Commercial Waxes

H. Bennett

Commercial Waxes
Second Edition

ISBN: 978-0-8206-0156-4

Chemical Publishing Company:
www.chemical-publishing.com
www.chemicalpublishing.net

First Edition: Chemical Publishing New York

Printed in the United States of America

preface to the second edition

The warm reception granted to the first edition of this book by readers and reviewers alike was ample reward for the editor's efforts in compiling the volume.

As the opinion was unanimous that the book filled a long-felt gap in the technical literature and was widely accepted as a source of information by chemists and other technical workers, it appeared quite justified to bring out a new, up-to-date, revised and enlarged edition.

Since the publication of the first edition, the applications of natural and synthetic waxes were further expanding in a great many industrial fields, new wax formulations and wax and resin blends have been developed, and new data have accumulated in the editor's files. All this information is included in this revised and enlarged edition to save time and effort for busy technical workers. Some of these data were widely scattered in the technical literature while others were not published before.

Thanks are due to those who took the trouble to point out errors and ambiguities in the first edition and grateful acknowledgment is made to those experts who have contributed much new and useful material to this new edition and who prefer to remain anonymous.

preface to the first edition

Natural waxes and waxlike substances have been used from time immemorial foı candles, waterproofing, sealing, decorative, preservative, and other purposes. Synthetic waxes and waxlike compounds, developed in the last decade, are finding increasing use and are supplanting natural waxes for many applications. Since different waxes have different properties, each one is used to obtain certain end results. Whereas certain waxes are used alone, as paraffin wax in waxed papers, other uses require combinations of different waxes, resins, etc., to obtain a desired result.

Since there is no one source of information available on the waxes used in industry, the editor was asked to prepare a book on commercial waxes for the chemist, engineer, superintendent, purchasing agent, and student. It is hoped that this book will be helpful to these and others who are interested in commercial waxes and wax compositions.

Much of the data included in the book is from the writer's own laboratory and has never been disclosed. Grateful acknowledgment is made to Allan B. Olsen, both for information supplied by him and his constructive criticism of the manuscript. The following, among others, were good enough to furnish useful data: H. M. Abbott, H. W. Kraaz, H. C. Eaton, Hercules Powder Co., Monsanto Chemical Co., Givaudan-Delawanna, Inc., General Electric Co., Century Stearic Acid & Candle Works, E. I. du Pont de Nemours & Co., Inc., Carbide and Carbon Chemical Corp., Armour & Co., Socony-Vacuum Oil Co., The Texas Co., and Glyco Products Co., Inc.

H. BENNETT

table of contents

introduction

If we look up the definition of "wax" in WEBSTER'S NEW INTERNATIONAL DICTIONARY (1953 Edition), we find that this common term is not clearly defined and people do not always mean the same thing when they talk about waxes. Even the chemical definition of wax is not clear-cut. Originally, the word "wax" was used to cover beeswax alone. Later it was extended to all substances of animal or vegetable origin which had a similar chemical composition to beeswax, that is, which consisted of esters of higher fatty acids with monohydric alcohols, some free fatty acids, and hydrocarbons. This definition would cover also some of the oils, such as sperm oil, which is called a "liquid wax."

A broader and more widely accepted definition of wax is based on its physical properties. According to this, wax is a plastic, slippery solid which is easy to melt. This definition embraces the mineral waxes which usually consist of higher hydrocarbons, certain fats, such as bayberry wax and Japan wax, which are glycerin esters of fatty acids, and synthetic waxes, such as fatty alcohols, fatty acids, and esters of polyethylene glycols.

As this book is directed to a large audience—chemists, engineers, manufacturers, technicians, executives, salesmen, and students—it will be based on the last definition of wax, that is, all waxlike substances will be treated regardless of their chemical composition.

For industrial purposes, waxes may be classified according to their origin or source as follows:

1. Mineral
2. Vegetable
3. Animal

4. Insect
5. Miscellaneous
6. Synthetic or Manufactured
7. Compounded

As Nature is not a chemist but an artist, natural waxes, as all natural products, will show variations in their physical properties, such as color, melting point, specific gravity, etc. The differences may be traced to soil, climate, age, oxidation, impurities, adulteration, and other causes. Therefore, it is not surprising if there will be discrepancies in the physical constants of one and the same natural wax within this book or as compared to other publications. For most commercial uses, small variations are not critical and acceptance or rejection of a wax for a certain application can be safely based on the data listed.

The great majority of applications of waxes in industry is based on one or several of the properties listed in alphabetical order as follows:

Adhesiveness or stickiness
Adsorption and absorption
Bacterial content
Blocking (self-adhering)
Bursting strength
Candle power
Clarity
Cold flow
Color
Compatibility with other ingredients
Corrosiveness
Crystalline structure
Crystallization from solution
Density or specific gravity
Dielectric strength
Edibility
Elasticity

Emulsifiability or dispersibility
Feel or hand
Film-surface continuity and topography
Flash point
Flexibility
Fluorescence
Form (amorphous or crystalline)
Fracture
Gelling or thixotropy
Hardness
Heating value
Homogeneity
Hygroscopicity
Iridescence
Length (when melted)
Lubricity
Melting (or solidification) point or range
Migration from composition
Odor
Opacity
Overprinting or coating
Penetration into adjacent materials
pH
Plasticity and ductility
Polishing properties
Polymerization
Power factor
Pressure (behavior under)
Purity
Reactivity
Refractive index
Resistance to fungi and bacteria
Saponifiability
Setting time

Sheen and reflectivity
Slipperiness
Solubility
Stability to electrical discharges
Stability to light
Static (electrical) state
Sublimation
Surface tension
Taste
Temperature and shock resistance
Tenacity or coherence
Tensile strength
Thermal conductivity
Toxicity
Ultraviolet screening
Viscosity (molten or in solution)
Volume change with temperature
Water repellency
Water-vapor transmission

As a single wax very seldom displays all of the properties needed for a specific application, waxes are usually compounded with each other and/or with resins, elastomers, fillers, and other substances to modify their properties.

In selecting a wax for a certain application, in addition to the previously listed properties, the following factors may also have to be considered:

Availability
Price
Ease of handling (bulk, slab, flakes, or powder)
Legal restrictions (health, safety, or patents).

abbreviations

AMP	American Melting Point
Amp.	Ampère
A.P.I.	American Petroleum Institute
ASTM	American Society for Testing Materials
Av.	Average
Bé.	Baumé
B.P.	Boiling point
C.	Centigrade
c.	Cold
cal.	Calorie
calcd.	Calculated
cc	Cubic centimeter
cm	Centimeter
cm^3	Cubic centimeter
D.	Dispersible
d.	Density
D.H.	Dispersible hot
Dist.	Distillation
dr.	Dram
E.	Engler
EMP	English melting point
F.	Fahrenheit
F.F.A.	Free fatty acid
F.P.	Freezing point
g	Gram
G.	Gels
gr.	Grain

H. ....Hot
hr. ....Hour

I. ....Insoluble
in. ....Inch
Inf. ....Infinite
ins. ....Insoluble

kg ....Kilogram

l ....Liter
lb ....Pound

m ....Meter
Max. ....Maximum
Min. ....Minimum
ml ....Milliliter
mm ....Millimeter
Mol. Wt. ....Molecular weight
M.P. ....Melting point

n ....Refractive index
No. ....Number

oz ....Ounce

P. ....Partially
P.S.H. ....Partly soluble hot
Pt. ....Point
pt ....Pint

q.s. ....A quantity sufficient to make

rd ....Specific optical rotation
Ref. ....Refined

S. ....Soluble cold
SAE ....Society of Automotive Engineers
Sap. ....Saponification
sec ....Second
S.H. ....Soluble hot
Sl. ....Slightly
Sol. ....Soluble
Soln. ....Solution

Sp. Gr. ....Specific gravity
S.u.v. ....Saybolt universal viscosity
S.X. ....Clearly miscible in certain proportions

Techn. ....Technical
Temp. ....Temperature

Unsap. ....Unsaponifiable matter

Visc. ....Viscosity
Vol. ....Volume

Wt. ....Weight

Symbols

~ ....Approximately

> ....More than

< ....Less than

± ....Plus or minus

CHAPTER 1

natural waxes

MINERAL WAXES

Under mineral waxes, those commercial waxes will be treated which are derived from the earth as such or obtained from a mass of earthy origin by refining. They occur in petroleum, bituminous coal and slate and consist of unsaponifiable hydrocarbons, nearly all saturated. The melting point of these waxes increases in the order of their molecular weights, disregarding any contaminants which they may contain.

Paraffin Wax (Paraffine Wax)

Paraffin wax is the most widely used commercial wax. It is mainly derived from the high-boiling fractions of petroleum. It is commonly sold in slabs or blocks which exhibit crystallinity and various degrees of translucency. There is often a difference in melting point between the outside and inside of a slab because of nonhomogeneity. The crude paraffin waxes, containing varying amounts of volatile ingredients, are odorous and greasy. The refined grades are tasteless, odorless, harder and less greasy.

The annual production of paraffin wax in the United States is between 500,000,000 and 600,000,000 lb. Not a single substance, but a group of related substances is marketed under this name. Considering the volume of yearly production, there is surprisingly little published literature dealing with the composition, manufacture and use of this commodity. It is known that paraffin wax is not a pure chemical compound, but a variable mixture of organic compounds. This probably explains the reluctance of investigators in the field of pure science to examine paraffin wax thoroughly. The investigation of the more practical or commercial aspects of production and utilization of paraffin wax has been conducted in the laboratories of suppliers or consumers. The more specific results of these studies are generally regarded as trade secrets and as such are not published in technical journals.

There are perhaps two dozen authoritative technical papers dealing with paraffin wax which have appeared in the past 10 or 25 years. Certain worth-while data were obtained by the authors of these papers which have, to a large degree, established the composition of paraffin wax. In addition, the laboratories of Socony-Vacuum Oil Co. have been engaged in studying paraffin wax for 5 or 10 years. The results of these investigations have as yet not been collected and presented in the form of a report.

Refined paraffin wax is the most commonly used grade of paraffin wax and has the following properties:

Melting Point (ASTM)	122–124°F.
Color (Saybolt)	25–30
Flash Point	395°F.
Penetration at 32°F.	16
77°F.	22
115°F.	221
Saponification Value	0
Acid Value	0
Viscosity at 212°F.	39
Structure	Crystalline
Specific Gravity	0.880–0.915

Soluble in: benzol, ether, chloroform, carbon disulfide, carbon tetrachloride, turpentine, petroleum, fixed oils

Insoluble in: water, cold alcohol

Uses: candles; waterproofing; sealing; lubricating; food, plant, fruit and vegetable protection; paper; polishes; cosmetics; crayons; pharmaceutical ointments and salves; electrical insulation, etc.

General Characteristics

Paraffin wax is a white, translucent, waxy, tasteless and odorless solid, consisting of a mixture of solid hydrocarbons chiefly of the methane (paraffin) series obtained from petroleum. Its specific gravity is about 0.880–0.915 at room temperature and its melting point about 110–150°F. (43.3–65.6°C.).

Paraffin wax is sold in various grades, which differ from one another chiefly in melting point. Generally, each grade covers a range of melting points of 2°F., for example 125–127°F. Melting points are given by either of two accepted standards: American melting point (AMP) and ASTM melting point, which is often called English melting point (EMP). At present, the American melting point is a calculated value based on the ASTM melting point, the first being arbitrarily taken as 3°F. higher than the second. The method for determining this value has the ASTM designation D87–37 and involves cooling molten wax in a cooling bath at a constant rate of temperature drop. The melting point of the wax is taken as the temperature at which the wax first shows a minimum temperature change. This method is based on the fact that, on solidification, the wax gives up its latent heat of fusion and the rate of temperature drop is considerably reduced.

The melting point of a wax, in itself, is no measure of quality or chemical composition, since it is derived from the melting points of the constituents which influence one another in various ways. Other characteristics of the wax must also be considered in determining quality. These include oil content, light stability, tensile strength, etc. Their relationship to the applicability of the wax will be discussed later in greater detail.

In general, the so-called *refined waxes* have an oil content of about 0.5% or less. In general, they are reasonably stable to darkening, when exposed to light, and their tensile strength is 160–400 lb per sq in. when measured on the Perkins Tensile Tester. These methods for evaluating and testing the uniformity of waxes are conventional. Unfortunately, many users of waxes consider these characteristics as the sole measure of their quality for all applications. The existing relationship between quality and applicability is, by no means, so clear cut as is frequently assumed, so that, regarding applicability, erroneous conclusions are drawn far too often.

The largest percentage of the paraffin wax on the market is of the "refined" type. As mentioned previously, this type of wax contains a small percentage of oil, is hard, and has been refined to a point where it is tasteless, odorless and stable to light. The bulk of the remainder consists of crude scale wax. This type of wax is sold either as white crude scale wax or as yellow crude scale wax. Both are similar except for one step in the refining procedure for bleaching the wax. The bulk of crude scale wax sold in the market is of the white type.

Crude scale wax differs from refined wax in several important details. First, it contains appreciably more oil (on the average 3%). For this, and other reasons, its quality is poorer than that of refined waxes of comparable melting points. The high oil content of crude scale wax is reflected in a low tensile strength (40–100 lb per sq in.) and also in a softer and more crumbly structure. Due to the lesser degree of refining of these waxes, they are not entirely tasteless and odorless and their color stability is questionable. There are wide variations in the overall quality of crude scale waxes, largely because no standards or specifications (except melting point) are established for them. This condition is also due to the large number of petroleum refiners who produce crude scale waxes and to the variations in the crudes from which they are derived.

Between refined waxes and crude scale waxes, there is a third type which is sold as "semirefined" wax. This designation has even less meaning, so far as quality is concerned, than the term "white crude scale wax." Just as crude scale waxes may vary in quality, so may semirefined waxes. In spite of a relatively uniform price, these grades will differ appreciably, depending on the supplier and the source. Frequently, semirefined waxes will approach fully refined waxes in quality, but just as frequently, they may be little, if any, better than crude scale waxes. Consequently, a purchaser of these grades, unless dealing directly with a producer, will have no real guarantee of uniformity.

Semirefined waxes are produced by a group of small refiners whose main business is the refining of wax. These concerns purchase crude scale wax on the open market and sweat it to a more highly refined grade. Depending entirely on the base crude scale wax, and on their ability to conduct this operation, they will offer waxes varying widely in quality. As a group, semirefined waxes will have melting points in the range 120–135°F., oil contents of about 0.6–3%, and tensile strength of 40–200 lb per sq in.

Match wax is a low-melting (104–116°F.) refined paraffin wax. Reclaimed paraffin wax is often available at a lower price. This refers to wax that has been scraped off or melted out of paper, textiles, etc. It is usually darker in color and may have different kinds of foreign impurities and decomposition products.

Occurrence in Nature

Petroleum is a natural resource found in various parts of the world. It is now generally conceded that petroleum was formed during a vast period of geologic time by the decomposition of organic refuse, largely of marine origin, aided by conditions of low temperatures and enormous pressures. An analysis of petroleum, freed from extraneous matter, shows that it is composed mainly of hydrogen and carbon in amounts ranging from 11–14% and 83–87%. Small quantities of sulfur, oxygen and nitrogen are frequently present. The elements contained in petroleum are in chemical combination, as are most of the elements in nature. It is a peculiarity of carbon that it has the ability to enter into a vast number of chemical combinations with itself and other elements and to be a part of all living organisms. Because of the last fact, early chemists distinguished between carbon compounds and all others, classifying the former as "organic" and the latter as "inorganic."

Most of the compounds contained in petroleum are hydrocarbons. They range from gases to solids with a predominance of liquid members. Relatively few pure chemical compounds have been isolated from petroleum, but thousands actually exist. This is due to the multiplicity of combinations which are possible between hydrogen and carbon atoms.

Petroleum is composed essentially of four distinct types of hydrocarbons that differ from one another in chemical structure. The four generic classes of hydrocarbons, contained in petroleum, are known as paraffins, olefins, naphthenes, and aromatics.

Paraffins or saturated, open-chain hydrocarbons are those in which the carbon atoms are linked together with single bonds, the remaining valences of the carbon atoms being satisfied with hydrogen atoms. Methane (CH_4) is the simplest paraffin hydrocarbon, followed by ethane ($CH_3.CH_3$), propane ($CH_3.CH_2.CH_3$), etc.

These compounds are straight-chain hydrocarbons, in which the carbons are linked together in a chain with no branches or side chains. A branched-chain paraffin hydrocarbon is represented by isobutane, having the formula

$$\begin{array}{c} CH_3{-}\underset{\displaystyle |\atop \displaystyle CH_3}{CH}{-}CH_3. \end{array}$$

Obviously, normal butane ($CH_3.CH_2.CH_2.CH_3$) has the same composition as isobutane, but is an entirely different compound, due to a distinct difference in arrangement of the carbon atoms. It can be seen that as the number of carbon atoms increases, the number of possible arrangements also increases, but at an even more rapid rate. This accounts for the complexity of petroleum products and the difficulty in isolating pure compounds from them.

Olefinic hydrocarbons are closely related to the paraffins in that they have an open chain. However, they differ in that, at some point in the chain, two carbon atoms are joined together by means of a double (or more rarely a triple) bond. Such products are said to be "unsaturated," meaning that under certain conditions they will take up hydrogen or other elements. Olefines can be converted into paraffins by a reaction which is known as "hydrogenation." It consists of the introduction of two hydrogen atoms into the molecule at the double bond and its conversion to a single bond.

Naphthenic and aromatic hydrocarbons are cyclic, i.e., their carbon atoms form ring structures. Simple members of these series are cyclohexane (C_6H_{12}) and benzene (C_6H_6).

Naphthenes are saturated hydrocarbons (no double bonds) containing one or more ring nuclei. In some naphthenes paraffinic side chains are attached to the carbon rings.

Aromatic hydrocarbons are characterized by the presence of a benzene ring in the molecule. These too may have paraffinic side chains attached to the benzene ring. The relationship between naphthenes and aromatics is about the same as between paraffins and olefins.

Petroleum crudes, occurring in various parts of the world, differ from one another in the percentage of the four classes of hydrocarbons which they contain. In so-called Pennsylvania crudes, paraffinic hydrocarbons predominate, whereas in California crudes naphthenes and aromatics constitute the greater portion. Crudes from mid-continent sources, in general, are of mixed base and contains substantial quantities of all four types.

Manufacture

As previously noted, petroleum fractions are complex mixtures of many ingredients. Their properties depend, in general, on two factors: the average molecular weight, and the average configuration. The differences in properties are used for the separation of one fraction from another.

Crude petroleum is first subjected to distillation to break it up into fractions varying in boiling point and viscosity. Gaseous light ends, gasoline, naphtha, kerosene, and gas-oil fractions are removed in this order. From paraffinic crudes, there is then obtained a fraction known as wax distillate, or paraffin distillate, which consists of a mixture of light-bodied oils ranging from 40–300 seconds S.u.v.* at 100°F., and various waxy fractions of about the same boiling range. Generally, fractions having boiling points slightly higher than that of the wax distillate cut are not distilled, but remain as a residuum and are further processed to prepare heavier bodied lubricating oils.

In the wax distillate, wax and oil must be separated for use. The wax distillate will have a pour point of about 80°F., which indicates that below this temperature it is, at best, a semifluid mixture of wax and oil with poor lubricating properties. Most oils are dewaxed to a pour point of 25°F., or lower, for use as lubricants. Similarly, satisfactory waxes for most uses must be deoiled to relatively low oil contents. Separation is based either on the differences in melting points of the wax and oil or on differences in the solubilities in a solvent. It should be emphasized, at this point, that there is a very close relationship

* Saybolt universal viscosity.

between paraffin wax and a paraffin oil having, for example, a viscosity of 100 seconds S.u.v. at 100°F. Both have approximately the same boiling range and volatility and, when heated to above the melting point of the wax, have comparable viscosities. Care in handling of molten paraffin wax is, therefore, essential for at elevated temperatures, its tendency to be oxidized and the loss due to evaporation are little less than those of paraffin oil. Paraffin wax differs from the oil primarily in that it has a nearly straight-chain structure, whereas the oil is more branched.

To separate wax from oil, the wax distillate is processed by either of two methods. The older procedure involves the so-called "sweating" operation, whereas the more modern one utilizes crystallization from a solvent mixture, and is known as solvent pressing. Wax distillates will vary in wax content over a range of 6–25%. A 15% average wax content may be expected in the distillate.

The Sweating Process

The first step in the separation of the wax distillate, pressing, consists, essentially, of filtering solid wax from liquid oil at low temperatures. Dewaxed oil and slack wax are produced in this operation. Often the paraffin distillate is diluted with dewaxed oil to facilitate handling. This is particularly necessary if the wax content is high, since otherwise, the chilled mass will be thick and difficult to pump from the chillers to the filter presses.

Frequently, the distillate is treated with a mild acid to lighten its color and remove highly polymerized impurities. After the acid treatment, the distillate is washed with water, neutralized with soda, and washed again. Finally, it is chilled at 13–15°F. to precipitate the higher-melting, waxy hydrocarbons. The temperature to which the distillate is chilled is dependent on the pour point which is desired in the dewaxed oil. For a 20°F. pour point, the above temperature is necessary. For lower pour points, the pressing temperature must be lowered accordingly. The chilled distillate is pumped to a bank of filter presses in which the crystalline wax is filtered from the dewaxed oil. This yields slack wax containing about 35% oil. The dewaxed oil is finally resolved into various paraffin oils and other light lubricating stocks by distillation and conventional refining procedures.

The slack wax, which melts at about 107°F., must be subjected to further processing, the first stage of which may be a second mild acid treatment. The molten slack wax is charged to sweaters which are generally of pan design. A conventional sweater consists of a room or oven in which a stack of ten to fifteen or more horizontal trays is housed, each equipped with a gravity drain. The slack wax is allowed to solidify over water, whose surface has been adjusted to be on a level with a wire mesh in each tray. When the wax is completely solid (wax temperature of about 80°F.), the water is withdrawn and the slack cake, 5–10 in. thick, remains suspended on the screen. The sweater is then heated at a rate of about 2°F. per hour by circulating warm water through banks of tubes embedded in the wax and along the walls of the oven. Heat causes beads of oil to form on the surface of the wax, resembling sweating. This accounts for the name of the process. The oil and lower-melting wax are allowed to drain away as the temperature increases. Finally, a cake of high-melting waxes remains and is melted and segregated. This is known as yellow crude scale wax which averages 5% oil and has a melting point of 122–124°F. The oil and low-melting waxes constitute the foots oil which is used as cracking feed stock for the manufacture of gasoline.

The crude scale wax is given another mild acid treatment, washed, neutralized, and again washed. Then it is ready to be resweated to refined wax. The subsequent resweatings are frequently highly complicated and generally involve the recycling of the lower-melting fractions (intermediates) to produce a series of three or four grades of refined wax. For example, waxes melting at 130–132°F., 126–128°F. and 122–124°F. may be produced. Some refiners will make a higher-melting grade and others one melting as low as 118–120°F. These waxes contain only 0.4%, or even less, oil and are used for blending to intermediate-melting waxes. Finished waxes are finally prepared by blending to melting point and filtering through bone char, bauxite, or some other medium to produce water-white products.

Many of the white crude scale waxes on the market were treated in a similar manner, but, due to their nature, refining could not make them meet the specifications. In other words, in spite of resweating from yellow crude scale wax, they still contain 2% oil, have low tensile strengths, are not entirely color stable and thus must be sold as white crude scale wax. Other refiners merely bleach yellow crude scale wax to obtain the same result. Such white crude scale waxes are inferior to those which have been resweated.

Solvent-Pressing Process

In this method of producing paraffin wax, the raw material is the same wax distillate as used in the sweating process. In general, it is pressed to slack wax by conventional procedures. In some cases, a solvent-pressed slack is used, whose oil content is much lower than usual. The slack wax is dissolved in a suitable solvent, which may be petroleum naphtha or a mixture of benzol and methyl ethyl ketone. In one process, three parts of slack wax and two parts of solvent are chilled to 45°F. and filtered in a wax press, as before, to yield a crude scale wax containing only 3% oil. This crude scale wax is further processed by repeated solvent pressing or recrystallization to yield suitable base waxes. The quantity of solvent and the temperature employed are by no means fixed, but are dependent on the yield and melting point of the desired wax. Intermediate fractions are recycled, but this step is not so complicated as in sweating because of the greater degree of fractionation obtained. For the same reason, only two or three blending waxes are produced, these being drier, less ductile and, in general, more oil-free than sweated waxes.

Composition

Paraffin wax, as stated, is a mixture of a great number of hydrocarbons. Small amounts of liquid "impurities" are present in the form of oil. The solid constituents are saturated paraffinic hydrocarbons.

Frequently, it is said that these hydrocarbons are all of the straight-chain type and that no branched-chain products are present. Considering that the oils from which they are separated are highly complex and contain a wide variety of branched-chain hydrocarbons, this would be indeed surprising, if true. Much work of an exploratory nature was done by numerous investigators, but it was not until 1931 that the exact composition of paraffin wax was established. The final solution may be credited to Ferris, Cowles, and Henderson.[1] The bulk of the contents of this section is derived from their publication, although reference will be made to the work of other investigators as well.

As mentioned previously, paraffin wax is derived from the wax-distillate fraction of crude petroleum and is separated from other lower-melting waxes lost in foots oil and sweater oil. In addition, higher-melting waxes, which are sold as microcrystalline, amorphous, or petrolatum waxes, are derived from heavier lubricating-oil fractions. A definite relationship exists between all of these waxes, as shown by Buchler and Graves.[2] These investigators found that by isolating pure, oil-free fractions from waxes of these types and plotting a melting point-refractive index curve, a substantially straight line was obtained. Certain deviations from the straight line occurred at higher melting points, but by recrystallization from a solvent, small amounts of an impurity were lost and then the melting point-refractive index curve for the purified waxes became a straight line.

It has also been observed that paraffin wax crystallizes in both plates and needle-type crystals, depending on the conditions of crystallization. One theory explained this as being due to differences in viscosity of the medium and also to the rate of cooling. This explanation discounted the fact that differences in composition had any bearing on the type of crystal formation. Buchler and Graves attributed the differences in the crystal system entirely to the presence of the impurity mentioned before, which they called "soft wax."

In the work of Ferris, Cowles and Henderson, a slack wax from a mid-continent crude was selected for study. It was completely deoiled by successive recrystallizations at about 5°F. and fractionally distilled into narrow cuts. These were purified into twenty-six separate waxes by repeated recrystallization and recombination of fractions having the same melting points. At the completion of purification, waxes were obtained which, on further crystallization, were practically unchanged in character. It was, therefore, assumed that practically pure components had been produced.

Then types of crystals which were formed when these pure

[1] *Ind. Eng. Chem.* **23**, 681 (1931).
[2] *Ibid.* **19**, 718 (1927).

waxes were crystallized from a solvent were carefully studied. Three different types were found. Most of these waxes formed plate-type crystals; another group crystallized as needles, and the remainder assumed forms that could not be identified. The authors called the last group "malcrystalline," for lack of a better name. Further studies of these fractions showed that, no matter what solvent was used, each individual fraction invariably yielded the same type of crystals. These observations effectively disproved the theory that differences in crystal system are due to the crystallization medium.

In order to study the significance of these three types of crystals, the physical characteristics of all of the pure waxes were determined. By plotting these characteristics, the relationships between melting point and refractive index, melting point and boiling point, melting point and density, and melting point and molecular weight were found. In each case, three curves were obtained which corresponded exactly to the three types of crystals. All of these observations show conclusively that paraffin wax contains three series of compounds, each of which has a characteristic crystal form, irrespective of the medium from which it is precipitated. It is now generally recognized that the rate of cooling merely affects the size of the crystals formed and not their type. Storage of wax, containing small crystals, may result in the development of larger crystals, but can never cause a change in their type. This can be accomplished only by removing from the wax, or adding to it, an impurity which is capable of impressing its form on the basic type.

It can be concluded that the plate-type crystals represent straight-chain hydrocarbons and constitute the basic fraction of paraffin waxes. The needle and malcrystalline type crystals are probably Buchler and Graves' "soft wax," and are assumed to be branched-chain hydrocarbons. Microcrystalline waxes are composed largely of these types, although high-melting paraffin waxes will sometimes be found to contain small amounts of the needle type. These conclusions have been substantiated by further work in the laboratory on certain selected fractions believed to be rich in "soft wax."

The authors also determined the relative percentages of these three types which might be expected in fractions of different boiling points. Their approximate values are given in Table 1.

Table 1

CRYSTAL TYPES IN MICROCRYSTALLINE WAXES

Boiling Point, °F. at 10 mm	Type of Crystals		
	Plates	Malcrystalline	Needle
180	100	0	0
200	99	1	0
220	98	2	0
240	91	7	2
260	78	12	10
280	54	17	29
300	37	27	36

It can be seen that in the highest-boiling fraction, the malcrystalline and needle types are present in a greater total amount than the plate type. These waxes yield a much lower melting point for a given boiling point. For this reason, they are usually almost completely lost in paraffin refining. These types probably represent Buchler and Graves' "soft wax" and also constitute the bulk of microcrystalline or amorphous waxes obtained from higher-boiling distillates and residues.

Since pure components are never isolated in practice, the authors conducted additional work to determine the effect of mixtures of the three components on crystallization behavior. One phase of this investigation involved the determination of relative solubility of the three types in various solvents. They found that:

1. Needle-type waxes are much more soluble in any solvent than plate-type waxes of the same molecular weight, and that the malcrystalline type was intermediate between the two.
2. For all solvents investigated, waxes of the same melting point, regardless of type, have about the same degree of solubility.

Mixtures of the three types, each having the same melting point, were prepared and their crystallization behavior was noted. It was found that the needle type was strong enough to impress its form on the plates if both types were allowed to crystallize simultaneously from the same solution. However, if crystallization was such that the needle type was kept in solution, either by high temperature or excessive solvent, the plate type formed its characteristic crystals.

Similar work showed that the malcrystalline wax is even more powerful than the needle type in impressing its form on plates and on mixtures of plates and needles. Thus, any blends with the malcrystalline type will yield malcrystalline formations, unless this impurity is kept from the sphere of activity by higher solubility.

One practical utilization of these differences in crystallization behavior is found in the manufacture of paraffin wax by the sweating process. It is well known that a wax distillate, with good pressing characteristics, contains plate-type crystals and that a slack wax, with good sweating characteristics, will crystallize as large, interlacing needles. In both cases, apparently the same wax is crystallized, but the conditions of crystallization are far different. In paraffin distillates, the percentage of oil is relatively high, thus holding the needles entirely in solution and allowing plates to form. However, in chilling slack wax, which contains the needles dissolved in the oil, to a solid cake, both types are precipitated simultaneously and the needles impress their form on the plates.

It has also been recognized by refiners that, if the wax distillate contains too much high-boiling constituents, it will not press satisfactorily. It was also found that, as the boiling point increases, the percentage of the malcrystalline-type wax also increases very rapidly. A long-cut paraffin distillate will contain excessive quantities of malcrystalline waxes which precipitate simultaneously with the plates and impress their form on them, making the distillate nonpressable.

It follows from this discussion that paraffin wax contains only paraffinic hydrocarbons having the generic formula C_nH_{2n+2} and cyclic hydrocarbons are absent. It is desirable to know what ranges of molecular weights, or lengths of carbon chains, are present. Buchler and Graves isolated many fractions believed to be relatively pure chemical compounds. In paraffin wax, these ranged from $C_{18}H_{38}$ to $C_{32}H_{66}$. Work with petrolatum waxes (microcrystalline waxes) disclosed the presence of hydrocarbons from $C_{34}H_{70}$ to $C_{43}H_{88}$. Clark and Smith[1] report hydrocarbons ranging from $C_{21}H_{44}$ to $C_{57}H_{116}$, but it appears that they are including high-melting paraffin waxes (up to 150°F. and higher) in their discussion. These investigators also claim, as a result of X-ray studies, that the number of different straight-chain hydrocarbons in the wax is small and seldom exceeds ten. The number of branched-chain hydrocarbons could not be determined, but is probably somewhat higher.

Practical Evaluation

It has often been stated that the distribution of paraffin wax, by uses, is roughly 80% for the coating and impregnating of paper and paper products, 10% for candles, and 10% for other applications. For the lack of any definite figures to prove otherwise, these proportions must be accepted as at least a good approximation. The 80% used in paper includes both refined wax and white crude scale wax. However, since the requirements of scale wax are not nearly so exacting, and, in general, represent lower-quality merchandise, only refined waxes will be considered in this discussion of quality and applicability.

The suppliers of refined paraffin waxes have adopted certain tests for controlling their production. Because of the more or less continuous nature of manufacture and the rapidity with which these tests can be conducted, they have served as a means of maintaining fair uniformity of quality. Such tests include determination of the following characteristics:

Melting point
Tensile strength
Color
Oil content
Reaction with sulfuric acid

In addition to these, there is a test which measures the degree of fluorescence exhibited by a wax when exposed to ultraviolet light.

Unfortunately, many consumers of refined wax have adopted these determinations as specifications for their purchases. They

---

[1] *Ind. Eng. Chem.* **23**, 697 (1931).

demand high tensile strength, low oil content and water-white waxes and assume that all these are an absolute necessity for satisfactory performance. When waxes fail to meet these arbitrary specifications, they are either rejected or viewed with suspicion and are held responsible for any failure which may occur in the finished product. In many ways, this is deplorable practice, although the exacting requirements of the consuming trade have probably been responsible for the advances in quality of refined waxes which have been made during the past decade.

Because of the complex nature of paraffin wax, a study of the correlation of physical characteristics and chemical composition with performance has been extremely difficult. There are practically no published data of this nature. In many cases, it has been impossible to arrive at definite conclusions regarding the performance characteristics, but frequently certain trends are indicated. Thus, it can be shown that several of the accepted tests are designed to determine performance, but others are not definitely related to it.

Of all paper-coating and impregnating applications, the manufacture of bread-wrapping paper, or other high-gloss wax-coated stocks, is probably the most exacting. Other uses for wax may have the same requirements to a lesser degree, but few demand such universal excellence. These requirements include: no tendency to block at ordinary temperatures, stability to discoloration, satisfactory sealing strength, ability to retain gloss, good ductility and no appreciable change on aging. Each of these items will be discussed separately, giving possible means of controlling production and maintaining uniformity.

Blocking

Blocking is a term used throughout the paper-converting trade and refers to the tendency of two sheets of waxed paper to adhere to each other. It is encountered where sheets of waxed paper are stored in stacks or piles and where a tightly wound roll is exposed to elevated temperatures. Various degrees of blocking are possible, all of which are primarily the result of either fusion of the wax from both sheets or localized adhesion of wax particles. The extreme case, from which the term is derived, is one where a practically solid cake, or block, is formed.

Milder forms of blocking result in the formation of pock marks in the surface caused by localized blocking or a general marring of the waxy surface due to a mild over-all tendency to form a wax bond between two layers of paper. As such a difference in severity exists, and since a solid block practically precludes re-running the paper, all forms of blocking are objectionable and should be avoided.

Blocking is due to complete or incipient fusion of wax from the surfaces of both layers of paper and is influenced by temperature, time and pressure. As any of these three increases, so does the tendency to block. Since time and pressure are somewhat beyond control (except in the case of the user), the effect of temperature must be overcome. This is accomplished, most simply, by the use of higher-melting waxes when elevated temperatures of storage are anticipated. Carefully controlled laboratory data have shown that, with waxes from the same source, blocking is almost entirely due to a low melting point. No safe temperature limits can be established for general use because of the effect of pressure and time, but an increase of about 5° in melting point is equivalent to a 10° increase in permissible storage temperature. This approximation holds only over the range of 125—135°F., in which most domestic refined waxes fall. Practically all converters purchase higher-melting grades during summer months and those catering to the southern trade ought to make certain that lower-melting waxes are not used on stocks shipped to such points.

In addition to melting point, there are two other factors which influence blocking in a lesser degree. The first of these is oil content, which, if too high, will tend to increase blocking. However, the quantity of oil required for so adverse an effect is excessive and would be objectionable for other reasons.

The second factor is the ductility of the wax. As previously shown, two waxes of identical melting point may have entirely different melting point distribution. Thus, in one of them, there may be low-melting fractions which are softer than usual. The presence of these constituents will result in a wax of greater ductility which is more easily deformed by constant pressure. Such waxes tend to block at slightly lower temperatures than harder, more brittle waxes of equal melting point.

Discoloration (Light Stability)

Many grades of paraffin wax will darken appreciably if allowed to stand in direct, or even indirect, sunlight for any length of time. This property of a wax was often attributed to its crude source and was assumed to be characteristic of all waxes derived from the same crude. However, careful investigation shows that this is not the case. The pure solid constituents of paraffin wax are remarkably stable to ultraviolet light, and will discolor only if subjected to it for prolonged periods or at elevated temperatures. However, the oil or liquid constituent is far from stable. It is found that, even in the presence of only a small (0.5) percentage of oil in a wax, this oil may be so unstable that its oxidation is sufficient to cause the entire mass of wax to become noticeably darker.

Light stability is an obvious requirement in waxes. Waxed paper must not yellow to be salable. Other articles, including candles, are even more degraded by discoloration and often deteriorate and become unsightly.

Two methods are available for improving light stability: removal of the oil, or refining until it is color stable. The first method is impracticable, so that wax producers have to resort to additional refining. After adequate treatment, even if 0.5% oil is left, it will not be harmful. Thus, the relation between color-stable and color-unstable waxes is the same as that between two theoretically oil-free waxes, to one of which a small percentage of white oil has been added, and to the other similar quantities of an unrefined paraffin distillate.

It follows from this that the quantity of oil present in a wax, in itself, is not a measure of color stability. The type of the oil must also be considered, for, if the oil is color stable, even excessive quantities will not be detrimental to color stability.

In view of the relationship between light stability and oil content, it would perhaps be well to explain the term "oil content." A determination of the so-called "expressible oil and moisture" was given tentative recognition by the ASTM. This method measured the quantity of liquid (chiefly oil and low-melting wax) squeezed from a sample of wax under standard conditions at 1000 lb per sq in. pressure. Due to poor reproducibility, this test has been abandoned and an improved method suggested instead. The new test actually measures the quantity of oil present in the wax by recrystallizing the wax from a solvent under prescribed conditions and weighing the oil held in solution. The oil is defined in terms of pour point to make certain that excessive quantities of wax are not removed at the same time. Highly reproducible results can be obtained by this method, which shows that refined waxes consistently contain less than 0.5% oil. As determined by the old "expressible oil and moisture" method, values ranging from 0–1.0% were often reported for the same wax.

Ductility

In general, refined waxes are described as being brittle and nonductile, especially when compared with microcrystalline or amorphous waxes. However, when this property is considered over a narrow range, it soon becomes evident that wide differences exist among various waxes. It is well known that, to a mild degree, refined waxes are thermoplastic below their melting point, and changes in temperature have a pronounced effect on ductility. Conversely, at any given temperature, lower-melting commercial waxes are, in general, more ductile than those of higher melting point, probably because they are closer to the liquid state. Besides these variations, there is also the fact to consider that a measurable difference in ductility exists between waxes of the same melting point when subjected to the same conditions of temperature and pressure.

The importance of this difference can be illustrated by the performance of two waxes used for coating paper milk bottles with the Excello machine. This machine immerses the paper container in a bath of molten wax, coating it inside and out, and delivers the coated article to a conveyor which carries it through a chilling tunnel. During cooling, drainage from the side walls to the bottom occurs, building up a thick cake of wax which is not completely congealed. Immediately, after leaving the conveyor, the container is filled with milk at 38°F., the filling time being only a few seconds. When the less ductile of the two waxes is used, the impact of the cold stream on the plastic mass of wax sets up strains, which result in open cracks, and allow leakage of the milk to the paper board. In one test, 4% of all bottles filled was classed as "leakers." With a more ductile wax, the strains are counteracted by cold flow of the wax and cracks are practically nonexistent. In a similar controlled test, the more ductile wax produced a fraction of 1% of "leakers."

Another practical application is, for example, the use of waxes for coating frozen-food containers which are subjected to temperatures below 0°F. Even papers creased at ordinary temperatures show that more ductile waxes cause less flaking than others.

A test has been developed to evaluate ductility. This is somewhat similar to a tensile-strength determination in that carefully prepared briquettes are broken, but has the important difference that constant loads are applied at controlled temperatures and are chosen so as to cause a measurable elongation of the briquette. Instead of obtaining a break in 10–15 seconds, as in tensile-strength measurements, times as long as several hours are required for breaking.

Attempts have been made to ascertain a correlation between ductility and other characteristics of refined waxes, such as tensile strength, oil content, sealing strength, etc., but apparently there is none. Instead, it appears that the most important factor is the distribution of the component parts with respect to melting point. Waxes with wide ranges of melting points of the ingredients are much more ductile than those with narrower ranges. Also, it appears that the evenness of distribution has an important effect on ductility. Waxes having a very even distribution, as contrasted with those composed of a blend of narrow cuts, will be noticeably more ductile. It is possible, therefore, for a consumer to prepare a highly ductile synthetic wax by blending together equal, or nearly equal, portions of several grades of wax.

The ductility of paraffin wax is also dependent on temperature, but different waxes are not always affected in the same manner. In general, ductility will increase, as temperature is increased, reaching a maximum at some point below the melting point, and above that, it will decrease as the temperature is further raised. The maximum ductility is apparently associated with the transition point of the wax, which occurs 40–50°F. below its melting point and is the result of a change in crystal structure.

Investigators of paraffin wax have noticed that, at temperatures below the apparent solidification point of the wax, changes in crystal structure frequently take place. The temperature at which this occurs is known as the transition point. Not all waxes exhibit transition points and most waxes, if quickly chilled to below the transition point, will show no change unless stored at slightly elevated temperatures where crystal reformation may occur. Waxes, having uniform melting-point distribution, particularly over a wide temperature range, will show no transition point, or at best an indistinct one. Waxes of uneven distribution invariably have the most pronounced transition points and show the greatest changes in ductility at these temperatures.

The following phenomena, which occur when waxes having transition points are cooled slowly, may be of interest. When a large wax mass is cooled and reaches the so-called melting point, latent heat of fusion is released and the temperature of the wax remains constant or increases slightly until solidification is complete. Below this temperature, cooling proceeds again at a regular rate. However, a large quantity of wax remains as an isotropic viscous fluid for some degrees below the solidification point. When this fluid congeals, contraction commences and air begins to enter the wax. This point probably represents the final solidification of all constituents. At a still lower temperature, transition takes place, which is accompanied by a definite change in volume caused by contraction of certain solid portions of the wax, an evolution of heat and a release of the air which was dissolved during the prior cooling. The release of heat causes a change in the rate of cooling and is used for determining the transition point. In some cases, the change in rate is pronounced, whereas in others, it is indefinite. Following transition, the wax again continues to cool at a regular rate. Frequently, a microscopic study of these cooling phenomena will show that a portion of the wax actually becomes liquefied and recrystallizes long after it appears to have become permanently solid. The quantity involved is quite small but is spread uniformly throughout the mass of wax.

The practical aspects of the transition point in waxes are not definitely established as yet. However, it appears that waxes showing indistinct transitions, or no transition points, are more translucent than those in which the transition is marked. It might also be reasoned that the absence of a transition point would be indicative of a more satisfactory wax for moisture-proofing, since there would be no fractures between crystal boundaries due to contraction. This is another way of saying that the more ductile waxes, which are those having no transition points, but having a uniform distribution of components, are more satisfactory for this purpose, due to their ability to be deformed without fracture.

An additional observation of interest is that molded articles, such as candles, can be made most readily from waxes with sharp transition points. It is found that these waxes show a greater contraction on cooling and thus have less tendency to adhere to the mold than those with indistinct transition points.

Oxidation Stability

In nearly all practical applications of paraffin waxes, they must be melted at one time or another. This is particularly true in the paper industry, which consumes the bulk of the production. For melting waxes, many types of equipment are employed, but, in general, they consist of iron or steel assemblies heated with steam, direct flame, or electricity. As steam constitutes the commonest source of heat and since even exhaust steam has a temperature of at least 212°F., the temperature of the molten wax is frequently about 200 or 210°F. At such elevated temperatures, the oxidation of the wax will proceed with great rapidity, unless extreme care is exercised.

Oxidation of paraffin wax is dependent on a number of variables encountered in usual practice. These are temperature, time, surface of exposure, and contact with metals and impurities. It is surprising, how little attention is paid to these factors in both the design and operation of commercial equipment. As previously stated, paraffin wax is closely related in chemical composition to low-viscosity paraffin oils. Few operators would expose light oils, such as these, particularly if used as lubricants, to conditions often encountered by paraffin wax. Yet they expect the wax to be unharmed by careless treatment. This is probably due to the erroneous assumption that a solid is much more stable than a liquid.

Paraffin waxes will show a gradual increase in stability as the melting point increases. This is a reasonable observation, since products of higher molecular weight, or higher viscosity, are present in these grades. However, the difference in stability is not sufficient to permit a high-melting wax (such as one of 135°F. AMP) to be treated with less care than one melting at a lower temperature (122°F.). Crude scale waxes are less stable than refined waxes of equal melting point and, in general, will deteriorate about twice faster than these more refined products. Most measurements of oxidation are relative, but are usually based on a darkening of color, lowering of melting point, loss of tensile strength, and the formation of products, such as aldehydes, ketones, organic acids and esters, most of which have a pronounced odor. Initially, oxidation probably causes a breakdown of the small oily fraction of the wax, but eventually it results in degradation of the solid constituents as well.

The effect of temperature and, for that matter, of all other variables is also dependent on other conditions of exposure. Under normal conditions, little oxidation occurs at 150°F., but when a temperature of 200°F. is reached, serious oxidation begins and above this temperature, it proceeds at a rapid rate. From a practical viewpoint, arrangements should be made to hold temperatures below 200°F., if this is consistent with the job being accomplished. Best results are obtained with thermostatic control, whatever the source of heat may be. Low-pressure steam should always be used for heating to avoid local overheating and consequent oxidation.

Similarly, the time of exposure to conditions of oxidation must be held at a minimum. Naturally, the longer a molten wax is held at elevated temperatures, the more extensive will be the oxidation. Equipment which is not used 24 hours a day should be shut down during idle periods. This should be usual practice in all plants and should be followed for all week-end shutdowns. Losses of time, in reheating solid wax, will be more than compensated for by savings in wax and a maintenance of uniform high quality. In this connection, it should also be mentioned that waxing equipment should be cleaned periodically to remove accumulations of oxidized wax. The presence of highly oxidized material in fresh wax, aside from contaminating the good wax, will catalyze further oxidation and accelerate further degradation. Where the quantity of wax used is large, cleaning may be less frequent. The simplest means of control is to resort to cleaning when the odor of the wax bath has become distinctly rancid.

The third variable affecting the rate of oxidation is the surface of exposure of the molten wax. Since it is the quantity of oxygen reacted with the wax that is important, reduction of area of contact between air and wax is desirable. Samples stored in containers of about the same surface area showed that oxidation in the open is four times as rapid as that in a closed container. Equipment used for handling molten wax should always be designed to provide for the least possible exposed surface area. Other precautions, which should be observed, are to reduce to a minimum any wax spray or foam. Drippings from coating rolls usually cannot be controlled, but often lines deliver wax to heating pans above the surface and cause spattering. Extending the lines below the surface will correct this fault. Spraying paraffin wax is always conducive to excessive oxidation, since the surface area is tremendously increased. If other means of application are available for a given job, they should be chosen in preference to spraying.

In many chemical reactions, the presence of certain metals markedly affects the speed and extent of the reaction. Oxidation of wax is no exception and the type of metal used in the equipment is, therefore, of considerable importance in preventing oxidation. Generally ferrous metals are satisfactory, and copper-bearing metals are not. There is apparently a difference among the various ferrous metals, since cast and wrought iron are less deleterious than stainless steel. Similarly, copper is much less desirable than brass and bronze. In order of desirability, the metals usually encountered may be listed as follows: cast and wrought iron, carbon steel, stainless steel, aluminum, nickel, brass, bronze, and copper. From this, it is apparent that fittings made of copper and copper-bearing metals should be eliminated from waxing equipment wherever possible. Replacements made of iron or steel should always be installed if at all possible.

Samples of refined paraffin, white crude scale, and microcrystalline waxes have been subjected to storage at several temperatures and in contact with metals. In addition, several of these waxes have been stored with no surface of exposure. From the data obtained, the following conclusions can be drawn.

1. Oxidation of petroleum waxes causes deterioration. This is evidenced by darkening, development of acid and saponifiable bodies, loss of tensile strength, lowering of melting point and loss of hardness.

2. High-melting-point waxes are more stable than those of lower melting point. This is particularly true of the microcrystalline waxes which are only slightly affected even at 250°F. for 7 days.

3. Refined paraffin waxes are more stable than crude scale waxes of similar melting points since the more oxidizable fractions have been removed during refining.

4. Refined paraffin and crude scale waxes stored at 150°F. exhibit little deterioration even after 21 days of exposure. Serious oxidation does not occur until a storage temperature of 200°F. is encountered. Above this temperature, oxidation is very rapid.

5. The surface of exposure has a pronounced effect on oxidation. In actual practice, it should be reduced to a minimum. This refers to agitation, aeration, spraying, and construction of equipment.

6. There is a period in the oxidation of waxes in which little change, other than darkening, occurs. Eventually, drastic oxidation commences with resultant increases in acid number, loss in tensile strength, and lowering of melting point. This is found when storing paraffin waxes at 250°F., at which temperature, it requires 3—5 days for serious oxidation to begin.

7. Drastic oxidation of microcrystalline waxes does not commence at 250°F., for a considerably longer period than for paraffin waxes. Less care is, therefore, needed in the handling of these waxes.

Antioxidants for Paraffin Wax[1]

Baths of paraffin wax, such as those used commercially for waterproofing, sizing or impregnating paper, tend to foul and

[1] Vore, U. S. Patents 2,325,085 and 2,325,167 (1943).

cause discoloration or undesirable odors due to oxidation of the paraffin, particularly after long continued use at elevated temperatures. Certain other substances, even when present in minute quantities in a bath of paraffin waxes or paraffin oils, prevent such oxidation and permit use of the bath for long periods of time without fouling.

Natural and artificial asphalts and pitches and their derivatives and asphaltlike substances, such as gilsonite, manjak and other asphaltites and petroleum residue asphalts, in extremely small percentages, will prevent oxidation of a paraffin-wax bath when the bath is operated at 235–240°F. It is understood that the percentage of antioxidant required increases in accordance with the temperature of the bath and at the higher temperatures, the increase is much more rapid.

Table 2 gives examples of such antioxidants:

Table 2

PREVENTING THE OXIDATION OF PARAFFIN BATHS

Antioxidant	Approximate % required to prevent oxidation of an industrial paraffin bath at 230–240°F.	Approximate % required to prevent oxidation of an industrial paraffin bath at 160–210°F.
Egyptian Asphalt	0.020	0.010
Cuban Gilsonite	0.040	0.020
Petroleum-Residue Asphalt	0.006	0.003
Isthmian Asphalt	0.020	0.010
Manjak	0.020	0.010
Utah Gilsonite	0.020	0.010
Fatty-Acid Pitch	0.500	0.250
Cottonseed Pitch	0.100	0.050
Stearin Pitch	0.500	0.250
Crude Montan Wax	0.040	0.020
Crude Ozokerite	0.100	0.050
California Asphalt	0.200	0.100
Carnauba Wax	0.600	0.300
Candelilla Wax	0.600	0.200
Esparto Wax	0.800	0.400

In addition to the antioxidants listed in Table 2 it has been found that about 0.6% ceresin breaks down the crystalline structure of the paraffin and tends to make the wax harder and more

*Table 3**

VARIATIONS OF PHYSICAL CONSTANTS IN DIFFERENT SHIPMENTS OF PARAFFIN WAX

M.P., °F.	ASTM Ball and Ring Softening Point, °F. Regular Min.	Regular Max.	Sloping Ring Min.	Sloping Ring Max.	Hardness Standard ASTM Needle Penetration 82°F. Min.	82°F. Max.	90°F. Min.	90°F. Max.	100°F. Min.	100°F. Max.	Tensile Strength lb per ¼ sq in. Area 90°F. Min.	Max.	Elongation or Stretch (mm) 90°F. Min.	Max.
Domestic 124–126	113.9	127.2	120.0	128.8	137	44	½ sec.	98	—	—	0	11.1	0	750
Domestic 125–127	121.1	129.1	127.6	136.2	90	21	2 sec.	236	½ sec.	2 sec.	20.2	45.0	½	2
Domestic 128–130	121.5	131.5	129.1	135.0	35	14	117	42	1 sec.	2 sec.	15.7	40.4	¼	45
Domestic 133–135	126.6	138.2	127.8	140.4	—	—	39	16	1 sec.	50	19.6	52.2	¼	¾
Domestic 135–137	127.1	136.2	136.2	139.5	—	—	42	18	3 sec.	140	29.8	63.3	¼	¾
Domestic 138–140	129.2	136.9	137.2	141.8	—	—	24	12	166	41	17.9	49.0	⅛	½
Imported 125–130	127.9	132.0	132.7	134.0	18	14	47	27	2 sec.	294	30.6	38.2	¼	¾
Imported 130–135	131.3	133.5	137.0	139.4	—	—	23	19	189	105	43.0	54.1	½	¼
Imported 133–135	130.8	137.2	136.4	141.0	—	—	26	15	207	32	35.9	59.7	¼	½
Imported 135–137	136.5	140.2	141.8	142.8	—	—	18	13	41	28	38.2	46.9	¼	½
Imported 138–140	136.6	142.1	142.2	147.4	—	—	15	10	29	19	37.8	61.8	¼	½
Imported 140–145	142.8	142.8	—	—	—	—	12	12	20	20	35.8	35.8	¼	¼

* The lowest and highest tests for each grade of paraffin wax are listed. The maximum and minimum tests were taken from a total of 426 tests made on different samples of paraffin waxes during 5 years.

water resistant. A bath of paraffin wax melting at 133—138°F., containing about 0.6% ceresin and about 0.04% petroleum-residue asphalt, is particularly suitable.

Data on Different Paraffin Waxes[1]

Paraffin waxes, from different sources, have very similar compositions. Differences in the melting points of commercial paraffin waxes are due to differences in their oil content. Paraffin wax apparently retains oil as a solid solution of oil in wax. The melting points of the solid solutions lie between the melting points of the components. Due to the existence of solid solutions, paraffin wax does not have a sharp melting point.

The solubility of paraffin wax in oil increases very markedly at temperatures approximately 10°C. below the melting point of the wax. The low-melting waxes are more soluble in oil than the high-melting waxes. Mixtures of high- and low-melting waxes have solubilities intermediate between the solubilities of the components. The increase in solubility of the mixtures in oil at temperatures approximately 10°C. below their melting point is as marked as for pure waxes.

Table 4

VARIATION OF THE MELTING POINT OF PARAFFIN WAX WITH THE OIL CONTENT[2]

Wax	Pressed Oil, %	Refractive Index at 25°C.	Melting Point °C.
"Parowax"	0.16	1.4481	51.7
Refined Wax	0.28	1.4487	50.0
Refined Wax	0.25	1.4485	50.0
Refined Wax	0.17	1.4481	51.7
Refined Wax	0.16	1.4479	51.7
Refined Wax	0.09	1.4476	54.4
Refined Wax	0.08	1.4475	54.4
Scale Wax	2.58	1.4570	48.3
Scale Wax	2.56	1.4568	48.3

The melting point under pressure of 1–200 atm. can be calculated from the following formula:

$$t_p = t_o + 0.029776\ (P\text{-}1) - 0.0000523\ (P\text{-}1)$$

[1] Wilson and Wilkins, *Ind. Eng. Chem.* **16**, 9 (1924).

[2] L. D. Myers and G. Stegeman, *Ind. Eng. Chem.* **20**, 640 (1928).

The Effect of Oil on the Plastic Properties of Petroleum Waxes[1]

Oil has a considerable effect on the properties of petroleum wax, particularly the microcrystalline waxes. Thirty blends of technical white oil with three typical waxes were examined with respect to melting point, penetration, crystallinity index, plastic point, ductility, tensile strength, compressive strength, and flexibility. Equipment and testing procedures were developed for most of the tests although several of the test methods were only modifications of standard methods. Equipment and procedure for a flexibility test have been developed in which a wax stick is bent through a 60° angle until rupture occurs. Compressive-strength tests were found to be more dependable than tensile-strength tests. Oil appears to enter the space between the interlocking crystals of paraffin wax, altering the structure only by dissolving some of the wax, whereas oil enters and dilutes the continuous phase of microcrystalline waxes, gradually altering the properties of the wax and possibly precipitating additional microcrystals. Only with extremely hard or high-melting point microcrystalline waxes does oil improve the plasticity or flexibility.

Properties of Wax–Oil Blends

Table 5 shows the melting-point and penetration tests of black oil blends, and the same data are given for white blends.

Table 5

MELTING POINT AND PENETRATION OF BLACK OIL AND MICROCRYSTALLINE WAX BLENDS

Percentage Oil	White 170 Blends		Brown 170 Blends		White 190 Blends		Brown 190 Blends	
	M.P.	Penetration	M.P.	Penetration	M.P.	Penetration	M.P.	Penetration
None	165	18	165	18	187	4	187	4
5	164	22	164	22	187	7	187	7
10	163	28	163	28	186	10	186	10
20	161	47	161	47	184	17	184	19
30	159	74	158	76	182	27	182	30
40	156	123	155	142	180	40	180	48
50	153	250	152	240	178	56	177	73

[1] W. L. Nelson and L. D. Stewart, *Ind. Eng. Chem.*, 41, 2231 (1949).

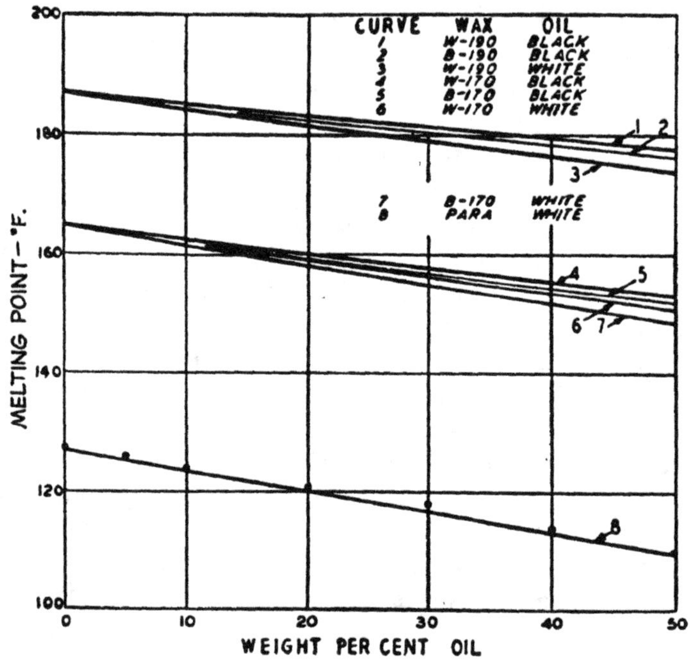

Figure 1. MELTING POINT OF WAX—OIL MIXTURES

Analysis of Wax—Oil Properties

The melting point of waxes is reduced in a nearly linear manner by the addition of oil (see Figure 1). This indicates that the oil merely reduces the relative amount of crystalline phase in proportion to the amount of oil added. The steeper slope of the brown-wax blends indicates that the brown waxes inherently contain more continuous phase than the other microcrystalline waxes. The position of the white-oil blends below corresponding black-oil blends may be explained by the lower viscosity of the white oil. The more rapid rate of reduction of melting point for the paraffin-oil blends is partly explained by the dissolving of paraffin wax by the oil. Also, the oil acts as a lubricant and does not adhere to the thermometer bulb as do the other wax-oil blends.

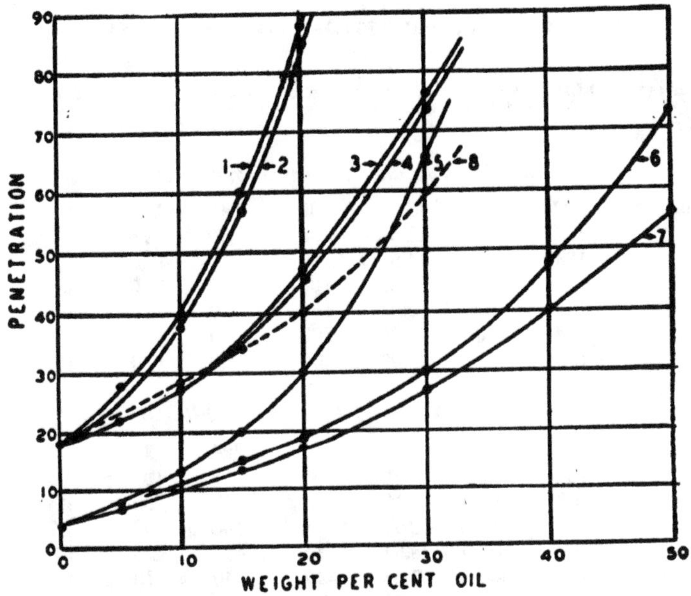

Figure 2. PENETRATION CURVES OF WAX—OIL MIXTURES

Curve	Wax	Oil
1	B–170	White
2	W–170	White
3	B–170	Black
4	W–170	Black
5	W–190	White
6	B–190	Black
7	W–190	Black
8	Paraffin	White

The penetration curves of Figure 2 exhibit the same general arrangement or relationship to one another as the melting-point curves of Figure 1 except that the paraffin-wax curve has a flatter slope than most of the others. But for the dissolving that occurs when oil is added to paraffin wax, the structure of the wax remains substantially the same.

Table 6 shows the results of all tests on white oil blends and Figures 3–7 show the same data plotted as curves.

Table 6

PHYSICAL PROPERTIES OF WAX—OIL BLENDS

Percentage of Oil in Blend	M.P. °F.	Penetration 1/100 cm.	Contraction %	Plastic Point	Tensile Strength lb/sq in.	Compressive Strength lb/sq in.	Ductility cm	Flexibility, Number of Bends
Paraffin Wax and Technical White Oil								
None	127	18	13.30	100	140	216	0	0
5	126	23	12.50	100	10	75	0	0
10	124	29	11.50	100	0	0	0	0
20	121	40	10.50	100	0	0	0	0
30	118	60	9.05	100	0	0	0	0
40	114	200	7.25	100	0	0	0	0
50	100	—	6.00	100	0	0	0	0
White 170 Wax and Technical White Oil								
None	165	18.0	11.00	70	224	232	1.0	86
5	164	27.5	10.70	50	128	132	1.3	112
10	163	38.5	10.10	35	80	80	1.3	60
20	161	86.0	9.17	–5	40	40	1.7	40
30	159	209.0	8.64	–34	24	24	1.8	30
40	155	300.0	7.55	–47	10	10	1.4	—
50	151	—	6.86	–52	—	—	—	—
Brown 170 Wax and Technical White Oil								
None	165	18	10.55	45	176	184	1.5	220
5	164	28	10.34	35	84	88	1.5	130
10	163	40	10.28	25	50	48	1.5	92
20	161	88	9.16	–20	16	16	1.5	64
30	158	215	8.18	–43	—	—	1.4	40
40	154	—	7.49	–58	—	—	1.0	—
50	149	—	6.79	–65	—	—	0.9	—
White 190 Wax and Technical White Oil								
None	187	4	11.40	118	300	720	0	0
5	186	8	11.40	110	226	416	0	0
10	185	12	11.04	100	180	282	0.3	40
20	183	30	10.52	84	96	120	0.7	92
30	180	66	10.24	30	48	56	1.0	80
40	177	161	8.76	–20	28	26	0.8	30
50	174	300	8.08	–43	16	16	0.6	16

The dotted contraction curves marked *A* in Figure 3 were computed from the total contractions (solid lines) of the blends by subtracting the contraction of the oil that had been added. The

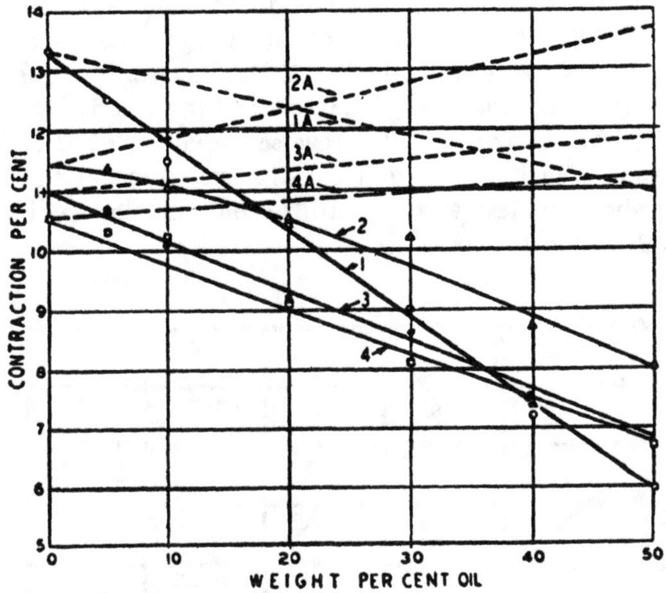

Figure 3. PERCENTAGE CONTRACTION OF WAX—OIL MIXTURES

Curve	Wax	Oil	Curve	Oil Free
1	Paraffin	White	1*A*	Paraffin
2	W–190	White	2*A*	W–190
3	W–170	White	3*A*	W–170
4	B–170	White	4*A*	B–170

contraction due to cooling and solidification of wax, C_w per cent, in terms of total contraction, C_t per cent, contraction due to cooling of oil, C_o per cent, and the percentage of oil, P, is:

$$C_w = \frac{100\ C_t - PC_o}{100 - P}$$

C_t and C_0 were measured experimentally and C_0 is approximately 2% for 60°F.

The addition of oil in all cases reduces the total contraction (Figure 3) of the wax blends. The apparent contraction of the wax part of the microcrystalline wax blends (dotted lines) increases as oil is added, but decreases for paraffin-oil blends. The dotted lines for the contraction of wax alone are smoothed data, presented only to indicate the general behavior of the wax portion of the blend. The behavior of the microcrystalline-wax blends appears to indicate that additional crystalline substance separates as oil is added, because more and more contraction due to crystallization occurs. The reverse behavior of the dotted paraffin-wax line indicates that oil dissolves some of the paraffin wax and thus less wax crystallizes and less contraction due to crystallization occurs.

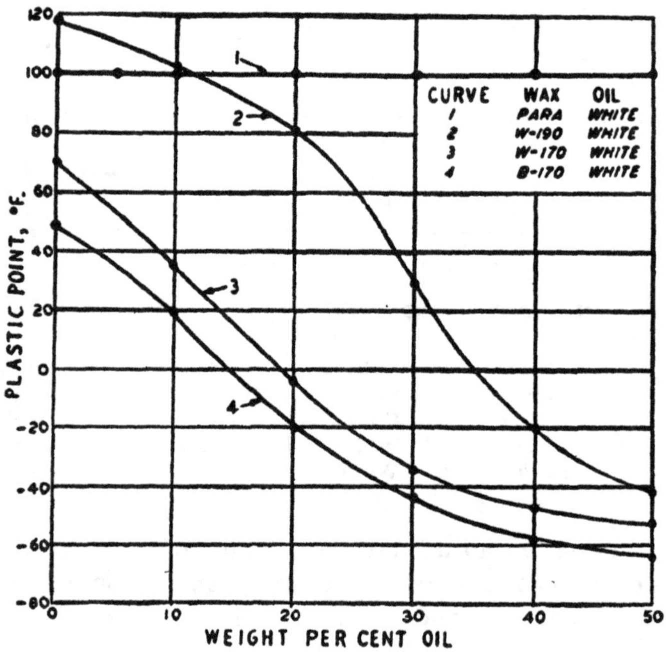

Figure 4. PLASTIC POINT OF WAX—OIL MIXTURES

The plastic point curves of Figure 4 are of interest primarily because the addition of oil did not alter the plastic point of the paraffin wax. The plastic points of the microcrystalline waxes are sharply reduced by the addition of oil and the three waxes generally arrange themselves in the order of their melting points. In microcrystalline waxes, the oil appears to enter and to alter the properties of the continuous phase, whereas in paraffin wax the original crystalline structure continues to dominate the behavior of the blend.

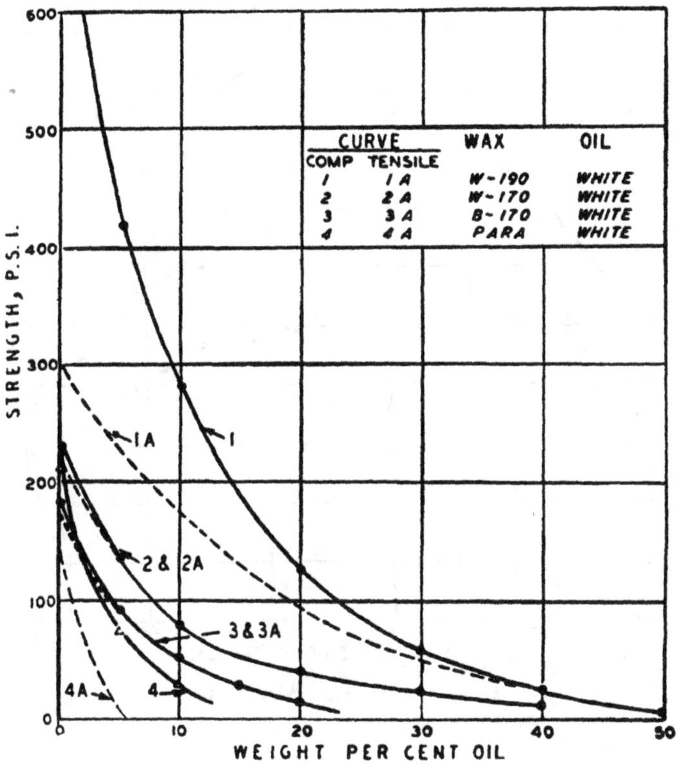

Figure 5. TENSILE AND COMPRESSIVE STRENGTH OF WAX–OIL BLENDS

The typically plastic White 170 and Brown 170 waxes have nearly the same tensile and compressive strengths, and the White 190 wax shows the same when enough oil has been added (30–50%) to soften its inherently plastic continuous phase. The tensile strength of paraffin-wax blends is always lower than the compressive strength (see Figure 5).

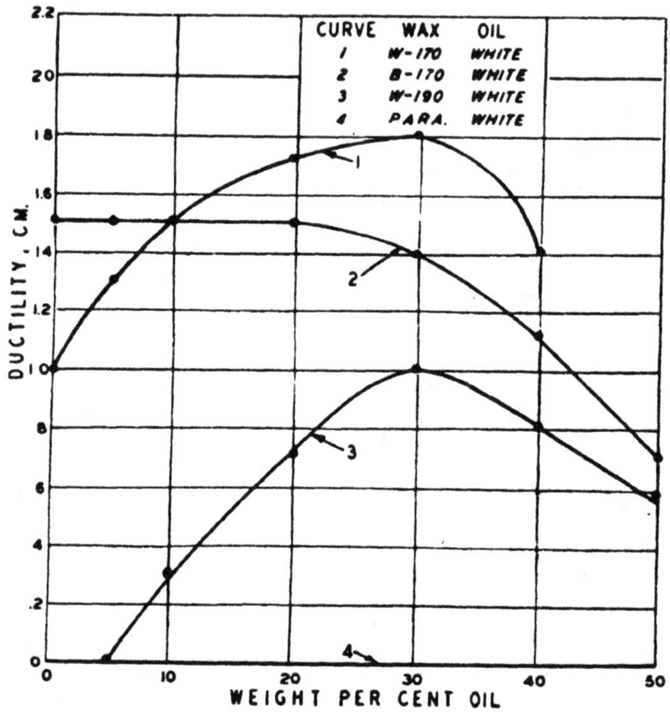

Figure 6. DUCTILITY OF WAX—OIL BLENDS

The ductility of White 170 and 190 waxes increases to a maximum at about 30% oil, that of Brown 170 wax is reduced by the addition of oil, and none of the paraffin-wax blends exhibits ductility. The continuous phase of waxes W—170 and W—190 is apparently so hard at 77°F. that the oil dilutes it to a phase that is more plastic or ductile, whereas the continuous phase of wax B—170 is already sufficiently soft so that oil dilutes it to a phase that is too thin to exhibit maximum ductility (see Figure 6).

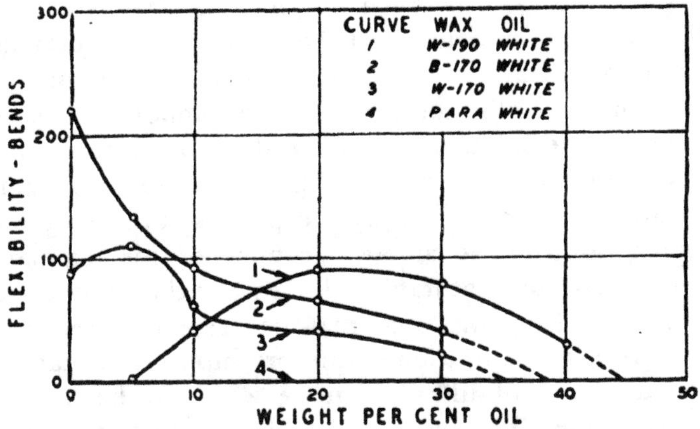

Figure 7. FLEXIBILITY OF WAX–OIL MIXTURES

Flexibility at 77°F., as shown in Figure 7, is not imparted to paraffin wax by additions of oil. Oil increases the flexibility of white microcrystalline waxes to a maximum and then, on larger additions, the adhesiveness of the continuous phase is inadequate for continued flexure. The flexure results are similar to the ductility results except that less oil is required in attaining maximum flexibility. The flexure test is a more severe one and, therefore, requires a somewhat stronger or harder matrix than that used for ductility.

Conclusions

Oil alters the melting point and percentage contraction of waxes in nearly direct proportion to the amount of oil. Softness is increased by the addition of oil as evidenced by penetration, plastic-point, and compressive-strength tests. Paraffin wax is partly dissolved by oil, whereas the crystalline portion of some microcrystalline waxes may be increased by the addition of oil. Apparently, additional crystalline substance separates from microcrystalline waxes as the continuous phase of these particular waxes is diluted with oil.

Microcrystalline wax-oil blends appear to exhibit a noncrystalline continuous phase throughout. Paraffin wax contains substantially no amorphous phase and oil merely dissolves some wax and pervades the space between interlocking crystals. The noncrystalline phase dominates the behavior of the microcrystalline waxes and the oil appears to enter only the continuous phase. The exact nature of the continuous phase cannot be ascertained from the test results, but the behavior after contraction indicates that this phase contains wax as well as oil and in an amount larger than can be explained by the plain solubility of wax in oil. Explanations for the large wax content of this phase are the existence of gels, supersaturated solutions, or the great solubility of the lowest-melting wax that may be present in the original mixture. Also, crystallization inhibitors suppress crystal formation and thus, on solidification, more wax may be left in the continuous phase. The behavior of brown waxes as compared with white ones of the same melting point, in all tests but the ductility test, appears to indicate a larger continuous phase. Colored petroleum substances are generally considered to act as inhibitors of crystallization. Oil imparts ductility and flexibility to hard microcrystalline waxes and the harder the wax structure the more oil is required to impart these properties.

Compressive-strength tests are more reproducible and sensitive than tensile-strength tests in indicating the structure and the plasticity of waxes. The predominance of the continuous phase of typical plastic substances permits them to exhibit the same tensile as compressive strength. Nonplastic waxes tend to shatter during the compression test whereas plastic waxes flow.

The flexibility test presented here is thought to be the best method, of those investigated, of indicating plasticity. It involves a combination of compressive strength and pliability. The compressive strength and plastic point results indicate that the maximum flexibility of a given wax is attained when these two properties are adjusted so that the ratio of the compressive strength to the melting point less the plastic point is about equal to unity:

$$\frac{C.S.}{M.P. - P.P.} = 1 \text{ (approximately)}$$

Two conditions seem necessary for the existence of a plastic wax: 1. small crystals which do not interlock, and 2. a continuous noncrystalline phase throughout which the crystalloids are dispersed. When these two conditions exist, plasticity is governed by the ratio of crystalline to noncrystalline matter as affected by inhibitors, oils, petrolatums, etc.; the melting point and the melting range of the wax components; and the temperature at which the wax is used as compared to the melting point.

Table 7

PROPERTIES OF FIVE COMMERCIAL MICROCRYSTALLINE WAXES BY SPECIAL TEST METHODS

	A	*B*	*C*	*D*	*E*
Color	Amber	Brown	Yellow	White	Brown
Melting Point, °F.	156.0	156.0	162.0	163.0	158.0
Penetration, 0.01 cm	40.0	37.0	31.0	25.0	29.0
Contraction, %	10.3	10.7	11.0	11.4	10.0
Plastic Point, °F.	13.0	20.0	48.0	55.0	15.0
Tensile Strength, lb/sq in.	78.0	80.0	100.0	116.0	140.0
Compressive Strength, lb/sq in.	78.0	84.0	102.0	120.0	140.0
Ductility, cm	1.6	1.6	1.6	1.6	1.8
Flexibility, number of bends	164.0	168.0	160.0	150.0	200.0

Several test methods are required for exploring commercial applications and many of the methods must yet be standardized. The properties of five additional commercial waxes, without the addition of oil, are shown in Table 7. They may be of assistance to commercial users of wax and may help in the formulation of specifications and test methods.

Solubility of Petroleum Waxes in Various Solvents

Frequently, it is desired to apply petroleum waxes in solution and data concerning their solubility at definite temperatures are essential. In addition, petroleum waxes are used extensively for coating purposes. These waxes are waterproof, but they are often called on to resist other solvents as well. In such cases, waxes of minimum solubility are required. Solubility data are also of value in separating waxes from other substances, since a solvent may be available which will preferentially dissolve one component.

Table 8

Solubility of Petroleum Waxes in Various Solvents

Solvent	Wax in Solution, % 60°F.	70°F.	80°F.	90°F.	100°F.	110°F.	Pounds of Wax per 100 lb Solvent 60°F.	70°F.	80°F.	90°F.	100°F.	110°F.
133–135°F. AMP Refined Paraffin Wax												
Sovasol No. 6	4.4	8.5	16.0	27.5	43.0	61	4.6	9.2	19.0	38.0	75	156
Turpentine	4.6	9.5	19.2	31.8	46.0	63	4.8	10.6	24.7	47.0	85	170
Pine Oil	0.6	1.4	2.9	6.2	13.4	29	0.6	1.5	3.0	6.6	16	41
125–127°F. AMP Refined Paraffin Wax												
Sovasol No. 3	16.0	26.0	41.0	61.0	81.0	90	19.0	35.0	70.0	156.0	425	—
Sovasol No. 4	13.0	23.0	37.0	56.0	76.0	88	15.0	30.0	59.0	127.0	316	—
Sovasol No. 5	10.0	19.0	32.0	49.0	69.0	83	11.0	25.0	47.0	96.0	222	—
Sovasol No. 6	7.9	14.5	26.5	42.0	61.0	81	8.6	16.5	36.0	72.0	156	—
Sovasol No. 5-A	12.0	21.0	33.0	50.0	70.0	84	14.0	27.0	50.0	100.0	233	—
Benzene	10.0	18.0	35.0	56.0	77.0	90	11.0	21.0	54.0	127.0	335	—
Carbon Tetrachloride	10.0	15.0	31.0	48.0	68.0	83	11.0	18.0	45.0	92.0	210	—
Ethylene Dichloride	0.2	0.8	1.6	7.0	30.0	81	0.2	0.8	1.6	7.5	43	—
Perchloroethylene	10.0	15.0	26.0	42.0	63.0	82	11.0	18.0	35.0	72.0	170	—
Trichloroethylene	12.0	20.0	32.0	50.0	68.0	83	14.0	25.0	47.0	100.0	210	—
Halowax 1000	2.0	4.0	8.0	13.0	31.0	62	2.1	4.2	8.7	15.0	45	—
Turpentine	9.2	17.5	29.2	43.5	60.0	78	10.0	21.2	41.0	77.0	150	—
Pine Oil	1.1	2.5	5.4	11.6	25.2	55	1.1	2.6	5.7	14.0	34	—
S/V 907	2.7	5.1	9.5	17.5	33.0	55	2.8	5.4	10.5	20.0	50	—
Petrolatum, M.P. 120°F.												
Sovasol No.6	10.5	19.3	33.0	51	72	—	11.7	25.0	50.0	104	260	—
Turpentine	12.0	22.0	36.0	54	74	—	14.0	28.0	56.0	117	284	—
Pine Oil	1.8	4.0	8.5	19	40	85	1.8	4.2	9.2	25	67	—
S/V 907	4.5	8.2	16.0	30	55	89	4.8	8.9	19.0	43	122	—

Table 9

SOLUBILITY OF PARAFFIN WAX (M.P. 53°C.)

1 cc of	Dissolves at 20°C. mg
Ethyl Acetate	1.1
Acetone	1.2
Alcohol (96%)	1.9
Ether	83.4
Petroleum Ether	200.0
Ligroin	244.0
Chloroform	246.0
Benzene	285.0
Carbon Tetrachloride	317.0

The solubility of paraffin wax in Pliolite increases with increase in temperature to the melting point of the wax. At the melting point, they become miscible in all proportions. These mixtures (Pliowax) are tougher and give more flexible and lustrous films than the wax alone.

Table 10

SOLUBILITY OF PARAFFIN WAX (M.P. 124.5°F.) IN ORGANIC SOLVENTS

Solvent	Temperature °F.	Grams Wax per 100 cc Solvent
Cyclohexanol	87	0.8800
	68	0.2940
	58	0.1420
	47	0.0823
Isopropyl Alcohol	85	0.2400
	66	0.0600
	52	0.0240
	39	0.0094
n-Propyl Alcohol	80	0.2140
	60	0.0535
	46	0.0214
	32	0.0086
tert-Butyl Alcohol	80	0.2920
	58	0.0730
n-Butyl Alcohol	77	0.3600
	59	0.0900
	45	0.0360
Fusel Oil	80	0.4620
	58	0.1150
	42	0.0460

Table 10 (continued)

Solvent	Temperature °F.	Grams Wax per 100 cc Solvent
tert-Amyl Alcohol	72	0.5520
	53	0.1380
	42	0.0552
Diacetone Alcohol	97	0.2560
	78	0.0640
	65	0.0256
Ethyl Alcohol	98	0.3580
	83	0.0895
	72	0.0358
Pentasol	83	0.8280
	61	0.2070
	50	0.0828
n-Propyl Formate	79	1.0480
	65	0.3160
	47	0.0790
Isopropyl Formate	75	0.4120
	57	0.1030
	45	0.0412
n-Butyl Formate	75	0.8980
	57	0.2250
	46	0.0898
	37	0.0475
Methyl Acetate	85	0.2780
	66	0.0695
	57	0.0278
	35	0.0086
Phenyl Acetate	96	0.3500
	75	0.0875
	62	0.0438
Ethyl Oxalate	112	0.6600
	93	0.1650
	80	0.0660
Chloroethyl Carbonate	70	0.5240
	50	0.1310
	39	0.0524
β-β-Dichloroethyl Ether	90	0.2800
	70	0.0700
	57	0.0280
Trichloroethylene	59	2.8600
	36	0.9160
	7	0.2290
Methyl *n*-Hexyl Ketone	65	1.0640
	48	0.2660
	38	0.1064
	30	0.0546
Isobultyl Methyl Ketone Hexone	83	2.0700
	72	0.8280
	46	0.2020

Table 10 (continued)

Solvent	Temperature °F.	Grams Wax per 100 cc Solvent
Cyclopentanone	74	0.5160
	57	0.2290
	45	0.0516
Methyl Cyclohexanone	83	3.0800
	54	0.3520
	35	0.0880
	22	0.0352
Dipropyl Ketone	61	0.8820
	43	0.2210
	33	0.0882
	23	0.0414
Cyclohexane	62	7.7600
	50	2.4480
Benzol (90%)	32	0.7440
	18	0.1860
Cresol	96	0.6080
	75	0.1520
	62	0.0608
Tetralin	56	2.5100
	43	1.1840
	26	0.3940

Table 11

SOLUBILITY OF REFINED PARAFFIN WAX (M.P. 130–132°F.) IN ORGANIC SOLVENTS

Solvent	Temperature °F.	Grams Wax per 100 cc Solvent
Butane	68	3.6700
	43	0.5480
	45	0.5960
	42	0.4880
	14	0.0660
	-4	0.0150
Propane	39-40	0.2380
	45	0.2600
	23	0.0548
	31	0.0824
	6	0.0189
	9	0.0189
	8.5	0.0189
	17.5	0.0330
	-2 to 2	0.0105
	-10	0.0056

Table 11 (continued)

Solvent	Temperature °F.	Grams Wax per 100 cc Solvent
Cylinder Stock Naphtha (P.B. 128–348°F.)	39	0.5000
	44	0.7500
	27	0.2500
	17	0.1250
	60	2.4900
	50	1.2500
	40	0.6200
	6	0.0610
	−10	0.0200
35% Acetone + 65% Benzol	62	0.507
	41	0.0995
	21	0.0205
	12	0.0100

Table 12

SOLUBILITY—GRAMS PARAFFIN WAX PER 100 CC SOLVENT

Temperature °C.	Pentane	Hexane	Heptane	Octane	Isodecane
0	—	2.77	1.37	0.99	—
5	—	3.69	2.18	1.69	0.94
10	5.11	4.81	3.55	2.90	1.44
15	6.94	6.07	5.06	4.24	2.74
20	9.53	8.31	7.18	5.93	4.98
25	17.16	16.23	14.36	11.66	9.17

Table 13

SOLUBILITY—GRAMS PARAFFIN WAX PER MOL SOLVENT

Temperature °C.	Pentane	Hexane	Heptane	Octane	Isodecane
0	—	3.61	2.01	1.60	—
5	—	4.81	3.22	2.73	1.84
10	5.83	6.28	5.23	4.67	2.84
15	7.92	7.91	7.48	6.84	5.40
20	10.87	10.83	10.57	9.58	9.80
25	19.48	21.33	21.06	18.81	18.03

The solubility of a solid paraffin in hydrocarbons of low molecular weight increases rapidly with a rise in temperature from that of isodecane to that of pentane. The increase in solubility with rise of temperature is more rapid with the higher-molecular-weight solvent. This can be seen by dividing the solubilities at 10°C. into the solubility at 25°C., the ratio being 2.3, 3.8, 4.0, 3.9, and 6.6 for pentane, hexane, heptane, octane, and isodecane.

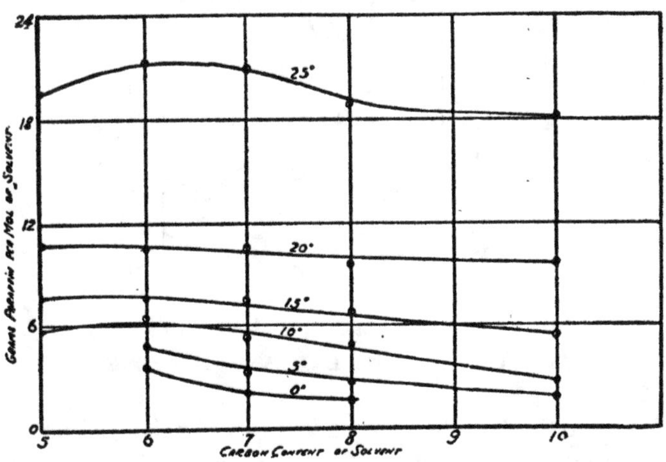

Figure 8. SOLUBILITY OF PURE PARAFFIN WAX IN HYDROCARBONS[1]

Paraffin wax (M.P. 56°C.) is dissolved in solvents at 19°C. as follows: benzene 9.34%; toluene 9.74%; xylene 9.90%; benzene (at 20°C.) 11.82%. The amount of paraffin wax required to cause gel formation decreases with the viscosity of the solvent as shown by the addition of paraffin to benzene, xylene, chloroform, turpentine and carbon disulfide. The solubility varies with the isomeric form. Tests with various mineral solvents show that minimum viscosity is associated with maximum adsorption.[2]

---

1 P. Weber and H. L. Dunlap, *Ind. Eng. Chem.* **20**, 384 (1928).
2 M. A. Rakuzin, *Petroleum Z.* 21, 1489 (1925).

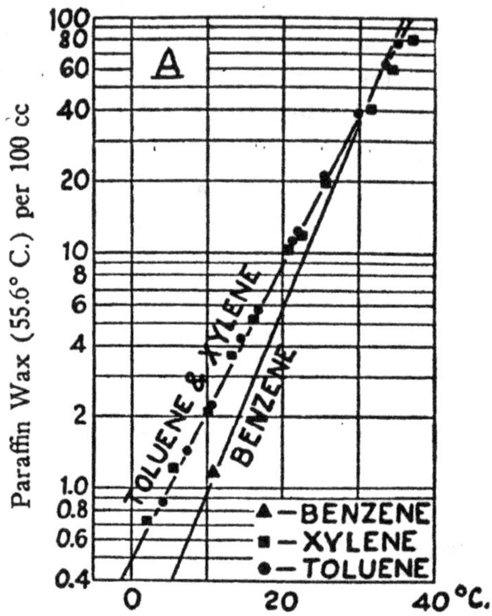

Figure 9. SOLUBILITY OF PARAFFIN WAX IN TOLUENE–XYLENE AND BENZENE

For any particular solvent, the solubility is the greater the lower the melting point of the paraffin wax and for any particular paraffin wax, the solubility decreases with increasing density of the solvent. Benzene of 0.677 density dissolves about ten times as much paraffin wax as machine oil, of 0.908 density. In the liquid state, the paraffin waxes are miscible in all proportions with all petroleum fractions, benzene, and isoamyl alcohol. As the solubility of the paraffins increases more rapidly at higher temperatures, a given lowering of temperature will cause the deposition of more paraffin wax at a high temperature than at a low temperature. The difference between the temperature at which paraffins first separate on cooling a solution and the solidifying point of the paraffins is the greater the lower the temperature of deposition, that is, the lower the concentration of the solution. The difference is 1–3°, 3–7°, and 7–20° for solutions containing more than 15 g, 15–5 g, and less than 5 g per 100 g of solvent. The higher the freezing point of the solution, the greater is the proportion of wax which separates out on cooling to the freezing point. During solidification, the whole of the solvent is absorbed by the solid paraffins, which reduces the rise of temperature and thus, at low temperatures, a small separation of paraffins is sufficient to cause solidification. The quantity of absorbed solvent is dependent on the size of the crystals, which again depends on the previous treatment of the solution.

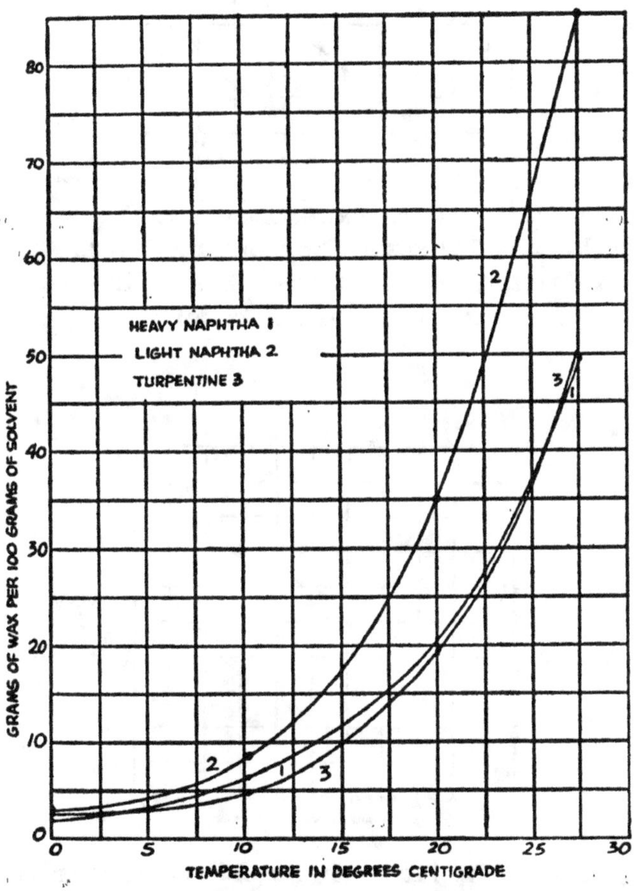

Figure 10. TEMPERATURE–SOLUBILITY CURVES FOR PARAFFIN WAX IN NAPHTHAS AND TURPENTINE[1]

[1] Figures 10–16 are reprinted from J. W. Poole, *Ind. Eng. Chem.* 23, 171 (1931).

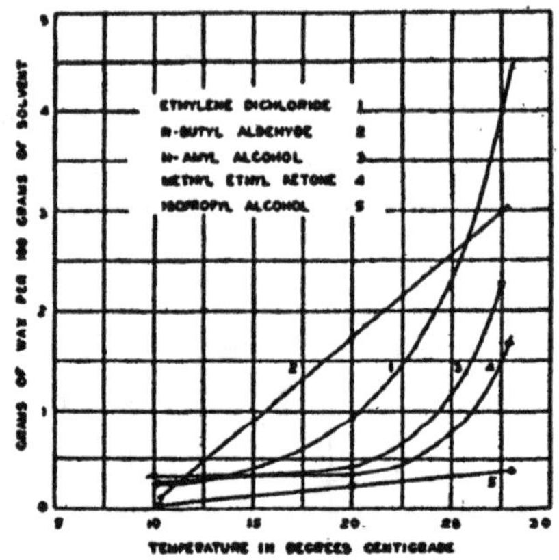

Figure 11. TEMPERATURE–SOLUBILITY CURVES FOR PARAFFIN WAX IN ETHYLENE DICHLORIDE *n*–BUTYL ALDEHYDE, *n*–AMYL ALCOHOL, METHYL–ETHYL KETONE, AND ISOPROPYL ALCOHOL[1]

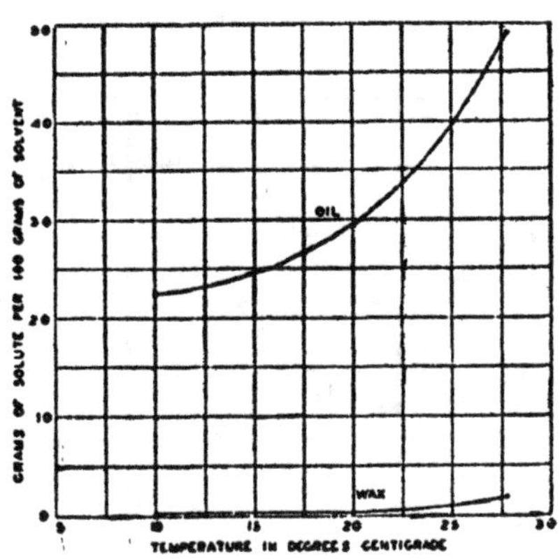

Figure 12. TEMPERATURE–SOLUBILITY CURVES FOR PARAFFIN WAX AND MINERAL OIL IN METHYL–ETHYL KETONE[1]

[1] Figures 10–16 are reprinted from J. W. Poole, *Ind. Eng. Chem.* 23, 171 1931).

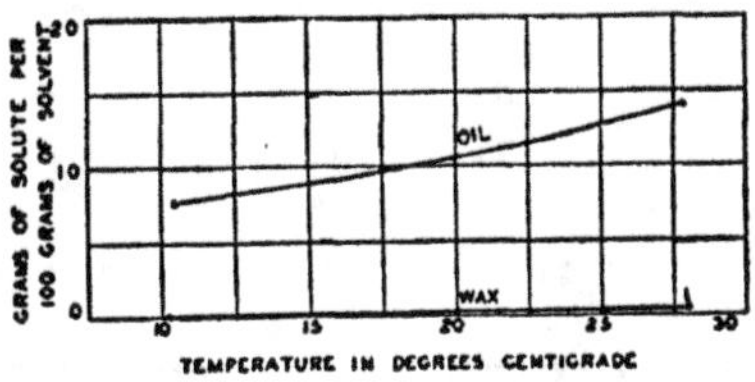

Figure 13. TEMPERATURE–SOLUBILITY CURVES FOR PARAFFIN WAX AND MINERAL OIL IN ISOPROPYL ALCOHOL[1]

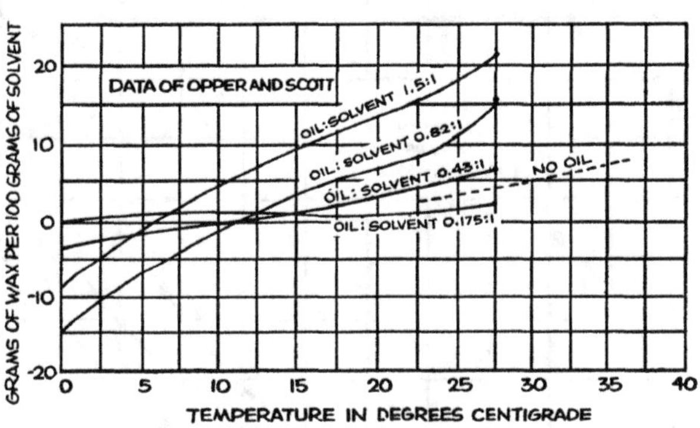

Figure 14. TEMPERATURE–SOLUBILITY CURVES FOR PARAFFIN WAX IN VARIOUS SOLUTIONS OF MINERAL OIL AND BUTANOL[1]

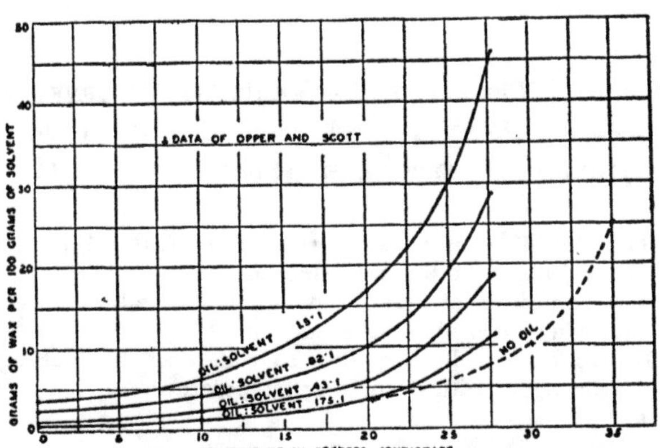

Figure 15. TEMPERATURE–SOLUBILITY CURVES FOR PARAFFIN WAX IN VARIOUS SOLUTIONS OF MINERAL OIL AND BUTYL ACETATE[1]

[1] Figures 10–16 are reprinted from J. W. Poole, *Ind. Eng. Chem.* **23**, 171 (1931).

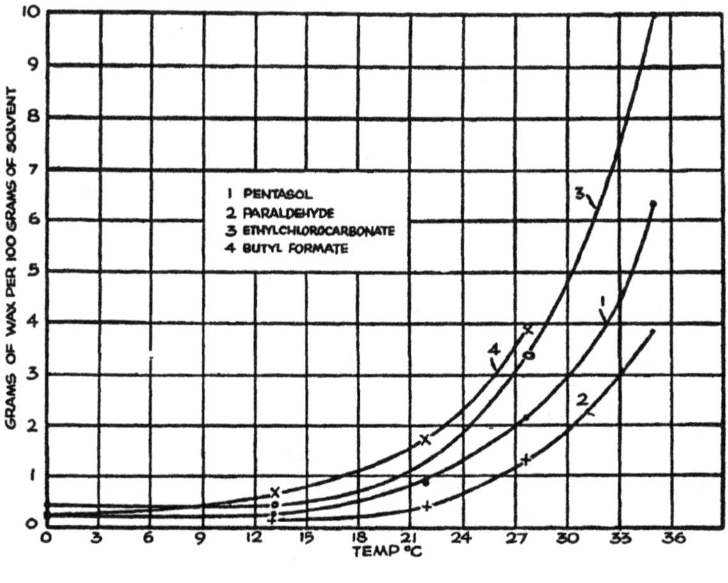

Figure 16. TEMPERATURE—SOLUBILITY CURVES FOR PARAFFIN WAX IN PENTASOL, PARALDEHYDE, ETHYL CHLOROCARBONATE, AND BUTYL FORMATE[1]

1. The solubility of paraffin wax in mineral oil increases as the melting point of the wax decreases.
2. The solubility of paraffin wax in mineral oil decreases with increasing viscosity of the solvent.
3. Difference of solubility due to difference in melting point of wax decreases with decreasing temperature.

[1] Figures 10–16 are reprinted from J. W. Poole, *Ind. Eng. Chem.* **23**, 171 (1931).

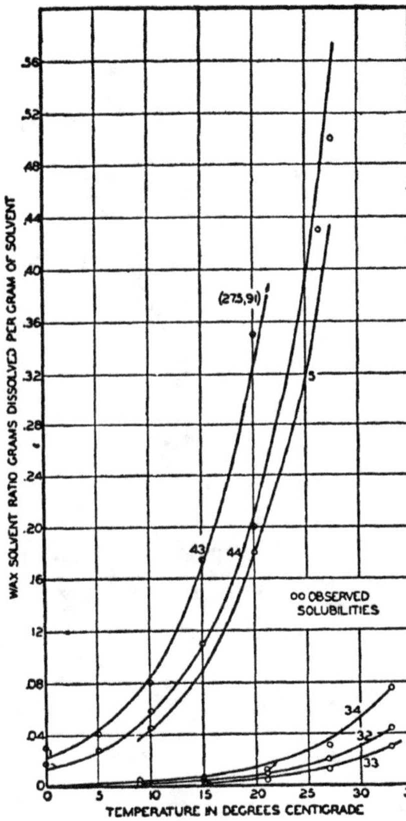

Figure 17. TEMPERATURE—SOLUBILITY CURVES FOR WAX IN VARIOUS SOLVENTS SHOWING AGREEMENT BETWEEN OBSERVED AND CALCULATED VALUES[1]

(Curves represent calculated values) 5, chlorobenzene; 32, pentasol; 33, ethyl chlorocarbonate; 34, butyl formate; 43, light naphtha (27.5 and 0.91 are calculated coordinates; 27.5 and 0.85 are observed); 44, heavy naphtha.

[1] J. W. Poole and T. A. Mangelsdorf, *Ind. Eng. Chem.* **24**, 1217 (1932).

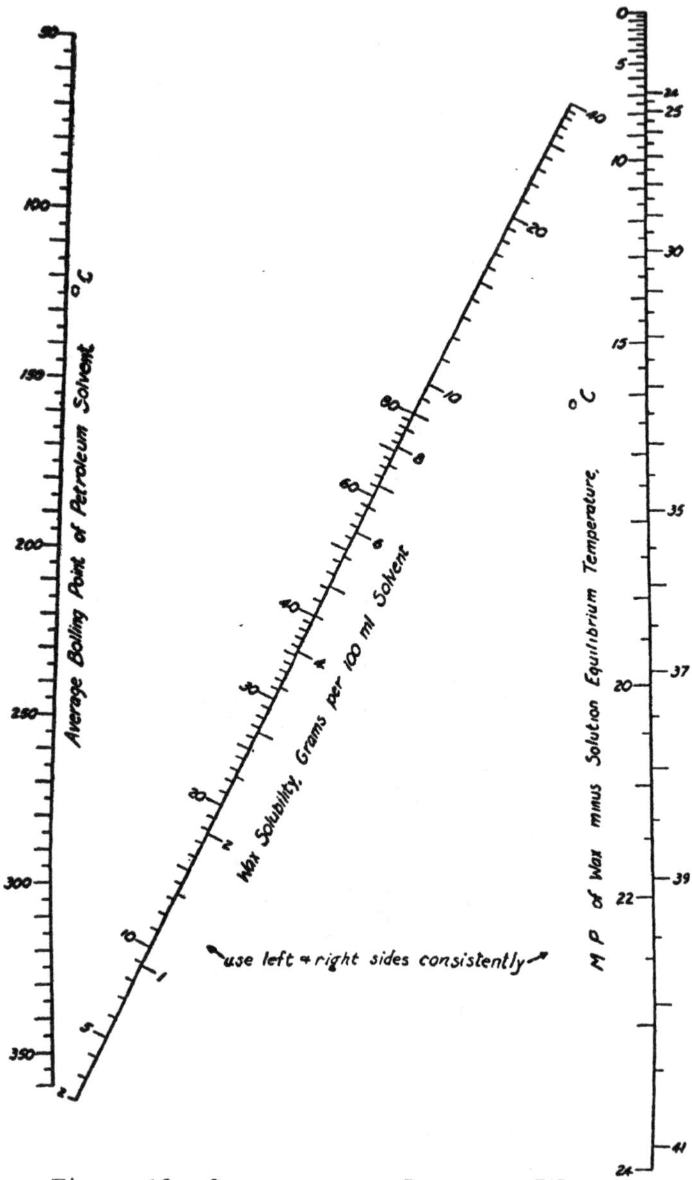

Figure 18. SOLUBILITY OF PARAFFIN WAX IN PETROLEUM FRACTIONS[1, 2]

[1] D. S. Davis, *Ind. Eng. Chem.* 32, 1293 (1940).
[2] A. Berne-Allen and L. T. Work, *Ibid.* 30, 811 (1938).

The use of the chart of Figure 18 may be illustrated as follows: What is the solubility of paraffin wax in petroleum when the melting point of the wax is 45°C., the temperature of the solution is 26°C., and the average boiling point of the pure distillate is 260°C.? The difference between the melting point and the solution temperature is 45–26, i.e., 19°C. Connect 19 on the right-hand scale with 260 on the left-hand scale and read the solubility on the inclined axis as 31 g wax per 100 ml solvent. Similarly, the chart reveals that 100 ml of a distillate, boiling at 190°C., will dissolve 5 g of paraffin wax when the melting point of the wax is 60°C. and the temperature of the solution 23°C. Note that one reads solubilities on the right-hand side of the inclined scale when temperature differences are read on the right-hand side of the temperature-difference scale and conversely.

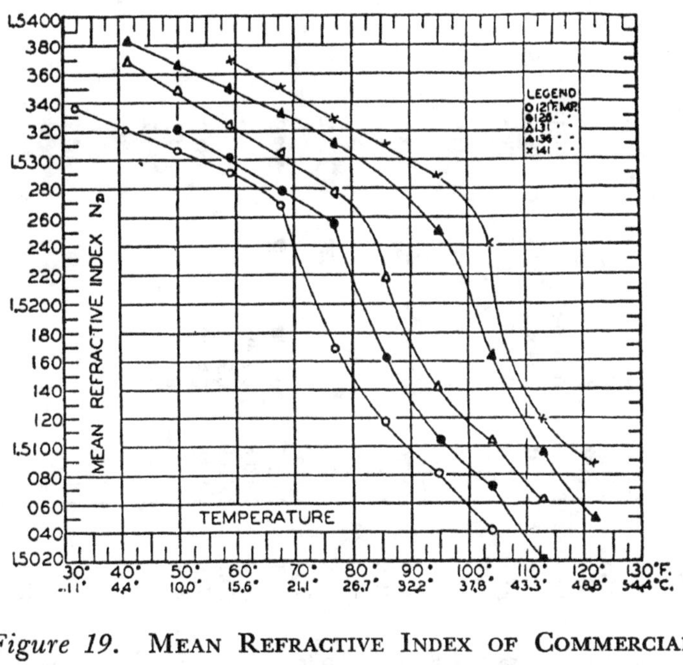

Figure 19. MEAN REFRACTIVE INDEX OF COMMERCIAL PARAFFIN WAXES IN THE SOLID STATE[1]

[1] J. M. Page, *Ind. Eng. Chem.* **28**, 856 (1936).

Table 14

Data on Purified, Oil-Free Petroleum Waxes[1]

Melting Point				Observed					Calculated			
°F.	°C.	n_D^{84}	d^{84}	Mol. Wt.	Mol. Refractivity	% C	% H	Formula	Mol. Wt.	% C	% H	Mol. Refractivity
					Paraffin Wax							
80.5	27.0	1.4133	0.7347	255	86.5	84.55	15.04	$C_{18}H_{38}$	254	85.04	14.96	86.9
88.5	31.4	1.4155	0.7397	271	92.0			$C_{19}H_{40}$	268			89.6
97.5	36.4	1.4173	0.7417	287	97.6			$C_{20}H_{42}$	282			94.1
101.0	38.3	1.4183	0.7440	292	99.1			$C_{21}H_{44}$	296			98.8
105.5	40.8	1.4191	0.7458	294	99.6	84.86	14.98	$C_{21}H_{44}$	296	85.10	14.90	98.8
107.0	41.7	1.4191	0.7466	298	100.8			$C_{21}H_{44}$	296			98.8
114.2	45.7	1.4210										
116.7	47.1	1.4217	0.7513	330	111.2	84.75	14.91	$C_{23}H_{48}$	324	85.18	14.82	107.9
118.7	48.2	1.4221										
121.0	49.4	1.4226	0.7526	337	113.5			$C_{24}H_{50}$	338			112.6
126.5	52.5	1.4240										
127.5	53.0	1.4242										
128.0	53.3	1.4244	0.7558	357	121.2	85.31	14.75	$C_{25}H_{52}$	352	85.21	14.79	117.2
131.5	55.3	1.4252										
133.5	56.4	1.4257	0.7580	367	123.9			$C_{26}H_{54}$	366			121.8
137.2	58.4	1.4266	0.7607									
138.5	59.1	1.4269	0.7609	385	129.9			$C_{27}H_{56}$	380			126.3
144.2	62.3	1.4285	0.7642	410	137.7			$C_{29}H_{60}$	408			135.5
149.5	65.3	1.4297	0.7660	430	144.7	85.22	14.74	$C_{31}H_{64}$	436	85.32	14.68	144.7
156.0	68.9	1.4315	0.7687	456	153.7			$C_{32}H_{66}$	450			149.3
					Petrolatum Wax							
159.7	71.0	1.4323	0.7706	474	159.6			$C_{34}H_{70}$	478			159.6
164.5	73.6	1.4333	0.7727	492	164.4	85.45	14.78	$C_{35}H_{72}$	492	85.36	14.64	163.2
166.7	74.8	1.4339	0.7739	502	168.9			$C_{36}H_{74}$	506			169.9
169.2	76.2	1.4347	0.7753	520	174.9			$C_{37}H_{76}$	520			172.4
170.7	77.1	1.4351	0.7759	539	181.3			$C_{38}H_{78}$	534			177.0
173.7	78.8	1.4359	0.7771	555	186.7	85.13	14.62	$C_{39}H_{80}$	548	85.40	14.60	181.6
177.2	80.7	1.4366	0.7784	572	192.3			$C_{41}H_{84}$	576			190.8
181.2	82.9	1.4375	0.7803	586	194.9			$C_{42}H_{86}$	590			195.4
182.7	83.8	1.4378		603		85.54	14.47	$C_{43}H_{88}$	604	85.43	14.57	200.0

[1] Buchler and Graves, *Ind. Eng. Chem.* 19, 725 (1927).

Table 14 (continued)

Melting Point °F.	°C.	n_D^{84}	d^{84}	Observed Mol. Wt.	Observed Mol. Refractivity	Observed % C	Observed % H	Formula	Calculated Mol. Wt.	Calculated % C	Calculated % H	Calculated Mol. Refractivity
					Slop	*Wax*						
132.2	55.7	1.4255		360		84.87	14.83	$C_{26}H_{54}$	366	85.23	14.77	
136.2	57.9	1.4263	0.7600	385	129.7			$C_{27}H_{56}$	380			126.3
138.5	59.2	1.4270										
139.0	59.4	1.4270	0.7614	394	133.0			$C_{28}H_{58}$	394			130.9
140.5	60.3	1.4275										
142.5	61.4	1.4280										
144.5	62.5	1.4284										
145.5	63.1	1.4287	0.7644	405	136.5	85.18	14.81	$C_{29}H_{60}$	408	85.29	14.71	135.5
148.5	64.7	1.4294										
150.5	65.9	1.4300	0.7677	431	145.1	85.19	14.74	$C_{31}H_{64}$	436	85.32	14.68	144.7
152.0	66.7	1.4303										
153.5	67.5	1.4306	0.7677	437	147.5			$C_{31}H_{64}$	436			144.7
156.0	68.5	1.4312	0.7684	451	152.1			$C_{32}H_{66}$	450			149.3
157.0	69.4	1.4316	0.7695	460	155.0			$C_{33}H_{68}$	464			153.9
160.0	71.1	1.4323	0.7706	470	158.1			$C_{33}H_{68}$	464			153.9
161.5	72.0	1.4327										
162.2	72.3	1.4329	0.7715	483	162.7	85.22	14.65	$C_{34}H_{70}$	478	85.35	14.65	163.1
168.0	75.6	1.4342	0.7741	508	171.6			$C_{36}H_{74}$	506			167.7
176.5	80.3	1.4365	0.7785	563	189.2			$C_{40}H_{82}$	562			186.1
182.0	83.3	1.4379	0.7810	600	201.8	85.21	14.65	$C_{43}H_{88}$	604	85.43	14.57	199.9
					Rod	*Wax*						
165.0	73.9	1.4334	0.7728	493	165.9			$C_{35}H_{72}$	492			163.2
167.5	75.3	1.4340	0.7739	516	173.6	85.40	14.55	$C_{37}H_{76}$	520	85.38	14.62	172.4
169.5	76.4	1.4345	0.7743	524	176.4			$C_{37}H_{76}$	520			172.4
171.0	77.2	1.4349	0.7757	527	177.2			$C_{38}H_{78}$	534			177.0
172.7	78.2	1.4353	0.7761	545	183.3	85.38	14.68	$C_{39}H_{80}$	548	85.40	14.60	181.6
174.2	79.0	1.4358	0.7771	550	185.0			$C_{39}H_{80}$	548			181.6
176.7	80.5	1.4363	0.7780	569	191.3			$C_{40}H_{82}$	562			186.2
180.5	82.5	1.4372		575		85.39	14.60	$C_{41}H_{84}$	576	85.41	14.59	190.8

It has been found that the melting point of the wax has a profound effect on its solubility in any given solvent. The higher-melting waxes are less soluble than those of lower melting point.

Table 15

SPECIFIC REFRACTION OF COMMERCIAL PARAFFIN WAXES IN THE LIQUID STATE

M.P. of Wax °F.	°C.	n	d	Temp. t°C.	Mol. Wt.	Approx. Formula	r_D At t°	r_D At 20°C.	r_D Calcd.
121	49.4	1.4349	0.7780	60	332	$C_{23.55}H_{49.1}$	0.3353	0.3343	0.3341
131	55.0	1.4355	0.7788	60	340	$C_{24.1}H_{50.2}$	0.3352	0.3342	0.3338
126	52.2	1.4352	0.7800	60	355	$C_{25.2}H_{52.4}$	0.3349	0.3339	0.3340
136	57.8	1.4370	0.7815	60	372	$C_{26.4}H_{54.8}$	0.3351	0.3341	0.3337
141	60.6	1.4294	0.7690	85	376	$C_{26.7}H_{55.4}$	0.3353	0.3337	0.3337

Table 16

SPECIFIC REFRACTION OF COMMERCIAL PARAFFIN WAXES IN THE SOLID STATE[a]

Temperature °F.	°C.	r_D of Wax Melting at: 121°F. (49.4°C.)	126°F. (52.2°C.)	131°F. (55.0°C.)	136°F. (57.8°C.)	141°F. (60.6°C.)
30	-1.1	0.3434	0.3405	0.3413	0.3400	—
40	4.4	0.3429	0.3403	0.3410	0.3410	—
50	10.0	0.3430	0.3401	0.3412	0.3403	0.3415
60	15.6	0.3437	—	0.3410	0.3405	0.3414
70	21.1	—	—	0.3411	0.3405	0.3413
80	26.7	—	—	—	—	0.3414
100	37.8	—	—	—	—	0.3414

Light Sensitivity of Paraffin Wax

Paraffin wax, containing more than 0.1% oil, is unstable to ultraviolet rays. Paraffin wax treated with Fuller's earth and exposed to ultraviolet light is light stable.[1]

[a] Calculated from curves.

[1] V. I. Shile, *Neftyanoe Khozyaistvo* **25**, 59 (1933).

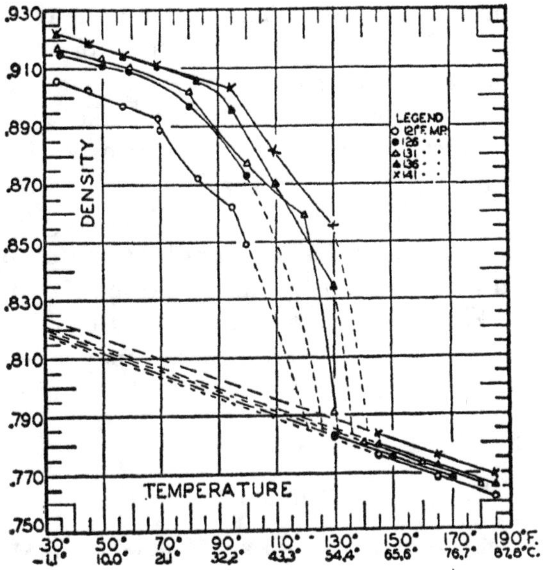

Figure 20. DENSITY OF COMMERCIAL PARAFFIN WAXES

Table 17

THE TRUE DENSITY OF AND VOLUME PER CENT IN SOLID COMMERCIAL PARAFFIN WAXES

M.P. of Wax		r_D at 20°C.	Temperature		n (Obsvd.)	d_1 (Obsvd.)	d_2 (Calcd.)	d_2—d_1	Volume % Air
°F.	°C.		°*F.*	°*C.*					
121	49.4	0.3343	50	10.0	1.5306	0.900	0.925	0.025	2.7
			60	15.6	1.5289	0.897	0.922	0.025	2.8
126	52.2	0.3342	50	10.0	1.5321	0.911	0.927	0.016	1.8
			60	15.6	1.5297	0.909	0.924	0.015	1.7
131	55.0	0.3339	50	10.0	1.5348	0.912	0.932	0.020	2.2
			60	15.6	1.5323	0.909	0.928	0.019	2.1
136	57.8	0.3341	50	10.0	1.5366	0.917	0.934	0.017	1.9
			60	15.6	1.5347	0.914	0.931	0.017	1.9
141	60.6	0.3337	60	15.6	1.5366	0.914	0.935	0.011	2.3
			70	21.1	1.5344	0.911	0.932	0.011	2.3

Tensile Strength and Density of Paraffin Wax[1]

The tensile strength of paraffin wax increases rapidly from 30 to about 50°F., when it begins to fall off; a maximum is

[1] W. F. Seyer and K. Inouye, *Ind. Eng. Chem.* 27, 567 (1935).

reached at about 0°. Cooling the wax from 15 to —6° over a period of 10 hours does not reduce the strain or flaws any more than rapid cooling. This decrease of tensile strength is due to tiny cracks. Above 30°F. the wax behaves like a plastic solid, because of liquefaction of some of the hydrocarbons. Below 25°F., the density of paraffin wax is a linear function of temperature. If density is plotted against tensile strength, the curve is not linear. It is evident that the tensile strength and the density are affected by the nature of the crystal matrix formed, since paraffin wax is known to contain three types of crystals: plates, needles and macrocrystals. Occluded air may also be a contributing factor.

Table 18

SPECIFIC GRAVITY OF THREE PARAFFIN WAXES
(M.P. 60–130°F.)

Temperature °F.	Temperature °C.	118.5°F. (48°C.)	123.2°F. (50.7°C.)	127.2°F. (52.9°C.)
60	15.56	0.906	0.907	0.920
70	21.10	0.899	0.903	0.913
80	26.70	0.885	0.890	0.905
90	32.20	0.882	0.887	0.893
100	37.80	0.874	0.868	0.878
110	43.30	0.858	0.856	0.876
115	46.10	0.841	0.830	0.870
120	48.90	0.801	0.802	0.848
125	51.70	0.802	0.801	0.818
130	54.40	0.780	0.785	0.803

Crystal Forms of Paraffin Wax

Certain paraffin waxes crystallize in two forms which differ in crystal symmetry, the less symmetrical form being nearly triclinic and the more symmetrical form monoclinic. The first exhibits, under the microscope, individual or twin crystals, whereas the second shows needlelike crystals in a fibrous structure.[1]

[1] Yannaquis, *Ann. combustibles liquidas* 9, 295 (1934).

Crystal Growth

Small amounts of Paraflow inhibit excessive growth of crystals and change the shape from the needle type to small equiaxial grains of uniform size.[2]

Transition Effects in Paraffin Wax[3]

When different temperatures are plotted against time, a minimum marks the melting point, a maximum, probably, the final solidification point and a second minimum, in some waxes, a transition point at which internal forces modify the crystalline structure, with release of energy. This second minimum marks the maximum rate of change, that is, from the α- to the β-condition. Isoparaffins show no transition points. Some waxes exhibit no transition point unless cooling is rapid, others, none at all. Waxes, blended for a particular melting point, tend to be more homogeneous in structure and more stable in temperature change if they have no transition point. Translucency and a transition point accompany each other. The transition change leads to contraction in crystalline boundaries and this is unfavorable where wax is used to prevent the passage of water vapor, as in electric condensers, or waxed papers; it is an advantage, however, when the wax is to be molded.

The transition from needle-shaped prisms to rhomboid plates or leafy masses occurs at 10–15° below the melting point. The crystal form depends on the solvent used, the rate of cooling and the wax used. Wax dissolves 7–15% of its own volume of air, at ordinary temperature.[4]

Fibrous Paraffin Wax

Paraffin wax, after drawing by hand or extrusion through a die, is shown by its X-ray diffraction effects to have a distinct fiber structure. The fiber axis may have either of two crystallographic directions, depending on the conditions of formation, but, in both cases, the long axes of the paraffin molecules are perpendicular to the fiber axis.[5]

---

[2] G. H. B. Davis, *Nat. Petroleum News* **24**, 32 (1932).
[3] C. R. Scott-Harley, *J. Inst. Petroleum* **25**, 238 (1939).
[4] J. A. Carpenter, *J. Inst. Petroleum Tech.* **12**, 288 (1926).
[5] H. Rivsan, *Mem. Coll. Sci. Kyoto Imp. Univ.* **A19**, 279 (1936).

Table 19

HEAT CONTENT AND HEAT OF SOLUTION OF PARAFFIN WAXES (M.P. 110, 125, AND 140°F., IN BTU/LB)[1]

Temp. °F	Heat content of wax			Effective heat content of wax in solution			Heat of solution			Temp. °F.
	Melting point			Melting point			Melting point			
	110°	125°	140°	110°	125°	140°	110°	125°	140°	
0	−15	−13	−11	62	65	68	77	78	79	**0**
5	−13	−11	−9	65	68	71	78	79	80	5
10	−11	−9	−8	67	70	73	78	79	81	10
15	−8	−7	−6	70	73	76	78	80	82	15
20	−6	−5	−4	72	75	78	78	80	82	20
25	−3	−2	−2	75	78	81	78	80	83	**25**
32	0	0	0	78	81	84	78	81	84	32
35	+1	+1	+1	79	82	85	78	81	84	35
40	4	3	+3	82	85	88	78	82	85	40
45	7	6	5	85	88	91	78	82	86	45
50	10	8	7	87	90	93	77	82	86	**50**
55	13	11	9	90	93	96	77	82	87	55
60	16	13	11	92	95	98	76	82	87	60
65	19	16	13	95	98	101	76	82	88	65
70	23	19	16	98	101	104	75	82	88	70
75	27	22	19	100	103	107	73	81	88	**75**
80	32	27	22	103	106	109	71	79	87	80
85	38	31	25	105	108	111	67	77	86	85
90	44	36	29	108	111	114	64	75	85	90
95	51	42	33	111	114	117	60	72	84	95
100	60	49	38	113	116	119	53	67	81	**100**
105	73	57	43	116	119	122	43	62	79	105
110	119	65	49	119	122	125	0	57	76	110
115	121	76	55	121	124	127	0	48	72	115
120	124	90	62	124	127	130	0	37	68	120
125	127	130	71	127	130	133	0	0	62	**125**
130	130	133	81	130	133	136	0	0	55	130
135	132	135	94	132	135	138	0	0	44	135
140	135	138	141	135	138	141	0	0	0	140
145	138	141	144	138	141	144	0	0	0	145
150	141	144	147	141	144	147	0	0	0	**150**
160	146	149	152	146	149	152	0	0	0	160
170	152	155	158	152	155	158	0	0	0	170
180	157	160	163	157	160	163	0	0	0	180
190	163	166	169	163	166	169	0	0	0	190
200	169	172	175	169	172	175	0	0	0	**200**
220	181	184	187	181	184	187	0	0	0	220
240	193	196	199	193	196	199	0	0	0	240
260	205	208	211	205	208	211	0	0	0	260
280	217	220	223	217	220	223	0	0	0	280
300	230	233	236	230	233	236	0	0	0	**300**
320	243	246	249	243	246	249	0	0	0	320
340	256	259	262	256	259	262	0	0	0	340
360	269	272	275	269	272	275	0	0	0	360
380	282	285	288	282	285	288	0	0	0	380
400	295	298	301	295	298	301	0	0	0	**400**

[1] *Thermal Properties of Petroleum Products*, U. S. Bureau of Standards, Pbl. 97, 45 (1929).

Heat of Fusion

Paraffin wax of an average molecular weight of 400 has a molecular heat of fusion of about 21,000 cal.[2] Its latent heat of fusion is 38.9—43.9 cal., increasing with rise of specific gravity, solidifying point and molecular weight.[3]

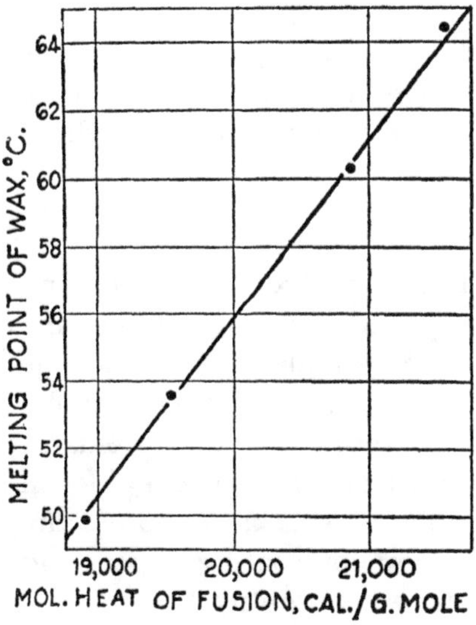

Figure 21. HEAT OF FUSION OF PARAFFIN WAX vs. MELTING POINT[4]

Table 20

MOLECULAR WEIGHTS AND ANILINE POINTS

ASTM Melting Point °F.	ASTM Melting Point °C.	Apparent Oil %	Observed Mol. Wt.	Aniline Point °C.	Calculated Mol. Wt.	Corresponding Hydrocarbon
121	49.4	0.5	332	116.7	330	$C_{23.55}H_{49.1}$
126	52.2	0.5	340	117.8	342	$C_{24.1}H_{50.2}$
131	55.0	0.3	355	119.0	359	$C_{25.2}H_{52.4}$
136	57.8	0.3	372	121.0	370	$C_{26.4}H_{54.8}$
141	60.6	0.3	376	122.2	375	$C_{26.7}H_{55.4}$

The apparent Poisson constant of paraffin wax is 0.50.

[2] Sakhanov, *Neffyanoe Slantsevoc Khoz.* **6**, 820 (1924).
[3] G. Von Kozicki, *Chem. Ztg.* **42**, 147 (1918).
[4] A. Berne-Allen and L. T. Work, *Ind. Eng. Chem.* **30**, 813 (1938).

Electrical Properties

The electrical conductivity of paraffin wax rises to a maximum at 25°C., then falls to a minimum at 35°C., and again rises sharply at 40°C. In the softening zone (40–49°C.), paraffin behaves like amorphous polar substances. "B.P." paraffin wax has a conductivity of 0.00030 at 20°C.

Property Changes

Exposure of paraffin wax to oxidation (air or chemical) or high-frequency discharges reduces its hardness and lowers its melting point.

Preservation

Autoxidation of paraffin wax, at ordinary temperature, is slight and slow and the products formed diffuse into the air before detection is possible. Small quantities of hydroquinone, malachite green, or sulfur prevent the oxidation. Paraffin wax is used for filling up spaces in explosives so that premature explosion does not occur; for converting explosives to a phlegmlike state to make them less susceptible to shock; for protecting metals against powder and explosives; for coating papers and cartons used to protect fireworks against humidity.

Wetting

Paraffin wax is "wetted" with water containing 0.01% or more Saponin.

The addition of fatty acids lowers the interfacial tension of paraffin wax against water.[1]

Expansion

The coefficient of expansion by heat of solid paraffin wax is about 0.005108. Paraffin wax obtained from lignite tar has approximately the same coefficient of expansion but, in the molten state, the value is almost double. The variation of the specific gravity of solutions of paraffin in solvents of high density, can be expressed by the rule of mixtures, but, in the case of solvents of low density, one observes certain irregularities in the neighborhood of the point of fusion. The density of solutions, at tempera-

[1] Bhatnagar and Gardner, *J. Soc. Chem. Ind.* **39**, 185 (1920).

tures above the fusion point of paraffin wax, may be approximately calculated by the rule of mixtures.[2]

Methods for Increasing or Decreasing Special Properties of Paraffin Wax

Length

By length is meant ductility or the ability to distort wax without fracture. Paraffin waxes, in general, are quite brittle and must be compounded with large quantities of an additive to improve them in this respect. Such additives should be quite miscible with paraffin wax, such as microcrystalline waxes or petrolatum. The use of oil alone is not satisfactory since it makes the waxes quite crumbly. If large quantities of rubberlike products, such as Vistanex, are used, the ductility of the wax can be improved somewhat. Other resins and hard vegetable waxes cannot be used for this purpose.

Microcrystalline waxes, however, are quite ductile in themselves and are easier to modify to improve ductility. Since microcrystalline waxes have an affinity for oil, they can be made more ductile by the incorporation of oil. A similar effect can usually be obtained with petrolatum. Most resins, unless they are soft, tend to harden the wax rather than to make it more ductile.

Body or Viscosity

The body or viscosity of paraffin wax can be increased by incorporating a highly viscous substance which is soluble, or partially soluble, in paraffin wax and forms a colloidal dispersion. Such additives are rubber, resins, metallic soaps, Vistanex, etc.

A decrease in viscosity can be obtained by blending the wax with a substance with which it is quite miscible and which has a lower viscosity than the wax in question. For example, the viscosity of a microcrystalline wax can be reduced by blending it with a paraffin wax or a low-viscosity mineral oil.

Adhesion

The adhesion or adhesiveness of a paraffin wax can best be increased by adding a somewhat tacky resin. The resin most

---

[2] Anon., *J. pétrole* **21,** 3 (1921).

commonly used for this purpose is rosin or one of its derivatives, such as ester gum or hydrogenated rosin. The increase in adhesiveness is practically proportional to the quantity of additive. If the resin is too soluble in the wax, the effect on the adhesiveness is relatively slight. Frequently, the adhesiveness of a mixture of wax and rosin can be improved by the use of small amounts of oil which increases the ductility of the mixture.

Slipperiness

In order to increase the slipperiness of the wax, it has to be hardened. With microcrystalline waxes, this can be done best by the use of large amounts of paraffin wax, or by adding substances which become insoluble when the wax solidifies. Carnauba wax or other vegetable waxes and various hard resins are suitable for this purpose. With paraffin wax, the problem is much less complicated, since it is quite hard and slippery. Small amounts of carnauba wax are usually quite effective in hardening the surface of paraffin wax. Sometimes, it is possible to increase slipperiness of paraffin wax by adding some oil which is generally repelled by the paraffin crystals and, therefore, lies on the surface as an oily film. Such mixtures, however, are usually objectionable because of an increased crumbliness and a loss in tensile strength.

Tensile Strength

The tensile strength of waxes is usually important only in the case of refined waxes. There are numerous additives which can be used to improve tensile strength, but one of the best is microcrystalline wax. Stearic acid and ozokerite are used for this purpose. It will be found that the tensile strength curve for a blend of refined and microcrystalline waxes is quite erratic and exhibits two maxima and one minimum. The maxima occur at about 5% and 25% of the microcrystalline wax, whereas the minimum is at about 10%.

Penetration or Wetting

Penetration or wetting properties of paraffin waxes are of interest only when they are molten. Then the waxes may be regarded as oils, so that additives, which are normally employed for oils to improve their wetting properties, will also work for the waxes. The list of agents is quite long, but the use of small amounts of fatty matter, such as stearic acid or high-titer glycerides, is a simple means of improving wetting. It will usually be found that 5% or less of these additives is all that is required to obtain best wetting properties. Excessive penetration, especially in paper, is prevented by the addition of carnauba wax or Acrawax C, which precipitate on cooling and form a solid barrier against liquid penetration.

Waterproofness

Practically any paraffin wax, when used as a continuous film, is moistureproof and waterproof. The main difficulty is to maintain a continuous film. For example, if paper is coated with a brittle, refined wax to produce a continuous film, this film will be found to be practically impervious to moisture. However, the wax film is easily broken by creasing or by shock chilling. Cracks in the wax film serve as channels through which moisture can pass. The problem, therefore, is one of improving the ductility of the wax to resist cracking or flaking. As a consequence, improvements in moistureproofness and waterproofness are related to improvements in ductility. It may be stated, however, that microcrystalline waxes are used extensively for improving waterproofness, because of their inherent ductility.

Shrinkage

Changes in the specific gravity of waxes with temperature are fundamental characteristics and cannot be greatly modified without using large quantities of additives. In general, it can be stated that the higher melting the wax, the greater will be its shrinkage on cooling from the liquid to the solid state. Conversely, the more closely the wax approaches an oil, the less will be its change in volume on cooling over the same temperature range. Sometimes, it is possible to improve waxes by adding substances which by themselves would exhibit little change over the critical temperature range, but the problem is, by no means, a simple one.

Paraffin Wax—Bitumen Data

Paraffin waxes, added to a series of bitumens,[1] lower the drop

[1] V. T. Cerchez, *Mon. pétrole roumain* **39**, 1535 (1939).

point; tube wax and especially ceresin raise it. Paraffins affect the Kraemer-Sarnow softening point variably, depending on the consistency of the bitumen; ceresins raise it; tube wax raises it very slightly or has no influence. The ring-and-ball softening point is lowered by paraffin wax, but raised by ceresins and tube waxes. All three kinds of wax cut down the ductility of the bitumens. The paraffins, again, have a variable effect on the penetration at 25°C., but both ceresins and tube waxes increase it. No influence on hardening point or stability can be found with any of the waxes. As to sensitiveness to heat, the paraffin additions increase it, the ceresins and tube waxes reduce it. In general, the ceresins are beneficial, the tube waxes have the least effect and paraffin wax lowers the physical constants of bitumens.

Compatibility of Additives

Propionic and hydroxystearic acids, which are not compatible with paraffin wax, may be made compatible by dissolving them in a small amount of stearic acid.

Opaque Wax[1]

Heat the wax, for example, paraffin wax, at 5–50°F. above its melting point, and add 0.2–10%, and preferably 0.5–2% of a froth stabilizer[2] to the molten wax. Then cool the wax mixture to a temperature within about 20°F. above its solidifying point, and blow vigorously with air, gaseous carbon dioxide, nitrogen or any other suitable gas. Allow the wax to solidify during the blowing. The air or gas occluded or dissolved in the melted wax forms minute bubbles when the wax solidifies which imparts a fine, glossy, opaque appearance to the wax. The presence of a froth stabilizer permits the reheating of the wax to a temperature above its melting point and retaining a greater portion of the minute bubbles on resolidification. A colloid mill or other mechanical device may be used to assist in producing the desired froth.

Opaque wax-coated paper and opaque candles may also be obtained by dissolving oxygen, air, carbon dioxide or gaseous

[1] U. S. Patent 2,185,046.

[2] Some froth stabilizers are alkaline sulfonic acid soaps, sodium stearate, alkyl xanthates, sodium and potassium soaps, higher fatty alcohols, ammonia and amine soaps.

hydrocarbons, such as propane, butane, etc., in the molten wax under pressure, just before the wax is to be applied to the paper or to be poured into the candle molds. As the wax cools, the pressure is released. This causes the occluded gas to effuse from the wax, thus forming minute gas bubbles which are encased in the solidified wax, giving an opaque appearance.

For this purpose, the use of a liquefied, normally gaseous, hydrocarbon is highly efficient. The liquefied hydrocarbon, such as propane, butane or their mixtures, may be dissolved in the wax under pressure. As the pressure is released, the hydrocarbons will vaporize, forming bubbles in the wax, and, at the same time, cool the wax to its solidification temperature.

This process of preparing opaque waxes is usually applied to mineral waxes and paraffin waxes having melting points of 110–160°F. Paraffin waxes having melting points of 120–145°F. resist breaking of the emulsion or foam on reheating. The air-wax mixture can be made by comminuting or powdering the wax and extruding or molding the product, preferably after it has been warmed and softened to a degree which will permit its coherence. By incorporating a stabilizing agent, at the same time, the wax-air mixture may be remelted without destroying the opacity of the wax. Coloring matter, such as pigments and oil-soluble dyes, may be added to the comminuted or emulsified wax to obtain any desired tint.

Rubber—Wax Compositions

Rubber is incorporated into hydrocarbon waxes, like paraffin wax or ozokerite, by first shredding or cutting into very small pieces and adding, in controlled quantities, to the molten wax while mixing slowly and continuously. The rubber swells and dissolves very slowly. Depending on the size of the pieces, temperature and type of mixing, the time of solution may be anywhere from hours to days.

Starting with a solution of rubber in a solvent or rubber latex, the time of solution is considerably reduced, but then the solvent or water must be driven off.

Such compositions are very viscous and stringy when melted. A nonviscous product may be made by using oxidized or de-polymerized rubber. This is attacked much less by benzol, fats and oils and does not absorb odors so readily as the usual rubber-wax composition.[1]

Hardening Wax

In paper coating, where paraffin wax is used for waterproofing, certain waxes are melted with paraffin to prevent excessive penetration and to stiffen the finished paper. Waxes, which are soluble in hot paraffin wax and insoluble below its setting point, are used, e.g., carnauba wax or Acrawax C.

MICROCRYSTALLINE WAXES

The designations of microcrystalline and amorphous waxes are used synonymously, although the first is a more accurate designation. These waxes differ from refined paraffin wax in crystal size and structure and in that they are tougher, more flexible, and have a higher tensile strength and melting point. They are also more adhesive and less lustrous and greasy. They bind solvents, oils, etc., much better than paraffin wax and thus prevent their sweating-out.

It should be noted in general that:

1. High penetration value and/or high refractive index are indicative of flexibility.
2. Waxes of high penetration value generally have more "tack."
3. Flexibility is not a function of melting point.
4. Oil content influences flexibility only to a limited degree.

The Differences between Microcrystalline and Paraffin Waxes*

Paraffin and microcrystalline waxes are both separated from crude petroleum. However, the process of manufacture and the resulting products are quite different.

As crude petroleum is subjected to distillation by heating in a still at atmospheric pressure, the following products are removed in the order of their increasing boiling points: light petroleum gases, gasoline, naphtha, kerosene, gas oil, paraffin wax distillate, light, neutral lubricating-oil fractions, and a residue in the still which will not distill overhead at atmospheric pressure without decomposition. Paraffin wax is separated from the paraffin-wax

[1] Lanigan, U. S. Patent 2,322,242 (1943).

* Bareco Oil Co., Tulsa, Okla.

distillate by a relatively simple process, while the microcrystalline wax, which cannot be distilled without decomposition, is separated by a complex series of solvent separations from the residue remaining in the still.

Paraffin wax, which has very little affinity for oil, is separated from the paraffin-wax distillate by cooling and filtering in a common plate-and-frame filter press where the wax is retained on canvas filter leaves and the distillate oil passes through the canvas and is removed. Oil remaining in the wax filter cake is removed by a sweating process which involves casting the wax in thin sheets and gradually raising the temperature to a point slightly below the melting point of the wax. In this process, the remaining oil runs out of the wax sheet leaving a paraffin wax which usually contains less than 0.5% oil.

Microcrystalline wax is present in the residual fraction from the still along with heavy residual lubricating oil and asphalt. Asphalt must be removed from the residue by conventional sulfuric acid treatment or one of the newer selective solvent processes before the wax can be separated. The microcrystalline wax is then removed as crude petrolatum from the residual lubricating oil by any one of several dewaxing processes which involve dilution with an organic solvent, chilling at 20—40°F., and separation of the crude petrolatum in high-speed centrifuges. Microcrystalline wax has a great affinity for oil and thus the crude petrolatum still contains 40—70% oil. Therefore, it is then mixed with another portion of organic solvent, usually a different solvent from the one used in the initial dewaxing step, and heated to dissolve the wax and oil. The blend is then cooled to precipitate the wax which is separated from the oil and solvent on a filter. The solvent used for deoiling microcrystalline wax must be polar, whereas for deoiling paraffin wax other types of solvent may also be used. Since the microcrystalline wax holds oil very strongly, the last solvent step is usually repeated to give a wax with lower oil content.

The oil content of the microcrystalline waxes varies with the grade of wax but is usually 2—12% as contrasted with 0.5% in paraffin wax.

Chemically, both paraffin and microcrystalline waxes consist of saturated hydrocarbon.

Paraffin and microcrystalline waxes are both long-chain compounds but microcrystalline waxes have much higher molecular weights. The common commercial paraffin waxes have molecular weights of 360–420, which means that the average molecule in these hydrocarbons contains 26–30 carbon atoms. The molecular weight of high-melting point paraffins may run as high as 600. The molecular weight of commercial microcrystalline waxes is 580–700, that is, the average molecule contains 41–50 carbon atoms. This molecular weight difference accounts for the fact that the paraffin waxes can be distilled at atmospheric pressure while the microcrystalline waxes decompose when distillation is attempted.

Extensive X-ray diffraction work has indicated differences in the chemical structure of the paraffin and microcrystalline wax molecules in addition to their molecular weight. Paraffin waxes have been found to consist mainly of straight-chain molecules

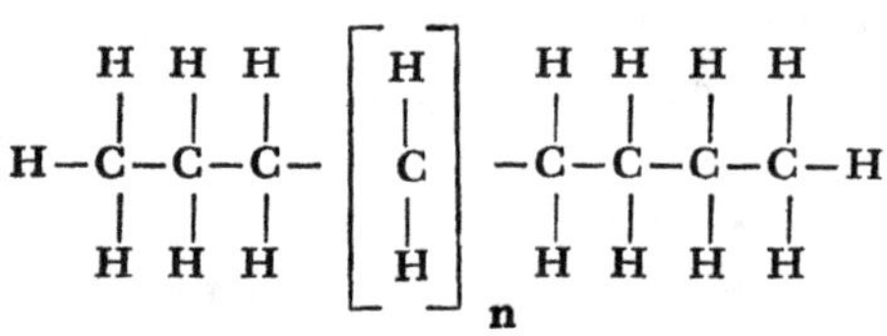

and a lesser amount of molecules containing branched chains.

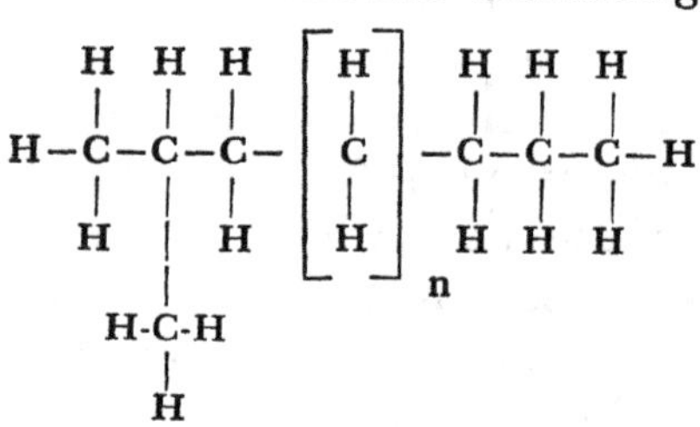

In some paraffin waxes, a very small quantity of aromatic hydrocarbons may also be present. The microcrystalline waxes contain some straight-chain molecules, but a large proportion of branched-chain molecules. The branched chains in the microcrystalline waxes are probably located at random along the carbon chain while in paraffin wax, they are near the end of the chain. There are indications that microcrystalline wax also contains a much higher concentration of ring-type compounds than paraffin wax.

Microscopic examination of paraffin and microcrystalline wax shows a striking difference. Paraffin wax has large, well-formed crystals both when crystallized from the melted wax and from solvents. Microcrystalline wax forms small irregular crystals from the melted wax and from solvents, no well-formed crystals of any size are obtained.

The melting point of commercial paraffin waxes is, in general, much lower than that of microcrystalline waxes. Paraffin waxes usually melt at 126—134°F., and microcrystalline waxes at 145—195°F. The melting point, however, is not sufficient for distinction, as paraffin waxes melting up to 170°F. are avaliable and microcrystalline waxes have been prepared with a melting point as low as 130°F.

Paraffin waxes are brittle while microcrystalline waxes are tough, resisting fracture. Most grades of microcrystalline wax are plastic and tend to flow under compression, while all grades of paraffin wax will shatter under compression.

Oil added to most grades of microcrystalline wax will increase its plasticity. Addition of oil to a paraffin wax will not yield a plastic product but one with greatly reduced strength which tends to crumble when handled.

Paraffin wax is transparent while microcrystalline wax is opalescent.

Films of paraffin and microcrystalline waxes of the same color and thickness transmit different amounts of light and thus paraffin wax films appear more transparent than microcrystalline-wax films.

Paraffin wax will form a stronger gel in turpentine than microcrystalline wax of the same melting point and solvent ratio.

If a glass rod is dipped into a solution containing a precipitate of microcrystalline wax in mineral oil and is withdrawn, the solution will drain leaving only a thin layer of oil on the rod.[1] If the solution contains a precipitate of paraffin wax, crystals will remain on the rod, because the surface tension of the paraffin wax crystals is less than that of the oil while that of the microcrystalline-wax crystals is greater than that of the oil.

Paraffin wax contracts to a greater extent when passing from the liquid state to the solid state than do microcrystalline waxes.[2] A test to differentiate between paraffin and microcrystalline waxes has been based on this behavior.

The Saybolt Universal viscosity at 210°F. for microcrystalline

[1] Jones and Blachly, *Ind. Eng. Chem.* **21**, 318 (1929).
[2] Kinsel and Phillips, *Ind. Eng. Chem.* **17**, 152 (1945).

waxes is usually 70—100 seconds, while that of paraffin wax 40—50 seconds.

The refractive index at 212° for paraffin waxes is usually 1.430—1.433 and for microcrystalline waxes 1.435—1.445.

The tensile strength of paraffin wax of very low oil content, (0.1% or less) is high, but the addition of 0.5 to 1% oil will reduce the tensile strength greatly (up to 50%). The reduction of tensile strength of microcrystalline wax with the same addition of oil will be about 5%. This phenomenon is due to the fact that oil weakens the structure of paraffin wax, but it is so strongly absorbed by microcrystalline waxes that a more plastic product results which will give plastic flow under tensile stress.

The extensibility is much greater with microcrystalline wax than with paraffin wax. Addition of oil will increase the extensibility of most microcrystalline waxes, but does not affect that of paraffin wax.

Small rods of most microcrystalline waxes cast will bend without breaking at temperatures considerably below their melting point. Paraffin waxes cast in the same way usually break at or near their melting point.

Table 21

PHYSICAL PROPERTIES OF SOME MICROCRYSTALLINE WAXES

	Brands				
Properties	Quaker State Microwax	Barnsdall Special Wax	Syncera Wax	Socony No. 2300	Be Square Special
Melting Point °F.	145–146	160–165	155–160	155	160–185
Specific Gravity at 60°F.	0.90–0.92	0.92–0.94	0.912	—	0.92–0.94
Specific Gravity at 210°F.	0.78–0.80	0.80–0.82	—	—	—
Viscosity at 210°F. (Saybolt)	75–85	75–100	51	65	75–100
Flash Point °F.	500 min.	500 min.	425	495	500
Fire Point °F.	575 min.	575 min.	—	550	575
Saponification Number	0.05–0.10	<2	0	—	<2
Acid Number	0.1 max.	0.1–0.2	0	—	0.1–0.2
Iodine Number					1.5
Color	white to amber	white to amber	yellow	yellow to brown	white to black
Penetration Number (ASTM) (Petrolatum method –77°F.; 100 g wt.)	—	5–10	—	20–25	5–20

Table 22

PROPERTIES OF COLWAXES

Colwax	No. 15	No. 21	No. 29	No. 58	No. 68	No. 70	No. 71	No. 85	No. 87	No. 89
Melting Point (Drop Method), °F.	175–177	180–182	189–191	185	189–191	180–185	170–175	175–180	160–165	145–150
Softening Point, °F. (ASTM E28–39T)	170–175	177–181	187–191	185–190	185–190	170–175	160–165	170–175	140–145	130–135
Specific Gravity at 60°F.	0.93–0.94	0.93–0.94	0.93–0.94	0.93–0.94	0.93–0.94	0.92–0.93	0.88–0.90	0.88–0.90	0.88–0.90	0.88–0.90
Solubility in Carbon Tetrachloride, %	100	100	100	100	100	100	100	100	100	100
Penetration, 100 g, 5 sec. 77°F.	55–65	31–37	6–12	30–40	8–12	10–25	10–25	15–30	15–30	15–30
Saybolt Universal Viscosity at 210°F. sec.	60–90	90–120	125–150	75–100	75–100	50–75	50–75	50–75	40–65	40–65
Flash Point (COC), °F.	355–380	420–450	585–610	475–500	475–500	435–460	425–450	425–450	410–435	410–435
Fire Point (COC), °F.	400–415	460–500	660–685	600–625	600–625	485–510	470–495	465–490	445–470	445–470
Iodine Number, Honus Maximum	4.0	4.0	4.0	2.5	4.0	0.5	0.5	0.5	0.5	0.5
Saponification Number, Maximum	2.0	2.0	2.0	2.0	2.0	1.0	1.0	1.0	1.0	1.0
Acid Number, Maximum	1.0	1.0	1.0	0.5	1.0	0.5	0.5	0.5	0.5	0.5
Ash, %	1.0	1.0	1.0	0.5	0.7	0.2	0.2	0.2	0.2	0.2
Sulfur, %	0.25	0.25	0.25	0.1	0.2	0.1	0.1	0.1	0.1	0.1
Color	brownish black	brownish black	brownish black	brown	brownish black	amber	amber	amber	amber	amber

Table 23

Physical Properties of Microcrystalline (Amorphous) Waxes

Name or Grade	ASTM Penetration at 77°F.	Refractive Index at 80°C.	ASTM Melting Point Drop Method	Oil Content	Flexibility: Room Temperature	Flexibility: at 34°F.	Flexibility: at 20°F.	Taste	Odor 200°F.	Remarks
Benowax C	21	1.4508	140.2	4.01–3.99	good	good	good	bad	burnt	very sticky
Micro Wax, Dark Amber	22	1.4491	145.4	1.26–1.56	good	good	good	poor	oily	—
Micro Wax, White	23	1.4451	146.2	0.66–0.93	good	good	good	none	none	—
Standard Oil Syncera, 145/50 Orange	19	1.4341	146.4	—	good	brittle	—	good–very little	slight	hard
Standard Oil Syncera, 145/50 Yellow	17	1.4330	146.4	0.88–1.06	good	brittle	—	good–very little	slight	hard
Standard Oil Syncera, 155/60 Yellow	19	1.4382	155.8	—	brittle	—	—	good	slightly oily	hard
Aristowax 160/65	14	1.4362	160.4	2.53–2.55	brittle	—	—	none	slight propane	—
Deep Rock, 160 Lemon Pale	33	1.4480	161.2	11.19–10.87	good	fair	brittle	poor	kerosene	mealy texture
Wadhams 2305	31	1.4461	162.7	5.40–5.27	good	good	good	poor	slightly oily	sticky
Standard Oil Superla, 165/70 Yellow	12	1.4446	165.8	—	brittle	—	—	good	slight	—
Wadhams Petrosene A	23	1.4448	167.8	8.61–8.01	good	brittle	—	good	slightly oily	—
Barnsdall 160/65, Amber	16	1.4481	176.2	6.76–6.56	good	brittle	—	none	good	—
Barnsdall 160/65, Plasticized to 32°F.	24	1.4500	177.0	—	good	good	good	none	good	—
Barnsdall 160/65, Plasticized to 40°F.	21	1.4500	177.4	—	good	brittle	—	none	good	—
Barnsdall 160/65, Plasticized to 50°F.	17	1.4500	178.1	—	good	brittle	—	none	good	—

Table 24

CROWN WAXES*

General Properties	Crown 200	Crown 500	Crown 700	Crown 1035
ASTM Petrolatum M.P., °F. (D 127–30)	192.5	194.4	194.8	196.8
ASTM Paraffin M.P., °F. (D 87–42)	184.0	184.5	188.5	190.2
Softening Point, °F. (Crown Cork and Seal Method)	162.0	164.0	173.0	179.0
Penetration 100/5 at 77°F. (ASTM D5–25)	8	7	3	2
Viscosity, SSU at 210°F. (ASTM D446–39)	77.3	76.8	77.0	78.5
Color, NPA Designation (ASTM D155–45)	brown-black	extra lemon-pale	extra pale	extra pale
NPA Number	8	2.5	2	2
Fracture	irregular	irregular	irregular	irregular
Acid Number	nil	nil	nil	nil
Iodine Number	nil	nil	nil	nil
Acetyl Value	nil	nil	nil	nil
Saponification Number	nil	nil	nil	nil
Specific Gravity, 25°C.	0.9445	0.9326	0.9415	0.9421
Conradson Carbon Residue, %	0.28	0.12	0.03	0.02
Ash, % by Weight	0.02	0.01	0.01	0.01
Molecular Weight	780	804	872	890
Molecular Structure	branched paraffins C–50 to C–65			
Refractive Index at 203°F.	1.4427	1.4413	1.4401	1.4390
Oil, % by Weight	12.4	10.4	2.3	0.6
Heat of Fusion, Btu/lb	70.5	70.5	70.5	70.5
Specific Heat, Btu/°F. lb	0.545	0.545	0.544	0.545

* Petrolite Corporation.

Compatibility of Microcrystalline Waxes

These waxes are compatible in all proportions with other mineral waxes, most vegetable waxes and resins. This property has made them valuable as agents for increasing melting point and hardness and for decreasing blocking, tack, sweating and other undesirable qualities in wax blends.

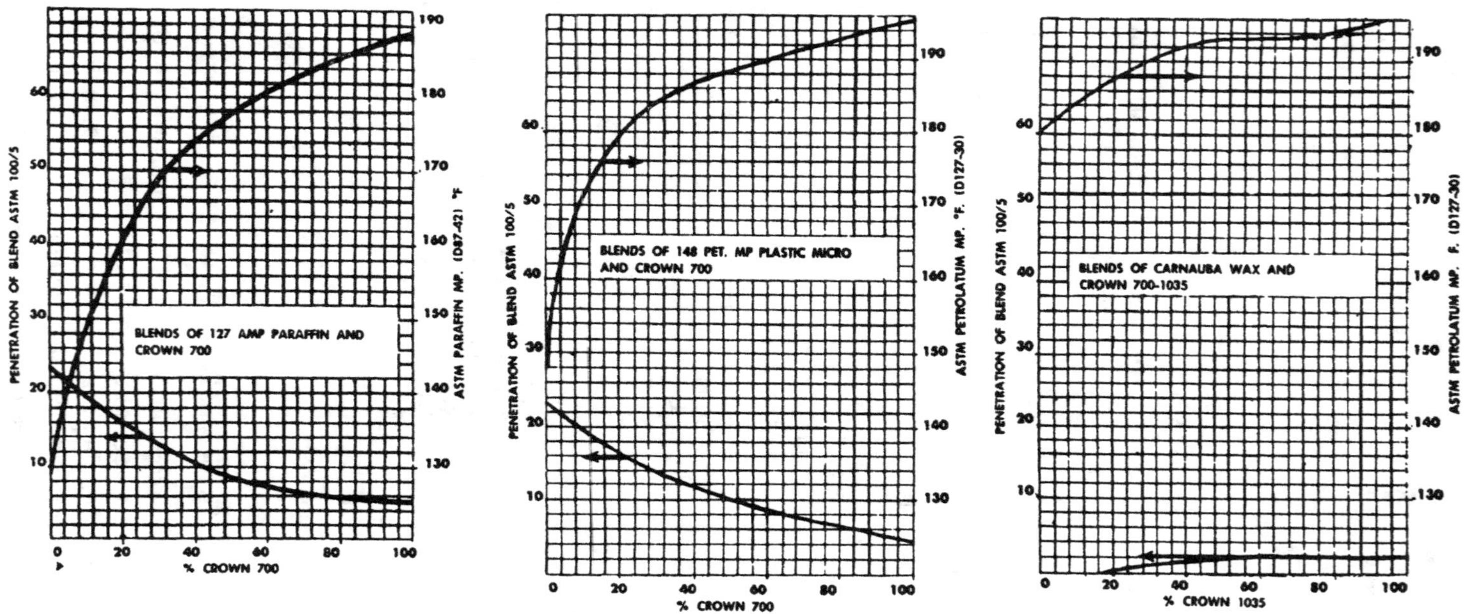

Figure 22. THE EFFECT OF CROWN WAXES ON THE MELTING POINT AND PENETRATION OF WAX BLENDS

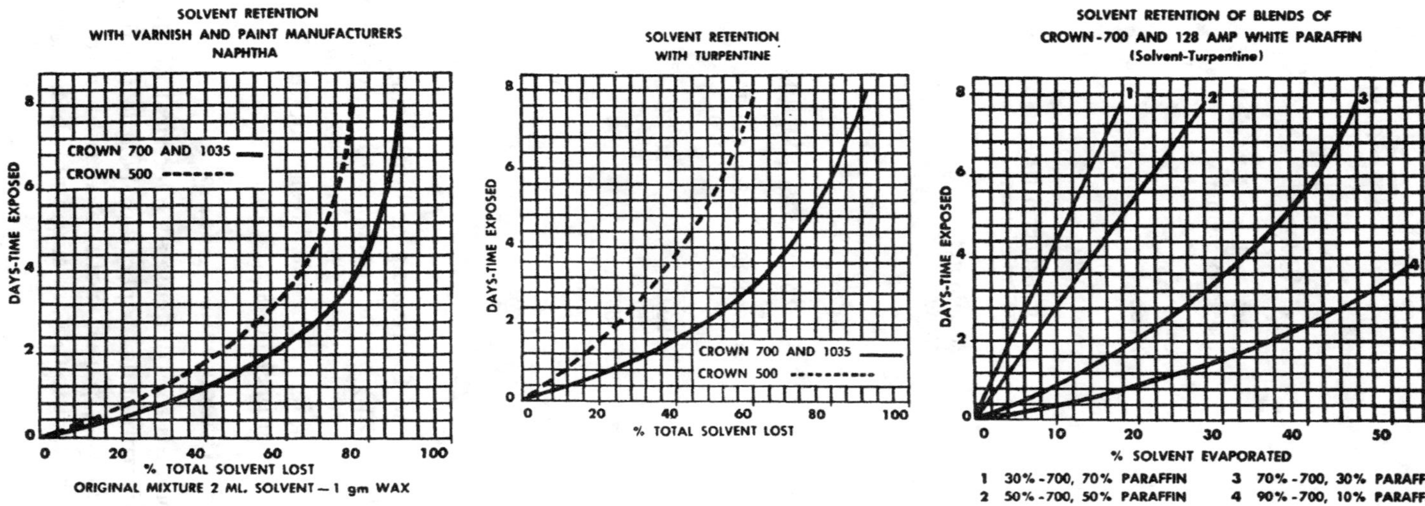

Figure 23. THE SOLVENT RETENTION OF CROWN WAXES*

GEM WAXES

Color	From Amber 5 NPA to Yellow 2 NPA
Melting Point	180°F. minimum
Penetration at 77°F. 50 g 5 sec	20 minutes
Odor	None

PEARL WAXES

Color	From Amber 5 NPA to Yellow 3 NPA
Melting Point	180°F. minimum
Penetration at 77°F. 50 g 5 sec	30 minutes
Odor	None

Table 25

PROPERTIES OF S/V SOVOWAXES 100, 105, 110

Properties	Sovawax 100	Sovawax 105	Sovawax 110
Melting Point, °F. ASTM, minimum	160	160	160
Needle Penetration at 77°F.	20/30	20/30	20/30
Saybolt Viscosity at 210°F.	85/95	85/95	85/95
Flash Point, °F. (COC)	500	500	500
Odor	OK	OK	OK
Color, ASTM	3–4 dil.	2–3	2 max.
Flexibility at 0°F.	OK	OK	OK

These waxes are used for coating or laminating, their application being restricted only by the special hardness or melting-point requirements of isolated uses. They are recommended for all applications requiring inert, chemical-resistant, durable, adhesive, flexible coatings or impregnants.

General Uses of Microcrystalline Waxes

Potting compounds for condensers; (electrical insulating) compound base; waterproofing paper, box-board, textiles, leather, wood, etc.; laminating paper, cloth, etc.; rubber compounding; slushing compounds; auto-top dressings; polishes; pattern making (sheet and fillets); binder for pipe coverings; filler for packings; molding-wax base.

Drum Linings

Waxes, especially of the microcrystalline variety derived from petroleum, are widely used for this purpose. The application of the wax is quite simple. The drum is heated to a temperature that will give the thickness desired; heated, molten wax is poured into the bung, the opening is closed and the drum rolled in such a way as to bring the molten wax into contact with the entire inner surface, the excess is then poured out or removed by other means. Any desired thickness of evenly distributed film can be obtained over the inner surface, with considerable reinforcement at the corners. The quantity required will vary from ½ lb to 3 or 4 lb, depending on the desired thickness. Adhesion will be excellent with the proper wax, if the interior of the drum is absolutely dry.

Vat Linings

Wood or steel vats can be lined with wax in a number of ways. The surface to be coated must be absolutely dry and preferably warm. The wax may then be applied by trowelling (hot, plastic), flushing, mopping or spraying. In the case of mopping or trowelling, it is good precaution to apply a preliminary thin spray coat and then the second coat while the surface is still warm enough to insure good adhesion. The quantities required to produce coats of different thickness are listed in Table 26.

These are approximate figures, but will serve as guides and multipliers. They are for vertical tanks, with only the bottoms and vertical sides waxed.

To protect articles against dehydration and water from the outside, wax is an ideal coating.

Wax coatings for packages are widely accepted. A final dip coat was given many items of supply during World War II, even though the article had been wrapped in some sort of waxed cover. The purpose was not only to make the wax coating heavier, but also to repair any minute leaks left inadvertently, such as failures at seals, laps or folds, or injuries of the wrap by

Table 26

WAX QUANTITIES FOR VAT LININGS

Dimensions Diameter and Height, ft	Volume gal	Coatings Thickness, in.	Quantity of Wax, lb
5 × 5	785	¼	119
5 × 10	1,469	¼	214
10 × 5	2,937	¼	285
10 × 10	5,875	¼	475
10 × 20	11,750	¼	855
20 × 10	23,501	¼	1,140
20 × 20	47,002	¼	1,900

handling. These coatings were produced by dipping, although a satisfactory spraying technique could undoubtedly be developed.

The original method was rather crude, as usually each end was dipped separately, but proved to be very useful. There has been a great deal of thought since on the development of dip and spray coats, compounded on a wax base and having the essential properties of wax, such as flexibility, water and water-vaporproofness, lack of toxicity, odor and taste, good appearance, nonblocking, and easy applicability.

Miscibility of Various Additives with Microcrystalline Waxes

In many applications of microcrystalline waxes, it is essential to know their miscibility with various resins and other additives. The solubilities of a number of products in Petrosene C (microcrystalline wax) were determined on a semiquantitative basis. Table 27 presents data indicating the solubility-temperature relationship for these mixtures. In all cases, 10% of additive was employed so that no attempt was made to determine the maximum solubility. It is expected that the solubility of any resin in paraffin wax will be greater than in Petrosene C and in other microcrystalline waxes it will be of the same order as in Petrosene C. Table 27 is divided into two sections, one listing the additives soluble up to 10% in Petrosene C at all temperatures above its melting point and the second listing those which are insoluble or only partially soluble.

Table 27

MISCIBILITY OF VARIOUS ADDITIVES WITH PETROSENE C (MICROCRYSTALLINE WAX)

Percentages Used: Petrosene C—90%; Additive—10%.

Additives Completely Miscible above and at Melting Point

Product	Product
1. Cumar Resin P-25	9. Chromium Oleate
2. Amberol—Grade ST-137	10. Diglycol Stearate
3. Opalwax 10	11. Lead Linoleate
4. No. 1 Singapore Damar Gum	12. Copper Stearate
5. Batavia Damar A/D	13. Manganese Naphthenate
6. Santo-Resin	14. Manganese Resinate
7. Burgundy Pitch	15. Calcium Stearate
8. Calcium Stearate	16. Zinc Stearate

Additives Insoluble or Partially Miscible

Product	Limit of Miscibility, °F.
1. Cumar Resin V 1½	above 250
2. Cumar Resin P-70	above 240
3. Cumar Resin MH 1½	above 230
4. Amberol M-16	at least above 300
5. Amberol M-82	at least above 300
6. Resin XR9517	above 270
7. Resin BR8900	above 230
8. Residuum Pitch	at least above 300
9. Acacia Gum	at least above 300
10. Gum Arabic	at least above 300
11. Gum Kauri	at least above 300
12. Cobalt Linoleate	at least above 300
13. Nuba No. 2	at least above 300
14. Nubalene Resin	at least above 300
15. Nuba No. 3	at least above 300
16. Resin No. 465	at least above 300
17. Nuba No. 1	at least above 300
18. Cobalt Resinate	at least above 300
19. Aluminum Oleate	at least above 300
20. Aluminum Naphthenate	at least above 300
21. Calcium Oleate	at least above 300

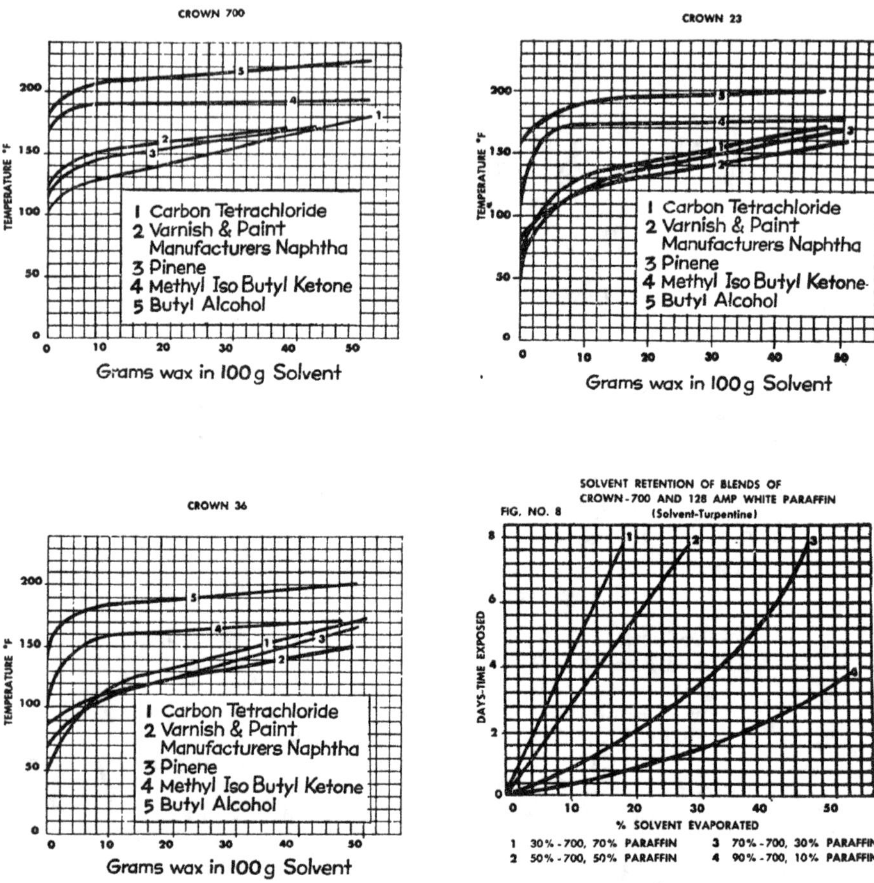

Figure 24. SOLUBILITY OF CROWN WAXES IN COMMERCIAL SOLVENTS

THERMAL EXPANSION AND MELTING POINT OF WAXES

Pure crystalline substances are very often identified by their melting point. Waxes, being complex mixtures, do not exhibit sharp melting points, but there is a gradual transition from the liquid to the solid state. This gradual softening results in an abnormally high apparent coefficient of expansion while the wax is still in the solid state. This characteristic of waxes has a bearing on their performance in many applications. The curves in Figure 25 compare this effect for a crystalline paraffin, carnauba wax, and

Crown 700. The breaking point in the curves may be interpreted as the liquefication temperature in each case.

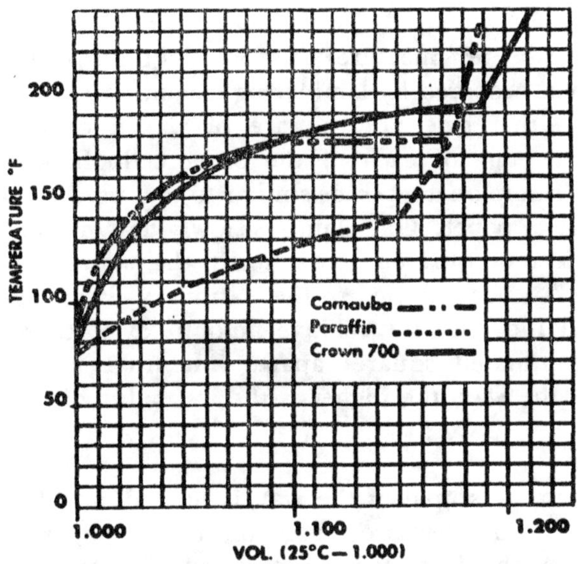

Figure 25. THERMAL EXPANSION AND MELTING POINT OF WAXES

Table 28

ELECTRICAL PROPERTIES OF MICROCRYSTALLINE WAXES

	Crown 1035			Crown 500		
			Dielectric Constant			
°C.	60	10^3	10^6	60	10^3	10^6
80	2.46	2.30	2.47	2.38	2.49	2.48
60	2.37	2.39	2.44	2.42	2.53	2.50
40	2.31	2.25	2.44	2.56	2.74	2.74
20	2.16	2.09	2.45	2.62	2.79	2.79
			% Power Factor			
80	2.20	2.05	0.11	0.11	0.04	0.12
60	2.10	1.98	0.08	0.04	0.03	0.08
40	1.69	1.64	0.08	0.22	0.04	0.20
20	1.79	1.51	0.33	0.06	0.03	0.17

Potential Breakdown (V/Mil)

°C.	20	40	60	80
Crown 1035	2050	710	550	675
Crown 500	1500	780	625	420

Volume Resistivity

Crown 1035	3.8×10^{15} ohms-cm 22°C. (55% humidity)
Crown 500	6.4×10^{15} ohms-cm 22°C. (55% humidity)

Laminating Properties of Microcrystalline Waxes

Adhesion Test

The adhesion test measures the force required to tear apart a test specimen prepared by laminating a square of unwaxed cellophane on either side of a square of waxed cellophane. An unlaminated edge of each of the outer cellophane squares is placed in a clamp on a machine similar to that used for measuring the strength of paper. One clamp is slowly drawn away from the other by a motor-driven screw until the squares have been separated. A lever attached to the stationary clamp moves along a scale and indicates the relative amount of force required to tear the laminated squares apart. The higher the adhesive strength of the wax, the higher is the test value obtained.

Bleed Test

In this test, a sample of the wax is melted, and five droplets (6—8 mm in diameter) are placed on a filter paper (Whatman No. 1). When the droplets have solidified, the paper is placed in an oven at 130°F. for 24 hours. The average value (in millimeters) obtained by subtracting the diameter of each droplet from the diameter of the oil ring that surrounds it is recorded as the bleeding number. The less the tendency of the wax to bleed, the lower is the number obtained.

The adhesion and bleed properties of Multiwax ML 445* as compared to those of other microcrystalline waxes on the market are presented in Table 29.

Table 29

ADHESION AND BLEEDING OF MICROCRYSTALLINE WAXES

Wax	ASTM M.P	ASTM Needle Penetration	Adhesion at 77°F.	Bleed Test
Multiwax ML 445	163	30	33	1
Wax A	175	12	1	0
Wax B	156	21	2	4
Wax C	155	24	6	5
Wax D	167	27	5	6
Wax E	164	34	43	4
Wax F	158	34	32	4
Wax G	157	36	35	5

* Petroleum Specialties Inc., New York City.

The bleed values might indicate that low needle penetration has been overstressed as a criterion of the suitability of a wax for laminating. It has been held that waxes with relatively high penetration have a greater tendency to strike through the laminating paper. These values do not support this theory.

It may be noted from Table 29 that there are some waxes that have very high adhesion at the expense of their bleed characteristics (Wax E, for example). Others have good bleed properties, but low adhesion (Wax A, for example). Obviously, the most suitable laminating wax would be one that has a combination of relatively high adhesion and relatively low bleed characteristics. Such a combination which depends on the proper selection of raw materials and the manufacturing process is represented in Multiwax ML 445, a wax which is used in increasing quantities as a paper laminant.

Microcrystalline-Wax (Bareco-Wax) Blends[1]

Blends were prepared by heating the components together on a water bath until melted and then stirring until solidification took place. Melting points were determined by the capillary method. Open capillaries were used. They were filled with equal amounts, immersed in the water bath to the same depth and heated at the same rate for all determinations. The temperature at which the wax began to flow, under the hydrostatic pressure, was taken as the melting point.

The Bareco 160 petro-wax used had the following properties:

Specific Gravity at 60°F.	0.92–0.94
Specific Gravity at 212°F.	0.80–0.82
Melting Point (ASTM D-127-30)	160–165
Saybolt Universal Viscosity, 212°F.	75–100
Flash, C.O.C.	above 500°F.
Color	white
Iodine Number	less than 1
Saponification Number	less than 2
Maximum Acid Number	0.1
Ash	trace
Odor	none
Taste	none
Penetration at 77°F.	15/25

[1] Ice and Harris, *J. Am. Pharm. Assoc.*, Apr. 1943.

Table 30

MELTING POINT OF WHITE PETROLATUM BLENDED WITH VARYING PERCENTAGES OF BARECO 160 WAX

Wax, %	Petrolatum, %	Point, °F.
0	100	110
1	99	112
2	98	116
3	97	119
4	96	121
5	95	122
10	90	134
15	85	140
20	80	144
25	75	148
30	70	150
35	65	152
40	60	154
100	0	168

Table 32

MELTING POINT OF ANHYDROUS LANOLIN BLENDED WITH VARYING PERCENTAGES OF BARECO 160 WAX

Wax, %	Anhydrous Lanolin, %	Melting Point, °F.
0	100	106
1	99	108
2	98	110
3	97	112
4	96	120
5	95	126
10	90	146
20	80	156
100	0	168

Table 31

MELTING POINT OF LARD BLENDED WITH VARYING PERCENTAGES OF BARECO 160 WAX

Wax, %	Lard, %	Melting Point, °F.
0	100	92
1	99	94
3	97	98
5	95	120
10	90	134
20	80	154
100	0	168

Table 33

MELTING POINT OF WHITE PETROLATUM WITH VARYING PERCENTAGES OF WHITE BEESWAX

Beeswax, %	Petrolatum, %	Point, °F.
3	97	110
5	95	112
10	90	115
20	80	119
100	0	124

Table 34

MELTING POINT OF WHITE PETROLATUM WITH VARYING PERCENTAGES OF YELLOW BEESWAX

Yellow Wax, %	Petrolatum, %	Melting Point, °F.
3	97	112
5	95	114
10	90	122
20	80	129
100	0	142

Specific Gravity of Microcrystalline Waxes at Various Temperatures

When microcrystalline waxes solidify, they contract and their specific gravity changes which has a bearing on their performance in many applications. In the impregnation of electrical or other equipment, for example, where voids are filled with molten wax, it is frequently essential to determine what contraction will occur on solidification so that adjustments may be made to avoid loss in waterproofing value. In addition, it may be necessary to determine what volume a given quantity of solid wax will occupy when melted and heated to a definite temperature.

In order to obtain this information, representative samples of microcrystalline waxes were held at definite temperatures and their specific gravities measured. The data so obtained are given in figure 26. Cerese Wax AA, Petrosene B, and Product 2305 were tested. These are the intermediate colored waxes in each of the three series. Since the other waxes differ from these only in color, their specific gravities may be considered comparable. Due to slight variations from batch to batch, other samples of any wax will probably show slightly different values from the figures noted. However, based on the available data, these variations are not expected to exceed 0.005, except for samples of widely different melting points, in which case, the greatest divergence will be at temperatures near the melting point.

It is interesting to note that, in the solid state, Cerese Wax AA has the highest specific gravity, whereas in the molten state, it has the lowest. Petrosene B is intermediate between Product 2305 and Cerese Wax AA.

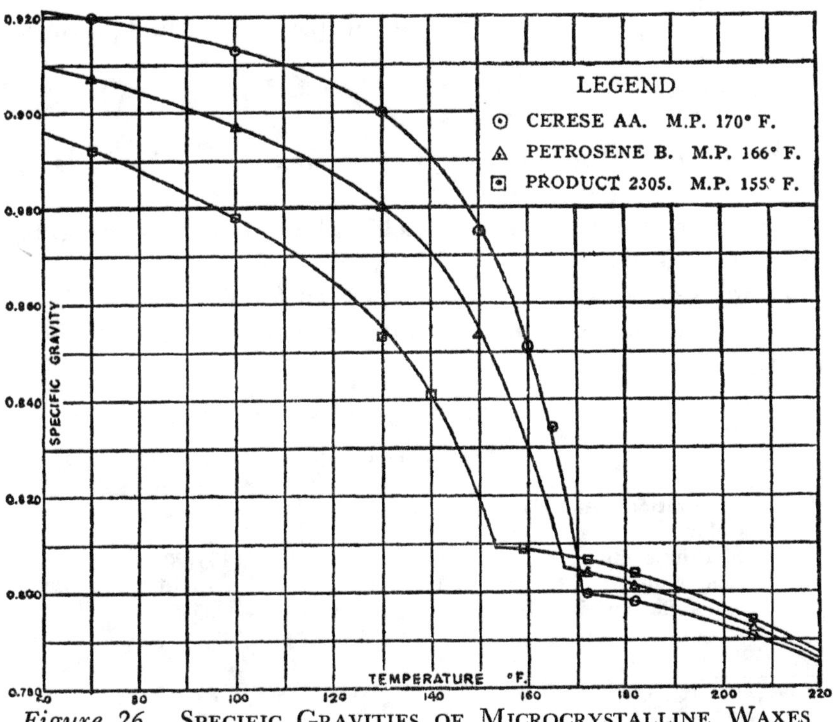

Figure 26. SPECIFIC GRAVITIES OF MICROCRYSTALLINE WAXES

Viscosity of Microcrystalline Waxes

The viscosity of microcrystalline waxes, when melted, is greatly lowered by small additions of ordinary paraffin wax.

Oxidized Microcrystalline Waxes

Oxidized microcrystalline waxes are prepared by blowing air through heated wax in the presence of a catalyst, often cobalt or manganese soaps.

Estawax[1] is an oxidized microcrystalline wax.

Properties

	Estawax 20	Estawax 25
Color, NPA	3 Max.	3 Max.
Melting Point, °F. (ASTM E28–42T)	210–220	210–220
Penetration (100/5/77)	2 Max.	2 Max.
Viscosity, centipoises at 300°F.	450–550	200–300

Uses: Phonograph-record component; paste and liquid polishes.

Cardis One[2]

This is an oxidized microcrystalline wax with a hardness of 1–2 (ASTM penetration).

It is easily emulsifiable and yields films of the highest gloss and wear resistance.

Its emulsions dry to highly scuff-resistant films.

It gives translucent and opaque emulsions of the highest quality.

Properties

Penetration (ASTM)	1–2
Color	4–5
Melting Point	195–200°F.
Acid Number	13–15
Saponification Number	55–60

1 Petrolite Corp.
2 Warwick Wax Co.

PETROLATUM WAX

Petrolatum (mineral jelly), freed from its liquid and semi-liquid hydrocarbons, yields an amorphous, white, translucent, plastic wax. It is commercially available under the name of Protowax or Fybrene. Its plastic and noncrystalline properties make it very useful. It also has good oil-holding capacity and prevents leakage or sweating out of oils. Thus, it can replace ceresin and ozokerite, where their low melting point is not undesirable.

Properties

	White Fybrene Wax	*Olive Fybrene Wax*
Specific Gravity at 60°F.	0.896–0.899	0.896–0.899
API Gravity	26.0–26.5	26.0–26.5
ASTM Melting Point, °F.	128–133	130–135
ASTM Consistency (Method D-217)	50–85	30–50
ASTM Consistency (Method D-5)	150–300	75–200
Flash Point, °F.	460–470	460–470
Fire Point, °F.	520	520
Saybolt Vis. at 210°F.	50–55	50–55
Color	white	olive green
Ash, %	0.03	0.03
Saponification Number	0.3	0.3

OZOKERITE (IMPORTED)

Ozokerite is an amorphous hydrocarbon wax occurring naturally, in the earth, in Poland, Austria, Russia, Utah, and Texas among other places. It occurs in Miocene formations, near petroleum deposits. It is separated from earthy matter by melting in boiling water and then drawing off the upper wax layer. The crude wax is purified by treatment with sulfuric acid and bleaching earth to obtain lighter-colored products.

The commercial grades of ozokerite are:

1. Unbleached (Black)
2. Single bleached (Yellow)
3. Double bleached (White)

Ozokerite is characterized by its excellent absorbing power for oil, grease and certain solvents. In this respect, it is superior to paraffin wax which tends to crystallize from mixture and allows fluid components, present in the composition, to sweat out. It is waterproof, alkali- and acid-resistant, only fuming sulfuric and concentrated nitric acids attack it. Low temperatures do not make it so brittle as paraffin wax. It is not so slippery as paraffin wax and actually has a slight "tack." It becomes harder on aging. It is less resistant to actinic rays than paraffin wax. Ozokerite sets more rapidly and shrinks more than beeswax. To offset these properties Venice turpentine, or another plasticizer, is often added.

The chemical constitution of ozokerite is not definitely known. It appears to be a mixture of solid saturated and unsaturated hydrocarbons of high molecular weight, with some liquid hydrocarbons and oxygenated bodies.

Properties

Melting Point	149–176°F
Specific Gravity	0.85–0.95
Acid Number	0
Saponification Number	0
Iodine Number	7.8–9.2
Sulfur	0.2%
Ash	0.16%
Color	white, yellow, green, black
Refractive Index (60°C.)	1.440
Dielectric Constant	2.37–2.55
Effective A.C. Conductivity	10–24
Volume Resistivity	7–40
Penetration at 77°F. (100 g, 5 sec)	14

Soluble in: benzine, benzol, turpentine, kerosene, toluol, xylol, mineral spirits, carbon disulfide, isopropyl ether, trichloroethylene

Insoluble in: ethyl and methyl alcohols.

Table 35

SOLUBILITY OF OZOKERITE

Solvent	Grams of Ozokerite[1] Dissolved by 100 g Solvent
Carbon Bisulfide	12.99
Petroleum Ether, Boiling up to 75°C.	11.73
Turpentine, B.P. 158–166°C.	6.06
Xylene, B.P. 135–143°C.	3.95
Toluene, B.P. 108–110°C.	3.83
Chloroform	2.42
Benzene	1.99
Ethyl Ether	1.95

[1] Pawlewski and Filemonewicz, *J. Chem. Soc.*, Abstr. 82, 1889.

Adulterations

Ozokerite is often adulterated with paraffin wax to reduce its cost. It is also sophisticated with carnauba wax, resins and asphaltum to increase its melting point or hardness.

Ozokerite—Paraffin Blends

Yellow Ozokerite M.P., 75°C. } mixed with { Paraffin Wax M.P., 50°C.

Proportions	New M.P., °C.
4 Ozokerite 1 Paraffin Wax	73.5
4 Ozokerite 2 Paraffin Wax	71.7
4 Ozokerite 3 Paraffin Wax	70.5
4 Ozokerite 4 Paraffin Wax	69.7

White Ozokerite M.P., 75.7°C. } mixed with { Paraffin Wax M.P., 58.3°C.

Proportions	New M.P., °C.
4 Ozokerite 1 Paraffin Wax	74.4
4 Ozokerite 2 Paraffin Wax	73.2
4 Ozokerite 3 Paraffin Wax	72.5
4 Ozokerite 4 Paraffin Wax	72.0

Uses: high-grade candles, colored lead pencils, finishing the heels and soles of shoes, shoe polishes, electrical insulation, acid-proof coatings for electrotypers' plates, waxing floors, acidproof bottles, waterproofings, cosmetics.

American Ozokerite

American ozokerite is now produced in large quantities and of qualities equal to the imported. Midcontinent* (Oklahoma, etc.) stocks (long fiber) of 135—150°F. melting point are used as starting material. By centrifuging, a very plastic grade, melting up to 165°F., is obtained. By vacuum distillation, harder grades, melting up to 220°F., are produced. These are bleached to obtain the so-called American, white ozokerite. This differs from the plastic grade in being free from oil and color.

Penetration Data on American Ozokerite

M.P., 145–180°F.	12–17	M.P., 185–220°F.	3–5

These figures were obtained with a penetrometer at 77°F., using a 100 g weight for 5 seconds.

* Pennsylvania and Gulf stocks are far less suitable.

CERESIN

Ceresin originally was a purified ozokerite. For economic reasons it was adulterated with paraffin wax to reduce its cost. Today commercial ceresin contains 50–80% paraffin wax. Some European ceresins also contain rosin and coloring matter to approach the properties of beeswax.

Commercial ceresin varies in color from yellow to white and melts at 130–170°F., depending on its purity. Its hardness, plasticity, etc., also vary with purity. The grades containing large percentages of ozokerite naturally approach ozokerite in properties.

Properties

(High-Grade Ceresin)

Melting Point	130–175°F.
Specific Gravity	0.88–0.92
Acid Number	0
Saponification Number	0–1
Iodine Number	7–10
Color	white, yellow
Refractive Index	1.4416–1.4465
Dielectric Constant	2.15–2.33
Effective A.C. Conductivity	1–12
Volume Resistivity	400 > 900

Soluble in the same solvents as paraffin wax.

UTAH WAX

This amorphous mineral (ozokerite-type) wax from Utah is available in a number of grades varying in hardness, melting point, and color.

Properties

(Crude)

Melting Point	73°C.	Acid Number	0
Cloud Point	76°C.	Saponification Number	0
Iodine Number	< 1	Sulfonation Residue	100%

Table 36

MEASUREMENT OF HARDNESS BY PENETROMETER WITH STANDARD NEEDLE, 100 G WEIGHT, 15 SECONDS TIME

Temperature °C.	Penetration mm	Temperature °C.	Penetration mm
20	0.43	50	2.2
30	0.64	60	3.8
40	1.27		

A comparison of these figures with similar measurements of other waxes shows that Utah wax is softened less by any given increase in temperature.

Dielectric Strength: in excess of 30,000 volts across a 1/10-in. General Electric standard gap. Retains a high value when melted. *Dielectric Constant:* (or permittivity) 2.03. *Power Factor:* 0.03. *Resistivity:* 36,000,000 megohms across a 1 cm cube. *Coefficient of Expansion* (Volume): 0.001 per degree Centigrade between 10–50°C.

Solubility: Slightly soluble in most organic solvents when cold. Solubility increases rapidly on heating, so that Utah wax becomes very soluble in practically all commercial organic solvents, with the exception of methyl and ethyl alcohol, glycol, and glycerine. Its best solvent is carbon disulfide.

Solutions made by heating retain a fluid character when cooled, if the concentration of wax is not too high. Otherwise, they set into pastes or gels of the nonsweating variety.

Uses: dance-floor wax, floor and furniture polishes, linoleum polish, shoe polish, leather dressings, rubber goods, candles, wax figures, carbon paper, wax bottles, waxed paper, electrical insulation, waterproofing fabrics, impregnation of wood, electrotypers' molding compound, pattern-makers' wax, waxing artificial flowers, wax crayons, salves and face creams.

MONTAN WAX

Montan wax is a hard, brittle, lustrous wax extracted from lignites, principally in Central Europe and also mined in Australia, New Zealand, Czechoslovakia, Russia, Great Britain and in the United States. The coal is granulated, dried and solvent extracted to remove the wax. The crude wax is usually further processed to produce an acceptable market product. One type of refining is by vacuum distillation, using steam and producing the so-called "double-refined" montan wax. In a second and more important refining process, the wax may be deresinified by solvent processing, and then treated by chromic acid oxidation. The crude wax is dark brown and the refined wax of a yellowish color. The so-called white montan wax is a mixture of refined montan wax with paraffin wax. Depending on the geographical occurrence, the resin content of the wax varies. For example, the German deposits are noted for their relatively low resin content, thus yielding a satisfactory product more economically.

Chemically, montan wax consists of monohydric alcohol esters and high-molecular-weight acids.

Properties

Melting Point	72–92°C.
Specific Gravity	1.00
Acid Number	15–85
Saponification Number	58–89
Unsaponifiables	35–45%
Iodine Number	7.75
Acetyl Number	11.2
Color	tan, brown
Dielectric Constant	2–68
Effective A.C. Conductivity	25–55
Volume Resistivity	270–630

Soluble in: benzol, carbon tetrachloride, chloroform, dichloroethylene, isopropyl ether, naphtha, tetralin, toluol, trichloroethylene, turpentine, xylol.

Montan wax solutions set more slowly than carnauba wax solutions and, therefore, should be filled into containers at a lower temperature. Pure montan wax can scarcely be scratched by the finger nail. It breaks with a conchoidal fracture. It is a good solvent (when melted) for basic dyes and shows good wetting and flow in oil solutions. It is the hardest of the natural nonvegetable waxes and is fairly resistant to oxidation.

The bleached or highly refined wax loses some of the listed characteristics, becoming softer and less emulsifiable.

Uses: in place of carnauba wax in many compositions because of its hardness, toughness, high melting point and emulsifiability;

polish, rubber, printing-ink, electrical insulating compositions, leather finishes and dressings, carbon papers, grease, phonograph records, waterproofing.

The wax from British lignite is a satisfactory substitute for montan wax while the peat wax is softer and has a lower melting point. The crude wax consists of a mixture of ester wax with a resinous and asphaltic matter. It is dark brown, hard, brittle and odorless. The properties of the crude wax are:

Melting Point	69–84°C.
Acid Number	29–34
Saponification Number	70–77
Ester Number	36–45

The resin content was 40%. In exceptional cases, the crude wax contained 50–95% resin.

Table 37

EFFECT OF THE TEMPERATURE ON THE YIELD AND PROPERTIES OF LIGNITE WAX[1]

(Devon lignite wax extracted by benzene)

	At 1 At. 80°C.	At 35 At. 250°C.	Commercial Montan Wax
Yield of Wax, % Dry Lignite	4·2	6·9	—
Melting Point	76–82°C.	58–68°C.	74–82°C.
Acid Number	34	25	33
Saponification Number	70	68	76
Ester Number	36	38	43
Ash, %	0·6	0·8	0·3

The use of a higher extraction temperature increased the yield of wax, but changed its properties. All the waxes were dark brown, but while the montan wax and the lignite wax extracted at 80°C. were hard and brittle, the wax extracted at 250° was relatively soft and was readily indented by finger pressure.

The color of the purified wax is ivory yellow. The color varies with the amount of oxidant added. If less oxidant is used, a brown wax is obtained while with more oxidant the wax has a green color. The best color is obtained by clarifying the wax with Fuller's earth.

[1] C. M. Cawley, and J. G. King, *J. Soc. Chem. Ind.*, p. 237, Sept. 1945.

The properties of the waxes obtained from the English and Scottish peats were similar in all respects. They were dark brown (almost black), and hard and brittle, but more readily fractured than lignite wax or montan wax.

Table 38

WAXES FROM ENGLISH AND SCOTTISH PEATS

No.	Yield of Crude Wax by Extraction with Benzene, % by wt. of Dry Peat	Melting Point of Crude Wax °C.	Properties of Crude Wax Acid No.	Saponification No.	Ester No.	Resin Content of Crude Wax, %
English Peats						
1	6·0	62–68	49	—	—	—
2	9·4	63–68	51	—	—	—
3	6·8	60–66	37	—	—	—
4	5·8	63–68	52	—	—	—
5	5·4	62–67	54	—	—	—
6	5·5	63–68	54	—	—	—
7	6·3	62–67	52	—	—	—
8	6·8	63–68	55	—	—	—
9	6·3	62–66	55	—	—	—
Bulked Samples 1–9	—	63–68	50	119	69	—
Scottish Peats						
10	4·8	67–74	51	119	68	—
11	7·6	62–72	45	112	67	25
12	4·3	59–64	44	106	62	—
13	3·2	63–75	51	103	52	—
14	3·2	68–73	43	107	64	28
15	11·8	64–72	44	122	78	20
16	7·6	64–72	45	113	68	20
17	2·7	68–77	55	120	65	—
18	7·8	60–71	47	106	59	20
19	8·6	64–74	47	116	69	—
20	5·2	63–75	59	122	63	26
Bulked Samples 10–20	—	64–73	48	113	65	23

They also differed from these waxes in having lower melting points (by about 10°) and higher acid and ester numbers. The properties of the various waxes are compared in Table 39.

Table 39

COMPARISON OF THE PROPERTIES OF DIFFERENT WAXES

	German Montan Wax	Devon Lignite Wax	English Peat Wax	Scottish Peat Wax
Melting Point	74—82°C.	73—83°C.	63—68°C.	64—73°C.
Acid Number	33	30	50	48
Saponification Number	76	75	119	113
Ester Number	43	45	69	65
Resin Content, %	14	40	—	23
Specific Gravity at 20°C.	1·05	1·04	1·03	—

The peat wax had a higher oxygen and a lower carbon content than lignite wax and montan wax. This property is associated with the higher acid and ester numbers of peat wax.

Table 40

YIELDS AND PROPERTIES OF CRUDE WAXES EXTRACTED FROM PEAT WITH DIFFERENT SOLVENTS

Scottish Peat No.	Solvent	Yield of Crude Wax % Dry Peat	Melting Point	Acid No.	Saponification No.
14	Light Petroleum, B.P. 100—120°C.	2·3	54—70	—	—
	Benzene	3·2	68—73	—	
	Benzene Alcohol Azeotrope (65/35 by vol.)	6·6	>86 S.O.*	—	— —
15	Light Petroleum, B.P. 100—120°C.	10·4	55—68	—	—
	Benzene	11·8	64—72	44	122
	Benzene Alcohol Azeotrope (65/35 by vol.)	18·6	>70 S.O.	62	155
17	Light Petroleum, B.P. 100—120°C.	2·4	70—78	—	—
	Benzene	2·7	68—77	55	120
	Benzene Alcohol Azeotrope (65/35 by vol.)	4·9	64—90	61	132
19	Light Petroleum, B.P. 100—120°C.	6·8	60—71	—	—
	Benzene	8·6	64—74	47	116
	Benzene Alcohol Azeotrope (65/35 by vol.)	12·8	65—90	—	—
20	Light Petroleum, B.P. 100—120°C.	4·1	54—74	—	—
	Benzene	9·0	>70 S.O.	55	152
	Benzene Alcohol Azeotrope (65/35 by vol.)	5·2	63—75	59	122

* S.O. = Softened only

VEGETABLE WAXES

BOTANICAL ORIGIN OF WAXES[1]

Palm Tree Waxes
- Carnauba
- Ouricury (Uricury)

Shrub, Cane and Grass Waxes
- Candelilla
- Rhimba
- Ocoxilla
- Banana
- Esparto
- Flax
- Hemp
- Cottonseed
- Kapok
- Cane
- Bamboo Leaf
- Coffee Berry
- Tea
- Palm
- Raffia
- Cocoa
- Opium
- Japan
- Bayberry
- Ocuba
- Apple
- Plum
- Rose
- Jasmine

Ficus and Tree Waxes
- Godang
- Cow Tree

Root Parasite Wax
- Blanophore

SOLUBILITY OF VEGETABLE WAXES

True vegetable waxes are soluble in most fat solvents, such as chloroform, ether or toluol, but are less soluble than fats. Since they contain no glycerides, they do not form acrolein on heating. They do not usually become rancid and are less easily hydrolyzed than fats. They can, however, be decomposed by prolonged heating with alcoholic potash. Because of the high molecular weight of their fatty acid components, their saponification numbers are lower than those of fats. They are also characterized by very high amounts of unsaponifiable constituents which consist of sterols and higher saturated alcohols, usually occurring as esters.[2]

CARNAUBA WAX (PALM WAX)

Carnauba wax occurs as a fine powder on the leaves of certain tropical palms. Most of it is produced in Brazil. The leaves are

---

[1] Ivanowsky, *Oil Colour Trades J.*, p. 918, Dec. 20, 1940.

[2] Hass, *Chemistry of Plant Products*, 4th Edition, p. 45. (New York, Longmans, Green & Co., 1928).

removed from the palm, dried in the sun, and then the powdery wax is brushed and scraped off and placed in vats of boiling water. The wax melts and floats to the top from which it is skimmed off. This crude wax, containing foreign matter, is dirty yellow, brown or green. It is refined by remelting in water and removing the impurities. Specially purified grades are produced by treatment with bleaching earths, chemicals, etc. The so-called white carnauba wax is a highly refined grade, containing some paraffin wax.

The commercial grades of carnauba wax, as sold in the United States, are:

No. 1 Yellow
No. 2 Yellow
No. 2 North Country
No. 3 North Country
No. 3 North Country, Refined
No. 3 Chalky

The first is the purest and lightest in color. Going down, the other grades are progressively darker and less pure.

Carnauba wax contains varying amounts of water. The refined grades contain 0.5 to 1.5%. The chalky grade usually contains 5% and some samples have tested up to 11%.

Carnauba wax is amorphous, hard, tough, lustrous and has a pleasing odor. It breaks with a clean fracture. Chemically, it

Properties

Melting Point	184–196°F.
Specific Gravity, 15°C.	0.990–0.999
Acid Number	4–9
Saponification Number	78–87
Unsaponifiable Matter	52–55%
Iodine Number	13.1–13.5
Acetyl Number	51–60
Color	yellow, green, gray, brown
Refractive Index, 60°C.	1.463
Dielectric Constant	2.67–4.20
Effective A.C. Conductivity	250–310
Volume Resistivity	0.5–4

is a mixture of hydrocarbons, higher alcohols and their esters, and possibly lactones with a small amount of inorganic matter. Its exact composition is not known. It is saponified by strong alkalies and is the hardest, highest-melting, natural commercial wax, except for some crude grades of ouricury wax. It is added to other waxes to increase their melting point, hardness, toughness, and luster and to decrease stickiness, plasticity and crystallizing tendencies.

Soluble in: hot alcohol, benzol, carbon tetrachloride, ether, dioxane, chloroform, trichloroethylene, isopropyl ether.

Uses: candles, polishes, lubricants, greases, floor and automobile waxes, insulating materials, carbon paper, chalk, matches, soaps, salves, phonograph records, plastics, cosmetics, protective coatings and paper coating.

Addition of large percentages of cumarone resins to carnauba wax increases its resistance to fracture (toughness) and makes it stringy when melted.

The melting point of carnauba wax is raised 3°C. by the addition of 1% Acrawax. Its flexibility is increased by the addition of 2% or more of beeswax, microcrystalline paraffin wax, or oleic acid.

Carnauba wax, dissolved in mineral oil, wets pigments and other insoluble particles quite well—better than most oils. Although a carnauba wax-oil mixture has good wetting properties, it sets fast (in thin layers) and does not penetrate rapidly into paper, textiles, etc. Solutions of carnauba wax precipitate more readily than those of other natural waxes on addition of nonsolvents. Evaporation of solutions of carnauba wax yields noncontinuous powdery layers rather than films.

Properties of Brazilian Grades of Carnauba Wax

In Brazil, seven types of carnauba wax are available, known as: 1. flower, 2. first, 3. medium clear, 4. medium purple, 5. "cauipe," 6. sandy and 7. fat.

Flower Wax

This is made in small quantities, seldom appearing in export lists; it is the best quality, with hardly any impurities. Its color is almost the same as that of the yolk of an egg.

	%		
Wax	99.32	Iodine Number	22.6
Moisture	0.48	Melting Point	84°C.
Impurities	0.20		

First Wax

It varies in color from dark to light with an insignificant percentage of impurities.

	%		
Wax	99.65	Iodine Number	24.3
Moisture	0.35	Melting Point	83°C.
Impurities	0.00		

Medium Wax

This is generally of a uniform color, varying from light yellow to light grey, according to the age of the tree and the treatments.

	%		
Wax	99.53	Iodine Number	21.6
Moisture	0.42	Melting Point	83°C.
Impurities	0.00		

Sandy Wax

This wax is light to dark grey and rough to the touch.

	%		
Wax	97.16	Iodine Number	16.7
Moisture	1.09	Melting Point	86°C.
Impurities	1.75		

Fat Wax

This is the darkest variety, almost black, containing a large percentage of impurities. Taken from the palm leaves, it is melted to drive off moisture.

	%		
Wax	96.91	Iodine Number	16.00
Moisture	0.59	Melting Point	85°C.
Impurities	2.50		

Carnauba-Wax Adulterations[1]

To detect possible adulterants, the Cleveland open-cup flash-point tester was used. This proved to be a very handy tool. The melting points of combinations of ingredients with carnauba wax could be artificially raised by heating with ethyl cellulose. The flash point would be lowered.

1 Bowers, *Soap and Sanitary Chemicals,* 113 (1949).

If the open-cup flash-point test[1] is followed carefully, reproducible results can be obtained. Care should be taken to have no drafts in the room which would dilute the fumes before they ignite. The sample should be heated at a constant rate of 9—11°F. a minute. These are the common sources of error.

A list of the highest and lowest flash-point tests of carnauba wax of various grades is given in Tables 41 and 42.

Table 41

FLASH POINTS OF CARNAUBA WAXES TESTED PRIOR TO 1942

Carnauba Wax	Flash Point, °F.
No. 1 Yellow	600–620
No. 2 Yellow	600–610
No. 3 N. C. Refined	585–595

Carnauba waxes tested since 1942 have shown a tendency to lower flash points as shown in Table 42.

Table 42

FLASH POINTS OF CARNAUBA WAXES TESTED AFTER 1942

Carnauba Wax	Flash Point, °F.
No. 3 N. C. Crude	575
No. 3 N. C. Refined	572
No. 3 N. C. Refined	572
No. 3 N. C. Refined	509
No. 3 N. C. Refined	536
No. 3 N. C. Refined	543
No. 3 N. C. Refined	Sputter 445 to 575 flashed
No. 3 N. C. Refined	536
Carnauba Wax	Flash Point, °F.
No. 3 N. C. Refined	554
No. 3 N. C. Refined	558
No. 3 N. C. Refined	568
No. 3 N. C. Refined	542
No. 3 N. C. Refined	554
No. 3 N. C. Refined	546
No. 3 N. C. Refined	564
No. 3 N. C. Refined	570

[1] *Standard Methods of Analysis of the Oil Chemists Society,* p. 41.

The low flash points in carnauba wax may be due to adulteration or to different methods of collecting and refining. In some formulas, these waxes do not make a stable emulsion, regardless of the usual formula adjustments. In other wax formulas, with more efficient emulsifiers, emulsions are formed, but the resulting waxes are sticky and, when stored at 120°F. for a few days, set up into jells. These results are shown in Tables 43 and 44. Table 43 shows the stickiness of two lots of wax of low flash points compared to waxes of high flash point made by the same formula.

Table 43

VARIATION OF TACK OF CARNAUBA WAX WITH THE FLASH POINT

Carnauba Wax	Pounds Pull	Temperature, °F.	Humidity, %
536	14	74	40
543	12	74	40
590	7	74	40
595	7	74	40

The method of measuring tack is as follows. Linoleum strips are coated with two coats of wax. The wax is dried overnight. A clean woven wool pad of polisher's felt is placed on the wax. A block of wood covers the felt. The block contains an eye screw. A 20-lb block of lead is placed on the wood cover. A 15-lb capacity spring scale is attached to the eye screw. Two minutes are allowed to lapse between putting on the weight and checking the pull in pounds necessary to move the felt pad.

The figures Table 44 shows some correlation between the shelf life and flash point. The same formula was used in each case. The expected life of this formula at 120°F. is 1 year or more.

Table 44

RELATION BETWEEN FLASH POINT AND SHELF LIFE

Flash Point of Wax, °F	*Shelf Life at 120° F.*
536	2 days
545	6 days
509	14 days
572	185 days
590	12 months

A correlation between both tack and shelf life and low flash points would depend on the ingredients causing the low flash point. These ingredients may vary slightly.

Table 45 shows flash points of various wax blends made with Carnauba wax.

Table 45

FLASH POINTS OF WAX BLENDS CONTAINING CARNAUBA WAX

			Flash Point, °F.
Refined Carnauba No. 3 N. C., alone			590
Refined Carnauba No. 3 N. C. with	5%	candelilla	575
	100%	candelilla	490
Refined Carnauba No. 3 N. C. with	5%	135° paraffin	527
Refined Carnauba No. 3 N. C. with	10%	180° paraffin	570
Refined Carnauba No. 3 N. C. with	10%	oxidized petroleum wax	582
Refined Carnauba No. 3 N. C. with	20%	oxidized petroleum wax	572
	100%	oxidized petroleum wax	539
Refined Carnauba No. 3 N. C. with	10%	Piccolyte S-100	543
Refined Carnauba No. 3 N. C. with	25%	Piccolyte S-100	503
Refined Carnauba No. 3 N. C. with	25%	Durez 219	540
Refined Carnauba No. 3 N. C. with	10%	wood I rosin	510

CANDELILLA WAX

The commercial, refined wax is hard, brittle and lustrous and has an aromatic odor. Different grades and shipments exhibit varying degrees of stickiness (on pressure, in solution, and in compositions).

Properties

	Standard	Double Refined
Melting Point	149–156°F.	158–162°F.
Specific Gravity, 15°C.	0.982–0.993	–
Acid Number	11–19	19.6
Saponification Number	46–66	58.6
Iodine Number	15–36	37.8
Ester Number	–	39.0
Ash	–	0.05%
Moisture	small percentage	0.73%
Unsaponifiable Matter	65–67%	–
Color	gray–brown	–
Refractive Index	1.4555	–
Odor	aromatic	–
Dielectric Constant	2.50–2.63	–
Effective Conductivity	19	–
Volume Resistivity	120	–

Chemically, candelilla wax is composed of a high proportion of hydrocarbons and resins with a small amount of lactones.

Candelilla wax is not quite so hard as carnauba wax and does not emulsify or saponify so readily. The commercial wax often contains a small amount of water which should be driven off by heating before use in nonaqueous compositions. It is very slow in setting in melted compositions and often does not reach maximum hardness for a few days. Its stickiness varies with its degree of refinement, different lots showing a variation. Mineral waxes and Acrawax C reduce its stickiness, but increase the aging and setting time. It is a good solvent for basic dyes. Additions of oleic or similar acids slow down crystallization and rapidly increase softness. It does not blend well with ouricury wax.

Soluble in: acetone, carbon tetrachloride, chloroform, turpentine, petroleum (forms a gel on cooling hot solutions in these solvents); alcohol-benzene (2:3 mixture).

Uses: in leather dressing, furniture and shoe polishes, candles, cements, varnishes, sealing wax, electrical insulating compositions, phonograph records, paper size, celluloid, rubber; in water- and insectproofing containers; in paint removers, as a soft wax stiffener, and as a substitute for carnauba wax and beeswax.

Properties of Standard-Grade Mexican Candelilla Wax

Melting Point (Drop)		66.0–71.0°C.
Specific Gravity at 15°/4°C.		0.9820–0.9856
Refractive Index—71°C.		1.4555
Saponification Number		47–64
Ester Number		40–47
Acid Number		12.3–20.6
Nonsaponifiable Matter		47–62
Hydrocarbons (Principally Hentriacontane and Tritriacontane)		40–60%
Resins		10–20%
Moisture and Volatile Matter		0.3–1%
Iodine Number		19–44
Color in Toluene Solution	Red	0.9–1.4
	Yellow	2.0–4.0
Myricilic Alcohol, Dihydroxymyricinoleic Acid, Lactones		30–40%
Reichert-Meissl Number		0.5–7.60
Electric Rigidity (Volts per mm of Section at 50 Cycles, Steel Spheres of 25 mm diameter)		23,500–45,400

Solubility

Alcohol	insoluble
Alcohol (Hot)	soluble
Acetone (Hot)	slightly soluble
Benzene (Hot)	soluble
Petroleum Ether (P.E. 30–75°C.) 20°C.	insoluble
Petroleum Ether (Hot)	soluble

JAPAN WAX

Japan wax is obtained from the berries of certain sumac trees in Japan and China. The berries are aged, crushed and, after the removal of the husks, are crushed again in heated presses to obtain a crude, green, soft wax. It is then refined by melting in hot water, skimming, and bleaching by sunlight. The two best-known brands are Kitagumi and Chickusan. This wax is often adulterated with varying amounts of local vegetable oils.

Japan wax is not hard but tough, malleable and sticky. It has a tallowy odor and becomes rancid, effloresces and darkens on aging. It breaks easily and the broken surfaces are dull.

Chemically, Japan wax is a fat (glyceride) and is composed of palmitin, palmitic and other high-molecular-weight fatty acids and their glycerides.

It contains 5.2—6.3% (sometimes 7.1%) dibasic acids of which one of the formula $C_{23}H_{44}O_4$ is predominant. Dibasic acids probably occur in Japan wax as mixed glycerides with oleic acid.

Properties

Melting Point	122–133°F.	(48–53°C.)
Specific Gravity	0.975–0.984	(0.975–0.992)
Acid Number	6–20	(18–20)
Saponification Number	216–236	(217–230)
Unsaponifiable Matter	1–1.5%	—
Iodine Number	4–16	(4.5–15)
Acetyl Number	27–31.2	—
Ratio Number	—	(10–11)
Hehner Number	—	(89–91)
Reichert-Meissl Number	—	(1–3)
Ash	0.02–0.08%	—
Refractive Index, 60°C.	1.450	—
Dielectric Constant	3.1–3.2	—
Effective A.C. Conductivity	33–34	—
Volume Resistivity	0.4–0.8	—

It is readily saponified by alkali to an emulsion.

When Japan wax is melted and allowed to solidify, its specific gravity and melting point are lowered. On aging, however, these constants tend to return to the normal standard values.

Soluble in: alcohol, carbon disulfide, chloroform, ether, benzol, petroleum ether, carbon tetrachloride, isopropyl ether, naphtha, pyridin, tetralin, toluol, xylol, trichloroethylene, turpentine.

Uses: wax matches, polishes, creams, substitute for beeswax, as an ingredient in the manufacture of special soaps, laundry glazes, textile finishes, and metal-drawing lubricants.

UCUHUBA WAX (UCUHUBA BUTTER, UCUHUBA RESIN, UCUHUBA TALLOW)

This wax is obtained from the ucuhuba plant, which grows in Brazil. It is of a yellowish color and of the consistency of butter. Its composition is as follows:

	%
Moisture	19.380
Albuminous Substances	11.409
Ether Extract	10.000
Cellulose	3.162
Ash	5.520
Nonnitrogenous Substance	50.529
Phosphorus (P_2O_5)	1.552

Properties

Melting Point	45°C.
Solidifying Point	40°C.
Saponification Number	219–221
Iodine Number	9–14
Acidity	17.5
Nonsaponifiable Matter	3.2

BAYBERRY WAX (MYRTLE WAX, LAUREL WAX, BAYBERRY TALLOW)

Bayberry wax is obtained from the berries of the myrica shrubs on the Atlantic coast of the United States. The berries are heated in water until the wax melts, which is then skimmed off and purified by remelting and filtering.

Commercial bayberry wax is a greenish-white, fatty substance, more unctuous and brittle than beeswax.

Chemically, bayberry wax is a fat consisting of stearin, palmitin, myristin and olein (glycerides).

Properties

Melting Point	102–120°F.
Specific Gravity	0.875–0.980
Saponification Number	206–217
Iodine Number	1.9–3.9
Color	green–white
Odor	faint
Taste	bitter

It saponifies readily with alkali solutions.

Soluble in: alcohol and turpentine (only slightly).

Uses: in candles, soaps, ointments, etc.; polishing leather; etching; and medicinals.

Ouricury Wax (Uricury Wax)

Ouricury wax is obtained from the exudation of the leaves of a palm called "ouricury," which grows in the northeastern states of Brazil. Chemically it is similar to carnauba wax. Ouricury wax, however, has a higher resin content, is more soluble in common solvents, has a lower fusion point, and has higher iodine and saponification numbers than carnauba wax. It is not compatible with certain waxes. On melting with other waxes, a resinous precipitate is formed. Its setting point (72–73°C.) is lower than that of carnauba wax (79–80°C.).

It differs from other vegetable waxes in that there is a lag of about 12° between its melting and setting points. Most vegetable waxes show a lag of 5–6° only.

Some grades contain as much as 18% foreign matter, which must be removed by filtering. Double-refined ouricury wax contains a much smaller amount of resin and foreign matter.

Properties

Melting Point	83–85°C.
Specific Gravity, 15°C.	0.970
Acid Number	5.5–10
Saponification Number	79–242
Iodine Number	8.6–12.8
Ester Number	73–92.5
Color	brown

Uses: as a substitute for carnauba wax in floor waxes, shoe polishes, miscellaneous polishing products, candles, etc.

COCOA BUTTER (CACAO BUTTER, OIL OF THEOBROMA)

Cocoa butter is a fatty material obtained from roasted cocoa beans, principally as a by-product of chocolate manufacture.

It is a yellowish white, greasy solid of a faint, agreeable odor and a bland, chocolaty taste. It is brittle below 25°C. It consists principally of glycerides of stearic, palmitic, oleic, lauric acids and other lower fatty acids.

Properties

Melting Point	30–35°C.
Specific Gravity, 100–25°C.	0.858–0.864
Saponification Number	188–195
Iodine Number	35–40
Refractive Index, 40°C.	1.4537–1.4578
Reichert-Meissl Number	0.1
Polenske Number	0.2

Soluble in: benzol, benzene, chloroform, ether and hot anhydrous alcohol.

Uses: cosmetics, pharmaceutical ointments, suppositories.

FIBER WAX (ESPARTO WAX, REED WAX)

Fiber wax is a dark-tan wax obtained from certain reeds and grasses (alfa and esparto grass) grown in Spain and North Africa. It is somewhat similar to candelilla wax but is less brittle and harder. When melted, it smells like decomposed straw. It has a grain structure, but much less pronounced than that of carnauba wax. When melted and cooled slowly, the surface is very smooth

Properties

Melting Point	75–77°C.	(61.2–74°C.)
Specific Gravity	0.978–0.994	(0.965–0.989)
Acid Number	23.9–30.2	(16.5–34.4)
Ester Number	40.6–45.9	—
Saponification Number	64–69.8	(61.4–75.7)
Iodine Number	15.0–19.2	(13.2–15.3)
Unsaponifiable Matter	16.8%	(66–72)
Ash	0.4–0.6	—
Color	brown	—

and flat. It contracts much less than carnauba wax. The purified commercial grade has an average melting point of 75.3°C.

Fiber wax contains about 65—70% hydrocarbon of the formula $C_{31}H_{64}$.

Fiber wax gives good wetting and penetrating properties in oil compositions and softer films than carnauba wax. It is a better solvent for basic dyes than carnauba. It can be plasticized by the addition of small amounts of amorphous waxes or oils.

Uses: replaces carnauba wax in polishes, shoe finishes, etc.

FIR-BARK WAX

A new fir-bark wax is obtained from the bark of the Douglas fir tree by steam distilling the solution of wax obtained by extraction with benzene of the ground bark.

COTTON WAX

Cotton wax is obtained from the so-called green cotton, in Arkansas, by solvent extraction. It is a hard, yellow, high-melting wax.

Properties

Melting Point	83–85°C.	(76–80°C.)
Specific Gravity	0.902	(0.9760–1.005)
Saponification Number	2.5	(50–76)
Acid Number	6.7	(21–28)
Iodine Number (Hanus)	12.75	(20–27)
Acetyl Number	—	(48–83)
Unsaponifiable Matter	—	(57–68%)
Acetyl Number of Unsaponifiable Matter	—	(115–124)

FLAX WAX

Properties

Melting Point	67–70°C.	(67.3–69.5°C.)
Specific Gravity	0.971–0.985	(0.936–0.985)
Saponification Number	78–83	(78.4–83.7)
Iodine Number (Hanus)	21–28	(21.6–28.8)
Acid Number	18–23	(17.5–23.8)
Unsaponifiable Matter	70–80%	(20%)
After Distillation with Superheated Steam		
Saponification Number	70	
Acid Number	33	
Iodine Number	18	

SUGAR–CANE WAX

Crude cane wax is extracted from cachaza, a by-product in the manufacture of raw sugar. Cachaza is the filter-press mud separated by sedimentation and filtration from the cane juice which is squeezed from the ripe cane. It consists of the suspended particles and the precipitate formed by the action of heat and chemical treatment on the colloid, and is removed in the clarification of the cane juice. The precipitate contains the cane wax that was originally on the surface of the cane stalk before it was crushed. The cachaza is removed from the cane juice, after chemical treatment and heating, by sedimentation and filtration.

When the filter cake is removed from the filters, it is mixed with a solvent which dissolves the cane wax as well as other soluble substances. The solvent is then evaporated, recovered and reused. The mass residue is waxy and contains about 50% sugar-cane wax, 30% sugar-cane fatty oil and 20% resinous matter.

In a refining process, the crude sugar-cane wax is dispersed in a solvent which readily dissolves the fatty-oil fraction and leaves the hard wax and resinous matter as a suspended solid. This slurry is filtered to remove the soluble fatty-oil fraction which is recovered as a green semifluid paste.

The sugar-cane wax and the resinous matter from which the sugar-cane fatty oil has been removed are now reslurried with additional solvent and heated until the cane wax melts. At this temperature, the resinous matter separates from the wax and is removed by decantation. The light layer containing the sugar-cane wax is then evaporated to remove all the solvent, leaving the refined cane wax which is cast into pans.

The characteristics of cane wax are similar to those of the other valuable hard natural vegetable waxes. The supply of this wax which could be obtained from the world cane crop would far exceed that of any other vegetable wax. The potential supply of hard wax from only two of the Cuban American Sugar-Cane Company's crushing plants is about 1,320,000 lb per crop year.

The wide variation in the constants is due to varying grades of purity, the higher figures referring to the pure and bleached wax.

Properties

Melting Point	68–80°C.	(82°C.)
Specific Gravity	0.963–0.985	(0.961–0.984)
Saponification Number	27–135	(40–57)
Acid Number	9.6–23.4	(4–12)
Iodine Number	19.8–31.5	(12.4–16.2)
Acetyl Number	55–95	(55–95)

Table 46

SOLUBILITY OF SUGAR-CANE WAX[1]

Solvent	%
Skelly-Solve B	0.67
Skelly-Solve C	0.71
Shell No. 8206	0.82
Shell No. 8182	0.73
Shell No. 8196	0.78
Benzol	1.06
Naphtha	0.73
Hexane	0.61
1,2-Dichloroethylene	>2.00
Cyclohexane	1.30
Methylhexane	0.93
Dioxane	0.29
Isopropyl Ether	0.52
Ethylene Dichloride	0.60
1,1,2,2-Tetrachloroethane	0.87
Trichloroethylene	>2.00
Butyl Chloride	1.17
Carbon Tetrachloride	1.24
Benzene	1.04
Toluene	1.43
Acetone	0.11

Nitroethane, nitromethane and 95% ethanol form gels at room temperature and dissolve less than 2% wax at the boiling point.

Table 47

ACID COMPONENTS OF SUGAR-CANE WAX

Acid	Percentage of Total Acids	Percentage of Wax
Caproic	0.6	0.3
Palmitic	27.7	14.7
Stearic	22.4	11.8
Oleic	41.5	22.0
Arachidic	3.3	1.8
Resinous	4.5	2.4

[1] R. T. Balch, *Sugar J.* 4, No. 6, 24 (1941).

Table 48

ELECTRICAL PROPERTIES OF DARK HARD SUGAR-CANE WAXES AND OTHER COMMERCIAL WAXES

Wax	Frequency kc/sec.	Dielectric Constant	Power Factor	A.C. Conductivity^{-1} mho–cm^{-1}	Volume Resistivity ohm–cm
Louisiana	1.0	2.93	103×10^{-4}	0.17×10^{-10}	
Sugar Cane	7.5	2.87	124×10^{-4}	1.5×10^{-10}	
	100.0	2.82	120×10^{-4}	19.0×10^{-10}	
					6×10^{15}
Cuban	1.0	2.65	96×10^{-4}	0.14×10^{-10}	
Sugar Cane	7.5	2.61	118×10^{-4}	1.3×10^{-10}	
	100.0	2.56	131×10^{-4}	19.0×10^{-10}	
					2×10^{15}
		*		*	*
Carnauba No. 2, N.C.	All	2.66–2.83		$4.9–5.3\times10^{-12}$	$60–120\times10^{13}$
Candelilla	samples	2.38–2.49		$6.0–6.4\times10^{-12}$	$>290\times10^{13}$
Crude Montan	measured	2.62–2.67		$29.0–31.0\times10^{-12}$	$>290\times10^{13}$
Crude Beeswax	at	2.87–2.88		$45.0–48.0\times10^{-12}$	$5–12\times10^{13}$
Chinese	1 kc/s	2.94–2.98		$125.0–137.0\times10^{-12}$	$1–2\times10^{13}$
Yellow Ceresin		2.16–2.24		$<0.37\times10^{-12}$	$>290\times10^{13}$
Refined Paraffin		2.21–2.24		$<0.37\times10^{-12}$	$>290\times10^{13}$

* Initial values.

Properties of Sugar-Cane Wax "Duplicane" 517

Melting Point	165–170°C.
Penetration	1 maximum
Color	brown
Acid Number	20–30
Saponification Number	65–75
Oil-retention Penetration	42 maximum

Table 49

SOLUBILITY OF SUGAR-CANE WAX (DUPLICANE 517) IN VARIOUS SOLVENTS

Weight % of Wax	*Solvent*	*Cloud point, °C.**
10	ethanol	insoluble
10	butanol	56°C.
10	acetone	insoluble
10	benzol	66°C.
10	ethyl ether	partially soluble
10	ethyl acetate	44°C.
10	carbon tetrachloride	28.5°C.
10	kerosene	40°C.
10	V M and P naphtha	38°C.
10	turpentine	34°C.
10	hydrogenated naphtha	32.5°C.

* The wax is completely soluble above the temperature shown.

Potential uses for the wax are in water-emulsion polishes, carbon paper and paste polishes; as pigment dispersers and for casting.

RICE-OIL WAX

Rice-oil wax is prepared commercially by hydrogenation of crude rice oil at high pressure and temperature. The dewaxing of the oil can be done by evacuated unglazed porcelain cylinders covered with filter cloth or by a De Laval and basket-type centrifuge.

The crude wax is purified by washing repeatedly with methyl alcohol, acetone, ether and finally chloroform to yield a substance which is an ester mainly composed of tetracosanoic acid and myricyl alcohol with small amounts of saturated acids and some unsaturated acids.

The commercial rice wax melts at 67–69°C. and has a saponification number of 142–144. The purified wax melts at 75–76°C. and has a saponification number of 72.8, iodine number of 12.6 and 48.71% unsaponifiable matter.

ANIMAL WAXES

SPERMACETI (CETIN; CETACEUM)

Spermaceti is obtained from the cavities in the head and from the fat of sperm whale. The sperm oil is chilled and the precipitated crude spermaceti is filtered off and refined by boiling with dilute caustic-soda solution, separated and washed with hot water until free from alkali. The melted, refined product is run into molds and allowed to crystallize.

Commercial spermaceti is white (translucent), shiny, somewhat greasy, brittle and crystalline. It has a faint odor and a bland taste. Chemically, it is composed of cetyl palmitate, cetyl alcohol, and esters of high-molecular-weight fatty acids.

Spermaceti is partially saponified by aqueous alkalies and forms emulsions.

Soluble in: boiling alcohol, ether, chloroform, carbon disulfide, fixed and volatile oils.

Uses: cosmetics and ointments.

Properties

Melting Point	107–120°F.	(41–52°C.)
Specific Gravity	0.945–0.960	(0.895–0.920)
Acid Number	0.5–1.1	(0.1–0.5)
Saponification Number	122–129	(118–135)
Unsaponifiable Matter	51–54	(47–55)
Iodine Number	2.5–3.9	—
Acetyl Number	2.7	—
Color	white (translucent)	
Refractive Index, 60°C.	1.440	—
Dielectric Constant	6–18	—
Effective A.C. Conductivity	>500	—
Volume Resistivity	0.0006–0.0003	—
Viscosity at 100°C.	—	(6.7–7.4)
Fatty Acids	—	(49–54)

INSECT WAXES

BEESWAX (WHITE WAX, CERA ALBA; YELLOW WAX, CERA FLAVA)

Beeswax is produced by the honey bee for building its honeycomb in the Americas, Africa, Europe and elsewhere. The honeycomb is melted and filtered, or boiled with water and then skimmed off to obtain a crude wax. The crude wax is refined by remelting and filtering with subsequent solvent extraction or bleaching with chemicals or sunlight. Beeswax produced by wild bees is considered better than that produced by domesticated bees, particularly for use in cosmetic creams.

Crude beeswax varies in color from a deep brown to a light yellow. The color is dependent on the type of flower from which the bee has taken the pollen, the age and care of the hive. Generally, the light-yellow crudes are of better quality because they can be bleached by natural processes. Crude beeswax is thoroughly washed, which process is known as refining. The refined wax is cast into approximately 1-oz squares, 1-lb bricks or 25-lb slabs, ready for the market. Wax to be bleached is shredded mechanically so that the maximum surface of the wax is exposed to the sun. These shavings are then placed into trays where, after exposure to sunlight, they turn white. The wax is then washed, cast into round discs, bricks or slabs, ready for use.

The specifications of U.S.P. beeswax are as follows: Melting point 61–65°C.; acid number 16–23, ester number 72–79, saponification number 85–101.

Beeswax is often sophisticated with different types of adulterants, including mineral matter. Among others, tallow, stearic acid, Japan wax, carnauba wax, rosin, paraffin wax and ceresin are important adulterants. The purity of beeswax is tested by the ratio of the ester number to the acid number. Even this value is not reliable if taken by itself. Waxlike mixtures can be obtained which, although free from beeswax, can be made to have the normal ratio, as, for example, a mixture obtained by melting together 37.5 parts of Japan wax, 6.5 parts of stearic acid, 6.5 parts of paraffin or ceresin wax. In buying beeswax, therefore, the safest course is to buy it from a reliable and recognized firm.

Beeswax, mixed with ceresin and some aromatic resins, is used for preparing artificial beehives for rearing bees in Europe.

Commercial beeswax is supplied in slabs or cakes in the following grades:

1. Crude
2. Refined
3. White, bleached

Chemically, beeswax is composed of myricyl palmitate, cerotic and homologous acids, with small amounts of hydrocarbons, cholesterol esters and ceryl alcohols, pollen and resins. The presence of resins, in undue amounts, makes bleaching difficult. The free fatty-acid content is an important factor in emulsifiability.

When beeswax is heated to 150–250°C., the acid number decreases, while the ester and saponification numbers increase. Continued heating causes reesterification or estolide formation with decrease of ester and saponification numbers.

Beeswax has an aromatic taste and does not adhere to the teeth or become pasty when chewed. It is plastic, when kneaded by hand, but hard and breaks with a snap when it is bent in the cold.

Soluble in: benzol, benzine, carbon tetrachloride, chloroform, dioxan, isopropyl ether, tetralin, turpentine, and in hot amyl alcohol and acetone. It is partially soluble in cold ether and hot alcohol. It mixes with fats, oils, resins and other waxes when melted with them. It is easily saponified by alkalies.

Uses: candle making, adhesive compositions, dressing and polishing leather, modeling fruits and waxes, wax polishes and finishes, transparent papers, pomades, toilet preparations, cosmetics, base for plasters, etc., in medicine, engraving and lithography, sizing and finishing textiles, chewing gums, food products, sweetmeats.

Table 50

CONSTANTS OF DIFFERENT GRADES OF BEESWAX

	Specific Gravity at 15°C.	Melting Point °C.	Acid No.	Saponification No.	Ester No.	Ratio No. (ester No. divided by acid No.)	Unsaponifiables %
Yellow Beeswax (Cera Flava)	0.958–0.970	62–64.0	17–23	87.0–97	70–80.0	3.3–4.0	50–56
White Beeswax (Cera Alba)	0.958–0.970	62–64.0	18–24	90.0–102	70–80.0	3.3–4.0	50–56
Extraction Beeswax (Unbleached)	0.953–0.957	61–62.5	23–27	92.0–95	66–70.5	2.4–3.0	50–56
Extraction Beeswax (Bleached)	0.970–0.984	69–72.5	22–30	91.5–104	69–77.5	2.5–3.3	50–56

Properties

Melting Point	145–158°F.
Specific Gravity	0.952–0.975
Acid Number	16.6–20.7
Saponification Number	90–96
Unsaponifiable Matter	52–56%
Iodine Number	4–12
Acetyl Number	15.1
Ester Number	72–78
Ester-Acid Ratio	3.6–4.3
Color	white, yellow, brown
Refractive Index, 75°C.	1.4398–1.4451
Odor	honey
Dielectric Constant	3.1–3.3
Effective A.C. Conductivity	70–86
Volume Resistivity	0.9–1.5

CHINESE WAX (INSECT WAX)

Chinese wax is deposited on the branches of a species of ash tree, in China, by certain insects. The wax is scraped off and refined by melting in hot water, skimming off, remelting, filtering and bleaching, if necessary.

Chinese wax is yellowish white, fairly hard, shiny, fibrous, and crystalline.

Chemically, it is composed of ceryl cerotate and minor amounts of other esters.

Properties

Melting Point	149–176°F.
Specific Gravity	0.926–0.970
Acid Number	3–13
Saponification Number	80–92
Unsaponifiable Matter	49–50%
Iodine Number	1.5–2
Color	yellow-white
Odor	very faint
Taste	very faint
Refractive Index, 40°C.	1.4566
Dielectric Constant	3.65–3.90
Effective A.C. Conductivity	190–210
Volume Resistivity	0.04–0.18

Soluble in: alcohol and ether (slightly). In benzol, isopropyl ether, toluol, xylol, trichloroethylene (completely).

Uses: in candles, furniture polishes and coatings for paper and cloth.

SHELLAC WAX

Shellac wax is a by-product in the production of alcoholic solutions of shellac. It precipitates out of such solutions as a haze or sludge and is filtered off and dried.

The commercial product is a hard, tough, brown, lustrous wax containing some shellac and impurities. It approaches carnauba wax in hardness and luster.

Properties

Melting Point	74–78°C.	(74–80°C.)	82°C.
Specific Gravity	0.971–0.980	(0.974–0.982)	0.972 (20°C.)
Acid Number	12–16	(10.9–25)	1.2
Saponification Number	120–126	(79–85)	46.7
Unsaponifiable Matter	72–76%	(68–72%)	—
Ratio Number	—	(7–8.6)	—
Ester Number	—	—	45.5
Iodine Number	—	—	1.25
Volatile Matter at 105°C.	—	—	0.2–0.4
Ash	—	—	0.12

It is used as a component of insulating material.

CHAPTER 2

manufactured and synthetic waxes

FATTY ALCOHOLS

Cetyl Alcohol

Commercial cetyl alcohol is obtained by "splitting" sperm oil or spermaceti, or by the reduction of cetyl palmitate or palmitic acid.

Properties

Melting Point	45–50°C.
Specific Gravity, 15°C.	0.810
Boiling Point	334°C.
Iodine Number	<3
Acid Number	0
Saponification Number	0
Unsaponifiable Matter	98–100%
Ash	0
Refractive Index, 60°C.	1.4345
Color	white
Form	crystalline
Heat of Fusion (cal./mol.)	8900

Cetyl alcohol is now obtainable as an odorless and tasteless product that is stable to alkali, acid, light and air. It is a good solvent for basic dyes.

Soluble in: alcohol, benzol, carbon disulfide, ether, glycol ethers, fats, mineral and vegetable oils.

Uses of Cetyl Alcohol

Cetyl alcohol is an emulsifier for water-in-oil type of emulsions. It causes mechanical adherence of the particles of water and when it is thus incorporated in viscous substances, such as petrolatum or hydrogenated oils, it will tend to stiffen the mixture, yielding an emulsion containing a larger percentage of water than would otherwise be possible.

Examples

Petrolatum alone will only take up 9–15% water. The addition of about 4% cetyl alcohol will permit the absorption of 39–52% water.

A mixture of 4% cetyl alcohol, 10% lanolin, and 86% petrolatum is able to take up 82–108% water.

1% cetyl alcohol more than doubles the water absorbing power of hydrogenated peanut oil, 75% water being taken up without cetyl alcohol and 170% being absorbed with it.

The addition of 3% cetyl alcohol to lard increases the amount of water absorbed from 8–245%.

Emulsions of the water-in-oil type, with large proportions of water, are less stable than those containing less water, but cetyl alcohol usually has a beneficial effect on the stability of these emulsions when it is used as an addition to the standard formulae, as it will act as an auxiliary emulsifying agent.

The proportions of cetyl alcohol to be incorporated in various types of water-in-oil emulsions will, as a rule, range from 1 to 5%, depending on the individual formula.

Since it acts to promote water-in-oil type emulsions with inert substances, cetyl alcohol alone cannot produce the reverse, i.e., oil-in-water emulsions. However, its oil-soluble hydrocarbon radical, when used with standard emulsifying agents, such as soaps, lauryl sulfates and wetting agents will give body and stability to oil-in-water emulsions, in general, and creams and lotions, in particular.

Example

A mixture of equal parts of mineral oil and water, containing about 1% soap, will soon separate into two more or less distinct layers; the addition of about 2% of cetyl alcohol (dissolved in the oil) will make a thick, creamy emulsion which will show only slight separation, on long standing.

Although the stabilizing action is most marked in lotions, a small amount of cetyl alcohol will definitely stabilize brushless shaves, vanishing creams, etc. Approximately 1% is usually sufficient and here again, it may be stated that cetyl alcohol acts as an auxiliary emulsifying agent.

Because of its ability to become hydrated and to assist the hydration of other substances, larger amounts of cetyl alcohol may make a cream softer and more translucent. An addition of 0.5–1% will hardly alter the physical appearance of the product, but still increase stability. Properly incorporated, cetyl alcohol will also cause a cosmetic cream to rub out more smoothly on the skin, as a result of the finer dispersion of ingredients.

LANETTE WAX

Lanette wax is a mixture of cetyl and stearyl alcohols.

Lanette wax SX consists of cetyl and stearyl alcohols, with a neutral emulsifier.

Lanette wax ester is a mixture of palmitic acid esters of cetyl and stearyl alcohols.

Lanette wax extra consists of higher fatty alcohols and saponifiable free fatty acid.

Properties

Melting Point	50°C.
Specific Gravity	0.81
Acid Number	0
Saponification Number	0
Ester Number	0
Acetyl Number	180–190

Soluble in: alcohol, benzol, ether.

Lanette wax, with soap sulfonated oils, aromatic sulfonic acids or other emulsifiers, acts as a stabilizer for emulsions. Lanette wax extra, with diluted alkali or soap, acts as an auxiliary emulsifier, particularly for paraffin- and ceresin-wax emulsions.

Lanette wax SX is neutral, melts at about 60°C. and is, by itself, an emulsifying agent. It disperses in hot water. On cooling, this dispersion thickens or gels. Secondary emulsifiers or stabilizers, such as glue, gelatin, methyl cellulose or gum arabic, are used with it to increase the stability of emulsions. If Lanette wax SX is heated above 100°C., it tends to lose its dispersing properties. Small amounts of this wax are said to give better tensile strength and elasticity to varnish films. When blended with lanolin, absorption bases are formed which can absorb large amounts of water. Certain drug and cosmetic emulsions, made with this wax, are more stable to weak acids, such as boric, carbolic and salicylic. It is also more stable to mild oxidizing and reducing agents than soap.

TECHNICAL STEARYL ALCOHOL (OCTADECANOL; "STENOL")

This is a white solid waxlike product; mixture of fatty alcohols, stearyl alcohol predominating.

Properties

Melting Point	54–57°C.
Specific Gravity	approx. 0.85
Distillation Range	300–375°C.
Hydroxyl Number	approx. 215
Calculated Molecular Weight	approx. 260
Iodine Number	<2

It is practically free from acids and esters.

FATTY ACIDS

STEARIC ACID

Commercial stearic acid is not a chemically pure compound, but a mixture of fatty acids in which stearic acid predominates.

Stearic acids are chiefly produced from saponified and distilled animal fatty acids. These fatty acids are usually composed of approximately 60% liquid and 40% solid acid. The bulk of the liquid acids is separated from the solid acids by hydraulic pressing. From the first pressing single-pressed stearic acid is obtained, from the second pressing, double-pressed stearic acids and from the third pressing, triple pressed stearic acids. Each successive pressing removes additional liquid fatty acids, thus increasing the purity and quality of the resulting solid stearic acid. The liquid acids present are unsaturated and the solid acids, saturated.

The stearic acid of commerce is frequently called stearin (stearine). This is a misnomer as stearin is a glyceride of stearic acid.

Table 51

PROPERTIES AND COMPOSITION OF COMMERCIAL STEARIC ACIDS

	Single-Pressed	Double-Pressed	Triple-Pressed
Titer	52.2–53.0°C.	53.5–53.9°C.	54.6–55.2°C.
Melting Point (Open Tube)	52.8–53.6°C.	54.1–54.5°C.	55.2–55.8°C.
Iodine Number (Wijs)	10–14	7–9	2–4
Free Fatty Acids (as Oleic)	103–104%	104–105%	104–105%
Acid Number	205–207	207–209	207–209
Saponification Number	207–209	208–210	208–210
Unsaponifiable Matter	0.50–0.60%	0.20–0.30%	0.10–0.20%
Average Molecular Weight	272–274	269–272	269–272
Specific Gravity, 60°/15°	0.853	0.855	0.857
Flash Point	400°F.	400°F.	405°F.
Fire Point	430°F.	425°F.	425°F.
Unsaturated Content	12–16%	8–10%	2–4%
Lovibond Color (Max.)	16 Y–3.0 R	4 Y–1.0 R	2 Y–0.5 R
Lovibond Color (Min.)	5 Y–0.6 R	2 Y–0.2 R	1 Y–0.1 R
Myristic Acid Composition	1–3%	1–3%	1–3%
Palmitic Acid	47–48%	49–50%	46–47%
Stearic Acid	28–33%	34–37%	46–48%
Palmitoleic Acid	trace	trace	nil
Oleic Acid	12–16%	8–10%	2–4%

Soluble in: alcohol, ether, chloroform, carbon bisulfide, carbon tetrachloride, methanol, pyridin.

The acetyl number should be close to 0.0. There should be very little spread between iodine and thiocyanogen numbers as no doubly unsaturated acids are present.

Properties of Pure Stearic Acid

Melting Point	69.3°C.
Iodine Number	0.0
Acid Number	197.5
Refractive Index, 80°C.	1.4300
Specific Gravity, 69.3°C.	0.847
Boiling Point	232°C. (15 mm)
Molecular Weight	284.28
Dipole Moment, 25°C.	0.77

Uses: cosmetics, ointments, polishes, soaps, candles, waterproofing, sun-checking compounds (rubber), emulsifiers, textile finishes, lubricants, greases, color solvents, crayons, plastics, soldering fluxes.

Stearic acid, produced by hydrogenation of fish oils, is used for many technical purposes. It is marketed in a number of grades having the following properties:

Titer °C.	Acid No.	Iodine No.
52	190	6
53	195	25
57	192	13
58	192	8

PALMITIC ACID

Palmitic acid occurs naturally as the glyceride in various vegetable and animal fats and oils. It is a hard, crystalline substance lying between myristic and stearic acids in the saturated fatty acid series.

Table 52

COMPARISON OF THE PROPERTIES OF COMMERCIAL AND PURE PALMITIC ACIDS

	Commercial Palmitic Acid	Pure Palmitic Acid
Chemical Formula	—	$C_{15}H_{31}COOH$
Mean Molecular Weight	258.0	256.25
Melting Point	58.0°C.	62.85°C.
Iodine Number (Wijs)	3.0	none
Neutralization Number	216.0	219.0
Unsaponifiable Matter	0.5%	none
Dipole Moment, 25°C.	—	0.77; 1.75[1]
Color	white	white
Odor	trace	none
Composition:		
Palmitic Acid	90.0%	—
Oleic Acid	4.0%	—
Stearic Acid	6.0%	—
Approximate Weight per U. S. Gallon	—	7½ lb

Uses: The metallic salts of palmitic acid are of particular interest in paints, varnishes, lacquers, inks, etc. Aluminum palmitate is one of the best thickening, suspension and flattening agents known. Zinc and calcium palmitate also find many applications in this field.

[1] Bhattacharya, *Indian J. Phys.* **16**, 396 (1942).

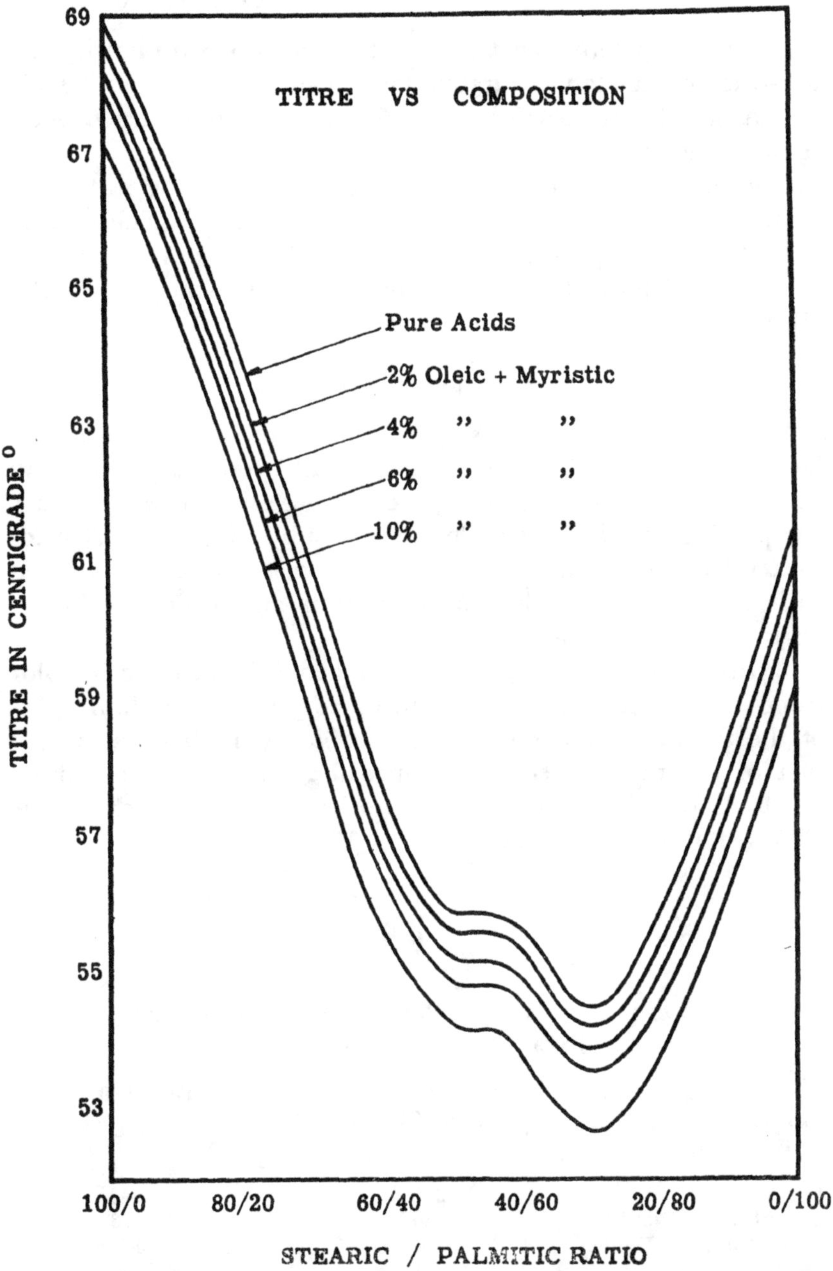

Figure 27. Titer of Pure Stearic and Palmitic Acid Blends

By mixing palmitic acid with myristic acid or with lauric acid, low-titer, completely saturated fatty acid mixtures can be obtained which are particularly suitable for preparing oxygen-stable greases.

The mono- and diglycerides are good emulsifying agents, in creams, lotions, brushless shaving compounds, etc. The sodium or potassium soaps are also useful for this purpose as normal changes in temperature do not materially influence their solubility.

MYRISTIC ACID

Myristic acid is found as the glyceride in vegetable oils and fats, in varying, sometimes rather small, percentages; e.g., in peanut oil, coconut oil, and some less accessible raw materials. By splitting the raw vegetable oils and subjecting the fatty acids to a fractional vacuum steam distillation, sometimes combined with fractional crystallization, filtration and deodorization, a fairly pure myristic acid is obtained.

Myristic acid is a hard, crystalline, fatty acid, ranging in color from white to light brown, according to purity. The faint odor of the technical grades is very agreeable, somewhat resembling that of ionone or sweet peas. Special requirements sometimes call for a practically odorless or colorless product, which can be obtained at an increased cost.

Table 53

COMPARISON OF THE PROPERTIES OF PURE AND TECHNICAL MYRISTIC ACIDS

	Pure	Technical
Formula	$C_{13}H_{27}COOH$	—
Purity	100	Approx. 80% Myristic Acid
Acid Number	246.1	237–243
Saponification Number	246.1	240–246
Iodine Number (Hanus)	0	7–13
Titer	—	42–46
Melting Point	53.8°C.	—
Specific Gravity	0.8622 54°/4°C.; 0.8584 60°/4°C.	0.869 at 50°C.

Table 53 (continued)

	Pure	Technical
Refractive Index	$n_D^{60}=1.43075$ $n_D^{76}=1.4218$	—
Molecular Refraction	68.77	—
Dipole Moment, 25°C.	0.77	—
Combustion Heat	cal./g=9.34; cal./mol.=2084.4	—
Boiling Range	196.5°C. at 15 mm 250.5 at 100 mm	ca. 220–230 at 20–30 mm
Color	white	light straw
Odor	none	characteristic, mild
Weight per Gallon	—	ca. 7¼ lb

Technical myristic acid is composed of:

	Approximate %
Lauric Acid	10
Myristic Acid	80
Palmitic-Stearic Acid	5
Oleic Acid	5

Soluble in: acetone, benzol, carbon tetrachloride, hexane, methyl alcohol, ethyl alcohol, butyl alcohol, methyl Cellosolve, butyl Cellosolve, *o*-dichlorobenzene, etc., and other hydrocarbons, alcohols and solvents of similar composition. It is insoluble in water.

DISTILLED FATTY ACIDS

Mixed fatty acids are offered in many grades.

Table 54

DATA ON MIXED FATTY ACIDS

Type	Titer, °C.	Free Fatty Acids	Iodine No.
Animal No. 1 Grade	40–42	100–102	50–60
Animal No. 2 Grade	38–39	99–100	50–60
Animal No. 3 Grade	35–37	98–99	60–70
Cottonseed	34–37	98–100	90–100
Coconut	22–25	128–130	14–16
Myristic	41–44	118–121	8–10
Soybean	22–26	96–100	130

SOLUBILITY OF FATTY ACIDS

Fatty acids are, in general, more soluble than the glycerides. They are very soluble in ether and soluble in chloroform, carbon tetrachloride, carbon disulfide, aromatic hydrocarbons, paraffin hydrocarbons, acetone and hot alcohol.

The glycerides are also very soluble in ether, but almost insoluble in alcohol.

Table 55

SOLIDIFICATION POINTS OF BINARY FATTY-ACID MIXTURES[1]

Palmitic-Stearic Mixture			Stearic-Arachidic Mixture		
C_{16} Mol. %	C_{18} Mol. %	Sol. Pt. °C. (Corr.)	C_{18} Mol. %	C_{20} Mol. %	Sol. Pt. °C. (Corr.)
0.00	100.00	68.85	0.00	100.00	74.35
2.81	97.19	68.15	2.80	97.20	73.65
6.05	93.95	67.20	5.85	94.15	72.80
9.77	90.23	66.20	7.65	92.35	72.30
14.00	86.00	65.00	13.03	86.97	70.80
18.64	81.36	63.80	14.09	85.91	70.55
18.85	81.15	63.72	16.71	83.29	69.80
24.33	75.67	62.20	21.49	78.51	68.30
31.29	68.71	60.10	23.46	76.54	68.00
34.16	65.84	59.20	26.76	73.24	67.00
38.85	61.15	57.70	28.90	71.10	66.50
39.82	60.18	57.28	30.74	69.26	65.95
42.33	57.67	56.70	35.03	64.97	64.80
44.19	55.81	56.52	38.24	61.76	63.90
45.65	54.35	56.25	39.16	60.84	63.75
47.73	52.27	56.00	42.03	57.97	63.25
49.54	50.46	56.02	43.38	56.62	63.15
50.26	49.74	56.00	46.69	53.31	62.75
52.87	47.13	56.00	46.72	53.28	62.70
56.11	43.89	55.85	48.69	51.31	62.50
58.65	41.35	55.70	50.46	49.54	62.30
62.24	37.76	55.45	52.00	48.00	62.20
65.00	35.00	54.90	54.05	45.95	62.15
65.36	34.64	54.85	55.82	44.18	62.15
68.06	31.94	54.45	58.30	41.70	62.05
68.30	31.70	54.42	60.60	39.40	61.85
68.98	31.02	54.39	62.78	37.22	61.72

[1] Schuette and Vogel, *Oil Soap,* July 7, 156 (1940).

Table 55 (continued)

Palmitic-Stearic Mixture			Stearic-Arachidic Mixture		
C_{16} Mol. %	C_{18} Mol. %	Sol. Pt. °C. (Corr.)	C_{18} Mol. %	C_{20} Mol. %	Sol. Pt. °C. (Corr.)
70.80	29.20	54.21	63.99	36.01	61.50
71.72	28.28	54.20	66.60	33.40	61.25
73.32	26.68	54.20	69.76	30.24	61.10
73.40	26.60	54.20	72.19	27.81	61.00
74.34	25.66	54.20	74.59	25.41	60.90
76.48	23.52	54.35	76.20	23.80	60.98
77.65	22.35	54.60	76.47	23.53	61.00
80.26	19.74	55.15	79.09	20.91	61.60
85.16	14.84	56.60	85.40	14.60	63.35
90.01	9.99	58.20	88.68	11.32	64.18
94.30	5.70	59.70	88.79	11.21	64.23
97.19	2.81	60.85	93.50	6.50	66.10
100.00	0.0	62.22	97.49	2.51	67.75
			100.00	0.00	68.85

Solidification points of fatty acids containing water are lowered as follows:[1]

Water	Solidification Point, °C.
Myristic Acid	
0.00	54.01
2.12	53.18
Palmitic Acid	
0.00	62.41
1.50	61.79
Stearic Acid	
0.00	69.20
1.28	68.74

Table 56

SOLUBILITY OF WATER IN MELTED FATTY ACIDS

	% Water	Solution at °C.
Myristic	1.70	53.2
Palmitic	1.25	61.8
Stearic	1.02	92.4

[1] Hoerr, Pool, Ralston, *Oil Soap*, **19**, 126 (1942).

Table 57

Physical Properties of Carbowaxes

	Carbowax Compound 1500	Carbowax Compound 1540	Carbowax Compound 4000
Physical Appearance	soft, waxy solid	medium-hard, waxy solid	hard, waxy solid
Density, g per cc at 20°C.	1.151	1.15	1.204
Melting Range	34–37°C.	40–45°C.	50–55°C.
Viscosity Range, Saybolt Universal at 210°F.	60–90 sec.	100–150 sec	400–600 sec.
Flash Point, Open Cup	430°F.	460°F.	535°F.
Freezing Points (°C.) of Mixtures Containing, by Weight:			
90% Water	— 1.0	— 0.5	— 0.5
70% Water	— 3.0	— 3.0	— 4.5
50% Water	—15.0	—10.0	—10.0
30% Water	+16.0	+20.0	+30.0
10% Water	+35.0	+37.0	+49.0
Solubility in Water, % by Weight at:			
0°C.	62	59	53
20°C.	73	70	62
40°C.	95	94	80
44°C.	100	–	–
45°C.	100	100	–
57°C.	100	100	100

Specifications

	Carbowax Compound 1500	Carbowax Compound 1540	Carbowax Compound 4000
Melting Point	not below 34°C.	–	not below 50°C.
Color, Pt-Co, 5% Aqueous Solution	not more than 100	–	not more than 50
Water Solubility, 5% Aqueous Solution	clear and substantially free from insoluble matter.		
Acidity, 5% Aqueous Solution	not more than 0.05%, calculated as acetic acid.		
Alkalinity, 5% Aqueous Solution	not more than 0.2%, calculated as sodium hydroxide.		

POLYGLYCOLS

CARBOWAX

"Carbowax" is the trade-name for certain solid, water-soluble, high-molecular-weight (> 500) polyethylene glycols that have a waxy appearance and feel.

Compatibilities of "Carbowaxes"

It will be seen from Table 58 that the compatibility of Carbowax 4000 is very limited, nitrocellulose and a few of the Bakelite resins being the only film-forming substances with which it is compatible in the proportions tested. The tests were run by adding a solution of Carbowax in alcohol, benzene or water to a solution of the resin in an appropriate solvent and examining a pour of the mixture on a glass plate for evidences of non-homogeneity. In many cases, these examinations were supplemented by an examination of hot melts of the resin and Carbowax in the absence of solvent.

Uses: base in cosmetics and ointments, a binding agent, size, softener, humectant, lubricant, preservative, and adhesive, especially for textiles, paper, leather, and other porous materials. It offers possibilities as a plasticizer for such substances as casein, gelatin, glue, cork, special printing inks, and for cellulose transparent films since it will not diminish their transparency. Carbowax Compound 1500 can be used with excellent results as a lubricant for wire drawing and metal extrusion operations. A small amount of Carbowax Compound in water, applied as a lubricant to molds in which rubber parts are formed, causes a perfect, quick release when the mold is opened. It imparts an attractive finish to the molded rubber product, and the molds require less cleaning when this releasing agent is used. It can also be used as a water-soluble lubricant in warp sizes for cotton yarn; as a scrubbing liquid in air purification; and as an anti-squeak fluid, with or without graphite. In addition, it shows promise in the formulation of carbon paper, because it is helpful as a wetting agent in wax mixtures, in belt dressings, and unvulcanized rubber compositions for self-sealing tires and tubes, and for transparentizing tracing-paper stock.

Carbowax Compound 4000 looks like paraffin wax, but is water soluble. It is not hygroscopic like Carbowax Compound 1500.

Table 58

COMPATIBILITY OF CARBOWAX 4000

	Ratio of Carbowax to Resin		
	5/1	1/1	1/5
Ethyl Cellulose	I	I	I
Methyl Cellulose	I	I	I
Polyvinyl Alcohol	I	I	I
Bakelite XR-3180		I	
Bakelite XR-4357		I	
Bakelite XJ-1434		I	
Bakelite XR-9396		I	
Bakelite XR-4357		I	
Bakelite XR-9392	I	C	C
Bakelite BR-3360	I	C	C
Bakelite BR-8900	I	I	C
Shellac		I	
Manila Loba B		I	
Carnauba Wax, No. 3 Refined		I	
Gum Damar		I	
Gum Accroides		I	
Gum Elemi		I	
Gum Mastic		I	
Batavia Damar		I	
Pale East India Macassar		I	
Gum Arabic	I	I	I
Raw Castor Oil	I	I	I
Tung Oil	I	I	I
White, Heat-bodied, Raw Linseed Oil	I	I	I
Lacquer Linseed Oil (A.D.M. No. 100)	I	I	1
Processed Linseed Oil (A.D.M.-O.K.O. M-17)	I	I	1
Paraffin Wax		I	
Ester Gum		I	
Rosin		I	
Algin S.P.		I	
Halowax No. 1012		I	
Uformite F-224		I	
Petrex No. 5		I	
Teglac Z-152		I	
Rezyl 1102		I	
Beeswax		I	
Mineral Oil		I	
Casein		I	I
Zein		I	
Polyvinyl Alcohol		I	
Nitrocellulose	1	SC	C

I = Incompatible
C = Compatible; completely homogeneous
SC = Partially compatible but with some evidences of inhomogeneity

Table 59

Properties of Glyceryl Esters

Product	Grade	Form and Color	Specific Gravity at 25°C.	M.P. or Solidif. P. °C.	Acid No.	Free Fatty Acids, %	Saponification No.	Iodine No.	pH 5% Aqueous Dispersion, 25°C.
Glyceryl (Mono) Laurate	S1062	soft solid, yellow	0.97	20–25	< 5	< 2.5	194–204	< 12	7.2–8.2
Glyceryl (Mono) Laurate	S547	soft solid, cream	0.97	23–27	< 10	< 5	175–185	< 12	8.0–8.6
Glyceryl (Mono) Oleate	S1096	soft solid, yellow	0.95	26–30	< 5	< 2.5	160–170	70–80	4.0–6.0
Glyceryl (Mono) Oleate	S1097	liquid, yellow	0.94	< 0	< 8	< 4	142–152	65–75	9.0–9.5
Glyceryl Oleo-Stearate	Aldo 40	soft solid, cream	0.96	34–37	< 5	< 2.5	165–175	50–60	7.0–8.0
Glyceryl (Mono) Ricinoleate	S1153	liquid, yellow	0.98	< 0	< 4	< 2	156–168	75–80	6.0–8.0
Glyceryl (Mono) Ricinoleate	S	liquid, yellow	1.01	< −30	< 4	< 2	125–135	65–70	9.0–9.5
Glyceryl (Mono) Stearate	Aldo 28	solid, cream	0.97	57–61	< 10	< 5	135–145	< 5	9.0–10.0 (3%)
Glyceryl (Mono) Stearate	Aldo 33	solid, white	0.97	57–61	< 5	< 2.5	160–165	< 5	7.0–8.0

Table 60

SOLUBILITY OF GLYCERYL ESTERS

Product	Grade	Water	Methanol	Ethanol	Acetone	Ethyl Acetate	Toluol	Naphtha	Mineral Oil	Vegetable Oil
Glyceryl (Mono) Laurate	S1062	ND	S	S	S	S	S	S	S	S
Glyceryl (Mono) Laurate	S547	D	SX	S	SX	S	S	S	S	S
Glyceryl (Mono) Oleate	S1096	ND	SX	S	S	PS	PS	PS	S	S
Glyceryl (Mono) Oleate	S1097	D	S	S	SX	SX	S	S	S	S
Glyceryl Oleostearate	Aldo 40	ND	SH	SH	SXH	PSH	SH	SH	SH	SH
Glyceryl (Mono) Ricinoleate	S1153	ND	S	S	S	S	PS	I	I	PS
Glyceryl (Mono) Ricinoleate	S	D	S	S	SX	S	S	I	I	S
Glyceryl (Mono) Stearate	Aldo 28	DH	SH	SH	SXH	SHX	SH	SH	SH	SH
Glyceryl (Mono) Stearate	Aldo 33	DH	SH	SH	SXH	PSH	SH	SH	SH	SH

Key to Table 60.

S = Soluble or miscible in all proportions
SX = Clearly miscible in certain proportions
PS = Poorly soluble, not completely miscible at any concentration
I = Insoluble or very slightly soluble
DH = Dispersible hot
D = Dispersible
SH = Soluble hot
ND = Not or poorly dispersible

POLYHYDRIC ALCOHOL FATTY ACID ESTERS

GLYCERYL STEARATES

The commercial, waxlike glyceryl esters are glyceryl tristearate, glyceryl distearate and glyceryl monostearate. The first contains as impurities small amounts of oleic, linoleic and linoleic esters. The second contains varying amounts of monostearate and the third, varying amounts of distearate. The mono- and distearates (water-dispersible grades only) contain varying amounts of soap.

Glyceryl Monostearate S (Aldo 28)

Properties

Color and Form	cream, waxlike solid
Melting Point	57–61°C. (capillary tube)
Odor	faint fatty
pH (3% aqueous dispersion at 25°C.)	9–10
Specific Gravity (20°/20°C.)	0.97

Soluble in: hot alcohol, hot hydrocarbons; dispersible in hot water.

Uses: protective coatings for edible hygroscopic powders, etc. Edible powders, crystals and tablets, which absorb moisture from the air, have a tendency to lump, cake, and decompose, on standing. A solution of Aldo 28 in hot alcohol, or other suitable solvent, is sprayed on the powder which can be tumbled in a heated barrel until all the solvent has evaporated. This leaves the product coated with a thin, protective film of Aldo 28 which disperses when the product is thrown into water. "Sun-checking" can be prevented by formulating polybutene-rubber combinations with Aldo 28, which not only improves light resistance, but also acts as a plasticizer and increases slightly the rate of vulcanization. By using glyceryl tristearate, the compound will become more rigid, but will still be flexible. Development of surface tack on aging can be eliminated from butadiene compounds (Vistanex), by compounding with Aldo 28. Apparently, the glyceryl monostearate sweats out of the compound at a similar rate to the development of tackiness and the two properties compensate each other so that the surface presents neither a tacky nor a waxy appearance and feel. *Cosmetics:* emulsifying agent for the manufacture of greaseless creams. *Edible and Pharmaceutical Emulsions:* for paste emulsions of edible oils and greaseless pharmaceutical ointments. *Lubricant:* integral lubricant for vinyls, particularly where a nontoxic edible agent is required; will not interfere with printing on vinyl sheeting; integral lubricant for ureaformaldehyde resins.

Glycerl (Mono) Stearate (Aldo 33)

Properties

Color and Form	light, white, waxlike solid
Melting Point	57–61°C.
Odor	faint
pH (aqueous layer when 3% is melted in water)	7–8
Specific Gravity, 25°C.	0.97

Soluble in: hot alcohol, hot hydrocarbons, hot oils.

Uses: in baked goods for longer shelf life, finer texture and constant volume; in candy making to reduce stickiness, to improve cutting and chewing qualities, and as an emulsifier for chocolate candies; in chewing gums as a plasticizer; in cosmetics as a water-in-oil emulsifier; in food and ice cream as an emulsifying agent; in margarine as an emulsifier and antispattering agent.

GLYCOL FATTY-ACID ESTERS

Ethylene glycol, diethylene glycol and propylene glycol derivatives will be discussed in this section. They are essentially the monoesters obtained from the reaction of one molecule of glycol with one molecule of fatty acid. The straight esters are water insoluble, but can be made water dispersible by adding small amounts of soap which will yield modified esters.

Glycol esters are widely used in industry as emulsifiers, stabilizers, plasticizers, lubricants and thickening agents.

Propylene glycol fatty-acid esters are nontoxic and edible. They are used in the food and pharmaceutical industries as emulsifiers and defoaming agents.

Table 61

PROPERTIES OF GLYCOL ESTERS

Product	Grade	Form and Color	Specific Gravity at 25°C.	M.P. or Solidif. p, °C.	Acid No.	Free Fatty Acid %	Saponification No.	Iodine No.	pH 5% Aqueous Dispersion, 25°C.
Diethylene Glycol (Mono) Laurate	Glaurin	Lq S	0.97	16–19	< 5	< 2.5	170–181	< 10	3.0–6.0
Diethylene Glycol Laurate	Diglycol Laurate S	Lq S	0.97	<12	< 6	< 3	165–175	< 9	8.7–9.4
Diethylene Glycol Oleate	Diglycol Oleate S	Lq A	0.93	< 0	80–90	40–50	160–175	70–85	7.5–8.5
Diethylene Glycol (Mono) Ricinoleate	S137	Lq A	0.98	< 0	< 10	< 5	140–150	65–71	3.0–5.0
Diethylene Glycol Stearate	Diglycol Stearate S	Ws W	0.96	54–59	95–105	47–53	170–180	< 6	6.8–7.5 (3%)
Ethylene Glycol (Mono) Laurate	S147	S-sd C	0.93	25–30	< 5	< 2.5	200–220	< 12	3.0–5.0
Ethylene Glycol (Mono) Ricinoleate	S152	Lq A	0.95	<—40	< 8	< 4	200–210	75–79	3.0–5.0
Ethylene Glycol (Mono) Stearate	S151	Ws C	0.96	57–60	< 6	< 3.0	196–201	< 8	3.0–5.0
Propylene Glycol (Mono) Laurate	S1003	Lq Y	0.93	< 12	< 5	< 2.5	155–165	< 10	9.0–10.0
Propylene Glycol (Mono) Stearate	Aldo 25	Ws W	0.94	43–46	< 10	< 5	143–150	< 7	9.2–10.0 (%)

Table 62

Solubility of Glycol Esters

Product	Grade	Water	Methanol	Ethanol	Acetone	Ethyl Acetate	Toluol	Naphtha	Mineral Oil	Vegetable Oil
Diethylene Glycol (Mono) Laurate	Glaurin	ND	S	S	S	S	SX	PS	SXH	SXH
Diethylene Glycol Laurate	Diglycol Laurate S	D	SX	S	SX	SX	S	S	S	SX
Diethylene Glycol Oleate	Diglycol Oleate S	D	S	S	SX	SX	S	S	SX	SX
Diethylene Glycol (Mono) Ricinoleate	S137	ND	S	S	S	S	SX	SX	PS	SX
Diethylene Glycol Stearate	Diglycol Stearate S	DH	SH	SH	SXH	SXH	SH	SH	SH	SH
Ethylene Glycol (Mono) Laurate	S147	ND	S	S	S	S	SH	S	S	PSC, SH
Ethylene Glycol (Mono) Ricinoleate	S152	ND	S	S	S	S	PS	SX	PS	PS
Ethylene Glycol (Mono) Stearate	S151	ND	SH	SH	SH	SH	SH	SH	SH	SH
Propylene Glycol (Mono) Laurate	S1003	D	S	S	SX	SX	S	S	S	S
Propylene Glycol (Mono) Stearate	Aldo 25	DH	SH	SH	SH	PS	SH	SH	SH	SH

Key to Tables 61 and 62.
FORM
COLOR
SOLUBILITY

Lq = Liquid
S-sd = Semisolid
Ws = Waxy Solid
A = Amber
C = Cream
S = Straw
W = White
Y = Yellow
S = Soluble or miscible in all proportions
SX = Clearly miscible in certain proportions
PS = Poorly soluble, not completely miscible at any concentration
I = Insoluble or very slightly soluble
DH = Dispersible hot
D = Dispersible
SH = Soluble hot
ND = Not or poorly dispersible

Diglycol Stearate S (Diethylene glycol monostearate)

Properties

Color and Form	white, waxlike solid
Melting Point	54–59°C. (capillary tube)
Odor	faint, fatty
pH (3% aqueous dispersion at 25°C.)	6.8–7.5
Specific Gravity 25°C.	0.96

Soluble in: hot alcohol, hot hydrocarbons, hot oils; dispersible in hot water.

Uses: antitack agent to prevent sticking of raw-rubber sheets; stabilizer for cosmetic lotions and creams; emulsifying agent for oils, solvents and waxes; lubricant for aluminum, nickel alloys, rubber, and vinyls; plasticizer and lubricant for paper and cardboard; dispersing agents for pigments and abrasives used in textiles and polishes; protective coating for hygroscopic powders and tablets; thickening agent.

SORBITOL STEARATES

Commercial sorbitol stearates consist chiefly of sorbitan and sorbide esters, formed during esterification. All of these esters contain varying amounts of mono-, di- and higher esters, along with small amounts of sorbitan.

Table 63

PROPERTIES OF SORBITOL STEARATES

Product	M.P. °C.	Sp. Gr.	Free Fatty Acids %	Acid No.	Saponification No.	Iodine No.	Odor	Color	Hardness	Solubility in— Water	Alcohol	Naphtha	Mineral Oil
Sorbitol Distearate	49–51	0.96	15–17	30–34	168–178	<5	faint	cream	3	D	PS	SH	SH
Sorbitol Disterate (A)	42–43	0.96	17–21	38–42	175–185	<5	faint	cream	3	I	PS	SH	SH
Sorbitol Monostearate (Arlex G-909)	45–55	0.98	—	<5	—	—	—	—	—	DH	SH	SH	—

Key
D=Dispersible
I=Insoluble
DH=Dispersible hot
PS=Partly soluble
SH=Soluble hot

PENTAWAXES

Pentawaxes are a series of waxy modified fatty-acid esters of pentaerythritol or polypentaerythritols. Some of their physical properties are listed in Table 64.

Table 64

PROPERTIES OF PENTAWAXES

Pentawax No.	Softening Point °C.	Flow Point °C.	Sward Hardness	Compatibility	Soluble in	Insoluble in	Suggested Use
145	60	81	18	X, O	O, W, E	A	emulsifier
177	70	73	22	X, O	O, H	W, A, E	polishes
217	63	65	55	X, O	O, H	W, A, E	carnauba substitute
218	54	63	10	X, O	O, H*	W	*beeswax
227	38	80	4	X, O	O, H, E	W, A	polishes
232	59	65	20	X, O	O, H, E	W, A	polishes
235	53	54	28	X, O	O, H	W, A, E	Japan wax substitute

Soluble or insoluble refers to the following substances at room temperature:
O = oil, W = water, A = alcohols, E = esters, H = hydrocarbons
Compatibility refers to the following, when melted together:
X = other waxes, including montan, beeswax, carnauba, opalwax, ceresin, paraffin
O = vegetable and mineral oils

* Pentawax 218 is soluble hot in all classes of organic solvents. It has an odor similar to that of beeswax. It does not shrink on solidifying.

HYDROGENATED OILS

Complete hydrogenation of cottonseed, coconut, peanut, soyabean, castor, fish, whale or other vegetable, fish or animal oils yields waxlike substances. The highest-melting commercial wax is produced from castor oil (Opalwax).

Uses: food, candy, baking, polish, cosmetics, molding, lubricants, ointments, etc.

Table 65

PROPERTIES OF HYDROGENATED OILS

Product	Titer °C.	Free Fatty Acids %	Saponification No.	Unsaponified %	Iodine No.
Hydrofol Glycerides					
(45–48)	42–45	< 3	190–195	—	25–35
(54–57)	51–53	< 3	190–195	—	< 6
(60–63)	57–60	< 1	185–190	—	25–30
(67–69)	64–66	< 2	185–190	—	< 6
(80–83)	77–80	< 2	176–178	—	6–8
Spermofol					
(45)	44–46	< 2	140	30–35	20–30
(52)	50–52	< 2	140	30–35	4–7
Cosmol Hard	48–50	< 1	115	50–52	2–5
Hydrogenated Cottonseed Oil	58–60	< 0.1	192–196	5	< 10
Hydrogenated Coconut Oil	43–44	0.08	246	—	6
Opalwax	86–88	< 2	175–185	—	2.5–8.5
Hydrogenated Tallow	59.5	0.7	103	0.2	1.5

OPALWAX (HYDROGENATED CASTOR OIL)

Opalwax consists principally of 12-hydroxy stearin (glyceryl trihydroxystearate) produced by catalytic hydrogenation of castor oil.

Properties

Odor	practically odorless
Color	pearl-white; clear when melted
Homogeneity	uniform composition, no extraneous matter
Fracture	conchoidal and amorphous
Luster	dull
Apparent Specific Gravity	0.98–0.99 at 20°C.
Melting or Solidification Point	86 to 88°C. (187 to 190°F.)
Hardness	quite hard
Dielectric Constant	approx. 12 at 30°C.; 22 at 60°C. (100 cycles)
Acid Number	< 2.0
Saponification Number	175 to 185
Iodine Number	2.5 to 8.5
Ash	< 0.005%
Hydroxyl Number	155 to 165

Opalwax is insoluble in solvents. None of the common solvents has been found to dissolve as much as 2% at temperatures up to

30°C. Toluene, carbon tetrachloride, and carbon disulfide are the most effective. Opalwax is dissolved, with discoloration, by concentrated sulfuric acid. It is disintegrated and partly dissolved by concentrated nitric acid.

It is only slightly compatible with paraffin wax, but more so with ceresin, ozokerite, montan and carnauba waxes, and many natural and synthetic resins, such as rosin, ester gum, coumarone-indene, n-butyl methacrylate, ethyl cellulose, alkyd, urea-formaldehyde and drying-oil types.

Molten Opalwax (100°C.), introduced, with rapid agitation, into about two and a half times its own weight of water, at 85°C., in which a small amount of a suitable emulsifying agent has been thoroughly mixed, in advance, will produce an emulsion that will remain stable when subsequently diluted and yield a continuous film on drying.

Uses: for impregnating and coating papers, fiberboard, leather, cork, textiles, etc., to make them grease-, oil-, and waterproof for electrical insulation; as a lubricant; also in the manufacture of candles, rubber-coated fabrics, polishes and finishes, carbon paper, inks, cutting oils. The addition of 20% naphthalene tetrachloride, behenone or stearone to hydrogenated castor oil (Opalwax) makes it approach carnauba wax in hardness, fracture and solvent retention.

Table 66

PROPERTIES OF HYDROGENATED OILS

Partly Hydrogenated Oils[1]

Oil	M.P. °C.	Congelation Point °C.	Butyrorefractometer Reading (40°C.)	Acid No.	Saponification No.	Iodine No.
Coconut	44.5	27.7	35.9	0.4	254.1	1.0
Sesame (Techn.)	62.1	45.3	38.4	4.7	188.9	25.4
Whale	45.1	33.9	49.1	1.2	192.3	45.2
Arachis	51.2	36.5	50.1	1.0	188.7	47.4
Sesame	47.8	33.4	51.5	0.5	190.6	54.8
Cottonseed	38.5	25.4	53.8	0.6	195.7	69.7

[1] Bomer and Leschyl-Hansen, *Chem. Rev. Fett-Harz-Ind.* 19, 218 (1912).

Table 67

WHALE OIL AT DIFFERENT STAGES OF HYDROGENATION[1]

Oil Stages	M.P. °C.	Congelation Point °C.	Acid No.	Saponification No.	Iodine No.	Molecular Equivalent of Fatty Acids
Original Oil	Fluid	Fluid	9.50	192.2	144.8	287.7
Artificial Tallow	47.5	38.1	9.88	183.7	56.9	296.4
Artificial Stearin	54.3	47.3	7.80	187.7	11.7	297.0
Hydrogenated Whale Oil	41.9	31.9	5.30	190.9	57.8	282.9

Table 68

COMPLETELY HYDROGENATED OILS[2]

Hydrogenated Oil or Fat	M.P. °C.	Iodine No.	Saponification No.	Fatty Acid, M.P., °C.
Tallow	62.0	0.1	197.7	64.0
Lard	64.0	1.0	196.8	62.0
Cacao Butter	63.5–64	0.0	193.9	65.5
Arachis	64–64.5	0.0	191.6	67.0
Cod Liver	65.0	1.2	186.2	59.0
Linseed	68.0	0.2	189.6	70.5
Sesame	68.5	0.7	190.6	69.5
Olive	70.0	0.2	190.9	71.0
Poppy	70.5	0.3	191.3	71.0
Almond	72.0	0.0	191.8	71.0

Table 69

PROPERTIES OF HYDROGENATED COCONUT AND COTTONSEED OILS

	M.P. °F.	Iodine No.	Saponification No.	Reichert-Meissl No.	Polenske No.	Refractive[3] Index
Hydrogenated Coconut Oil	107.2	2.2	250	8.5	15.5	1.4486 (40°C.)
Hydrogenated Cottonseed Oil	140.7	3.9	195	0.5	0.7	1.4468 (60°C.)

Hydrogenated vegetable oils have dielectric constants of 20–25.[4]

[1] Sandelin, *J. Soc. Chem. Ind.* **33,** 1097 (1914).
[2] Mannich and Thiele, *Arch. Pharm.* **26,** 36 (1916).
[3] Langwill, *Mfg. Confectioner* **21,** 15 (1941).
[4] Valter, *J. Tech. Phys.* (U.S.S.R.) **10,** 1970 (1941).

Table 70

Properties of Hydrogenated Sardine Oil and Tallow

	45° Incompletely Hydro-generated Sardine Oil		53° Fully Hydro-genated Sardine Oil		Fully Hydro-genated Tallow	
	Glycerides	Acids	Glycerides	Acids	Glycerides	Acids
Titer	—	44–46	—	52–54	—	56–58
Melting Point (open tube)°C.	43	44.5–46.5	54	52.5–54.5	57	56.5–58.5
Iodine Number	27–34	28–35	2–4	2–4	2–6	2–6
Free Fatty Acids as Oleic	2–5%	95–97%	2–5%	95–97%	3–5%	96–98%
Acid Number	4–10	189–193	4–10	189–193	6–10	191–195
Saponification Number	185–189	193–199	185–189	193–199	190–193	199–201
Unsaponifiable	0.8–1.0%	0.8–1.0%	0.8–1.0%	0.8–1.0%	0.6–1.0%	0.6–1.0%
Average Molecular Weight	896	286	896	286	878	280
Specific Gravity 60°/15°	0.873	0.895	0.876	0.854	0.856	0.881
Flash Point °F.	530	405	530	385	515	405
Fire Point °F.	580	440	585	425	595	430
Unsaturated Content	37%	38%	3%	3%	4%	4%
Lovibond* Color	15–35 Y 2.5–5 R	10–15 Y 2–3 R	10–35 Y 2–5 R	10–15 Y 2–3 R	10–35 Y 2–5 R	10–15 Y 2–3 R
Thiocyanogen No.	No data. Should be same as iodine on 53° and hydrogenated tallow, only slightly lower than iodine on 45°.					

* Lovibond readings on glycerides, 5¼ in. column; fatty acids, 1 in. column.

Table 71

COMPOSITION OF HYDROGENATED SARDINE OIL AND TALLOW

	45° Incompletely Hydrogenated Sardine Oil %	53° Fully Hydrogenated Sardine Oil %	Fully Hydrogenated Tallow %
Saturated Acids:			
Myristic	7	7	4
Palmitic	26	30	30
Stearic	18	24	62
Arachidic (C_{20})	8	21	—
Behenic (C_{22})	3	14	—
Unsaturated Acids:			
Myristoleic	0.2	trace	trace
Palmitoleic	4	trace	0.5
Oleic	6	0.5	3.5
Eicosenoic (C_{20})	14	1.0	—
Docosenoic (C_{22})	13	2.0	—

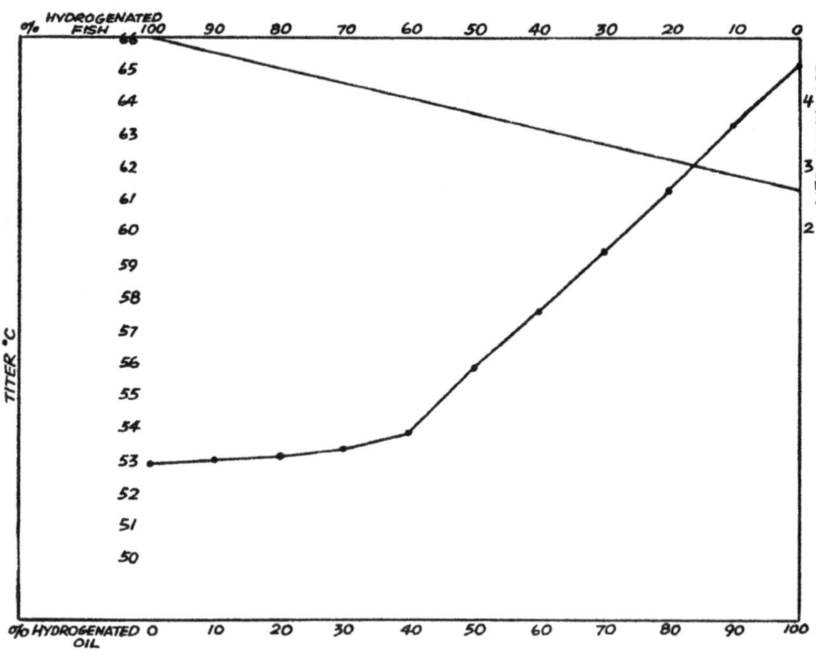

Figure 28. TITER AND IODINE NUMBER OF HYDROGENATED SOYBEAN- AND FISH-OIL BLENDS

Table 72

MELTING POINTS OF COMPLETELY HYDROGENATED OILS

Hydrogenated Oil	Melting Point, °C
Coconut	43–45
Palm Kernel	43–45
Herring; Cod	52–55
Whale	54–60
Palm	56–58
Cottonseed	62–63
Olive; Peanut	68–69
Soyabean; Linseed	69–71
Castor	86–90

CHLORINATED NAPHTHALENES

HALOWAX

Chlorinated naphthalene waxes are marketed under the name of Halowax. They will not support combustion, are good dielectrics, inert to alkali and acid and noncorrosive. They must be used with caution as they affect many individuals physiologically.

Soluble in: benzol, toluol, carbon tetrachloride; in general, in chlorinated or aromatic solvents.

SEEKAY WAX

A British chlorinated naphthalene made in different grades; color, yellow to black; melting point, 65–125°C. They are insoluble in water and soluble in most hydrocarbon solvents.

Uses: applied to wood, metal or other surfaces to prevent attack by acids, alkalies, and other corrosive liquids and fumes; plasticizer and toughener for gums, synthetic resins, waxes, etc.; moistureproof coatings; polishes; flameproofing; electrical insulation.

Toxic Effects

Poisoning from chlorinated naphthalene may take two forms: damage to the skin, producing acne, particularly of the face; and toxic jaundice, produced by either an acute or subacute necrosis

Table 73

PROPERTIES OF HALOWAXES

	Halowax No. 1012	Halowax No. 1013	Halowax No. 1014
Description (68°F.)	hard crystalline solid	hard semiamorphous solid	tough amorphous solid
Color	white to pale yellow	pale yellow	pale yellow
Flow Point (ASTM—Ball and Modified Ring, °F).	203–210	247–252	277–283
Specific Gravity (68°/68°F.)	1.53–1.59	1.65–1.71	1.75–1.81
Boiling Range (ASTM, °F.)	600–650	615–655	680–730
Flash Point, °F.	284	356	392
Fire Point, °F.	none	none	none
Viscosity	35 sec. at 212°F. (Saybolt)	31–35 sec. at 266°F. (Saybolt)	37 sec. at 302°F.
Maximum Acidity (mg of KOH Halowax)	0.1	0.1	0.1
Penetration (200 g load—5 sec. at 77°F.)	10–15	5–8	5–8
Volatility (g/sq in./hour at 221°F.)	0.01	0.004	0.001
Coefficient of Lineal Expansion in./in.:			
Liquid	0.00013	0.00012	0.00011
Solid	0.000082	0.000064	0.000049
Coefficient of Cubical Expansion cu in./cu in.:			
Liquid	0.00049	0.00026	0.00024
Solid	0.00027	0.00018	0.00010
Vapor Pressure, 130°F.	0.2–0.4	0.1–0.2	0.1–0.2

* Lovibond readings on glycerides, 5¼ in. column; fatty acids, 1 in. column.

of the liver, or by acute yellow atrophy of the liver. Twelve cases of dermatitis and one of acute atrophy of the liver among workers who had been exposed to the fumes or dust of chlorinated naphthalene were reported. If cases of poisoning are to be prevented, the wax must not be overheated and the fumes and dust must be properly evacuated by exhaust ventilation.[1]

ACRAWAXES

Acrawaxes are complex nitrogen derivatives of the higher fatty acids, some of which have very high melting points.

PROPERTIES OF ACRAWAX

Color and Form	hard, tan wax
Luster	high
Melting Point	94–97°C.
Flash Point	230°C. (open cup)
Specific Gravity, 24°C.	1.04

Soluble in: hot alcohol, hot toluol (forms gel on cooling), hot butyl acetate, hot turpentine; insoluble in water and mineral spirits.

Used as: antitack agent for vinyl resin films; for impregnating leather packings; in water- and oil-resistant greases.

Compatibility

9 parts Acrawax to 1 part of Congo, Acrawax B, Shellac and Candelilla Wax; 4 parts Acrawax to 1 part of Gum Manila Loba, Diolin, Ester Gum and Candelilla Wax; 1 part Acrawax to 1 part of Albacer, Aldo 33, Ester Gum, Rosin, Shellac, Stearic Acid, Acrawax B and Acrawax C.

ACRAWAX B

Properties

Color and Form	hard, light-brown wax
Luster	good
Melting Point	83–86°C.
Flash Point	235°C. (open cup)
Acid Number	<2
Specific Gravity, 25°C.	0.97

[1] E. Collier, *Lancet 72* (1943).

Soluble in: hot alcohol, hot toluol, hot butyl acetate, hot mineral spirits (forms gel on cooling) and mineral oil. Insoluble in water.

The properties of Acrawax B are very similar to those of Acrawax except that Acrawax B is compatible with paraffin wax and forms gels with mineral spirits and kerosene.

Used as: flatting agent for paints; lubricant and antitack agent for natural and synthetic rubbers.

Compatibility

9 parts Acrawax B to 1 part of: Acrawax, Ester Gum, Gum Dammar, Cumar and Shellac; 3 parts Acrawax B to 1 part of Flexo Wax C and Paraffin Wax; 1 part Acrawax B to 1 part of Albacer, Aldo 33, Ester Gum, Flexo Wax C, Candelilla Wax, Carnauba Wax, Beeswax, Paraffin Wax, Reso Wax B and Rosin.

ACRAWAX C

Properties

Color and Form { Lumps, Powder	hard, tan wax
Luster	high
Melting Point	140–143°C.
Flash Point	285°C. (open cup)
Specific Gravity, 25°C.	0.97
Moisture Absorption at 96°F. for 48 hours	0.3 g moisture per 50 g wax
Oxidation Potential	400–500 millivolts
Dielectric Strength	428 volts per mil
Insulation Resistance at 25°C.	68,000 megohms
Power Factor at 1,000 Cycles	0.0277

Compatibility

1 part Acrawax C to 1 part of Candelilla Wax, Carnauba Wax, Flexo Wax C, Glyco Wax S932, Paraffin Wax, Pentalyn, Rezo Wax B, Rosin, etc.

It is compatible with tricresyl phosphate which can be used with it as an efficient plasticizer.

Solubility

Acrawax C is insoluble in boiling water and low-boiling

organic solvents, such as alcohol, acetone, benzol, and carbon tetrachloride. It is soluble in hot high-boiling solvents, such as kerosene, toluol, Cellosolve,[1] turpentine, etc., at and above its melting point (140—143°C.). On cooling, the wax precipitates or from a turpentine solution a gel is formed.

Uses: as antiblocking or antitack agent in heat-sealing adhesives, plastics, synthetic and natural rubbers; antistatic agent; for raising the softening point of asphalts; in dental waxes; in electrical insulating coatings and cable fillers; as a lubricant in wire drawing, molded metal parts, metal coatings; plastic mold release; stop-off wax in electroplating; moistureproofing and antimar agent in paints, varnishes and lacquers; moistureproofing and nonblock agent for textiles; printing inks and bake finishes.

Acrawax C can be incorporated by three methods into lacquer sealers and top coats:

1. Acrawax C beads are dissolved in hot toluol and the solution added, with rapid agitation, to the nitrocellulose wetted with cold toluol.
2. Acrawax C atomized is added to an alcoholic solution of nitrocellulose and mixed thoroughly.
3. Acrawax C powdered is dispersed in the lacquer by mill-grinding technique.

IG WAXES

The IG waxes (made in Germany) cannot be readily classified from a chemical standpoint as they vary greatly in composition. Some are glyceryl, glycol or polymerized ethylenic esters of acids derived from montan wax and other sources. Some contain metallic soaps and other modifiers.

The grades mentioned in the following table are classified as follows:

Hard waxes:

IG Wax O and OP
IG Wax E and EG
IG Wax S and L
IG Wax Special

[1] Carbon and Carbide Chemical Co.

Table 74

PROPERTIES OF IG WAXES

IG Waxes	M.P.* °C.	Sp. Gr. 20°C.	Acid No.	Ester No.	Saponification No.	Unsaponified %	Color	Remarks
O	101–106	1.03	10–15	103–120	111–133	5–8		
OP	101–106	1.03	10–15	103–120	110–132	5–8	light yellow	hard and noncrystalline
E	79–82	1.01	17–25	140–156	158–178	5–8	light yellow	emulsifies well
EG	72–75	1.01	18–26	135–153	158–173	5–8		
S	80–83	1.01	10–30	142–157	160–175	7–10	light yellow	used in hardening paraffin wax
L	80–83	1.00	127–139	22–37	155–170	7–10		
Special	90–93	1.00	13–18	80–90	95–110	20–22		
CR	80–83	1.00	31–37	84–98	115–130	13–16		
B (Unbleached)	71–74	0.95	13–23	120–140	140–160	22–25		
B (Bleached)	70–72	0.96	66–86	86–106	163–183	21–24		
N	71–74	0.97	78–88	68–82	152–164	16–19		
N (New)	70–74	0.97	33–43	83–95	118–133	22–26		
Z	100–102	0.93	0	0	0	100	white	hard and tough
KP	80–82	1.02 (70°)	25–35	100–120	130–145	12–14	light brown	substitute for carnauba wax
V	50	—	0	10	—	—	white	high luster and solubility in turpentine

* Dropping point, Ubbelohde method.

IG Wax CR
IG Wax A

Soft waxes:

IG Wax B unbleached and bleached
IG Wax N and N new

IG Waxes O and OP are light-colored, solid waxes of high melting point. They have a conchoidal fracture and great hardness, are free from impurities, and always show similar properties, as indicated by their constants. Their outstanding property is their colloidal nature in the presence of organic solvents, as a result of which, they take up and bind large quantities of these solvents. In this respect, they surpass, by far, the natural waxes. The gels or jellies produced from IG Waxes O and OP with solvents, such as turpentine and turpentine substitutes, give a high gloss and are, therefore, suited, with other waxes and waxy substances, for the manufacture of boot polishes, leather dressings, floor polishes, and similar articles.

IG Wax E is a pale-yellow wax of crystalline structure. As it may be saponified and emulsified much more readily than IG Waxes O and OP, it is suited for the production of saponified and mixed-saponified preparations.

IG Wax EG is practically identical with IG Wax E, from a chemical point of view. It is a special grade which emulsifies easily and enables the production of thin liquid concentrated emulsions, especially in combination with Emulphor O.

IG Wax S is a crystalline wax of very light color and great hardness. Owing to its crystalline structure, it hardens softer waxes or waxlike bodies, such as paraffin wax, ozokerite, etc. It yields polishes of very high luster and free from haziness. Owing to its high acid number it can form soaplike combinations and emulsions with alkalies. It is also excellently suited for dissolving color bases and, therefore, is used in the manufacture both of boot polishes and carbon papers.

IG Wax L has practically the same properties as IG Wax S, but is slightly darker. It is, however, considerably harder than the S brand and, therefore, is used for polishing shoe trees and also for producing flatting pastes, stick polishes, etc.

IG Wax Special is chiefly intended for the production of black boot polishes. It makes these polishes resistant to variations of temperature, imparts a good surface, and forms surface rings, like carnauba wax, under suitable treatment.

IG Wax CR is similar to IG Wax Special in several respects, but it is a little more crystalline. It is very easy to saponify and can, therefore, be used for solid or liquid emulsions where the dark color is not a disadvantage.

IG Waxes B unbleached and B bleached are representatives of soft waxes (like beeswax). The color of the unbleached variety is brownish yellow, similar to that of beeswax; the bleached quality has an ivorylike color. They can be easily molded between the fingers at ordinary temperatures. They are soft and should be used for wax compositions which are required to be elastic and pliable. They are very easily saponified and emulsified and, therefore, suited primarily for the production of saponified and mixed-saponified preparations. They are also used in sizes, finishes, cosmetics and similar products.

IG Wax Z is chemically inert. It is completely colorless, It absorbs considerable quantities of solvent in the emulsions. Normally, emulsions of the water-in-oil type result, which are practically neutral and show no case of alkaline reaction. The color of the wax is practically white and it is a soft wax. Its specific properties make it suitable for the preparation of emulsified products, such as skin creams and mixed saponified polishing and lustering agents.

IG Wax N new is identical in properties with IG Wax N, with the only difference that it yields emulsions of the oil-in-water type under normal working conditions. It is used for mixed saponified shoe and floor polishes.

IG Wax Z is chemically inert. It is completely colorless, odorless and without taste, very hard, but not brittle, and of excellent luster; the melting point is above 100°C. Accordingly, a slight addition to paraffin wax, stearin or similar substances is sufficient for considerable hardening; it is used for increasing the stability to heat of paraffin candles, waxed papers, and wax flowers. Its application is, however, of particular value where, apart from the hardening effect, extreme purity is essential, as in producing impregnating masses for food and drinking cartons, etc.

Table 75

PROPERTIES OF MISCELLANEOUS SYNTHETIC WAXES

Wax	Color	Luster		Melting Point, °C.	Flash Point, °C.	Specific Gravity at 20°C.	Acid No.
Acrawax	tan	high		94–97	230	1.04 (24)	—
Acrawax B	light brown	good		83–86	235	0.97	<2
Acrawax C	tan	high		140–143	285	0.97 (25)	<10
Albacer	white	high	(flow point)	99–104	285–295	1.05	—
Ceramid	yellow	good		78–81	260	0.98	<12
Flexo Wax C	tan	dull		69–73	257	0.82	—
Flexo Wax C Lt.	cream	dull		62–64	230–240	0.82	—
Glyco Wax S932	white	good		60–64	328	0.96	—
Stroba Wax	tan	good		98–102	212	1.04 (29)	35–43
Wax S1167	black	dull	(flow point)	110–125	275	1.01	<12
Rilan Wax	cream	hard		84–85	—	0.98–0.99	25

	SOLUBILITY IN					
Wax	Water	Alcohol	Toluol	Naphtha	Mineral Oil	Vegetable Oil
Acrawax	IC	SH	SH, GC	PSH	PSH	SH
Acrawax B	I	SH	SH	SH, GC	SH	SH
Acrawax C	I	I	SH	SH	SH	SH
Albacer	I	I	SH	SH	SH	SH
Ceramid	I	SH	SH	SH	SH, GC	SH, GC
Flexo Wax C	I	I	SH	SH	SH	SH, GC
Flexo Wax C Lt.	I	I	SH	SH	SH	SH, GC
Glyco Wax S932	I	I	SH	SH	SH	SH
Stroba Wax	I	I	SH, GC	SH, GC	SH, GC	SH
Wax S1167	I	PSH	SH, PSC	SH, PSC	PSH	PSH
Rilan Wax			S	S		

Key: S = Soluble; D = Dispersible; H = Hot; I = Insoluble; G = Gel; C = Cold; P = Partially.

IG Wax V is a chemically inert product of high luster. It has an opaque, pure-white color. Its most pronounced property is an exceptional solubility in turpentine and turpentine substitutes. IG Wax V is not only soluble itself in high proportions in these solvents but also exerts a liquefying effect on other waxes which is maintained at low temperatures. It is suitable for the production of liquid floor polishes, either alone or in combination with the usual wax mixtures. The Gersthofen waxes are similar to the IG waxes.

MISCELLANEOUS SYNTHETIC WAXES

ALBACER

Properties

Color and Form	white, hard wax
Luster	high
Melting Point	99–104°C. (flow point)

Soluble in: hot hydrocarbons. The best solvent for Albacer is a mixture of 1 part ethylene dichloride, 6 parts toluol and 4 parts butyl acetate, which will dissolve up to 5% Albacer, most of it remaining in solution even at room temperature.

Uses: the addition of 5–10% Albacer to paraffin wax, for wax-coated paper, gives a finish which exhibits considerable "slip." This is of interest for wrapping sticky candies (caramels, etc.). The paraffin wax is heated to 100°C. and the Albacer, in small pieces, is added slowly with stirring until dissolved. Albacer, plasticized with up to 40% castor oil, is used as a paper coating for packaging polybutene (Vistanex-type) synthetic rubbers to prevent adhesion.

When Albacer is melted, it changes to a heavy-bodied non-gelatinous liquid which sets very slowly.

Albacer can be blended with certain synthetic resins, mineral and vegetable oils, and other waxes.

DIOLIN (OCTADECANEDIOL)

This white, small-lump wax powders with pressure.

Properties

Melting Point	65°C.
Molecular Weight	286
Boiling Point	172–186°C. at 0.5 mm
Specific Gravity	0.84 at M.P.
Hydroxyl Number	> 350
Iodine Number	< 4
Saponification Number	< 5
Acid Number	< 2
Insoluble in Water	

FLEXO WAX C

Properties

Color and Form	tan, amorphous wax
Luster	dull
Melting Point (Ball and Ring)	69–73°C.
Flash Point	495°F. or 257°C.
Fire Point	565°F.
A.P.I. Gravity, 60°C.	36.5
Viscosity, 210°F.	76 sec
Penetration, A.S.T.M. Needle, 77°F.	33

Flexo Wax C is a noncrystalline wax with high adhesive properties. Because of its amorphous character it is much more waterproof than paraffin wax.

Soluble in: hot hydrocarbons, toluol.

Uses: as an adhesive for cellophane, cellulose acetate, etc.; self-sealing waterproof coatings; greaseproofing; in nitrocellulose lacquers as waterproofing and antiblock agent; protective coating for nickel alloys; for increasing slip effect in nonmar enamels; stop-off wax for differential plating; water-repellent coatings for fabrics.

GLYCO WAX S932

Glyco Wax S932 is an edible waxlike ester melting at 60—64°C. It has a flash point of 328°C. and does not smoke or give off much volatile matter on heating to 260°C. It is a quick-setting wax with a softening point within 1°C. of its melting point.

Compatibility

1 part Glyco Wax S932 to 1 part of: Acrawax B, Acrawax C, Carnauba Wax, Ethyl Cellulose, Flexo Wax C, Japan Wax.

Uses: in greaseproof coatings for paper; plasticizer for chewing gum (4–8% based on the chicle in chewing gum); in solder fluxes.

STROBA WAX

Stroba Wax is a hard, tan wax melting at 98–102°C.

Compatibility

Stroba Wax is compatible with most waxes and resins.

Uses: in paints, varnishes and lacquers as a flatting agent; blended with paraffin waxes gives increased luster and melting point to paraffined papers; in paste polishes; a high-luster, non-smearing, rubbing paste wax can be made using 15 parts each Stroba Wax and a high-melting microcrystalline wax in 70 parts solvent. It will not sweat solvent at high temperatures.

KETONES, AMINES, AMIDES AND NITRILES

16–HENTRIACONTANONE

The average molecular weight of the commercial product is 454.7. It consists of 98% ketone divided in about the following proportions: 16–hentriacontanone 85 and 18–tritriacontanone 15.

The symmetrical aliphatic ketones are water-white in the pure state. The melting point of 16–hentriacontanone is 82.8°C. and the iodine number 0. Commercial samples have a light, lemon-yellow color, the melting point is 76°C. and the iodine number 2. The boiling range is 240–260°C. at 0.5 mm. It is very heat stable and does not decompose readily.

High-molecular-weight aliphatic ketones are insoluble in water. They are fairly insoluble in most organic solvents at room temperature, but quite soluble above the melting point of the ketone. Because of their high molecular weight, they are inert chemically and do not react readily. They can be changed to reactive derivatives by halogenating.

Table 76

SOLUBILITY OF STROBA WAX

(10 cc Solvent to 1 g of Stroba Wax)

	Hot, Temperature, °C.	Cold
Mineral Spirits	S–93	cools milky white, changing to a fine-texture translucent gel on standing
Turpentine	S–93	translucent gel
Toluol	S–93	translucent gel
Ethylene Glycol	I–130	transparent flocculent white solid
Diethylene Glycol	SS–130	same as 4 only more soluble
Ethylene Glycol Diacetate	SS–130	same as 4
Propylene Glycol	SS–130	same as 4
Triethylene Glycol	VSS–130	VSS
Cellosolve	I–115	I
Methyl Cellosolve	I–95	PS at BP affecting separation of wax into two components
Butyl Cellosolve	PS–97	white, finely divided suspension approaching gel
Cellosolve Acetate	I–95	I
Methyl Carbitol	PS–130	similar to 4
Butyl Carbitol	I–95	I
Carbitol Acetate	I–95	I
Methyl Acetate	I–55–BP	
Ethyl Acetate	I–75–80–BP	
Butyl Acetate	PS–93	PS–MS–SP
Methanol	I–60–BP	
Butanol	PS–93	PS–MS–SP
n-Hexanol	S–93	partial suspension
Octanol	S–95	milky white suspension
Methyl Amyl Alcohol	S–93	finely divided suspension
2-Ethyl Butyl Alcohol	PS–95	PS–MS–SP
Tetrahydrofurfuryl Alcohol	S–130	translucent to opaque gel
Synasol (Denatured Alcohol)	I–85–BP	slush (unstable)

PS = Partially Soluble; MS = Milky Suspension; SP = Solid Particles; SS = Slightly Soluble; VSS = Very Slightly Soluble; BP = Boiling Point.

Their solvent characteristics at elevated temperatures are excellent. The indicated fields of commercial utility are chemical synthesis and intermediates, waxes, polishes, moistureproofing agents and rust-preventing compositions, candle dipping, textile lubricating and softening agents.

These ketones do not deteriorate in storage and may be held indefinitely. The chief impurities are n-hexadecane and palmitic acid.

OCTADECYL–HEXADECYL AMINE

Average Molecular Weight	267
Percentage of Amine by Titration	<98
Specific Monoamine Mixture:	
Octadecylamine	94
Hexadecylamine	6

Properties

Color	water white to light yellow
Initial Boiling Point	340°C.
90% Over	348°C.
Final Boiling Point	350°C.
Nonvolatile Matter	trace
Melting Point	53°C.
Iodine Number	<2
Moisture	none

OCTADECYLAMINE

Properties

Molecular Weight	269.5
Color	water white
Boiling Point, 760 mm	348.8°C.
Boiling Point, 2 mm	166.1°C.
Vapor Pressure, 20°C.	<0.1 (0.1 mm at 115°)
Melting Point	55°C.
Specific Gravity	about 0.844
Weight per Gallon	about 7 lb

ARMIDS[1]

High molecular weight aliphatic amides are in large-scale commercial production and may be obtained in almost unlimited quantities.

[1] Armour and Company.

Table 77

CHEMICAL COMPOSITION OF ARMIDS

Code	Name	Formula
AD-1080	octanamide	$CH_3(CH_2)_6CONH_2$
AD-1100	decanamide	$CH_3(CH_2)_8CONH_2$
AD-1120-A	dodecanamide	$CH_3(CH_2)_{10}CONH_2$
AD-1140-A	tetradecanamide	$CH_3(CH_2)_{12}CONH_2$
AD-1160-A	hexadecanamide	$CH_3(CH_2)_{14}CONH_2$
AD-1180-A	octadecanamide	$CH_3(CH_2)_{16}CONH_2$
AD-1180-C	25%—hexadecanamide	$CH_3(CH_2)_{14}CONH_2$
	75%—octadecanamide	$CH_3(CH_2)_{16}CONH_2$
AD-1181-A	9—octadecanamide	$CH_3(CH_2)_7CH{:}CH.(CH_2)_7CONH_2$
AD-1181.5-C	20%—hexadecanamide	$CH_3(CH_2)_{14}CONH_2$
	50%—octadecanamide	$CH_3(CH_2)_{16}CONH_2$
	30%—octadecadienamide	$CH_3(CH_2)_4CH{:}CHCH_2CH{:}CH(CH_2)_7CONH_2$
TAD-1181.5-C	40%—substituted cyclic amide	$(CH_3)_2CH(CH_3)_2C_{14}H_{16}CONH_2$
	30%—octadecanamide	$CH_3(CH_2)_7CH{:}CH.(CH_2)_7CONH_2$
	30%—octadecadienamide	$CH_3(CH_2)_4CH{:}CHCH_2CH{:}CH(CH_2)_7CONH_2$

Table 78

PROPERTIES OF ARMIDS

Code	Molecular Weight	Boiling Point °C.	Melting Point °C. Minimum		Iodine No.	Purity %	Free Fatty Acid %
AD-1080	143	> 200—d	—	(106)*	1	95	< 5
AD-1100	171		98	(99)*	1	95	< 5
AD-1120-A	199	200, 12 mm	98	(102)*	1	95	< 5
AD-1140-A	227	217, 12 mm	98	(103)*	2	95	< 5
AD-1160-A	255	236, 12 mm	98	(106)*	3	95	< 5
AD-1180-A	283	251, 12 mm	98	(109)*	3	95	< 5
AD-1180-C	276	245, 12 mm	98	(100)*	3	95	< 5
AD-1181-A	281	> 200	76		90	95	< 5
AD-1181.5-C	280	> 200	60		120	95	< 5
TAD-1181.5-C	289	> 200	44		130	95	< 5

* Figures in parentheses are the melting points of the chemically pure amides.

All of the figures in Table 78 are for commercial productions. Normally, the melting point ranges from 99 to 102°C., although 98°C. is taken as a minimum.

The solubility data of Table 79 were obtained by work with chemically pure (C.P.) amides. Regular commercial productions will only approximate these figures, although they can be used as a reliable guide.

Table 79

THE SOLUBILITY OF AMIDES

	Methanol			*Ethanol*		
Amide	10°C.	30°C.	64.7°C.	10°C.	30°C.	78.5°C.
A	22.1	53.0	263	13.4	32.8	303
B	6.8	15.2	170	4.3	12.0	350
C	4.2	12.4	206	2.9	11.4	370
D	1.0	2.7	56	1.2	3.3	180
E	0.7	1.2	51	0.4	1.5	216
F	0.4	0.7	23	0.2	0.8	124

	Isopropanol			*n-Butanol*		
Amide	10°C.	30°C.	82.3°C.	10°C.	30°C.	100°C.
A	15.7	26.3	288	17.1	25.8	1300
B	6.0	10.9	380	3.5	9.1	—
C	3.7	11.4	357	2.2	10.1	3550
D	1.2	3.2	208	0.8	3.8	1150
E	0.4	1.8	229	0.3	1.9	1050
F	0.2	1.0	172	0.1	1.1	590

	Ethyl Acetate			*Butyl Acetate*		
Amide	10°C.	30°C.	80.1°C.	10°C.	30°C.	81.4°C.
A	2.2	5.7	115	2.6	6.0	740
B	0.5	2.2	123	0.9	2.9	—
C	0.5	2.5	133	0.4	3.0	2550
D	0.3	1.0	55	0.3	1.2	700
E	0.2	0.6	65	0.2	0.8	670
F	0.1	0.3	41	0.1	0.5	465

	Tetrachloromethane			*Nitroethane*		
Amide	10°C.	30°C.	76°C.	10°C.	30°C.	100°C.
A	—	0.4	42.5	1.0	2.1	1600
B	—	0.2	59.0	0.8	1.1	—
C	—	0.4	72.0	0.8	1.1	5100
D	—	<0.1	27.6	0.3	0.5	1500
E	—	0.1	46.5	0.2	0.3	1075
F	—	<0.1	28.4	0.1	0.2	61

Table 79 (continued)

	Benzene			Cyclohexane		
Amide	10°C.	30°C.	80.1°C.	10°C.	30°C.	81.4°C.
A	0.5	0.6	153	—	0.1	53
B	0.6	0.8	195	—	0.4	150
C	0.6	1.0	238	—	0.5	163
D	0.3	0.4	95	—	0.2	61
E	0.3	0.4	145	—	0.3	122
F	0.3	0.4	89	—	0.3	88

	Acetone			2-Butanone		
Amide	10°C.	30°C.	56.5°C.	10°C.	30°C.	79.6°C.
A	3.4	7.8	48.4	3.4	8.3	163
B	2.0	3.8	23.9	2.2	3.8	189
C	1.3	3.4	30.5	1.2	3.8	194
D	0.4	1.0	6.4	0.6	1.2	100
E	0.3	0.8	6.9	0.4	1.0	118
F	0.2	0.5	3.7	0.3	0.7	79

A—Octanamide. B—Decanamide. C—Dodecanamide. D—Tetradecanamide. E—Hexadecanamide. F—Octadecanamide.

All figures are gravimetric quantities per 100 g of solvent.

High-molecular-weight aliphatic amides are readily hydrolized to the corresponding fatty acid by boiling in aqueous alkali or acid solutions. This reaction is quantitative over a period of a few hours. Care must be used in planning chemical reactions, to avoid conditions that will favor hydrolysis of the amide, to the exclusion of the desired reaction product.

These amides are unstable above 170°C. in the absence of a catalyst, and at lower temperatures in its presence. This decomposition is simple, does not change the alkyl-chain length and produces equal molar proportions of the corresponding nitrile, fatty acid and ammonia.

These amides readily react with organic halogens, to give the corresponding substituted amide, accompanied by the liberation of hydrogen halide.

One or both hydrogens, attached to the nitrogen atom in the amide group, can be substituted when heated with active metals, some organic acids, and metal oxides. This reaction can be widely used for the application of amides as chemical intermediates and in the synthesis of other products.

They are already being used as solvents for organic dyes in carbon-paper manufacture. They are excellent neutral, nonionic textile-softening and sizing agents when applied from an aqueous dispersion containing ordinary emulsifying agents.

Because of their high melting points, amides are partially replacing carnauba, montan and other waxes in commercial waxing and polishing. They are also used, to some extent, in waterproofing textiles. Because of their high melting points, they should be excellent for candle dipping and for any use where a soft, plastic, high-melting wax is desirable.

OCTADECANAMIDE—HEXADECANAMIDE

The average molecular weight of this technically pure mixed product is 275. It consists of 95% amide divided in about the following proportions: octadecanamide 70 and hexadecanamide 30.

Pure aliphatic amides are water white. Commercial samples have a very light lemon-yellow color. The melting point is 99°C. and the iodine number 3. The specific gravity is about 0.8. The boiling point is 241—251°C. at 12 mm; there is decomposition at this temperature.

High-molecular-weight aliphatic amides are insoluble in water and almost insoluble in the usual organic solvents, at room temperature. They are quite soluble in most organic solvents at elevated temperatures and are themselves excellent solvents for a wide variety of high-melting resinous substances at temperatures above their melting points. They are fairly reactive chemically. Thermal decomposition begins at temperatures above 175°C. which converts half of the amide quantitatively into the corresponding nitrile, whereas the other half is converted into the corresponding free fatty acid. Strong acids and alkalies hydrolyze the amides to fatty acids. They can be condensed with formaldehyde to form substituted amides. The indicated fields of commercial utility are chemical synthesis, waterproofing textiles, soft, plastic high-melting waxes, emulsifiers, textile softening and lubrication, candle dipping.

Amides do not deteriorate in storage and may be held indefinitely. The chief impurity is free stearic acid.

OCTADECANE NITRILE

Properties

Molecular Weight 265.4

Color	water white
Boiling Point, 760 mm	357.4°C.
Boiling Point, 2 mm	174.6°C.
Refractive Index (Solid at 20°C.)	1.4405^{40}
Vapor Pressure, 20°C.	< 0.1 mm (0.1 mm at 121°)
Melting Point	41°C.
Specific Gravity, 41°C.	0.8147
Weight Per Gallon	about 7 lb

OCTADECANE–HEXADECANE NITRILE

Average Molecular Weight	263
Nitrile, %	99
Specific Nitrile Mixture:	
Octadecane Nitrile	94
Hexadecane Nitrile	6

Properties

Color	white to light yellow
Distillation:	
Initial Boiling Point	345°C.
95% Over	355°C.
Final Boiling Point	357°C.
Nonvolatile Matter	trace
Melting Point	39°C.
Iodine Number	< 3
Moisture	none

SANTOWAXES

These are solid hydrocarbons ranging in melting point from a little above room temperature to as high as 210°C. (410°F.).

Stability: boiling 10% sulfuric acid or caustic soda has no detectable effect on Santowaxes.

Uses: Santowax O—rust-inhibiting coatings for steel; Santowax M—moistureproofing compositions for paper; Santowax P—high-melting, moistureproof impregnation compositions for paper and strawboard.

Table 80

PHYSICAL PROPERTIES OF THE SANTOWAXES

Santowax	O	M	P
Color, NPA	< 1.8	2.2–4.2	0–1.25
Density, g/cc	1.14	1.164	1.236
Melting Point	Santowaxes do not have sharp melting points		
Begin to Soften	35°C. (95°F.)	75°C. (167°F.)	200°C. (392°F.)
Completely Liquid	50°C. (122°F.)	85°C. (185°F.)	215°C. (419°F.)
First Crystals	–	< 90°C. (< 194°F.)	209–213°C. (408–415°F.)
Hold Point	52–54°C. (125.6–129.2°F.)	83–85°C. (181–185°F.)	
Distillation Range ASTM D-20 (corrected)	330–345°C. (626–653°F.)	370–378°C. (698–712°F.)	381–388°C. (718–730°F.)
Flash Point	171°C. (340°F.)	207°C. (405°F.)	207°C. (405°F.)
Flame Point	183°C. (390°F.)	229°C. (445°F.)	238°C. (460°F.)
Viscosity at 210°F. (Saybolt Universal Seconds)	40.6	39.3	28.8 (225°C.)

Table 81

ELECTRICAL PROPERTIES OF SANTOWAXES

	Dielectric Constant	Resistivity—ohm/cm³ 100°C.	Resistivity—ohm/cm³ 250°C.
Santowax O	2.54	$8{,}200 \times 10^9$	–
Santowax M	2.62	$2{,}600 \times 10^9$	–
Santowax P	–	–	30×10^9

Table 82

COMPATIBILITY OF SANTOWAXES

Santowax M

Other Component	% Santowax M in Mixture 10	25	50	75	90
Paraffin Wax, 120/125°F.	incompatible with 2% or more Santowax M				
Barnsdall Wax, 160/165°F.		C	B	I	I
Ceresin	I		I		I
Carnauba Wax	C		C	C	I
Candelilla Wax	C	C	I		I
Montan Wax	I		I		I
Beeswax	I		I		I
Japan Wax	C	I	I		
Opal Wax	C	C	I		I
IG Wax O	I		I		I
Stroba Wax	I		I		I
Acrawax	C		C		B
Stearic Acid	I		I		I
Shellac	C		I		I

Table 82 (continued)

Manila Resin (Loba B)	C		I		I
Ester Gum	C		I		I
"N" Wood Rosin (Hercules)	C	C	I		I
Piccolite Resin, 85°C.	C		I		I
Staybellite Resin	B		I		I

Santowax O

	% Santowax O in Mixture				
Other Component	10	25	50	75	90
Paraffin, 120/125°F.	C		C		C
Barnsdall Wax, 160/165°F.	C		C		C
Beeswax	C	C	C		I
Opal Wax	C		I		I
Stearic Acid	I	I	I		I
"N" Wood Rosin	C	C	I	I	

Santowax P

	% Santowax P in Mixture				
Other Component	0.5	1–2	3	5	10–75
Paraffin Wax, 120/125°F.	I				
Paradene No. 1				C	I
Paradene No. 2		C	I		
"N" Wood Rosin					I

C = Compatible; I = Incompatible; B = Almost Compatible.

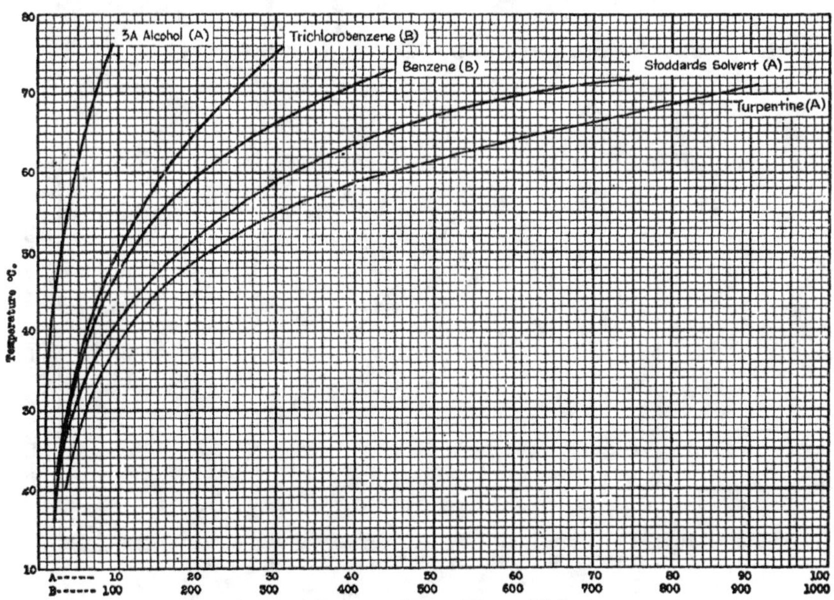

Figure 29. SOLUBILITY OF SANTOWAX M

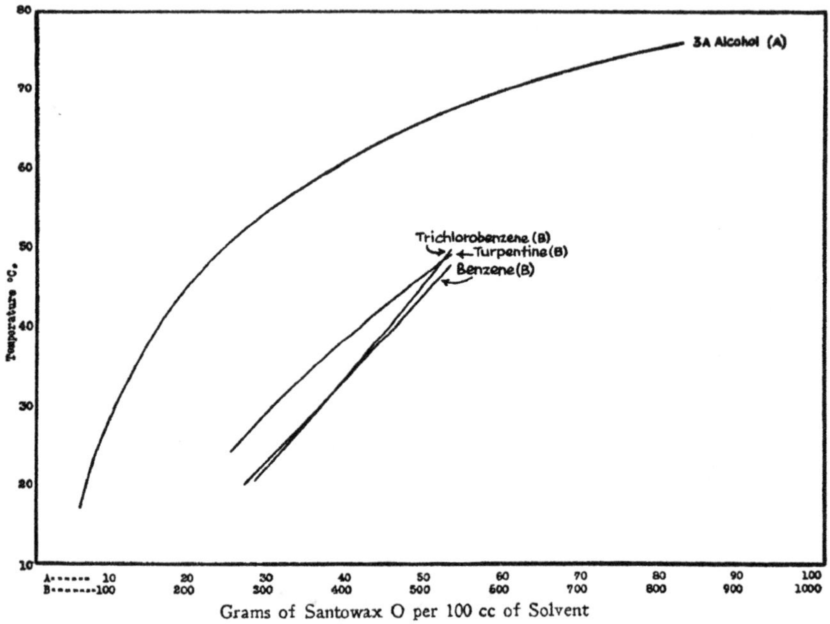

Figure 30. SOLUBILITY OF SANTOWAX O

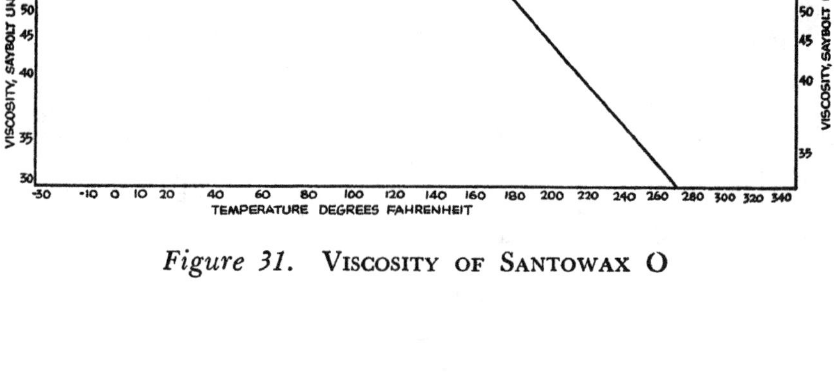

Figure 31. VISCOSITY OF SANTOWAX O

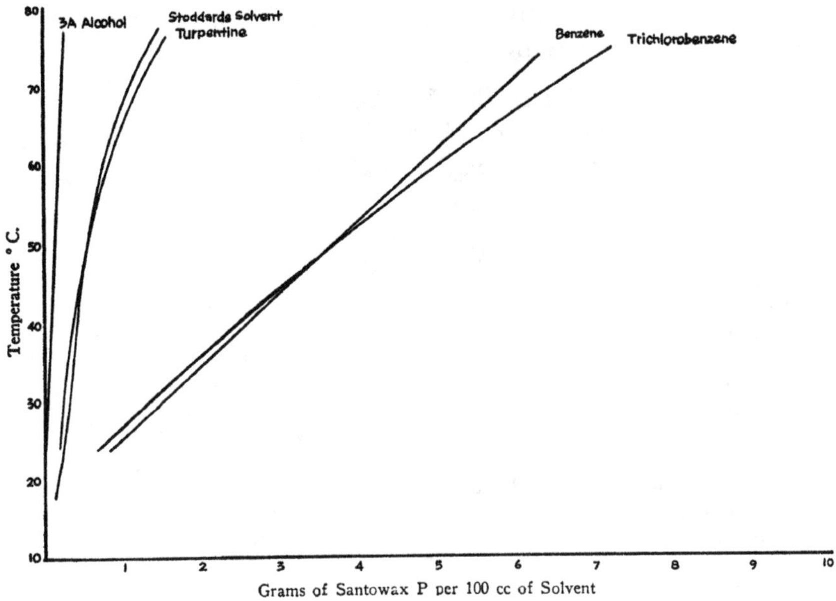

Figure 32. SOLUBILITY OF SANTOWAX P

Table 83
PROPERTIES OF SANTOWAX OSA AND SANTOWAX PSA

	OSA	PSA
Color in 25% n-butyl alcohol, NPA	1.0	2.0–2.2
Density g/cc	0.99 at 29.5°C.	1.04 at 28.9°C.
Refractive Index	—	1.58
Softening Point	59–59.2°C.	139.2–139.5°C.
Crystallizing Point	50–52.2°C.	135.5–136°C.
Flash Point	246°C.	268°C.
Fire Point	279°C.	318°C.
Decomposition Point	232°C.	193–204°C.
Acid Number	4.7	0–1

Penetration ASTM-D5-25 Load	50 g	100 g	150 g	200 g	50 g	100 g	150 g	200 g
cm × 10^{-2} at 25°C. for 60 sec	4.4	10.0	13.2	17.3	15.8	25.9	32.2	36.2

Solubility at °C.	25	75	25	75
Benzene	sol.	sol.	ins.	sol.
Ethanol	ins.	ins.	ins.	ins.
Naphtha	sol.	sol.	ins.	ins.
Trichlorobenzene	sl. sol.	sol.	ins.	sol.
Turpentine	sol.	sol.	ins.	ins.

ins. (insoluble) = less than 5% (by weight). sl. sol. (slightly soluble) = 5–10%. sol. (soluble) = more than 10%.

Uses: polishes, wax finishes, paints, cosmetics, paper coating, opacifier, sealing compounds.

SANTOWAX R (REGULAR)

This high-melting hydrocarbon wax has an unusual stability toward heat. Boiling 10% H_2SO_4 and boiling 10% NaOH have no detectable action on Santowax R.

Properties

Density at 25°C., g/cc	1.133
Color, NPA	4–5
Melting Point	Santowax R has no sharp melting point
Begins to Soften	60°C. (140°F.)
Completely Liquid	140°C. (284°F.)
Upper Hold Point	140–145°C. (284–293°F.)
Lower Hold Point	60–65°C. (140–149°F.)
Distillation Range D–20 Modified (Corrected)	364–418°C. (687–784°F.)
Flash Point	191°C. (375°F.)
Flame Point	238°C. (460°F.)

Table 84

ELECTRICAL PROPERTIES OF SANTOWAX R

Dielectric Constant		2.58
Resistivity, ohm/cm³ at	100°C.	$140{,}000 \times 10^9$
	135°C.	$35{,}000 \times 10^9$
	155°C.	550×10^9

Table 85

COMPARATIVE HARDNESS OF SOME WAXES
(Smaller figure denotes greater hardness)

Wax	Average penetration in cm $\times 10^{-2}$ in 60 seconds at 25°C. under load indicated. ASTM D–5–25. Load 50 gr	150 gr	250 gr
Santowax M	0	0	2.3
Santowax P	0	0	3.3
Carnauba Wax	0	2.3	4.7
Candelilla Wax	0	3.3	4.7
Santowax O	1.3	4.0	10.7
50% Paraffin Wax and 50% Santowax M	3.7	12.3	18.0
75% Paraffin Wax and 25% Santowax M	7.3	18.0	25.3
Ceresin Wax	9.3	15.7	26.0
95% Paraffin Wax and 5% Santowax M	10.7	21.0	29.7
98% Paraffin Wax and 2% Santowax M	11.7	26.0	32.7
Paraffin Wax	14.3	23.3	31.3
Santowax R, Aged 24 Hours	28.7	—	—
Santowax R, Nonaged	60.0	—	—

Table 86

SOLUBILITY OF SANTOWAX R

Solvent	Grams Santowax R per 100 ml of Solvent 25°C.	75°C.
3A Alcohol	0.008	0.6
Turpentine	1.200	9.0
Fuel Oil, A.P.I. 36° SUS 100°F.—38	1.660	8.4
Benzene	6.000	42.0
Trichlorbenzene	9.000	45.0
Stoddard's Solvent	0.700	8.0
Nitromethane	0.200	7.5
Nitroethane	2.000	15.0
1-Nitropropane	2.000	20.0
2-Nitropropane	2.000	20.0

Table 87

COEFFICIENTS OF THERMAL EXPANSION OF SANTOWAX R

Temperature Range °C.	Coefficient of Cubical Expansion Volume/Unit Volume/°C.	Coefficient of Linear Expansion Length/Unit Length/°C.
25–60	8.3×10^{-4}	2.8×10^{-4}
60–77	11.5×10^{-4}	3.8×10^{-4}
25–77	9.4×10^{-4}	3.1×10^{-4}

Table 88

TEMPERATURES FOR PREPARATION OF 20% SANTOWAX R SOLUTIONS

Solvent	Minimum °C. for Complete Solution (20 g Santowax R and 80 g Solvent)
Tetrachlorethane	53
Dioxane	58
Orthodichlorbenzene	58
Benzoyl Chloride	60
Monochlorbenzene	60
Trichlorbenzene (Mixed Isomers)	60
Benzyl Chloride	62
Benzene	64
Trichlorethylene	64
Pyridine	66
Chloronaphthalene	72
Furfural	72
Xylene	72
Bromonaphthalene	73
Diphenyl Oxide	76

Uses: impregnating paper to increase its moisture resistance and as a plasticizer in synthetic-resin-bodied paints.

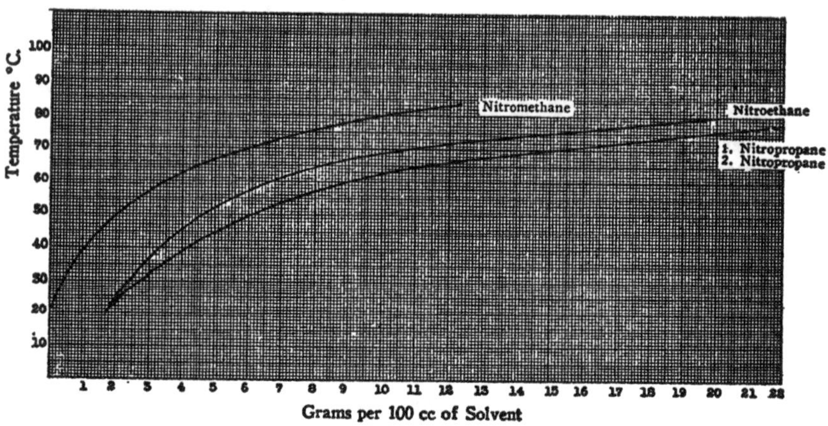

Figure 33. SOLUBILITY OF SANTOWAX R (REGULAR) IN NITROPARAFFIN SOLVENTS

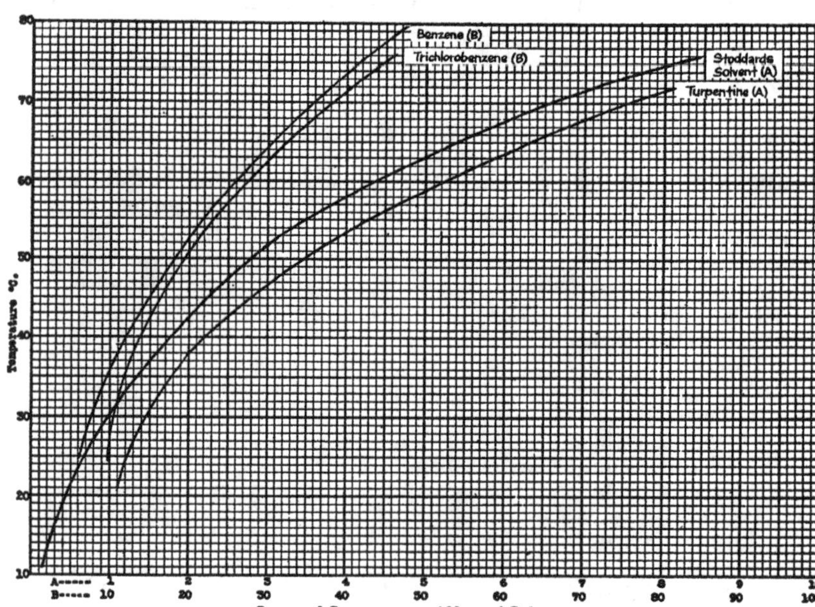

Figure 34. SOLUBILITY OF SANTOWAX R (REGULAR)

ALCOWAX

Alcowax is a low-molecular-weight waxlike solid derived from ethylene and is resistant to water and chemicals. It has a high melting point, but a moderately low melt viscosity. It is hard, white, translucent, tasteless, nontoxic and substantially odorless. It has low density, excellent electrical properties and is compatible with most waxes and waxlike substances. It offers resistance to most chemicals at room temperature.

Physical Properties

	Grade	
	No. 6	*No. 7*
Approximate Average Molecular Weight	2100	2100
Melting Point, °C.	97/102	102/106
Hardness, mm at 75°F.	0.3–0.5	0.2–0.3
200 g at 5 secs Flash, °F.	470–495	480–500

Solubility

Alcowax is only slightly soluble in turpentine, petroleum naphtha, xylene and toluene, at room temperature. At 180°F., it is soluble in xylene, toluene, trichlorethylene, turpentine, and mineral oils. It is practically insoluble in water at all temperatures and only slightly soluble in methyl acetate, acetone, and alcohol up to their boiling point. Above 225°F., Alcowax is miscible in all proportions with the solvents boiling above this temperature.

ALCOWAX M

This is an emulsifiable grade of Alcowax.

Uses: wax polish emulsions; leather finishes; textile size; glazing and embossing; latex dispersions; water paint.

ALCOWAX G

A low-cloud-point grade that offers better solubility, particularly in paraffin wax. It disperses more readily than other Alcowax grades.

Uses: in adhesives to metal; coating for electrical condensers; potting compounds; collapsible tubes.

CHAPTER 3

physical properties of waxes and wax compositions

INCREASING THE VISCOSITY OF MOLTEN WAXES

In certain hot-wax operations, it is desirable to increase the viscosity of the wax melt in order to decrease penetration, increase the coating thickness, facilitate pickup by rollers, brushes, etc., or for other reasons. To attain this end, the wax is melted and with further heating and mixing, a metallic soap[1] (e.g., aluminum stearate), rubber polybutenes, methacrylate polymers, synthetic resins, cellulose ethers, Acrawax, Strobawax, or other bodying agents are incorporated.

Table 89

STORMER VISCOSITIES OF WAXES AT 95°C.
(50 g; Baffled Cup)

	Stormer Seconds/100 Revolutions
Socony Paraffin Wax, 123–125°F.	9¼
Black Dallawax, 190°F. (Dallas Laboratories)	27¼
Candelilla Wax Prime A1 Grade	22
Z-52 Wax (Allied Asphalt)	15½
Utah Wax	24¼
Settled Crude Carnauba Wax N C No. 3	39¾
Settled Crude Ouricury Wax	80½
Norcowax 26 (Neatsfoot Oil Refineries)	35¼
Gulf Process Wax No. 4	14.0
Black Dallas Soft Wax	9¾

[1] Where the wax composition contains some free higher fatty acid (e.g., stearic acid) or its ester, a metallic soap is formed *in situ,* by mixing in a metallic oxide, hydroxide, or carbonate (e.g., magnesium oxide) and heating above 100°C., with stirring.

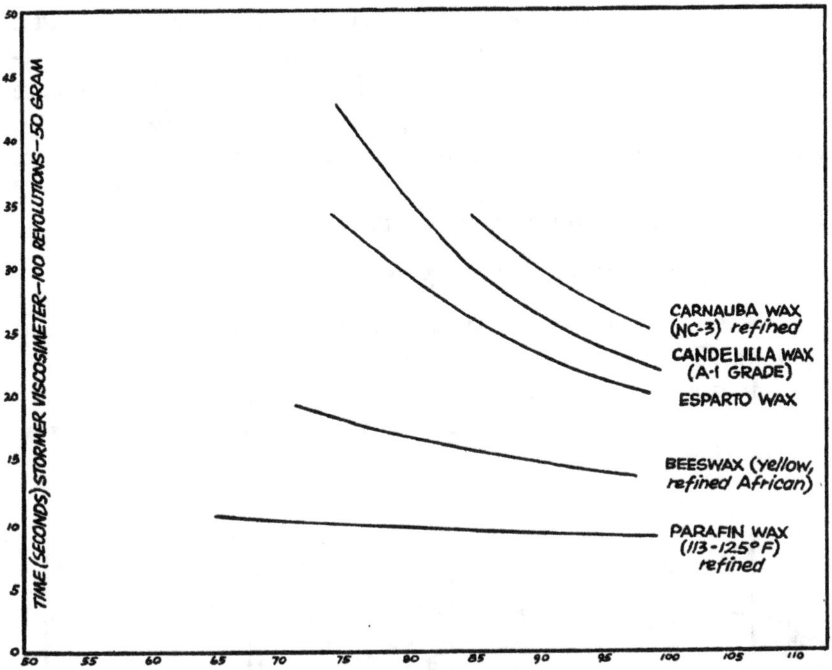

Figure 35. STORMER VISCOSITY OF WAXES

PENETRATION

Candelilla and paraffin waxes have very good wetting or penetrating qualities, when applied to paper. The addition of small amounts of Acrawax to certain wax or oil compositions often assists penetration. Increasing the temperature, lowering the viscosity or the addition of solvents will also increase penetration. In some cases, as in the use of insulating compounds, the mass is preheated or treated in a vacuum to remove moisture, air or other substances which would hinder the wetting or penetration of wax.

Where only a surface coating, with a minimum of penetration, is desired, the wax composition has to be applied at the lowest possible temperature. Other methods are to increase the viscosity; to apply to a chilled or damp surface; or the addition of a suitable, high-melting substance. Carnauba wax or Acrawax

C is used for this purpose. Either of these is added to paraffin wax, for example, and heated until just dissolved. On application, the high-melting waxes precipitate first, at a temperature above the melting point of the paraffin wax, and form a solid layer on the surface to be coated. This prevents penetration of the paraffin wax. As the higher-melting waxes are hard, they will stiffen the coating more than paraffin wax.

SURFACE TENSION OF WAXES[1]

Most waxes exhibit low surface tensions. Carnauba wax and opal wax give large contact angles with water and high-surface-tension organic liquids. Soft waxes of low surface tension are wetted more readily by liquids than hard waxes of higher surface tension.

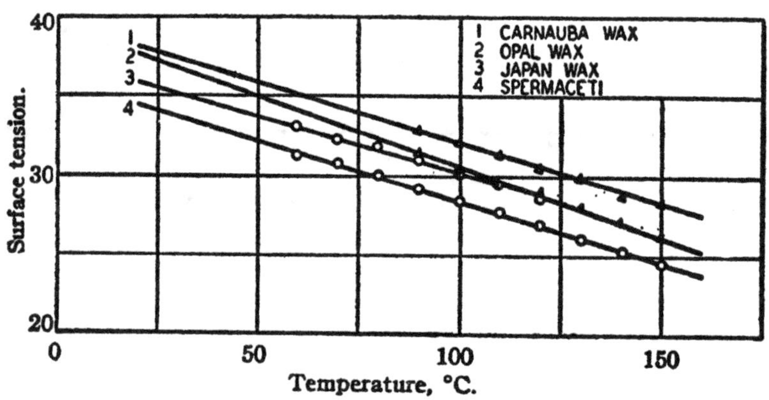

Figure 36. SURFACE TENSION OF WAXES

ADHESIVENESS

Most waxes are nonadhesive and slippery. Flexo Wax C (a hydrocarbon wax) is one of the few unblended waxes that have adhesive properties. It is only a weak adhesive, but is suitable for certain laminating and other uses. Waxes can be made by compounding with oils, natural and synthetic rubber, natural and synthetic resins and various organic compounds, such as

[1] F. E. Bartell and H. H. Zuidema, *J. Am. Chem. Soc.* 58, 1453 (1936).

Table 90

CONTACT ANGLES IN AIR AND INTERFACIAL CONTACT ANGLES FORMED BY LIQUIDS ON WAXES

	Carnauba Wax		Opalwax	
Liquid	θ ln	θ n3	θ ln	θ n3
Water	107°		119°	
Methylene Iodide	70°	128°	77°	123°
Acetylene Tetrabromide	65°	140°	72°	136°
α—Bromonaphthalene	57°	146°	65°	149°
Benzyl Alcohol	58°	142°	67°	134°
Ethylene Dibromide	45°	148°	35°	156°
Bromobenzene	49°	145°	50°	151°
Chlorobenzene	35°	153°	27°	159°
Toluene	26°	154°	26°	161°
Butyl Acetate	0–10°	166°	24°	153°
Amyl Alcohol	0°	149°	25°	120°
Benzene		152°		157°

Table 91

CONTACT-ANGLE MEASUREMENTS ON WAXES[1]

Wax	Temperature °C.	Contact Angle	Effective Adhesion Tension dynes/cm
Amorphous Paraffin Wax	31	110.2°	–24.9
Yellow Ozokerite	29	110.0°	–24.6
Bleached Ozokerite	29	113.2°	–28.4
Beeswax	30	107.6°	–21.8
Crude Carnauba Wax	28	122.8°	–39.0
Refined Carnauba Wax	27	125.3°	–41.6
Candelilla Wax	29	111.5°	–26.4
Spermaceti	26	110.2°	–24.9
Japan Wax	27	102.9°	–16.1
Crude Montan Wax		139.8°	–55.0

resin esters, cellulose esters, rubber derivatives, shellac, chlorinated diphenyls, etc.

The inclusion of Staybelite Esters in paraffin waxes increases their gloss, transparency, and adhesion to various supports. The

[1] R. N. Wenzel, *Ind. Eng. Chem.* 28, 992 (1936).

effect of 5% Staybelite Ester on the physical properties, as measured by hardness, of paraffin (M.R. 125°F.) is given in Table 90.

It will be noted that Staybelite Ester No. 10 causes relatively little change in the hardness or embrittlement of paraffin, its chief function being to add gloss, transparency and adhesion.

Staybelite Esters No. 1 and No. 2 show a definite softening action on paraffin, while Staybelite Ester No. 3 has powerful plasticizing action. On exposure at 0 to 3°C. the paraffin film modified with Staybelite Ester No. 3 showed no exudation of the plasticizer after storage for 24 hours. It increases the low-temperature flexibility of paraffin, to some extent, and may be used in preparing wax adhesives, sizes, coatings, and laminating formulations.

Hercules Staybelite Esters No. 1, 2, 3, and 10 show less tendency to sludge in molten paraffin blends than similar unhydrogenated rosin esters.

Table 92

PENETRATION* OF MODIFIED PARAFFIN WAX MELTING AT 125°F.

Type	At 32°F.	At 77°F.
Control (100% Paraffin Wax)	14	22
Staybelite Ester No. 10 (5%)	14	21
Staybelite Ester No. 1 (5%)	15	25
Staybelite Ester No. 2 (5%)	18	27
Staybelite Ester No. 3 (5%)	18	28

INCREASING COMPATIBILITY

Certain waxes are incompatible with other waxes, resins or other substances. Addition of a mutually soluble or thickening agent often produces compatibility.

Approximately 1% of aluminum stearate stirred into a mixture of melted waxes until the entire mixture is uniform will prevent separation of some waxes, e.g., 10% Acrawax and 89% paraffin wax are held together by this treatment. Considerable variation is possible. The viscosity of the melt is increased and it is not perfectly clear, but compatibility is maintained apparently indefinitely. This procedure may be of interest when applied to other incompatible waxes.

* As measured by ASTM Penetrometer.

RAISING THE MELTING POINT OF WAXES

The admixture of higher-melting substances does not necessarily raise the melting point of waxes proportionately. In some cases, the melting point is only increased slightly and in others, the melting point may even be reduced.

Carnauba wax or ozokerite is added to other waxes, particularly hydrocarbon waxes like paraffin, to increase the melting point. The addition of 1% Acrawax to carnauba wax raises the melting point of carnauba wax to 3°C.

Fatty-acid or fatty-acid-ester waxes are treated with small amounts, about 1.2%, of metallic oxides, hydroxides or carbonates at about 100°C. to increase their melting point, e.g., beeswax with magnesium oxide. The metal reacts with the fatty acid to form a metallic soap.

Ethyl cellulose increases the melting point, toughness, hardness and luster of waxes. Table 93 illustrates the effect of 25% low viscosity ethyl cellulose on various waxes.

Table 93

EFFECT OF ETHYL CELLULOSE ON THE MELTING POINT OF WAXES

	M.P. °C.	M.P. °C. with 25% Ethyl Cellulose
Glyco Wax S932	60–64	142.5
Glyceryl Monostearate	56–57	125.0
Diglycol Stearate S	54–59	110.0
Acrawax B	83–86	120.5
Acrawax	94–97	134.7

Table 94

EFFECT OF VARYING QUANTITIES OF ETHYL CELLULOSE ON THE MELTING POINT OF WAXES

Composition, %		Drop Melting Point, °C.		
Wax	Ethyl Cellulose	Purified Montan Wax	Bleached Beeswax	Stearic Acid
100.0	—	69.5	62.5	55.0
90.0	10.0	86.5	103.5	56.0
75.0	25.0	115.0	144.0	71.0
66.7	33.3	117.5	—	87.0
50.0	50.0	134.5	182.0	114.0

Table 95

EFFECT OF CARNAUBA WAX ON THE MELTING POINT OF WAXES

Carnauba Wax M.P. 85°C. % Added	Ceresin Wax M.P. °C.	Paraffin Wax M.P. °C.
0	72.7	60.10
5	79.1	73.90
10	80.56	79.20
15	81.60	81.10
20	82.53	81.10
25	82.95	81.10

RAISING THE MELTING POINT OF PETROLATUM

The maximum increase in melting point of petrolatum is usually obtained by a wax addition of 3-5%, for a variety of waxes. But 5% white beeswax increases the melting point 2°, 10% increases it 5°, 20% increases it 9°, obviously, the optimum amount is somewhere between 10 and 20% beeswax. An addition of 5% yellow beeswax increases the melting point 2°, 10% increases it 10° and 20% increases it 17°. That is, yellow beeswax is better than white beeswax for increasing the melting point of petrolatum.

Table 96

WAX SHRINKAGE

Wax	Volume when Melted	Volume when Cold (22°C.)	% Expansion on Melting
Carnauba Wax	26.0	20.2	29.9
Ouricury Wax	26.0	20.0	30.0
Fiber Wax	24.0	20.0	20.0
Paraffin Wax, M.P. 125°F.	23.5	22.2	5.9
Montan Wax	24.0	15.8	60.5
Yellow Utah Wax No. 4	28.0	24.0	16.6

Table 97

ANILINE POINT OF WAXES[1]

Beeswax	62–65
Carnauba Wax	78
Japan Wax	36–37
White Wax	61
Paraffin Wax	115–117

Table 98

WATER ABSORPTION OF WAXES AND OTHER SUBSTANCES

(After Six Months of Immersion)

Substance	Final Water Content %	Difference in Initial and Final Weights %
Candelilla Wax	0.13–0.20	0.00
Carnauba Wax	0.72–0.90	0.00
Montan Pitch	0.09–0.11	+0.22 to +0.27
Montan Wax	0.62–0.67	+0.02
Japan Wax	0.26–0.58	−1.02 to −1.4
Chinese Wax	0.75–1.16	−0.53 to −0.81
Bayberry Wax	6.1	−0.50
Beeswax	0.42–0.43	−0.13 to −0.20
Ozokerite	0.85–1.20	−1.0 to −1.3
Ceresin	0.04–0.05	0.00 to −0.05
Paraffin Wax	0.04–0.06	−0.43 to −0.66
Paraffin Wax	0.06–0.11	−0.66 to −0.71
Superla Wax	0.00–0.03	−0.02
Naudella Wax	0.50	+0.15
Zinsser's Insulating Wax	1.72–2.21	−0.03 to 0.00
Zinsser's Insulating Wax—Beeswax	1.5 -2.4	−1.4 to 2.4
Paraffin—Beeswax	0.11–0.54	−0.18 to −0.31
Paraffin—Rosin—Rosin Oil—Carnauba	0.32–0.33	−0.42 to −0.48
Paraffin—Rosin—Hydrated Lime	0.17	+0.50
Zinc Stearate	3.0	+0.45
Aluminum Stearate	2.7	+0.20
Blown Gilsonite and Residual Petroleum Oil	3.3	+0.63
Cumar Resin	0.41	−0.04 to 0.00
Rosin	0.46	−0.28
Manila Copal	0.41–0.60	0.00
Stearin Pitch	1.23–1.40	+0.30 to +0.48
Soft Blown Petroleum Asphalt	0.47–0.52	−0.02 to 0.00
Soft Blown Petroleum Asphalt—Rosin—Rosin Oil	0.44–0.56	−0.05 to 0.00
Spermaceti Wax	1.27–2.11	−0.37
Shellac	4.3 -5.1	−1.9 to 0.00
Halowax	0.20–0.28	−0.15
Stearic Acid	3.2 -4.0	−0.2 to −0.4

[1] C. G. Katrakis, *Praktika Akad. Athenon* 5, 311 (1930).

EFFECTS OF A WAX SIZING ON THE MOISTURE ABSORPTION OF INSULATING BOARD[1]

The results of wax sizing are shown in Table 99. Addition of 2% wax produces a board having a moisture absorption of 30.7% in ½ hour, which, although not satisfactory, shows definite sizing action. The absorption with 4.4% wax is 18.7%, which is still somewhat high. When 6% wax is used, the sizing is excellent, since the absorption is only 11.4%. The hydrogen-ion content of the white water (that is the water drained away from the board in the process of formation) varies between *p*H 7 and 8 or is nearly neutral.

Table 99

EFFECT OF A WAX SIZE ON THE MOISTURE ABSORPTION OF CORNSTALK INSULATING BOARD

Board No.	% Wax Used	% Moisture Absorption in ½ Hour
1	2.0	30.7
2	4.4	18.7
3	5.5	13.0*
4	6.0	11.4
5	7.1	9.8
6	8.8	11.0

* The board was only ¼ inch thick; other boards were ½ inch.

The wax emulsion used for sizing is made as follows: 300 cc water, 3 g triethanolamine, and 9 g stearic acid are boiled gently with stirring. To this 88 g of melted paraffin wax is added, with vigorous stirring. The wax sizing emulsion is added to the pulp in the usual manner, but is precipitated on the fibers by the addition of calcium chloride; 3 g of calcium chloride to 25 g of wax gives good results.

PLIOLITE–PARAFFIN WAX COMPOSITIONS

The addition of 5% Pliolite to paraffin-wax compositions improves its waterproofing qualities for paper. However, the Pliolite seems to hinder penetration so that, with certain methods

[1] Arnold, *Paper Ind.* 25, 156 (1943).

of treatment, its beneficial effect is cancelled by its lack of penetration. Oxidation inhibitors are ineffective in preventing the oxidation of paraffin wax in the presence of Pliolite. This prevents the use of Pliolite-paraffin compositions where the waxes must be held at a high temperature for a long period of time.

Table 100

REFRACTIVE INDICES (n_D^t) OF NORMAL SATURATED FATTY ACIDS[1]

Acid	20.0°C.	25.0°C.	30.0°C.	40.0°C.	45.0°C.	50.0°C.
Caproic	1.4170	1.4150	1.4132	1.4095		1.4054
Enanthic	1.4230	1.4209	1.4192	1.4155		1.4114
Caprylic	1.4280	1.4260	1.4243	1.4205		1.4167
Pelargonic	1.4322	1.4301	1.4287	1.4250		1.4210
Capric				1.4288		1.4279
Hendecanoic				1.4319		1.4248
Lauric					1.4323	1.4304
Tridecanoic						1.4328

Acid	55.0°C.	60.0°C.	65.0°C.	70.0°C.	80.0°C.
Caproic		1.4012		1.3972	1.3931
Enanthic		1.4073		1.4037	1.3993
Caprylic		1.4125		1.4089	1.4049
Pelargonic		1.4171		1.4132	1.4092
Capric		1.4210		1.4169	1.4130
Hendecanoic		1.4240		1.4202	1.4164
Lauric	1.4288	1.4267	1.4250	1.4230	1.4191
Tridecanoic	1.4310	1.4290	1.4272	1.4252	1.4215
Myristic	1.4329	1.4310	1.4291	1.4273	1.4236
Pentadecanoic	1.4348	1.4329	1.4310	1.4292	1.4254
Palmitic			1.4328	1.4309	1.4272
Margaric			1.4340	1.4324	1.4287
Stearic				1.4337	1.4299

Table 101

DENSITIES OF NORMAL SATURATED FATTY ACIDS AT 80°C.

Acid	d_4^t	Acid	d_4^t
Caproic	0.8751	Tridecanoic	0.8458
Enanthic	0.8670	Myristic	0.8439
Caprylic	0.8615	Pentadecanoic	0.8423
Pelargonic	0.8570	Palmitic	0.8414
Capric	0.8531	Margaric	0.8396
Hendecanoic	0.8505	Stearic	0.8390
Lauric	0.8477		

[1] A. Dorinson, M. R. McCorkle, A. W. Ralston. *J. Am. Chem. Soc.* **64**, 2739 (1942).

Table 102

ELECTRICAL CONSTANTS OF WAXES AND OTHER SUBSTANCES[1]

Dielectric Constant

Substance	Initial [a]	After Immersion [b]	Final [c]
Candelilla Wax	2.38–2.49	2.50–2.62	2.45–2.56
Carnauba Wax	2.66–2.83	3.84–4.19	2.82–2.83
Montan Pitch	2.60–2.62	2.63–2.68	2.60–2.62
Montan Wax	2.62–2.67	2.88–3.05	2.68
Japan Wax	2.96–2.99	3.15–3.20	2.91
Chinese Wax	2.94–2.98	3.62–3.89	2.95–2.97
Bayberry Wax	3.22–3.27	10.4	3.31
Beeswax	2.87–2.88	3.11–3.26	2.84–2.90
Ozokerite	2.37–2.43	2.50–2.55	2.40–2.47
Ceresin	2.16–2.24	2.29–2.32	2.28–2.29
Paraffin Wax	2.21–2.24	2.31–2.38	2.24–2.27
Paraffin Wax	2.19–2.24	2.37–2.55	2.26–2.30
Superla Wax	2.32–2.35	2.36–2.37	2.34–2.35
Naudella Wax	2.50	3.37	2.63
Zinsser's Insulating Wax	3.98–4.27	21	5.2 –6.5
Zinsser's Insulating Wax—Beeswax	3.21–3.40	14.0–14.2	3.64–3.83
Paraffin—Beeswax	3.50–3.55	3.55–3.61	2.53
Paraffin—Rosin—Rosin Oil—Carnauba	2.45–2.47	2.58–2.58	2.48
Paraffin—Rosin—Hydrated Lime	2.67	2.64–2.69	2.60–2.68
Zinc Stearate	2.58	5.83	
Aluminum Stearate	2.62–2.78	4.93–5.27	2.57
Blown Gilsonite and Residual Petroleum Oil	2.53–2.54	2.92–3.03	2.58
Cumar Resin	2.79–2.80	2.87–2.92	2.66
Rosin	2.69–2.76	2.81–2.87	
Manila Copal	3.05–3.09	3.22–3.38	2.63
Stearin Pitch	3.34–3.41	3.55–3.60	3.26
Soft Blown Petroleum Asphalt	3.07–3.12	2.90–3.03	2.96
Soft Blown Petroleum Asphalt—Rosin—Rosin Oil	3.07–3.22	2.94–3.04	2.88
Spermaceti Wax	3.60–3.75	6–17	3.74–3.89
Shellac	3.62–3.71	7.7 –>20	3.58
Halowax	3.63–3.83	4.24–6.37	3.97–4.23
Stearic Acid	2.37–2.41	>20	2.33–2.38

Effective A–C Conductivity

(Unit = 10^{-12} ohm cm)

Substance	Initial	After Immersion	Final
Candelilla Wax	6.0–6.4	18	9.3–9.4
Carnauba Wax	4.9–5.3	240–300	74–75

[1] J. A. Lee and H. H. Lowry, *Ind. Eng. Chem.* **19**, 304 (1927).

Table 102 (Continued)

Substance	Initial [a]	After Immersion [b]	Final [c]
Montan Pitch	20–21	25–26	24
Montan Wax	29–31	45–55	25
Japan Wax	9–12	33–34	13
Chinese Wax	125–137	189–210	120–123
Bayberry Wax	89–105	6000	110
Beeswax	45–48	70–85	56–58
Ozokerite	10–12	19–24	15
Ceresin	< 0.37	10–12	1.1
Paraffin Wax	< 0.37	3–7	0.74–0.80
Paraffin Wax	< 0.37	16–22	< 0.74–0.94
Superla Wax	4.2–4.6	2.1	0.37
Naudella Wax	16	110	16
Zinsser's Insulating Wax	48–60	600–2500	250–375
Zinsser's Insulating Wax—Beeswax	80–95	2000–2800	89–120
Paraffin—Beeswax	22–26	30–39	15–18
Paraffin—Rosin—Rosin Oil—Carnauba	23–24	30–32	18
Paraffin—Rosin—Hydrated Lime	3.2–3.9	18–20	8
Zinc Stearate	5.7	210	
Aluminum Stearate	13–17	74–130	12
Blown Gilsonite and Residual Petroleum Oil	14	126	22
Cumar Resin	9.0–9.7	15–17	1.0
Rosin	2.4–6.2	7.4–7.7	4.2–5.0
Manila Copal	7–15	21–23	17
Stearin Pitch	67–75	77–81	57
Soft Blown Petroleum Asphalt	35–43	32–42	30
Soft Blown Petroleum Asphalt—Rosin—Rosin Oil	36–43	56–58	33–35
Spermaceti Wax	6.0–6.6	500–100,000	10–12
Shellac	16–17	40,000–910,000	17
Halowax	4.4–7.4	24–40	3–4
Stearic Acid	4.4–12	> 130,000	5–6

Volume Resistivity
(Unit = 10^{13} ohms cm)

Substance	Initial	After Immersion	Final
Candelilla Wax	> 290	> 110	> 630
Carnauba Wax	60–120	0.3–4	270
Montan Pitch	> 290	75–160	> 1200
Montan Wax	> 290	> 270	> 610
Japan Wax	6–7	0.40–0.75	21
Chinese Wax	1–2	0.03–0.16	1–2
Bayberry Wax	2–5	0.00054	0.96
Beeswax	5–12	0.8–1.4	7–8

Table 102 (Continued)

Substance	Initial [a]	After Immersion [b]	Final [c]
Ozokerite	30–40	7–9	30–40
Ceresin	> 290	400–> 600	> 900
Paraffin Wax	> 290	150	> 900
Paraffin Wax	> 290	18	> 660
Superla Wax	> 460	> 560	> 1000
Naudella Wax	> 610	2.6	> 1000
Zinsser's Insulating Wax	> 490	0.00031	60–90
Zinsser's Insulating Wax—Beeswax	140–> 290	< 0.00031	130–150
Paraffin—Beeswax	20–27	8–10	31–33
Paraffin—Rosin—Rosin Oil—Carnauba	> 450	100–> 500	> 500
Paraffin—Rosin—Hydrated Lime	> 910	> 740	> 900
Zinc Stearate	640	0.0003	
Aluminum Stearate	400–> 900	4–12	> 1000
Blown Gilsonite and Residual Petroleum Oil	> 290	50–95	> 560
Cumar Resin	> 300	640	> 1000
Rosin	550	300–700	
Manila Copal	> 410	720	> 800
Stearin Pitch	14–18	6.5–7.0	39
Soft Blown Petroleum Asphalt	50–72	40–52	1000
Soft Blown Petroleum Asphalt—Rosin—Rosin Oil	14–18	7.9	26–33
Spermaceti Wax	3–4	0.00059–0.00031	9–15
Shellac	> 560	< 0.00031	> 1000
Halowax	4–9	0.4	0.1–0.7
Stearic Acid	3–7	< 0.00031	1–6

[a] After drying over calcium chloride.
[b] After 6-months' immersion in 3.5% sodium chloride solution.
[c] After redrying.

SPECIFIC HEAT OF PETROLEUM WAXES

The specific heat of petroleum waxes can be calculated from the following formula:[1]

$$S = 0.492 + 0.0009\,t$$

where t is the temperature in °C.

[1] C. M. Gateley, *Fuel* **12**, 26 (1933).

Table 103

MELTING POINTS AND SPECIFIC GRAVITIES OF COMMERCIAL WAXES (ARRANGED IN THE ORDER OF DECREASING MELTING POINTS)

Wax	Melting Point, °C.	Specific Gravity at 25°C.
Santowax P	209–213	1.100
Santowax PSA	136–139	1.040
Acrawax C	140–143	0.975
Santowax DF	125–130	0.990
Stroba Wax	98–102	1.040
Albacer (Flow Point)	99–104	1.010
Acrawax	94–97	1.040
IG Wax OP	91	1.020
Opalwax	87	0.980
Rilan Wax	84–85	0.985
Acrawax B	83–86	0.955
Crude Ouricury Wax	84	1.090
Santowax M	83–89	1.100
Carnauba Wax	81	0.995
Wax B—430	81	1.120
Refined Ouricury Wax	80	0.995
Crude Montan Wax	78	1.020
Fiber Wax	71	0.996
Crude Sugar-Cane Wax	70	0.993
Esparto Wax	70	0.993
Utah Wax No. 1	68	0.952
Flexo Wax C, Light	62–64	0.820
Crude Candelilla Wax	66	0.980

Table 104

SOFTENING POINT, HARDNESS, TENSILE STRENGTH AND STRETCH OF PARAFFIN WAX-CERESIN COMPOSITIONS

	Wax Mixture Used					
125–127°F. Paraffin Wax	330	275	220	165	110	0
138–140°F. Paraffin Wax	0	55	110	165	220	330
180–185°F. Imported Ceresin	2	2	2	2	2	2
ASTM Softening Point[1]—Regular	123.8	124.1	124.9	125.7	127.2	130.3
ASTM Softening Point[1]—Sloping	129.0	130.2	130.6	132.4	134.6	138.4
Hardness at 90°F.[2]	78	65	54	44	31	16
Hardness at 100°F.	2 sec.	2 sec.	2 sec.	2 sec.	2½ sec.	95
Tensile Strength at 90°F.[3]	21.7	22.9	25.7	28.0	28.6	26.5
Stretch at 90°F.[4]	28	1¼	1–	¾–	½–	¼

[1] ASTM Ball and Ring Method. [2] Penetrometer. [3] Pounds ¼ sq. in. cross section.
[4] In millimeters.

Table 105

SOFTENING POINT, HARDNESS, TENSILE STRENGTH AND STRETCH OF PARAFFIN WAX-CERESIN COMPOSITIONS

Domestic 133–135 °F.	Imported 133–135 °F.	Domestic 135–137 °F.	Ceresin	Softening Point[1] Regular	Softening Point Sloping	Hardness[2] 90°F.	Hardness[2] 100°F.	Tensile Strength[3] 90°F.	Stretch[4] 90°F.
—	100%	—	—	136.4	138.2	21	69	46.4	¼
—	100%	—	—	134.4	138.2	19	67	43.1	¼
73%	—	100%	1	133.3	138.0	27	206	48.8	½
73%	—	73%	1	136.0	138.4	21	154	53.5	½

[1] ASTM Ball and Ring Method.
[2] Penetrometer.
[3] Pounds ¼ sq in. cross section.
[4] In millimeters.

Table 106

MELTING POINT AND SPECIFIC GRAVITY OF VARIOUS WAXES

Wax	Melting Point, °C.	Specific Gravity at 25°C.
Diolin	65	0.920
Glyco Wax S932	60–64	0.926
Palmitic Acid	62.2	0.846
Refined Beeswax	60	0.954
Glyceryl Tristearate	58–60	0.920
Glyceryl Monostearate	56–57	0.970
Monostearin	56–57	0.898
Ozokerite	56	0.860
Steryl Alcohol	54–57	0.850
Triple-Pressed Stearic Acid	54	0.950
Diglycol Stearate S	54–59	0.960
Santowax OSA	50–59	0.990
Carbowax 4000	50–55	1.204
Ceraflux	50–51	0.880
Paraffin Wax	50	0.970
Cetyl Alcohol	45–50	0.810
Carbowax 1500	34–37	1.151

MELTING POINTS OF BINARY WAX MIXTURES[1]

The melting points are determined by the method devised by Koch, Hable, and Wrangel (*Ind. Eng Chem.*, Anal. Ed., **10,** 166, 1938). The apparatus consists of a brass plate (12 by 3 cm;

[1] W. J. Ellis, *Petroleum,* 116 (1945).

3 mm thick), and a number of small aluminium cylinders (5 mm diameter, 6 mm height), with one end slightly convex. The cylinders are fixed to the brass plate by their convex ends with the wax to be investigated. The plate carrying the cylinders is suspended in 2-liter beaker filled with 1.7-liter water which is slowly warmed (1°C. per minute) and gently agitated. The temperature at which one of the pellets falls from the plate is recorded as the melting point of the wax. Usually five tests were made with each mixture (variations did not exceed 1°C.) and the average recorded as the melting point. Each melting-point curve was determined by testing ten equally spaced mixtures; 1 g of each mixture was found to be adequate.

The results confirm, more or less, previously published data by other authors. Some of the waxes and wax mixtures, however, have been investigated for the first time. Gram wax is described as a wax from a plant grown in New South Wales; Synthetic 11165 (M.P. 82.5°C.) is a wax of British manufacture; and H.O.S. (M.P. 57.5°C.) is classified as a hydrogenated oleostearine, a waxlike by-product of the stearine production. The H.O.S. brand actually used is made in Australia and is known as Stearax. For lack of details regarding the last two waxes, their melting-point curves have not been reproduced. It can, however, be stated that the H.O.S. mixtures behave similarly to the paraffin-wax mixtures, the Synthetic 11165 mixtures to those with carnauba wax.

According to a proposal by Ivanovzsky, wax mixtures can be classed, with regard to the position of their melting-point curves, into four main groups. Wax mixtures of the first group have melting points which are higher than their corresponding simple-mixture values (which would be represented by the straight line connections of the two melting points). Wax mixtures of the second group are characterized by melting points lying below their simple-mixture values. Those of the third group are more or less in agreement with the latter, while wax mixtures of the fourth group are represented by melting-point curves crossing the mixture rule line.

Referring to the wax mixtures featured in the diagrams, the following classification can be made:—*Group* 1: Wool-wax mixtures with beeswax (2), gram wax (4), sugar-cane wax (5), shellac wax (6), and carnauba wax (7); paraffin wax mixtures with esparto wax (9), gram wax (10), sugar-cane wax (11), shellac wax (12), and carnauba wax (13); beeswax mixtures with gram wax (15), and carnauba wax (18). *Group* 2: Beeswax mixtures with paraffin wax (8); esparto wax mixtures with sugar-cane wax (19), gram wax (23), shellac wax (24), and carnauba wax (25); carnauba wax mixtures with shellac wax (27), and gram wax (28). *Group* 3: Sugar-cane wax mixtures with paraffin wax (11), gram wax (20), shellac wax (21), and carnauba wax (22); beeswax mixtures with esparto wax (14), sugar-cane wax (16), and shellac wax (17); shellac wax mixtures with gram wax (26). *Group* 4: Wool-wax mixtures with paraffin wax (1), and esparto wax (3).

It is of interest to note that there is often no relation between melting point and hardness. Although, for instance, a wide range of the paraffin wax mixtures with wool wax (1) have the same melting point, they differ widely in hardness. The hardness of wool-wax mixtures with sugar-cane wax (5) is, however, in many cases, much lower than would be expected from their melting points.

Following, the waxes are listed with rising melting points. The numbers refer to the melting-point curves in Figure 37, *the figures in parentheses to the melting points of the waxes. The abscissæ show the composition of the wax mixtures from* 0 *to* 100%, *in steps of* 10%. *In each diagram, the relatively lower-melting wax has its* 100% *abscissa in the left corner. The melting points are given as ordinates.*

For the sake of better grouping, reference to some of the wax mixtures had to be repeated in the following list.

1. *Wool Wax* (35) : *Paraffin Wax* (55)
2. *Wool Wax* (35) : *Beeswax* (62.5)
3. *Wool Wax* (35) : *Esparto Wax* (77.5)
4. *Wool Wax* (35) : *Gram Wax* (82.5)
5. *Wool Wax* (35) : *Sugar-Cane Wax* (72)
6. *Wool Wax* (35) : *Shellac Wax* (82.5)
7. *Wool Wax* (35) : *Carnauba Wax* (82.5)

1. *Paraffin Wax* (55) : *Wool Wax* (35)
8. *Paraffin Wax* (55) : *Beeswax* (62.5)

9. *Paraffin Wax* (55) : *Esparto Wax* (77.5)
10. *Paraffin Wax* (55) : *Gram Wax* (82.5)
11. *Paraffin Wax* (55) : *Sugar-Cane Wax* (72)
12. *Paraffin Wax* (55) : *Shellac Wax* (82.5)
13. *Paraffin Wax* (55) : *Carnauba Wax* (82.5)

2. *Beeswax* (62.5) : *Wool Wax* (35)
8. *Beeswax* (62.5) : *Paraffin Wax* (55)
14. *Beeswax* (62.5) : *Esparto Wax* (77.5)
15. *Beeswax* (62.5) : *Gram Wax* (82.5)
16. *Beeswax* (62.5) : *Sugar-Cane Wax* (72)
17. *Beeswax* (62.5) : *Shellac Wax* (82.5)
18. *Beeswax* (62.5) : *Carnauba Wax* (82.5)

5. *Sugar-Cane Wax* (72) : *Wool Wax* (35)
11. *Sugar-Cane Wax* (72) : *Paraffin Wax* (55)
16. *Sugar-Cane Wax* (72) : *Beeswax* (62.5)
19. *Sugar-Cane Wax* (72) : *Esparto Wax* (77.5)
20. *Sugar-Cane Wax* (72) : *Gram Wax* (82.5)
21. *Sugar-Cane Wax* (72) : *Shellac Wax* (82.5)
22. *Sugar-Cane Wax* (72) : *Carnauba Wax* (82.5)

3. *Esparto Wax* (77.5) : *Wool Wax* (35)
9. *Esparto Wax* (77.5) : *Paraffin Wax* (55)
14. *Esparto Wax* (77.5) : *Beeswax* (62.5)
19. *Esparto Wax* (77.5) : *Sugar-Cane Wax* (72)
23. *Esparto Wax* (77.5) : *Gram Wax* (82.5)
24. *Esparto Wax* (77.5) : *Shellac Wax* (82.5)
25. *Esparto Wax* (77.5) : *Carnauba Wax* (82.5)

6. *Shellac Wax* (82.5) : *Wool Wax* (35)
12. *Shellac Wax* (82.5) : *Paraffin Wax* (55)
17. *Shellac Wax* (82.5) : *Beeswax* (62.5)
21. *Shellac Wax* (82.5) : *Sugar-Cane Wax* (72)
24. *Shellac Wax* (82.5) : *Esparto Wax* (77.5)
26. *Shellac Wax* (82.5) : *Gram Wax* (82.5)
27. *Shellac Wax* (82.5) : *Carnauba Wax* (82.5)

4. *Gram Wax* (82.5) : *Wool Wax* (35)
10. *Gram Wax* (82.5) : *Paraffin Wax* (55)
15. *Gram Wax* (82.5) : *Beeswax* (62.5)

20. *Gram Wax* (82.5) : *Sugar-Cane Wax* (72)
23. *Gram Wax* (82.5) : *Esparto Wax* (77.5)
26. *Gram Wax* (82.5) : *Shellac Wax* (82.5)
28. *Gram Wax* (82.5) : *Carnauba Wax* (82.5)

7. *Carnauba Wax* (82.5) : *Wool Wax* (35)
13. *Carnauba Wax* (82.5) : *Paraffin Wax* (55)
18. *Carnauba Wax* (82.5) : *Beeswax* (62.5)
22. *Carnauba Wax* (82.5) : *Sugar-Cane Wax* (72)
25. *Carnauba Wax* (82.5) : *Esparto Wax* (77.5)
27. *Carnauba Wax* (82.5) : *Shellac Wax* (82.5)
28. *Carnauba Wax* (82.5) : *Gram Wax* (82.5)

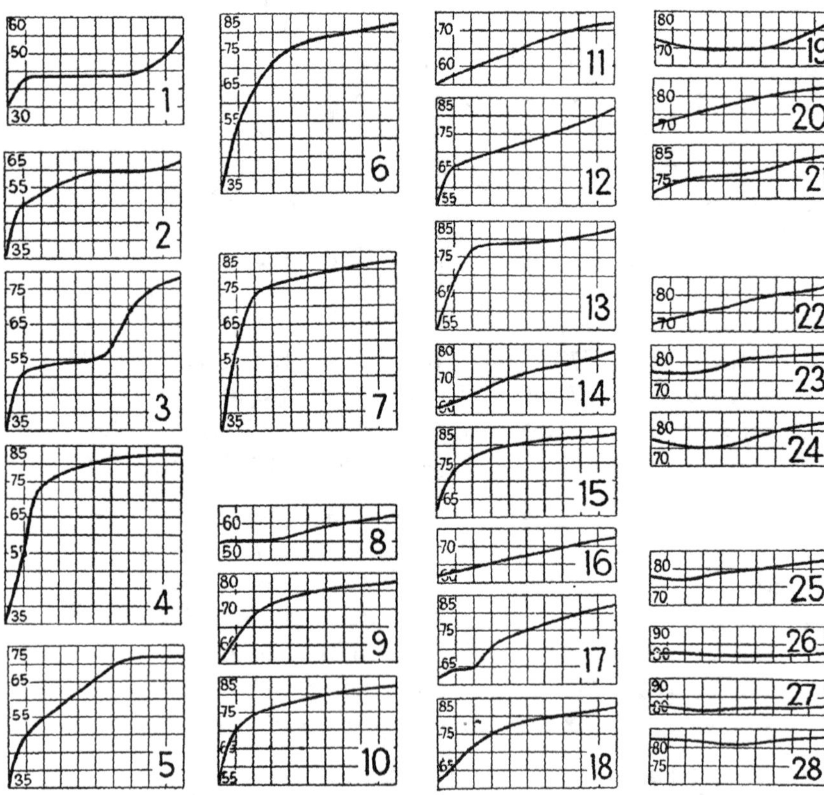

Figure 37. MELTING–POINT CURVES OF BINARY WAX MIXTURES

Table 107

Softening Point, Hardness, Tensile Strength and Stretch of Paraffin Waxes and Wax Compositions

Imported 125–130 °F.	Domestic 133–135 °F.	Imported 133–135 °F.	Domestic 135–137 °F.	Ceresin	ASTM Ball and Ring Softening Point Regular	ASTM Ball and Ring Softening Point Sloping	Hardness 90°F.	Hardness 100°F.	Tensile Strength 90°F.	Stretch 90°F.
—	—	—	100	—	133.2	136.7	23	165	31.7	½
—	100	—	—	—	132.2	136.9	29	258	52.2	½
—	100	—	—	—	131.3	136.0	30	266	48.9	½
—	100	—	—	—	133.8	138.0	22	191	51.8	½
—	73	—	73	1	136.6	138.8	21	110	44.7	½
—	73	—	73	1	135.5	138.8	29	133	46.0	½
—	146	—	—	1	136.3	139.2	25	136	51.6	½
—	146	—	—	1	136.8	139.6	24	87	45.5	½
100	—	—	—	—	132.8	134.4	36	2 sec.*	36.8	½
100	—	—	—	—	129.4	133.0	35	2 sec.*	36.1	½
100	—	—	—	—	129.6	133.2	33	2 sec.*	38.2	½
—	—	100	—	—	133.2	138.2	16	69	50.3	¼
—	—	100	—	—	133.2	139.9	17	62	47.2	¼
70	—	70	—	—	133.8	139.9	18	65	60.5	¼
70	—	70	—	1	134.2	136.7	19	160	54.4	½
70	—	70	—	1	134.2	136.1	22	211	54.1	½
70	—	70	—	1	134.5	136.7	22	152	53.4	½
70	—	70	—	1	132.1	135.2	22	232	49.6	½

* Time required for a needle to fall through the sample.

Table 108

Softening Point, Hardness, Tensile Strength and Stretch of Paraffin Waxes and Compositions

Paraffin Wax 128–130°F.	Paraffin Wax 133–135°F.	Added Wax	ASTM Ball and Ring Softening Point Regular	ASTM Ball and Ring Softening Point Sloping	Penetrometer Hardness 82°F.	Penetrometer Hardness 100°F.	Tensile Strength 90°F.	Stretch 90°F.
20		1 carnauba	147.6	154.0	26	125	40.2	39
20		1 carnauba	149.0	150.2	22½	133	31.3	41
110	67	2 carnauba	130.2	134.0	24	273	53.2	2–24
110	67	2½ carnauba	130.8	134.5	23	265	58.9	2–15
110	67	3 carnauba	131.3	135.4	22	228	58.6	13
110	67	4 carnauba	130.1	135.0	21	200	56.7	14
110	67	5 carnauba	131.3	138.2	19	170	56.8	15
110	67	6 carnauba	133.2	146.2	18	145	57.6	17
110	67	4 carnauba, 3 palm wax	131.8	137.0	15	196	31.3	36
73⅓	67	3 carnauba, 2 palm wax	132.0	136.8		165	55.1	10
20		1 Isco Wax	140.7	148.8		139	26.1	43
110	67	5 Isco Wax, 2 ceresin	132.1	139.0	19	146	51.5	21
110	67	5 carnauba, 2 palm wax	132.2	138.8		176	48.2	16
20		1 palm wax	127.3	132.0	28	281	26.1	35
10		1 palm wax	129.2	134.8	24	250	52.3	8
110	67	4 palm wax, 2 ceresin	131.1	134.6	20	146–200 (4 sec.)	45.4	2–10
110	67	4 palm wax, 4 ceresin	131.9	135.3	16	246	43.1	1–5
5	67	5 palm wax	131.8	137.4	18	222	55.1	1–5
110	67	0	132.2	142.9	35	141	64.3	10
11	67	0	134.5	137.5	20	223	50.7	½
11	67	1 ceresin	134.5	137.0	15	197	51.7	½

Table 108 (Continued)

Paraffin Wax 128–130°F.	Paraffin Wax 133–135°F.	Added Wax	ASTM Ball and Ring Softening Point Regular	ASTM Ball and Ring Softening Point Sloping	Penetrometer Hardness 82°F.	Penetrometer Hardness 100°F.	Tensile Strength 90°F.	Stretch 90°F.
22	134	1 palm wax	135.3	137.5	19	209	65.2	½
11	67	1 palm wax	134.1	137.5	19	194	66.1	½
11	67	2 palm wax	133.9	138.8	16	171	61.4	½
11	134	1 carnauba	134.9	137.7	14	129	74.6	½
11	134	1 carnauba 1 ceresin	135.3	139.1	12	85	70.7	½
11	134	1 palm wax 1 ceresin	136.3	138.4	16	102	65.1	½
11	134	2 palm wax 1 ceresin	135.9	138.5	15	87	61.5	½
22	134	1 carnauba	134.6	137.4	14	152	57.5	½
22	67	1 carnauba	133.5	137.2	19	206	71.8	½
20		1 carnauba	147.4	153.9	53	127	39.4	40
20		1 dark carnauba	138.1	151.0	55	127	36.9	41
	134	0						
	134	1 carnauba	135.8	139.5	16	66	71.7	½
11	134	1 carnauba	135.2	139.0	17	113	68.5	½
11	134	2 carnauba	135.6	139.0	13	78	65.1	½
22	134	1 carnauba	135.2	138.0	18	151	67.6	½
33	134	1 carnauba 1 ceresin	136.2	137.8	19	152	60.8	½
11	67	2 palm wax	136.5	138.8	19	138	69.1	½
11	134	2 palm wax	136.1	139.0	19	117	65.7	½
110	67	6	134.1	145.1	27	136	68.7	8
20		1 carnauba	144.8	153.9	36	119	48.1	20
	134	7 carnauba 2 ceresin	144.7	158.8	10	40	88.5	½

Table 108 (Continued)

Paraffin Wax 128–130°F.	Paraffin Wax 133–135°F.	Added Wax	ASTM Ball and Ring Softening Point Regular	ASTM Ball and Ring Softening Point Sloping	Penetrometer Hardness 82°F.	Penetrometer Hardness 100°F.	Tensile Strength 90°F.	Stretch 90°F.
20		1 crude montan	131.3	140.0	25	178	40.0	25
15		1 crude montan	134.1	146.0		155	35.0	34
10		1 crude montan						
20		1 IG wax S	130.0	133.6	24	125	45.4	2
20		refined No. 1						
		1 montan	129.7	132.0	24	295	25.8	1
20		refined No. 2						
		1 montan	128.7	128.8	32	3 sec.	22.4	30
20		1 candelilla	127.9	131.6	29	246	29.3	2–25
10		1 candelilla	130.9	136.2		236	27.5	42
20		1 esparto	129.8	132.2	31	272	26.9	37
20		1 Adurco No. 1	128.2	133.0	28	179	28.3	34
20		1 Adurco No. 3	127.7	131.0	30	249	24.2	34
20		1 shellac wax	129.8		33	4 sec.	24.8	34
20		1 shellac wax	131.1	142.2	29	285	27.5	35
20		1 opal wax	126.6	135.0		263	21.8	31
20		1 carnauba	145.0	153.0	44	121	39.8	33
110	67	6 carnauba	132.5	143.3		130	60.8	10
11	134	0	135.1	138.4	15	143	53.4	½
11	134	1 carnauba	136.0	138.4	13	92	65.0	½
22	134	1 carnauba	134.2	136.4	17	154	67.1	½
11	134	1 candelilla	135.8	136.5	15	136	57.1	½
22	134	1 candelilla	134.1	136.2	16	204	58.7	½
11	134	1 cumar	135.5	137.3	14	135	51.8	½
22	134	1 IG Wax	134.7	136.9	16	161	66.2	½
11	134	1 shellac wax	135.2	138.1	17	112	51.6	½
20		1 carnauba	145.0	153.0	44	121	39.8	33
10		1 carnauba	168.0	170.0	31	71	51.9	22
20		1 carnauba	143.6	154.5	46	109	37.6	35

Table 108 (Continued)

Paraffin Wax 128–130°F.	Paraffin Wax 133–135°F.	Added Wax	ASTM Ball and Ring Softening Point Regular	ASTM Ball and Ring Softening Point Sloping	Penetrometer Hardness 82°F.	Penetrometer Hardness 100°F.	Tensile Strength 90°F.	Stretch 90°F.
20		1 carnauba	147.1	152.5	45	138	42.5	30
10		1 carnauba	163.0	168.0	36	70	50.4	15
22	134		134.6	137.8	17	165	58.8	½
11	134		136.0	139.8	14	93	56.2	½
0	134		136.2	139.4	15	56	61.3	½
0	134		136.1	139.6	10	54	48.6	¼
0	134	2 carnauba	137.4	141.8	11	43	67.4	½
0	50	1 ceresin, 1 carnauba	135.7	141.0	11	42	69.2	½
0	134	1 ceresin, 5 carnauba	136.2	146.3	8	41	70.4	½
0	134	2 ceresin, 7 carnauba	150.3	157.0	7	41	97.0	½
0	14	1 carnauba	161.1	166.0	23	74	83.4	3
0	10	1 carnauba	165.5	170.0	18	53	91.6	2
20		1 carnauba	146.8	151.8	39	112	35.3	39.9
110	67	6 carnauba	131.7	144.0	30	148	53.7	16
110	67	2 carnauba	130.7	134.0	39	304	47.1	16
55	67	1 carnauba	132.9	135.4	30	273	54.1	2
33	67	1 carnauba	133.8	137.0	28	215	56.6	1
11	67	¾ carnauba	135.1	138.6	15	90	56.1	½
	100	1 carnauba	136.2	140.3	11	44	63.3	very slight
20		1 carnauba	146.7	154.4	4.9	117	41.0	46
110	67	6 carnauba	134.9	145.2	35	147	58.8	11
55	67	1 carnauba	132.4	136.2	34	278	57.1	1
33	67	1 carnauba	133.6	137.2	2.4	209	66.7	¾
22	67	1 carnauba	133.8	138.4	20	155	55.6	½
11	67	1 carnauba	135.1	139.9	13	91	58.8	⅓

Table 108 (Continued)

Paraffin Wax 128–130°F.	Paraffin Wax 133–135°F.	Added Wax	ASTM Ball and Ring Softening Point Regular	ASTM Ball and Ring Softening Point Sloping	Penetrometer Hardness 82°F.	Penetrometer Hardness 100°F.	Tensile Strength 90°F.	Stretch 90°F.
20		1 carnauba	138.4	151.6	47	139	37.8	38
110	67	6 carnauba	131.8	138.6	33	156	54.4	17
55	67	1 carnauba	132.1	135.2	29	269	52.7	1–2
33	67	1 carnauba	132.8	136.2	24	214	59.3	1
22	67	1 carnauba	134.2	137.4	18	164	58.7	½
11	67	½ carnauba	134.7	137.0	17	130	55.9	½
11	67	1 carnauba	134.6	138.2	13	97	58.2	½
110	67	6 carnauba	133.8	144.6	29	151	57.4	13
55	67	1 ceresin, 1 carnauba	132.8	136.8	30	232	53.4	1
22	67	½ ceresin, ½ carnauba	133.8	137.0	18	188	56.8	½
11	67	½ ceresin, ½ carnauba	135.0	139.2	13	116	60.2	½
33	67	1 palm wax, 1 carnauba	134.0	137.2	22	234	58.0	½
33	67	1 ceresin, 2 palm wax	133.2	136.5	21	216	51.3	½
11	67	2 palm wax	134.4	138.4	13	129	69.4	½
33	67	1 candelilla, 1 carnauba	134.8	138.5	21	189	49.1	½
88	22	2 candelilla, 2 carnauba	131.9	136.8	39	210	47.8	16

Table 108 (Continued)

Paraffin Wax 128–130°F.	Paraffin Wax 133–135°F.	Paraffin Wax 138–140°F.	Added Wax	ASTM Ball and Ring Softening Point Regular	ASTM Ball and Ring Softening Point Sloping	Penetrometer Hardness 82°F.	Penetrometer Hardness 100°F.	Tensile Strength 90°F.	Stretch 90°F.
110	67		6 carnauba	131.9	147.0	35	140	61.2	10
20			1 carnauba						
22	134		1 carnauba	134.8	137.7	17	163	64.5	½
22	134		1 ceresin	135.1	136.7	16	139	50.6	½
	134		5 carnauba, 1 ceresin	153.1	157.6	10	40	93.9	½
11	67	67	1 ceresin	137.8	140.2	12	41	52.4	¼
	134		1 ceresin	136.3	138.8	13	55	56.0	¼
110	67		6 carnauba	131.8	147.8	38	136	56.2	4
20			1 carnauba	147.1	152.5	45	138	42.5	3
10			1 carnauba	163.0	168.0	36	70	50.4	15
22	134		⅛ montan	134.6	137.8	17	165	58.8	½
22	134		1 montan	135.5	138.8	12	144	62.9	½
11	134		1 montan	136.0	139.8	14	93	56.2	½
	134		⅛ montan	136.2	139.4	15	56	61.3	½
	134		1 montan	136.1	139.6	10	54	48.6	¼
	134		1 montan+2 carnauba	137.4	141.8	11	43	67.4	½
	50		2 carnauba+1 ceresin	135.7	141.0	11	42	69.2	½
	134		5 carnauba+1 ceresin	136.2	146.3	8	41	70.4	½
	134		7 carnauba+1 ceresin	150.3	157.0	7	41	97.0	½
	14		1 carnauba	161.1	166.0	23	74	83.4	3
	10		1 carnauba	165.5	170.0	18	53	91.6	2

Table 109

Softening Point and Hardness of Rosin—Paraffin Wax Compositions

Wax Mixture			Softening	Penetrometer Hardness at			
Parts	Paraffin Wax	Parts Rosin	Point °F.**	70°F.	82°F.	95°F.	100°F.
	124–126°F. (Supplier A)						
100	"	0	122.4	37	158	Infinite*	Infinite
30	"	100	113.0	11	57	Infinite	Infinite
50	"	100	113.5	11	65	Infinite	Infinite
75	"	100	112.3	10	74	Infinite	Infinite
50	"	55	111.0	10	68	Infinite	Infinite
	(Supplier B)						
100	"	0	122.6	24	69	Infinite	Infinite
30	"	100	115.0	5	21	143	Infinite
50	"	100	114.6	7	30	169	Infinite
75	"	100	112.2	9	37	194	Infinite
50	"	55	114.0	10	42	220	Infinite
	128–130°F.						
100	"	0	129.2	17	32	145	Infinite
30	"	100	120.0	3	11	144	215
50	"	100	117.8	5	19	218	Infinite
75	"	100	117.9	7	23	129	Infinite
50	"	55	119.2	7	24	124	Infinite

* Sinks through in 5 seconds.

** ASTM Ball and Ring (Regular).

Table 110

SOFTENING POINT, HARDNESS, TENSILE STRENGTH AND STRETCH OF MICROCRYSTALLINE—AND PARAFFIN—WAX COMPOSITIONS

% Be Square Wax	0	1%	5%	10%	25%	50%	75%	100%
Parts Be Square Wax	0	3.5	17.5	35.0	87.5	175.0	262.5	100
Domestic 133–135°F.								
Paraffin Wax	220	346.5	332.5	315.0	262.5	175.0	87.5	0
Softening Point–Regular	129.1	127.0	131.5	131.0	135.9	163.2	178.5	187.4
Softening Point–Sloping	136.0	134.8	137.4	136.6	146.8	173.6	184.2	192.0
Hardness at 90°F.	30	36	32	26	20	11	9	8
Hardness at 100°F.	3½ sec.	3 sec.	269	119	46	25	18	11
Tensile Strength at 90°F.	36.6	35.6	36.8	58.6	63.3	68.6	72.3	56.4
Stretch at 90°F.	–½	½	–½	¾	½	–½	½	¼

Table 111

SOFTENING POINT, HARDNESS, TENSILE STRENGTH AND STRETCH OF MICROCRYSTALLINE—AND PARAFFIN—WAX COMPOSITIONS

Domestic 133–135°F. Paraffin Wax	150	150	350	347.4	347.4	347.4	347.4	347.4	347.4	347.4
Imported Ceresin	1	1	0	0	0	0	0	0	0	0
A.W.S. Crown Amber Wax	0	0	0	2.4	2.4	0	0	0	0	0
A.W.S. Gem Amber Wax	0	0	0	0	0	2.4	2.4	0	0	0
180–185°F. Barnsdall Wax	0	0	0	0	0	0	0	2.4	0	0
160–165°F. Barnsdall Wax	0	0	0	0	0	0	0	0	2.4	0
Utah Ceresin	0	0	0	0	0	0	0	0	0	2.4
% Amorphous or Microcrystalline Wax	0.66	0.66	0	0.69	0.69	0.69	0.69	0.69	0.69	0.69
Softening Point, Regular	127.0	127.0	127.4	126.2	127.0	127.1	126.9	127.1	129.4	126.3
Softening Point, Sloping	134.0	134.0	132.4	133.2	134.2	133.4	134.0	133.4	133.4	133.6
Hardness at 90°F.	34	34	36	38	37	34	33	32	33	37
Hardness at 100°F.	3½ sec.	3½ sec.	4 sec.	3½ sec.	2½ sec.	3 sec.	3 sec.	3 sec.	2½ sec.	2 sec.
Tensile Strength at 90°F.	37.9	37.9	32.3	34.6	30.1	27.3	26.1	36.7	38.1	36.2
Stretch at 90°F.	½	½	–½	½	¼	¼	–¼	½	–½	½

Domestic 133–135°F. Paraffin Wax	75	75	346.5	346.5	339.5	332.5	325.5	346.5	339.5
Imported Ceresin	1	1	3.5	0	0	0	0	0	0
A.W.S. Amber Microcrystalline Wax	0	0	0	0	0	0	0	3.5	10.5
A.W.S. Black Microcrystalline Wax	0	0	0	3.5	10.5	17.5	24.5	0	0
% Amorphous Mineral Wax	.54	.54	1.0	1.0	3.0	5.0	7.0	1.0	3.0
Softening Point, Regular	131.7	131.7	129.3	130.3	131.7	132.9	131.5	127.9	131.3
Softening Point, Sloping	139.4	139.4	135.6	135.4	136.2	139.2	140.1	135.0	136.6
Hardness at 90°F.	–16	–16	–27	–28	29	28	–26	29	28
Hardness at 100°F.	118	118	3½ sec.	4 sec.	273	262	162	4 sec.	272
Tensile Strength at 90°F.	39.4	39.4	44.1	38.3	38.7	42.1	43.3	38.2	36.2
Stretch at 90°F.	¼	¼	½	½	½	½	½	½	–¾

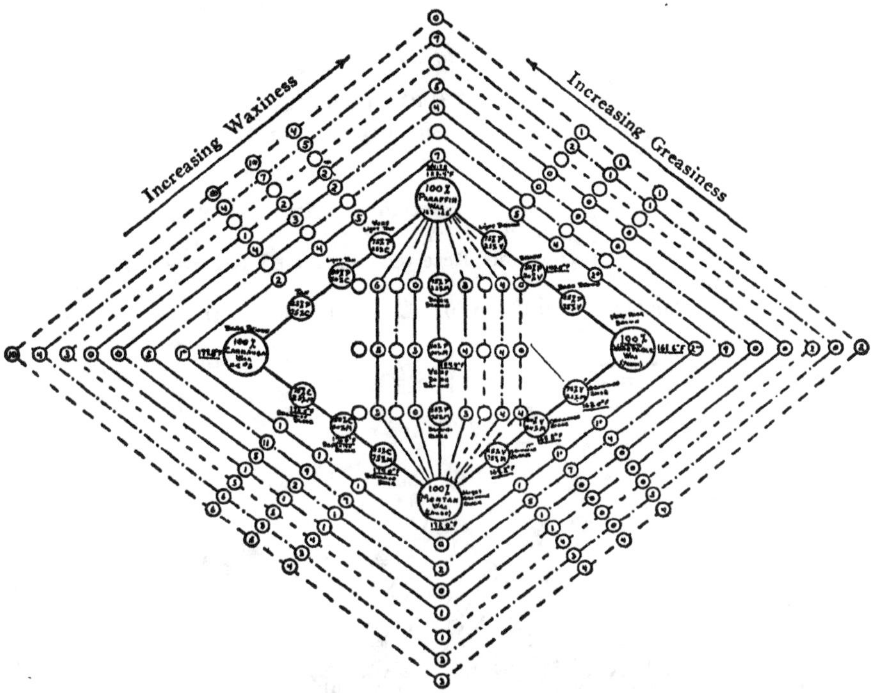

Figure 38. PHYSICAL PROPERTIES OF WAXES

Line 1
Composition of Waxes
Line 2
Scratch Hardness (0 = Hardest)
Line 3
Impact Hardness (0 = Most Brittle)
Line 4
Flatness of Surface (0 = Flattest)
Line 5
Smoothness of Surface (0 = Smoothest)
Line 6
Cross-Section (0 = Most Homogeneous)
Line 7
Graininess or Pittiness of Surface (0 = Least)
Line 8
Contraction or Cracking (0 = Least)

Table 113 gives the melting point of various percentages of wax when cut with oil. The oil, used in this case as being representative, is 1200/100 mineral oil.

Table 113 is useful when a reduced or lowered melting point is desired. Using an oil, a scale is established for the wax when a blend without reaction is achieved. Blends of waxes can be roughly estimated from this table and the melting points of the two waxes.

Table 112

BENZENE INSOLUBLES, SPECIFIC GRAVITY AT 25°C. AND CAPILLARY—TUBE MELTING POINTS OF WAXES AS RECEIVED AND AFTER BENZENE EXTRACTION

	% Benzene Insoluble	Specific Gravity As Received	Specific Gravity Extracted	Melting Point °F. As Received	Melting Point °F. Extracted
1. No. 3, Refined Carnauba Wax N.C.	0.18	0.9912	0.9916	180.5	183.0
2. No. 3, Crude Carnauba Wax N.C.	1.60	1.0097	0.9979	181.2	180.1
3. Refined Ouricury Wax	0.005	1.0025	1.0040	181.0	179.5
4. Crude Ouricury Wax	4.54	1.0321	1.0085	181.0	181.0

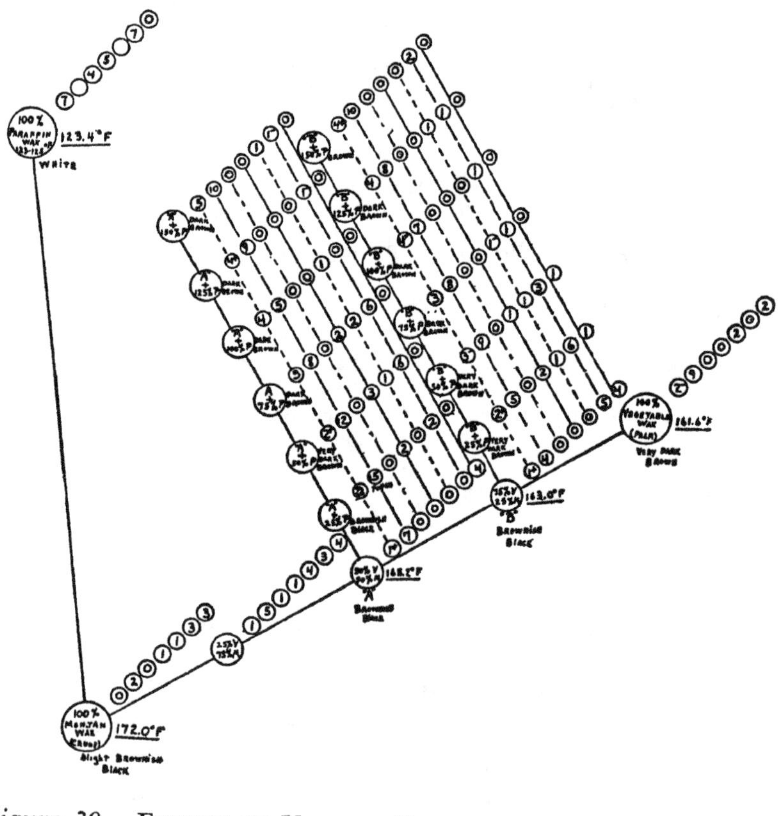

Figure 39. EFFECT OF VARIOUS PERCENTAGES OF PARAFFIN ON VEGETABLE WAX—MONTAN WAX MIXTURES

KEY TO CHART

Large Circle
Composition of Waxes
1st. Small Circle
Scratch Hardness (0 = Hardest)
2nd. Small Circle
Impact Hardness (0 = Most Brittle)
3rd. Small Circle
Flatness of Surface (0 = Flattest)
4th Small Circle
Smoothness of Surface (0 = Smoothest)
5th. Small Circle
Cross Section (0 = Most Homogeneous)
6th. Small Circle
Graininess or Pittiness of Surface (0 = Least)
7th. Small Circle
Contraction or Cracking (0 = Least)

Table 113

MELTING POINTS OF WAX—OIL COMBINATIONS

% Wax	Carnauba M.P. °C.	Ouricury M.P. °C.	Fiber M.P. °C.	Candelilla M.P. °C.	Utah M.P. °C.	Montan M.P.°C.
100	78	79	72	63	66	77
90	76	78	70	60	64	75
80	74	76	69	59	62	73
70	72	75	66	57	60	71
60	71	73	64	52	58	68
50	69	71	63	50	56	65
40	64	69	58	46	52	60
30	53	61	52	40	48	52

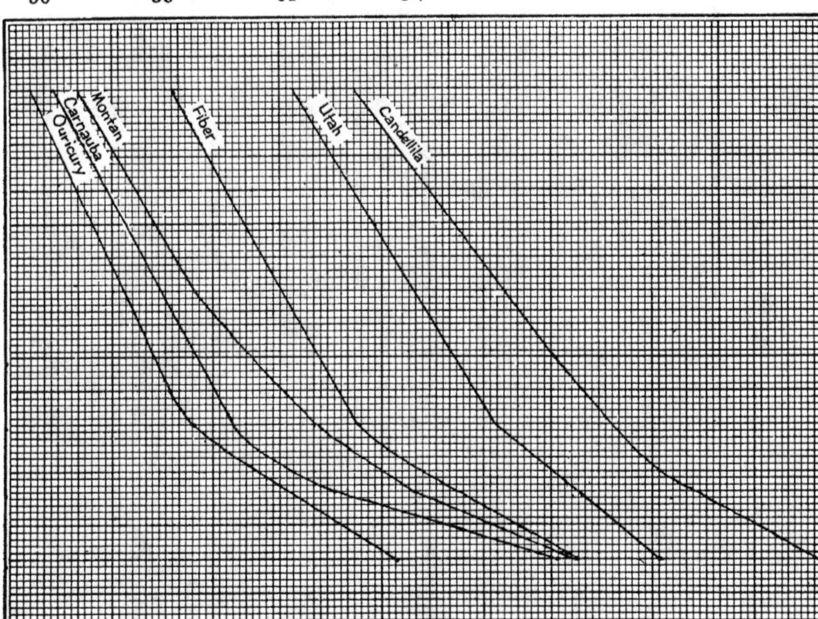

M.P. ° C. of Wax and Mineral Oil 1200/100 (S.A.E.) Mixtures

Figure 40. MELTING POINTS OF WAX—OIL MIXTURES

Table 114

WAX–OIL COMPOSITIONS

Wax	Mixed 50–50 with No. 752 Oil*	Mixed 50–50 with Sperm Oil	Mixed 50–50 with Half Sperm Oil Half 752 Oil
Bayberry Wax	Soft body similar to Philadelphia cream cheese, when worked up.	Smooth surface. Consistency and hardness of Roquefort cheese.	Softer than sperm, harder than 752-oil mix. Characteristics similar to 752-oil mix.
Candelilla Wax	Smooth surface. Takes up oil well. Hardness slightly greater than American cheese.	Smooth surface. Breaks with sharp fracture. Mixture harder than candelilla in 752 oil. Not so hard as carnauba in sperm oil.	Softer than sperm-, harder than 752-oil mix. Properties and appearance similar to sperm-oil mix.
Carnauba Wax	Smooth surface. Oil absorption excellent. Hard mix. Breaks in sharp fracture.	Smooth surface. Oil is taken up well. Breaks with sharp fracture. Slightly harder than carnauba–752 oil mixture.	Harder than 752-oil, softer than sperm-oil mixture. Properties and appearance similar to sperm-oil mix.
Esparto Wax	Smooth surface. Breaks in sharp fracture. Hardness not quite equal to carnauba in 752 oil. Forms homogeneous mixture.	Hardness practically equal to carnauba–sperm oil mixture. Breaks with sharp fracture.	Slightly softer and less brittle than 752-oil mix. Hardness about equal to sperm-oil mix.
Japan Wax	Quite soft. Consistency of amber petrolatum. Grainy appearance.	Smooth surface. Consistency of hard tallow. Oil is taken up well. Homogeneous mixture.	Harder than 752-oil, but softer than sperm-oil mix. Properties and appearance similar to sperm-oil mix.
Japan-Wax Substitute	Takes up more oil than natural Japan wax. Appearance soft, granular, and cheesy.	Surface smooth. Consistency of Roquefort cheese. Looks, when worked up, like mashed potatoes.	Softer than sperm-, but harder than 752-oil mix. Properties similar to 752-oil mix.
Ouricury Wax	Appearance and hardness of mixture is practically same as carnauba in 752 oil. Smooth surface. Breaks in sharp fracture (not quite so sharp as carnauba).	Smooth surface. Breaks with sharp fracture. Mixture harder than 752 oil-ouricury mixture. Takes up oil very well. Homogeneous mix.	Hardness and properties similar to sperm-oil mix.

* Mid-continent mineral oil (Gulf Co.), Saybolt viscosity 100 sec. at 100°F.

Table 114 (Continued)

	Mixed 50–50 with No. 752 Oil*	Mixed 50–50 with Sperm Oil	Mixed 50–50 with Half Sperm Oil Half 752 Oil
French, Yellow Ozokerite	Homogeneous mixture. Hardness of candles.	Homogeneous mixture. Slightly harder than 752-oil mixture.	Softer than sperm-, but harder than 752-oil mixture. Properties similar to 752-oil mixture.
Italian, Yellow Ozokerite	Surface semismooth. Consistency of dry, amber petrolatum. Homogeneous mixture.	Homogeneous mixture. About same hardness as French ozokerite with sperm oil.	Softer than sperm-, but harder than 752-oil mixture. Properties similar to sperm oil mix.
Montan Wax (Natural)	Smooth top. This wax takes up all of the oil; but a considerable amount can be forced out under pressure. Cheesy appearance. Hardness equal to overcooked fudge.	Smooth surface. Consistency and hardness of overcooked fudge.	Similar to both sperm-montan mixture and 752 oil-montan mixture. All three alike.
Utah Wax No. 1	Crystalline appearance. Dark brown. Hard and cheesy. Excess oil may be squeezed out by pressure.	Smooth surface. Consistency slightly harder than American cheese. Oil may be forced out under pressure.	Softer than sperm-oil mixture. Slightly harder than 752-oil mixture.
IG Wax E	Smooth surface. Breaks in sharp fracture. Takes up oil well. Hardness almost equal to that of carnauba.	Smooth surface. Breaks with sharp fracture. Hardness not equal to that of carnauba wax. Homogeneous mixture.	All three about equal in hardness and properties.
IG Wax S	Smooth surface. Takes up oil well. Mixture is slightly more brittle and drier than IG Wax E. Breaks with sharp fracture.	Smooth surface. Forms hard mixture equal to that of carnauba. Breaks with sharp fracture.	All three about equal in hardness and properties.
IG Wax KP	Smooth surface. Hard, dry, cheesy appearance, when broken up. Does not break in sharp fracture. Much softer than IG Wax S.	Smooth surface. Hardness slightly greater than that of candelilla in same mixture. Breaks with sharp fracture. Homogeneous mixture.	Softer than sperm-oil mixture. Harder than 752-oil mixture. Properties similar to sperm-oil mixture.

* Mid-continent mineral oil (Gulf Co.), Saybolt viscosity 100 sec. at 100°F.

Table 114 (Continued)

	Mixed 50–50 with No. 752 Oil*	Mixed 50–50 with Sperm Oil	Mixed 50–50 with Half Sperm Oil Half 752 Oil
Mannitan Monostearate	Not miscible.	Not miscible.	Not very miscible. Properties similar to sperm-oil mixture. Harder than 752-oil, but softer than sperm-oil mixture. Properties similar to sperm-oil mixture.
Normal Stearamide	Two layers. Top is soft and cheesy, the bottom somewhat harder.	Crystalline surface. Consistency dry, crystalline and semi-cheesy.	
Sorbitan Tetrastearate	Very soft, granular appearance. Consistency of cold bacon grease.	Smooth surface. Consistency of mashed potatoes but slightly harder.	Softer than sperm-, harder than 752-oil mixture. Properties of 752-oil mixture.
Beeswax	Smooth surface. Forms homogeneous mixture. Hardness equal to that of dark-green petrolatum.	Smooth surface. Homogeneous mixture. Consistency and hardness of hard tallow.	Softer than sperm-oil, harder than 752-oil mixture. Properties and appearance of sperm-oil mixture.
Spermaceti	Quite sugary and crystalline. Oil may be squeezed out readily. Consistency about same as that of paraffin wax in 752 oil.	Crystalline surface. Appearance granular and crystalline. Hardness about equal to paraffin wax in 752 oil.	Softer than sperm-oil mixture. Harder than 752-oil mix. Properties and appearance similar to sperm-oil mixture.
Paraffin Wax (125°F.)	Quite crystalline and sugary. Hardness not greater than that of Crisco.	Crystalline surface. Can be worked up into mixture which is equal in hardness to hot mashed potatoes. Oil may be readily squeezed out under pressure.	Softer than sperm-oil mixture. Harder than 752-oil mixture. Properties similar to sperm-oil mixture.

* Mid-continent mineral oil (Gulf Co.), Saybolt viscosity 100 sec. at 100°F.

Table 115

HARDNESS OF WAXES AND WAX COMPOSITIONS

Temperature of Wax	25°C.
Instrument	precision penetrometer with brass cone
Weight Used	brass cone plus 150 grams
Time	60 seconds
Values	penetration in centimeters $\times 10^{-2}$ (average of 5 to 10 readings)
Accuracy	about 5%.

	Wax Alone	50 Wax 50 Ink Oil	47.5 Wax 47.5 Ink Oil* 3 Aerosol OT
	Hardness	Hardness	Hardness
Carnauba Wax N.C. No. 3	0.5	23.2	21.2
Santowax M	0.3	>45.0	53.6
Code 200 (Werner G. Smith)	1.0	59.3	
Special Wax "A" (Werner G. Smith)	1.3	23.9	
Dallawax Black, 190°F.—(Dallas Laboratories)	7.6	> 60	74
Texas Paraffin Wax, 127°F., Refined	18.3		
Socony Paraffin Wax, 123–125°F., Refined	18.7		

Composition	Hardness	Composition	Hardness
50 Carnauba Wax 50 Paraffin Wax, 123–125°F.	7.8	50 Santowax M 50 Dallawax, Black, 190°F.	7.6
50 Texas Paraffin Wax 50 Stearic Acid	14.5	50 Code 200 50 Dallawax, Black, 190°F.	6.5
50 Paraffin Wax, 123–125°F. 50 Stearic Acid	14.7	50 Special Wax "A" 50 Dallawax Black, 190°F.	3.6
50 Carnauba Wax 50 Dallawax Black, 190°F.	2.2	50 Paraffin Wax, 123–125°F. 50 Dallawax Black, 190°F.	23.9

* Ink oil referred to in all cases is a Mid-continent mineral oil, 100 Saybolt sec. at 100°F.; 39 Saybolt sec. at 210°F.

Table 116

HARDNESS OF WAX COMPOSITIONS

	86% Wax 14% Carbon Black Hardness	43% Wax 43% Ink Oil* 14% Carbon Black Hardness
Special Wax A	0.10	
Code 150 (Werner G. Smith)	0.15	
Code 200 (Werner G. Smith)	0.00	
Code 1000 (Werner G. Smith)	1.00	
Utah Wax	0.00	
Z–51 (Allied Asphalt)	0.10	
Z–52 (Allied Asphalt)	0.70	
NC Wax (Foster D. Snell)	0.25	
HC Wax (Foster D. Snell)	0.25	> 4.0
LOH Wax (Foster D. Snell)	0.10	
Dallawax (Black, 190°F.)	0.60	> 4.0
Carnauba Wax NC No. 3		1.3
Norcowax C–37 (Neatsfoot Oil Refineries)		> 4.0
Norcowax No. 55 (Neatsfoot Oil Refineries)		> 4.0

* Mid-continent mineral oil, 100 Saybolt viscosity at 100°F.

Table 117

VISCOSITIES AND DENSITIES AT 95°C. OF WAXES AND WAX—OIL COMPOSITIONS

(Accuracy 1–2%; viscosity with Ostwald viscosimeter)

	Wax Alone			Wax in Ink Oil 50–50		
	Centi-stokes	Density	Centi-poises	Centi-stokes	Density	Centi-poises
Carnauba Wax NC No. 3	33.80	0.845	28.60	13.40	0.846	11.30
Code 200 Wax	31.20	0.896	28.00	11.00	0.845	9.30
Dallawax (Black) 190°F.	32.20	0.842	27.10	11.60	0.833	9.70
Special Wax A	25.20	0.865	21.80	11.20	0.823	9.20
Santowax M	4.28	1.038	4.44	3.67	0.888	3.26
Texas Paraffin Wax–127°F.	5.71	0.762	4.35	—	—	—
Socony Paraffin Wax–123–125°F.	3.58	0.756	2.71	—	—	—

Table 118
Wax–Oil Consistency for Inks

Wax	Color	M.P.	Hardness* Compared with Carnauba Wax	Mineral Oil (1 Part Oil 1 Part Wax)	Pigment Wetting	Flow in Ink
Acrawax	tan	94–97°C.	softer	much softer than carnauba	good	poor
Albacer	white	99–104°C.	softer	slow-blending, softer than montan	fair	poor
Opal	white	98°C.	softer	similar to Albacer	poor	poor
Esparto	brown	75°C.	slightly softer	slightly softer than carnauba	good	good
Candelilla	brown	68°C.	slightly softer	slightly softer than carnauba	good	good
Fiber	brown	70°C.	slightly softer	slightly softer than carnauba	good	good
Ouricury	brown	180°F.	slightly harder	slightly harder than carnauba	good	good
Sugar Cane		80°C.	slightly softer when pure	slightly softer than carnauba	good	fair
Wax No. 703	brown	180°F.	similar	harder than carnauba	good	good
Wax No. 648	brown	180°F.	similar	hard as montan	good	good
Wax No. 688	brown	143°F.	similar	soft	good	poor
Mineral Waxes						
"B"	white	158°F.	softer	good-blending	good	good
"B"	yellow	158°F.	softer	good-blending	good	good
Barnsdall Special		160–165°F.	softer	good-blending	fair	fair
Crown	black	190°F.	softer	good-blending	good	good
Crown	amber	200°F.	softer	good-blending	good	good
Gem		185°F.	very soft	good-blending	good	good

* Hardness rating is based on pure wax.
The oil taken for comparison is mineral oil 200/100°F. viscosity.

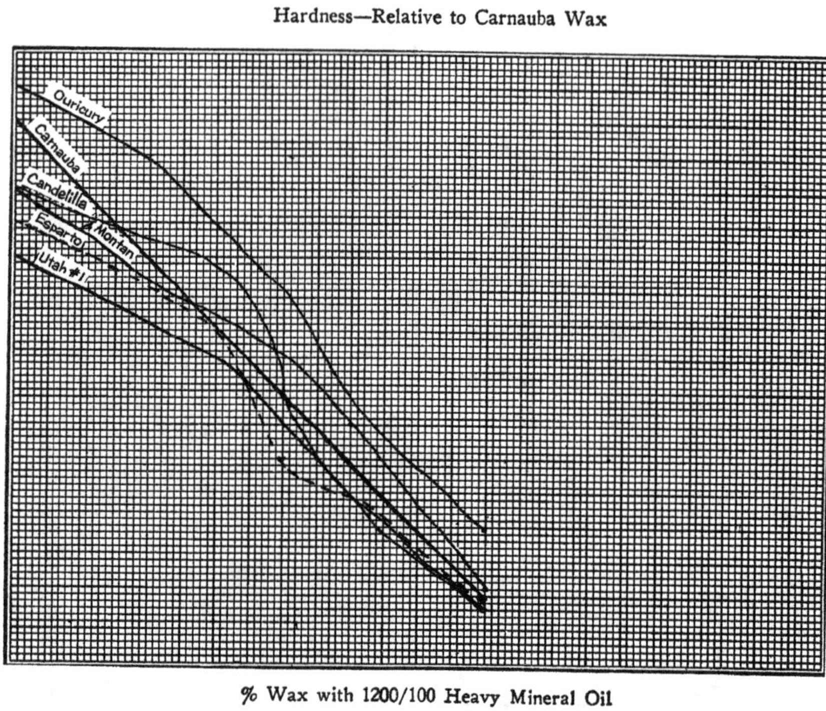

Figure 41. HARDNESS OF WAX–OIL COMPOSITIONS

Table 119

DENSITIES OF WAXES AND WAX–OIL COMPOSITIONS

	Gravimetric Method At 95°C. ± 1.0 g/cc	Hydrometer Method At 99°C. g/cc	At 100°C. g/cc
Carnauba Wax–NC No. 3	0.845	0.847	0.843
Ink Oil*	0.838	0.840	0.836
Socony Paraffin Wax, 123–125°F.	0.756	—	0.754
Dallawax Black, 190°F.	0.842	—	—
Santowax M	1.038	—	1.036
Special Wax A	0.865	—	0.861

* Mid-continent mineral oil, 100 Saybolt viscosity.

Table 120

VISCOSITIES AND DENSITIES OF WAX AND WAX–OIL COMPOSITIONS

Grams	Wax Composition	Centistokes	Density	Centipoises
50.0 50.0	Carnauba Wax NC No. 3 Socony Paraffin Wax, 123–250°F.	13.6	0.778	10.60
50.0 50.0	Special Wax A Dallawax Black, 190°F.	26.8	0.848	22.70
50.0 50.0	Carnauba Wax NC No. 3 Dallawax Black, 190°F.	31.5	0.776	24.40
50.0 50.0	Santowax M. Dallawax Black, 190°F.	10.4	0.918	9.60
50.0 50.0	Socony Paraffin Wax, 123–250°F. Dallawax Black, 190°F.	9.1	0.783	7.10
47.5 47.5 3.0	Carnauba Wax NC No. 3 Ink Oil Aerosol OT	13.8	0.828	11.40
47.5 47.5 3.0	Code 200 Wax Ink Oil Aerosol OT	11.4	0.863	9.84
47.5 47.5 3.0	Dallawax Black, 190°F. Ink Oil Aerosol OT	12.5	0.848	10.60

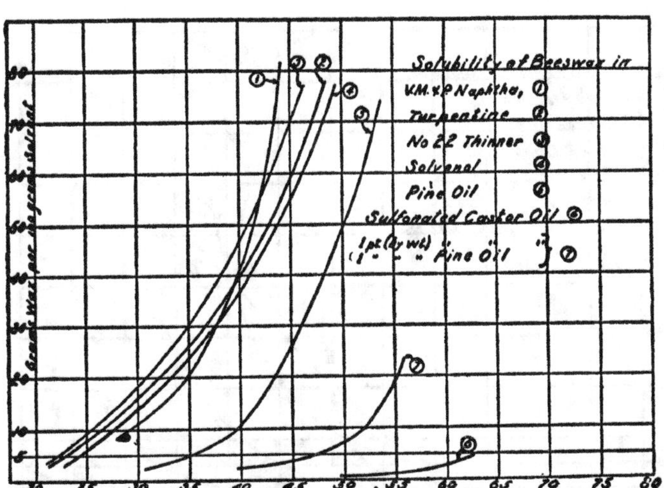

Figure 42. SOLUBILITY OF BEESWAX

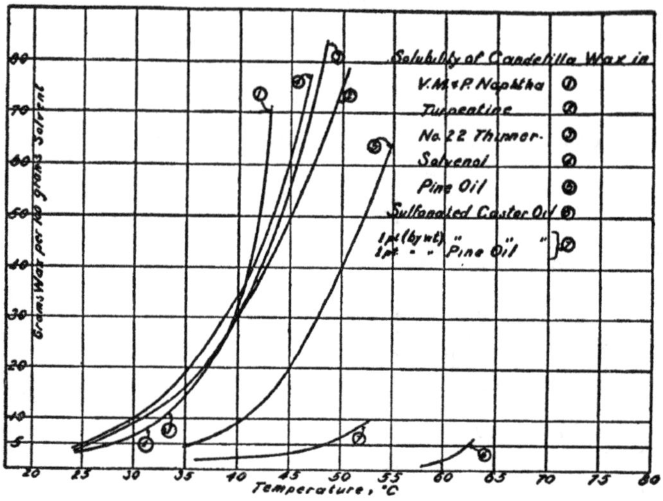

Figure 43. SOLUBILITY OF CANDELILLA WAX

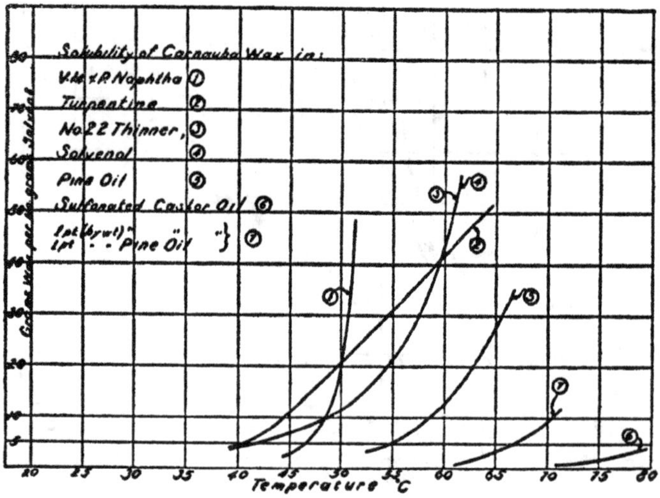

Figure 44. SOLUBILITY OF CARNAUBA WAX

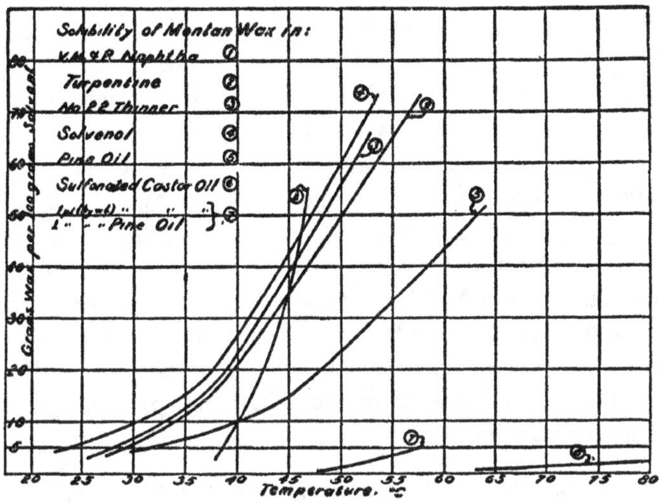

Figure 45. SOLUBILITY OF MONTAN WAX

SOLUBILITY OF WAXES[1]

Variations in data on wax solubilities maybe due to impurities different grade or source, temperature or other factors.

The waxes used in the experiments illustrated in Figures 41-44 were as follows:

Beeswax. Commercial grade; yellow, unbleached; setting point 62°C.

Candelilla wax. Commercial grade; light yellow; setting point 66°C.

Carnauba wax. Commercial grade; light gray with slight greenish tint; setting point 82°C.

Montan wax. Commercial grade; dark brown; setting point 69°C.

The following solvents were used in these experiments:

1. V. M. and P. naphtha. Standard grade petroleum hydrocarbons.
2. Turpentine. Standard, steam-distilled, wood turpentine.
3. No. 22 Thinner. Industrial solvent carrying approximately 55% terpene hydrocarbons and 45% gasoline.
4. Solvenol. Industrial solvent, consisting of approximately 97% terpene hydrocarbons and 3% saturated hydrocarbons.
5. Pine oil. Standard grade of Yarmor pine oil.
6. Sulfonated castor oil. Industrial grade carrying 1.46% total sulfur equivalent to 3.65% SO_3.
7. Equal parts, by weight, of pine oil 5 and sulfonated castor oil 6.

[1] O. A. Pickett, *Ind. Eng. Chem.* **21**, 767 (1929).

It is evident that solubility curves 1, 2, 3, and 4 are all relatively close together, with 5 somewhat lower. At lower temperatures, Solvenol, No. 22 Thinner, and turpentine are all better solvents for the waxes tested than V. M. and P. naphtha, whereas at higher temperatures, the naphtha is the best solvent.

Curves marked 6 show that sulfonated castor oil is a relatively poor solvent for waxes. Yet, it is known to be a good emulsifying agent in the dewaxing of textile fibers. Mixtures of pine oil and sulfonated castor oil are extensively used for such purposes; curves marked 7 partly explain why such mixtures are increasing in favor. The addition of pine oil increases the solubility of the wax, without appreciably decreasing the emulsifying power of sulfonated castor oil, since pine oil itself is a good emulsifying agent.

SOLUBILITY OF WAXES IN DIFFERENT SOLVENTS[1]

The behavior of candelilla wax, ozokerite and crude montan wax blended with paraffin wax suggests that these waxes contain large amounts of long-chain hydrocarbons which, being

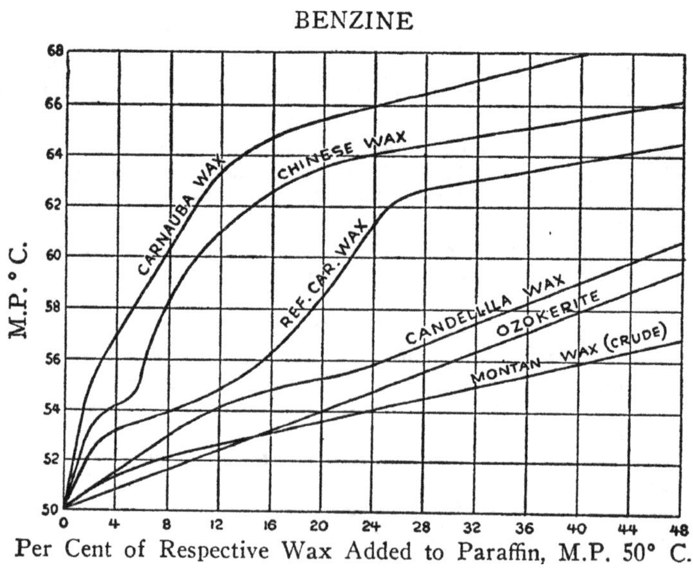

Figure 46. SOLUBILITY OF WAX BLENDS IN BENZINE

[1] C. F. Mason, *Chem. Industries* **38,** 254 (1936).

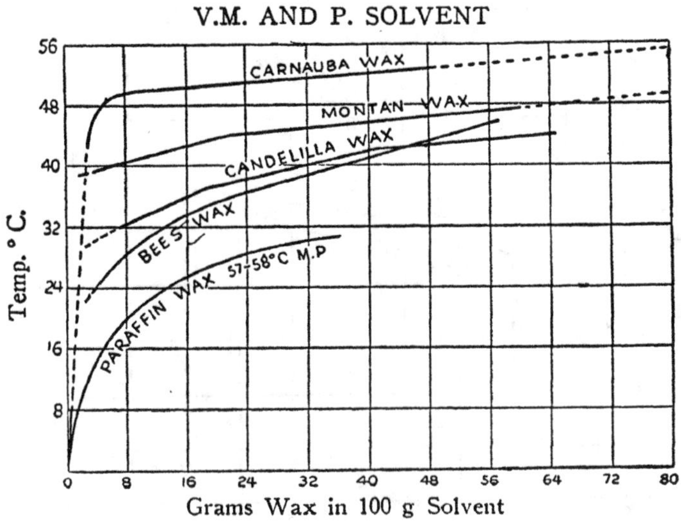

Figure 47. SOLUBILITY OF WAXES IN V. M. AND P. NAPHTHA

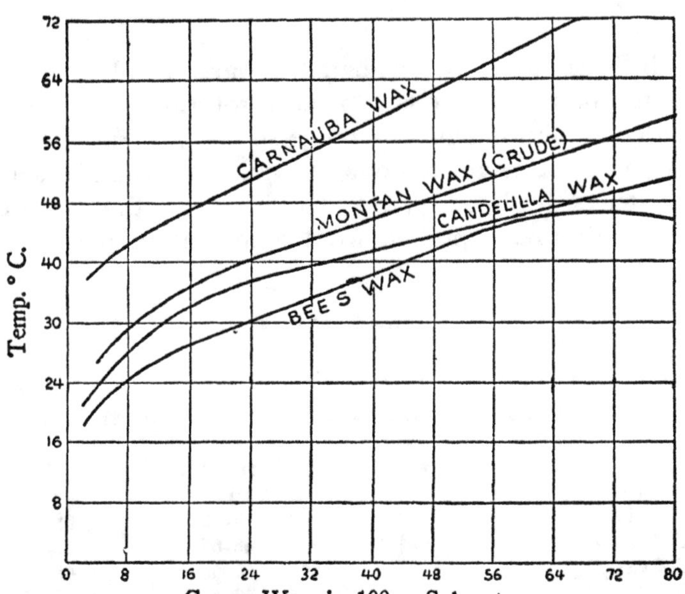

Figure 48. SOLUBILITY OF WAXES IN TURPENTINE

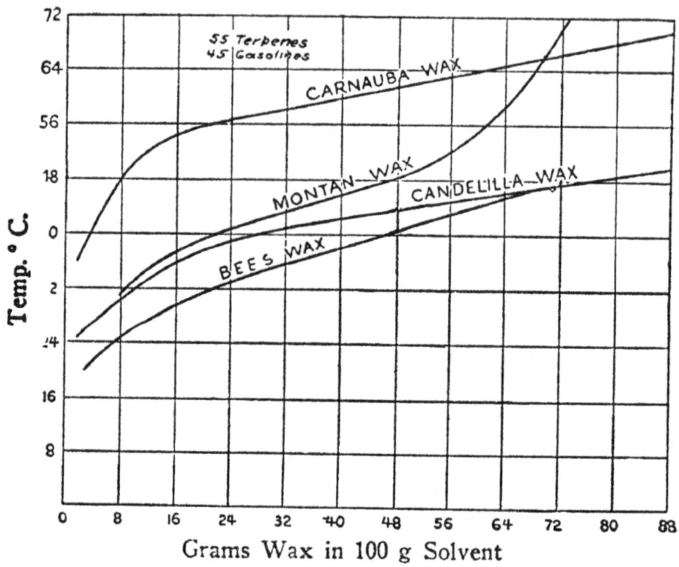

Figure 49. SOLUBILITY OF WAXES IN NO. 22 THINNER

similar in type to those of paraffin wax, dissolve mutually, obeying approximately the laws of ideal solutions. The presence of no eutectic mixtures in this range shows how futile the past attempts have been to compound waxes in volatile solvents which, after spreading, would dry bright without rubbing. A eutectic mixture, deposited from a solvent in a continuous

Table 121

CRYSTALLIZATION POINTS OF 10% WAX SOLUTIONS FROM RESPECTIVE SOLVENTS (SYNTHETIC WAXES)

Wax	Turpentine °C.	Heavy Benzine °C.
Rilan Wax	58–59	68–69
IG Wax Z	66–67	66–67
IG Wax C R	45–46	48–49
IG Wax B J	43–44	69–70
IG Wax S	47–48	55–56
IG Wax E	54–55	59–60
Shellac Wax	38–39	42–43

film, without crystals or amorphous particles which would reflect the light at one angle, would be the answer.

Comparing these values with those in the graphs of Figures 45-48 for 10% concentrations, Rilan Wax, IG wax E and Z are superior to carnauba in respect to the temperature of solution, and although they come out of solution at a higher temperature than carnauba wax, they do so in a finely divided colloidal or gelatinous condition which does not subside and, in addition, retards the settling of other natural waxes present. This is beneficial in making wax gels of low concentration.

Various, two-component wax blends were dissolved in benzine (25% wax content), and the percentage loss of the solvent was measured after 30 days of exposure to the atmosphere, under conditions identical with those in Table 121. The data are listed in Table 122.

Table 122

Paraffin Wax %	100	99.5	98	95	90	80	50	25	0
Added Wax %	0	0.5	2	5	10	20	50	75	100
Paraffin Wax M.P. 48 and 61°C.	34	—	35	—	40	—	50	—	66
Paraffin Wax 53°C. Ozokerite 77°C.	58	36	25	31	33	35	63	74	79
Paraffin Wax 53°C. Ozokerite 79°C.	58	—	—	—	20	30	—	—	83
Paraffin Wax 61°C. Ozokerite 79°C.	66	—	—	—	75	77	—	—	83
Paraffin Wax 53°C. Beeswax	58	—	—	—	25	—	10	—	88
Paraffin Wax 61°C. Beeswax	66	—	—	—	79	26	—	—	88
Paraffin Wax 48°C. Montan Wax A	34	—	—	—	50	—	38	—	49
Paraffin Wax 61°C. Montan Wax	66	—	—	—	64	—	54	—	48
Paraffin Wax 61°C. Carnauba Wax	58	—	46	—	—	—	50	—	87

The upper horizontal columns contain the percentage of paraffin wax and the percentage of the added wax; and under each, the figures, opposite the names of the ingredients, represent the percentage of benzine, lost by exposure for 30 days. Some values are very striking, e.g., the low loss of 20% solvent corresponding to 90% paraffin and 10% ozokerite, and the loss of 26%, corresponding to 80% paraffin and 20% beeswax. These experiments cannot be considered analogous to the behavior in a film, but they are indicative of how tenaciously certain mixtures retain the solvent.

Table 123

SOLUBILITY OF WAXES AT 25°C.
(Grams per 100 grams of solvent)

Wax	Naphtha (Distillation Range 300–400°F.)	Toluene
Acrawax	0.55	~0.00
Acrawax B	0.10*	1.20
Acrawax C	0.21	0.42
Beeswax USP	3.50	4.50
Candelilla Wax	3.60	> 1.00*
Carnauba Wax No. 2	0.30	0.40
Flexowax C	4.20	4.30
Light Flexowax C	4.40	4.00
Glycowax S932	0.90	> 5.00
Japan Wax	4.50	> 5.00
Bleached Montan Wax	2.40	2.30
Bleached Ozokerite	4.60	4.40
Paraffin Wax	> 5.00	> 5.25
Strobawax	0.40*	4.20
Wax S688	4.75	4.90

* Gels in 5% solution.

COMPATIBILITY OF PARAFFIN WAXES WITH OTHER SUBSTANCES

The compatibility of waxes varies greatly and there are only some general conclusions which will be given as follows.

Petroleum waxes are, in general, miscible with fatty waxes, such as carnauba and candelilla wax, if held in the molten condition. The solubility of vegetable waxes, particularly of those of high melting point, is quite low if the temperature of the molten wax is only slightly above its melting point. Long-time storage of such a mixture, at a temperature intermediate between the melting points of the two components, will cause crystallization of the higher-melting component and its alternate deposition. In the solid state, a two-phase system exists, with crystals of the higher-melting wax interspersed throughout the mass of the petroleum wax. This is probably responsible for the pronounced effect on hardness of carnauba wax when blended with paraffin wax. It will be found that if the higher-melting is entirely soluble in the lower-melting wax, it will have much less effect on the properties of the blend than if it is only slightly soluble as is carnauba wax. For example, if a petroleum wax of high melting point is blended with a lower-melting paraffin wax, the blend will be changed relatively little from the original paraffin, unless, of course, large quantities of the higher-melting wax are used.

A great many resins, both natural and synthetic, have been studied as blending agents for petroleum waxes. It will be found that many of the natural resins, including rosin, burgundy pitch and others are quite miscible with wax. Such mixtures are somewhat unstable in that deposition of the resin will frequently occur as a result of crystallization or oxidation. In the case of rosin, this can be somewhat overcome by the utilization of hydrogenated rosins, ester gum, or Acrawax C. It will be found once again that if the additive is to have any appreciable effect on the pertoleum wax, its solubility must not be too good.

Synthetic resins present a far more complex problem in that their chemical compositon is far more diversified than that of natural resins. It will be found, in general, that resins which contain considerable amounts of oxygen in the molecule are soluble with difficulty or insoluble. Those, which are composed largely of hydrocarbons, will be found to be compatible. In the second group are such products as polymerized terpenes, rubberlike polymers, such as Vistanex, certain vinyl chloride resins, etc. The solution of resins in molten wax is very closely related to the solution of resins in conventional solvents, and the same general rules will hold. It should be remembered, however, that petroleum waxes are quite paraffinic in nature and thus have much poorer solvency than do ordinary solvents. The well-established rule that materials of similar composition are miscible with one another is true, of course, also of compatibility.

The main purpose in adding a resin to a wax is to modify the properties of the wax. This modification might be an increase in melting point, improvements in ductility, greater adhesiveness, an increase in viscosity, etc. Once again, if the resin is too soluble, its effect will be neglible unless abnormally large quantities are used.

COMPATIBILITY OF MIXED FATTY ACIDS

Arranged in order of compatibility with resins, rubbers, etc., the fatty acids would line up approximately as follows: linseed, soybean, oleic, corn, cotton and saturated fatty acids, waxes.

The fatty acids are compatible with most of the synthetic resins before polymerization, at which time, they are sometimes added as modifiers. After polymerization, the resins are apt to be oil-resistant, the resistance being as follows:

Excellent: acrylate, methacrylate, phenol-formaldehyde, polyvinyl, urea-formaldehyde.

Good: casein, cellulose esters.

Poor: polystyrene, chlorinated rubber.

Resistance of the resins to fatty acids would be expected to be somewhat less good than to the neutral oils.

Although temperatures as high as 250°C. are sometimes required, all the natural resins except accroides and sandaric are completely compatible with the unsaturated oils and, therefore, would be compatible with the fatty acids.

The amount of fatty acids, milled into natural and synthetic rubbers, is usually not more than 5%; but higher quantities are compatible.

Table 124

COMPATIBILITY[1] OF WAXES WITH OTHER SUBSTANCES

Wax	Other Substance	Proportions[2]
Acrawax	Acrawax B	1:1
	Albacer	1:1
	Beckacite No. 1100	9:1
	Beckacite, Super No. 2000	9:1
	Candelilla Wax	2:1
	Carnauba Wax	1:1;1:99
	Cetyl Alcohol	3:2;1:1

[1] The compatibilities in this section refer to wax compositions at normal temperatures. Certain substances may be compatible at higher temperatures, but will separate on solidification of the wax.

[2] The first figure in the ratio indicates the proportion of wax and the second figure that of the other substance (which may also be a wax). The proportions given do not necessarily mean that other proportions are not compatible. They are only indicative that these proportions have been tried and found compatible.

Table 124 (Continued)

Wax	Other Substance	Proportions
	Congo Gum, Fused	9:1
	Diolin	4:1
	Ester Gum	1:1
	Ester Gum, Mannitan	1:1
	Ester Gum, Sorbitan	1:1
	Ethyl Cellulose	3:1
	Manila Loba A	9:1;4:1
	Monostearin	1:1
	Pentalyn	1:1
	Rosin	1:1
	Santowax M	1:9;1:1
	Shellac, Dewaxed	1:1;9:1
	Stearic Acid	1:1
	Teglac Z152	9:1;1:1
Acrawax B	Acrawax	1:1;9:1
	Albacer	1:1
	Beckacite No. 1100	1:1;9:1
	Beeswax	1:1
	Candelilla Wax	1:1
	Carnauba Wax	1:1;99:1
	Congo Gum, Fused	1:1;9:1
	Cumar	1:1;9:1
	Dammar	9:1
	Ester Gum	1:1;4:1
	Ester Gum, Mannitan	1:1
	Ester Gum, Sorbitan	1:1
	Ethyl Cellulose	3:1
	Flexo Wax C	1:1;3:1
	Glyceryl Tristearate	1:1
	Manila Gum	1:1;9:1
	Manila Loba A	4:1
	Microcrystalline Wax	1:1;1:3
	Monostearin	1:1
	Nypene Resin	3:1
	Paraffin Wax	1:1;1:9;2:1
	Paranol, Extra Hard	9:1
	Pentalyn	1:1
	Piccolyte S100	1:1
	Rosin	1:1
	Shellac, Dewaxed Orange	1:1;9:1
	Spermaceti	
	Teglac Z152	1:1
	Tristearin	1:1
Acrawax C	Aluminum Stearate	*
	Candelilla Wax	1:1
	Carnauba Wax	1:1

* Where no proportion is given, compatibility in certain proportions does exist, but the exact amount has not been noted.

Table 124 (Continued)

Wax	Other Substance	Proportions
	Ester Gum, Mannitan	1:1
	Ester Gum, Sorbitan	1:1
	Halowax	
	Microcrystalline Wax	1:1
	Nypene Resin	3:1
	Paraffin Wax	1:9½;1:2;1:1;2:1;9:1
	Pentalyn	1:1
	Phenolformaldehyde Resins	All proportions
	Piccolyte Resin S100	1:1
	Polymer No. 34	1:1
	Polyvinyl Butyral	
	Rosin	1:1
	Santowax P	3:1
	Scale Wax	1:9
	Teglac Z152	1:1
	Tristearin	1:1
	Vinylite	
Albacer	Acrawax	1:1;4:1
	Acrawax B	1:1
	Beeswax	9:1;7:3;1:1
	Candelilla Wax	7:1;4:1
	Carnauba Wax	9:1;7:3;1:1
	Cetyl Alcohol	4:1;5:1
	Congo, Fused	4:1
	Cumar	9:1;7:3;1:1
	Dammar	4:1
	Ester Gum	4:1
	Japan Wax	9:1;7:3;1:1
	Montan Wax	9:1;7:3;1:1
	Paraffin Wax	1:1;1:3;1:9
	Teglac Z152	9:1;7:3;1:1
Bayberry Wax	Beeswax	
	Cumarone Resins	
Beeswax	Acrawax B	1:1
	Bayberry Wax	
	Carnauba Wax	1:2;1:3;1:4;1:6
	Ceresin	
	Chinese Insect Wax	
	Chlorinated Paraffin	All proportions
	Cumarone Resins	$<30\%$
	Dammar	
	Diglycol Stearate S	
	Elemi	
	Ethyl Cellulose	$<50\%$
	Glyceryl Monostearate	
	Japan Wax	
	Methacrylate Polymers	

Table 124 (Continued)

Wax	Other Substance	Proportions
	Ozokerite	
	Paraffin Wax	1:1
	Petrolatum	3:97;1:4
	R. H. Resin	< 10%
	Rosin	
	Spermaceti	
	Staybelite Esters	
	Stearic Acid	
	Tallow	
	Wool Wax	
Candelilla Wax	Acrawax	1:2
	Acrawax B	1:1
	Albacer	1:9;1:7;1:4;1:1
	Cumarone Resins	
	Ethyl Cellulose	1:13;1:4
	Ouricury Wax	3:7;2:3;1:1;1:2;1:4
	Paraffin Wax	3:1;4:1;6:1;8:1;9:1
	Piccolyte Resin	4:1
	Santowax M	1:9;1:3
	Staybelite Resin	95:5
Carbowax	Bakelite XR–9392; BR3360	1:1;1:5
	Nitrocellulose	1:5
	Vinylite Resin AYAF	9:1
Carnauba Wax	Abalyn	
	Acrawax	1:1;99:1
	Acrawax B	
	Albacer	1:1;3:7;1:9
	Beeswax	2:1;3:1;4:1;6:1
	Candelilla Wax	2:1;1:4
	Cetyl Alcohol	All proportions
	Cumarone Resins	< 30%
	C. V. Wax	1:1
	Dammar	
	Durez Resin 409	
	Elemi	
	Esparto Wax	1:4;2:1;5:1
	Ethyl Cellulose	1:7;1:13;1:27;2:1;3:1
	Flexowax C	8:1;15:1
	Glyceryl Monolaurate	2:3
	Glyceryl Tristearate	
	Glyco Wax S932	2:23
	Hercolyn	
	Hydrogenated Vegetable Oils	2:23
	Methacrylate Polymers	
	Montan Wax	3:1;1:1;1:2
	Nypene	
	Ouricury Wax	3:2;4:1

Table 124 (Continued)

Wax	Other Substance	Proportions
	Ozokerite	2:1;3:1;15:1
	Paraffin Wax	1:1;1:5;4:1;5:1;6:1;15:1
	Petroleum, Dark Green	4:1;3:2;2:3;1:4
	Polymer No. 34	1:4
	Santowax M	1:9;1:1
	Staybelite Esters	1:1
	Staybelite Resin	97½:2½
	Stearic Acid	1:1;2:1
	Sugar-Cane Wax	1:3
	Utah Wax No. 1	1:2
	Utah Wax No. 4, Yellow	4:3;7:2;5:2;3:2;2:1
Ceresin (See also Paraffin Wax)	Carnauba Wax	
	Cumarone Resins	< 30%
	Paraffin Wax	
	Rosin	
	Rubber	
	Rubber Resins	
	Staybelite Esters	
Cetyl Alcohol	Acrawax	2:3;1:1
	Albacer	1:4;1:5
	Carnauba Wax	All proportions
	Cumarone Resins	
	Ethyl Cellulose	
	Opalwax	1:4;1:6
	Ouricury Wax	1:1;3:1
	Paraffin Wax	1:5
	Piccolyte Resin	
	Rubber Resins	
Chinese Insect Wax	Beeswax	
	Paraffin Wax	
	Rubber, Depolymerized	
Chlorinated Naphthalene	Acrawax C	
	Butyl Stearate	
	Chlorinated Diphenyl	
	Cumarone Resins	< 30%
	Gilsonite	
	Naphthalene	
	Polymerized Isobutylenes	
	Resins	
	Rosin	
	Rubber	
	Rubber, Depolymerized	
	Stearic Acid	
	Sulfur	
	Vinylite Resin AYAF	9:1
Dammar Wax	Paraffin Wax	1:3
Diglycol Stearate	Beeswax	
	Cocoa Butter	

Table 124 (Continued)

Wax	Other Substance	Proportions
	Paraffin Wax	
	Spermaceti	
Fiber Wax	Carnauba Wax	1:4;2:3;3:2;4:1
(Esparto Wax)	Ethyl Cellulose	1:99;1:9
	Montan Wax	3:1;1:1
	Ouricury Wax	1:2;1:1;3:1;4:1;9:1
	Paraffin Wax	1:1;3:1;1:3
	Petrolatum, Dark Green	4:1;3:2;2:3;1:4
	Utah Wax No. 1	4:1
	Utah Wax No. 4	3:1
Flexo Wax C	Acrawax B	1:1;1:3
	Carnauba Wax	1:8;1:15
	Chlorinated Paraffin	All proportions
	Hercolyn	9:1
	Lanolin	
	Microcrystalline Wax	All proportions
	Nevillite Resin No. 1	1:1
	Nevindene Resin	4:1
	Paraffin Wax	All proportions
	Petrolatum	All proportions
	Piccolyte Resin S100	3:7;7:3;1:1
	Protowax	All proportions
	Record Wax, Light	1:1
	Rosin	4:1
	Staybelite	4:1
Glyceryl Mono- stearate	Beeswax	
	Cocoa Butter	
	Ethyl Cellulose	3:1
	Glycol Monostearate	
	Paraffin Wax	
Glyceryl Tristearate	Acrawax B	1:1
	Acrawax C	1:1
	Carnauba Wax	
	Ethyl Cellulose	
Glyco Wax S932	Acrawax B	
	Carnauba Wax	23:2
	Cocoa Butter	
	Ethyl Cellulose	3:1
	Japan Wax	5:4;4:5
Halowax (See Chlorinated Naphthalene)		
Hydrogenated Oils	Acrawax B	1:1
(See also	Carnauba Wax	23:2
Opalwax)	Cumarone Resins	
	Ethyl Cellulose	
	Naphthalene Tetrachloride	
	Paraffin Wax	1:100

Table 124 (Continued)

Wax	Other Substance	Proportions
	Piccolyte Resin	
	Rosin	
	Stearone	
IG Wax OP	Paraffin Wax	1:1
IG Wax S	Paraffin Wax	
IG Wax Z	Paraffin Wax	
	Stearic Acid	
Japan Wax	Beeswax	
	Cocoa Butter	
	Cumarone Resins	
	Dammar	
	Elemi	
	Ethyl Cellulose	30%
	Glyco Wax S932	5:4;4:5
	Rosin	
	Santowax M	1:9
	Tallow	
Microcrystalline Wax (See Paraffin Wax)		
Monostearin	Acrawax	1:1
Montan Wax	Carnauba Wax	1:3;1:1;2:1
	Cumarone Resins	30%
	Dammar	
	Elemi	
	Esparto Wax	1:3;1:1
	Ethyl Cellulose	$<50\%$
	Nypene Resin	
	Paraffin Wax	3:97;2:3;3:1
	Staybelite Resin	17:3
	Stearic Acid	
	Tar	
Opalwax (See also Hydrogenated Oils)	Behenone	5:1
	N-Butyl Methacrylate	
	Carnauba Wax	
	Ceresin	
	Cetyl Alcohol	4:1;6:1
	Cumarone Resins	
	Diglycol Laurate	6:1
	Ester Gum	
	Ethyl Cellulose	
	Montan Wax	
	Naphthalene Tetrachloride	5:1
	Ozokerite	
	Paraffin Wax	1:1;4:1
	Rosin	
	Santowax M	1:9;1:3

Table 124 (Continued)

Wax	Other Substance	Proportions
	Staybelite Esters	1:1
	Staybelite Resin	92½:7½
	Stearone	5:1
Ouricury Wax	Candelilla Wax	7:3;3:2;1:1;2:1;4:1;5:1; 9:1
	Carnauba Wax	2:3;1:4
	Cetyl Alcohol	1:1;1:3
	Esparto Wax	2:1;1:1;1:3;1:4;1:9
	Microcrystalline Wax	3:2;1:1;2:1
	Ozokerite	15:1
	Paraffin Wax	4:1;3:1;1:1
	Utah Wax No. 1	5:3;6:1;2:1;1:1
Ozokerite	Beeswax	
	Carnauba Wax	4:1;1:2;1:3;1:15
	Cumarone Resins	< 30%
	Dammar	
	Elemi	
	Ouricury Wax	1:15
	Paraffin Wax	All proportions
	Rubber	
	Staybelite Esters	
	Rubber Resins	
Palmitic Acid (See also Stearic Acid)	Piccolyte Resin	
Paraffin Wax	Abalyn	1:1
	Acrawax B	9:1;1:1;1:2
	Acrawax C	1:9½;1:2;2:1;9:1
	Albacer	1:1;3:1;9:1
	Aluminum Stearate	2½–4½%
	Aroclor	
	Asphalt, Blown	
	Balsam, Oregon	
	Beeswax	
	Bitumens	
	Candelilla Wax	97:3;3:2
	Carnauba Wax	1:1;6:4;97:3
	Cetyl Alcohol	5:1
	Chinese (Insect) Wax	97:3;3:2
	Chlorinated Paraffin	All proportions
	Cumarone Resins	< 30%
	Dammar Resin	
	Dammar Wax	3:1
	Dammar Wax Residue	9:1;3:1
	Diglycol Stearate	
	Durez Resin No. 409	
	Durez Resin No. 51185	
	Durez Resin No. 219	

Table 124 (Continued)

Wax	Other Substance	Proportions
	Esparto Wax	1:1;1:3;3:1
	Ester Gum	9:1
	Glyceryl Monostearate	
	Hercolyn	50% or more
	Hycar OR	<2%
	Hydrogenated Oils	100:1
	IG Wax OP, S, Z	1:1
	Jelutong	
	Kavri Resin	
	Lanolin	
	Lard	
	Lead Stearate	3%
	Lewisol Rosin 21	
	Limed Resin	1:8.9
	Methacrylate Polymers	
	Microcrystalline Wax	All proportions
	Montan Wax	97:3;3:2;1:3
	Nevtex Resin No. 10	9:1;3:1;1:1
	Nuba (Resinous Pitch)	
	Nypene Resin	
	Opalwax	1:1;1:4;5:1
	Ouricury Wax	2:3;1:1;1:2
	Ozokerite	All proportions
	Petrolatum, Dark Green	2:3;1:4
	Piccolyte Resin	1:1;3:7;19:1
	Pliolite	
	Polybutenes	
	R. H. No. 4	1:3;3:1
	R. H. Resin	3–4%;3:1
	Rubber	
	Rubber, Chlorinated	
	Rubber, Depolymerized	
	Rubber Resins	
	Santowax M	1:1;3:1;19:1
	Shellac Wax	5:1
	Staybelite Esters	
	Staybelite Resin	1:1;19:1;97½:2½
	Stearic Acid	
	Stearone	1:9
	Velsicol Resins	All proportions
	Vistanex	19:1
Protowax (See Paraffin Wax)		
Santowax M	Acrawax	1:9;1:1
	Candelilla Wax	1:9;1:3
	Carnauba Wax	1:9;1:1
	Ester Gum	1:9;1:3

Table 124 (Continued)

Wax	Other Substance	Proportions
	Japan Wax	1:9
	Manila Resin	1:9
	Opalwax	1:9;1:3
	Paraffin Wax	< 2%
	Piccolyte	1:9
	Rosin	1:9;1:3
	Shellac	1:9
Santowax O	Beeswax	1:9;1:3;1:1
	Opalwax	1:9
	Paraffin Wax	All proportions
	Rosin	1:9;1:3
Santowax P	Acrawax C	1:3
	Paradene No. 1	1:19
Scale Wax (See Paraffin Wax)		
Spermaceti	Beeswax	
	Cumarone Resins	
	Ethyl Cellulose	< 50%
	Stearic Acid	
Stearic Acid	Acrawax	1:1
	Aluminum Stearate	
	Beeswax	
	Carnauba Wax	1:1;1:2
	Cumarone Resins	
	Ethyl Cellulose	1:1;< 50%
	Gum Elemi	
	Halowax	
	Hycar OR	< 2%
	Montan Wax	
	Paraffin Wax	
	Piccolyte Resin	
	R. H. Resin	< 10%
	Spermaceti	
Stearyl Alcohol	Cumarone Resins	
	Ethyl Cellulose	
	Piccolyte Resin	
Sugar-Cane Wax	Carnauba Wax	1:1;2:1;3:1
Tristearin (See Glyceryl Tristearate)		
Utah Wax	Carnauba Wax	2:1;1:2
	Esparto Wax	1:4
	Montan Wax	
	Ouricury Wax	3:5;1:6;1:2;1:1
	Ozokerite	
	Paraffin Wax	
	Rosin	

COMPOUNDED WAXES

Wax compositions, containing different waxes and/or other substances, are often employed for special effects, not obtainable otherwise. For example, in making paper milk-bottle caps or hoods, a composition of paraffin wax, carnauba wax and rosin is used.

Paraffin wax is extremely slow setting, when employed alone, is rather greasy and adheres to glass. Rosin sets fairly rapidly and is notably adhesive and sticky. Carnauba wax sets quickly and is substantially nonadhesive. It is surprising to find that the sticky qualities of rosin and the slow-setting properties of paraffin wax are so considerably modified by their interaction and that carnauba or another high-melting wax, even in the small proportion of about 10%, introduced into the mixture of rosin and paraffin, causes mild adhesion, substantial absence of greasiness, quick setting, strong binding effects, and excellent water resistance.[1]

[1] Ellis, U. S. Patent 1,987,288 (1935).

Table 125

DATA ON PARAFFIN WAX—CERESIN—RESIN MIXTURES

Resin	0	20% WW Rosin	20% Stay-belite	20% Resin P–88	20% Ester Gum	20% Hard Ester Gum
Paraffin Wax (Imported, 133–135°F.)	140.0	140.0	140.0	140.0	140.0	140.0
Ceresin Wax	1.0	1.0	1.0	1.0	1.0	1.0
Softening Point °F., Regular*	138.0	129.4	131.0	133.6	133.6	132.8
Softening Point °F., Sloping†	138.6	136.1	136.6	137.0	136.6	136.8
Wax Hardness, 70°F.†	10.0	6.0	7.0	6.0	6.0	6.0
Wax Hardness, 90°F.†	15.0	15.0	18.0	14.0	17.0	18.0
Wax Hardness, 100°F.†	50.0	48.0	89.0	58.0	56.0	51.0

* A.S.T.M., Ball and Ring method.
† Penetrometer.

Table 125 (Continued)

Resin	0	20% WW Rosin	20% Stay-belite	20% Resin P–88	20% Ester Gum	20% Hard Ester Gum
Tensile Strength, 90°F. in lb/¼ sq in.	45.6	56.5	75.6	49.8	69.2	74.3
Stretch, 90°F. in mm	–0.5	0.5	1.0	0.5	0.75	0.75

The following data were obtained from commercially impregnated milk bottle caps or hoods, using sulfite or kraft paper of about 0.025 in. thickness.

Wax % in Caps	83.8	88.6	87.6	86.6	83.3	86.6

Immersion in Water

16 hours 32°F. then 2 hours 70°F.

% Absorbed Water	18.4	14.9	18.4	18.2	21.6	20.5
Flare 18 hours, mm‡	1.2	1.2	1.8	1.7	1.8	2.0
Cap Hardness Test D	21.2	28.1	26.1	23.7	25.2	25.0
Cap Hardness Test B	10.4	17.9	15.7	14.0	12.9	12.5
Cap Hardness Test C	37.9	50.3	49.4	41.6	46.2	43.9
Cap Hardness Test DB	15.8	23.0	20.9	18.9	19.1	18.8

16 hours 32°F. then 2 hours 90°F.

% Absorbed Water	21.8	21.2	27.4	23.5	23.8	23.5
Flare 16 hours (mm)	1.1	0.5	1.0	1.0	1.0	1.0
Flare 18 hours (mm)	2.1	2.0	3.0	2.9	2.8	3.0
Cap Hardness Test A	24.0	23.0	13.8	23.2	19.0	18.0
Cap Hardness Test B	9.0	8.6	5.8	8.5	6.8	6.4
Cap Hardness Test C	34.2	37.9	20.9	36.3	30.3	27.8
Cap Hardness Test AB	16.5	15.8	9.8	15.9	12.9	12.2

‡ Flare refers to linear expansion of cap after immersion test.

WAX–ROSIN COMPOSITIONS

According to concentration and temperature, rosin-wax mixtures may be fluid, solid solutions, or mixtures in which there is a balance between solid solution and pure wax. Up to 25% wax, solid solutions are formed, which may be supersaturated with wax by quick chilling. The wax molecules gradually diffuse through the solid rosin, forming a turbidity which may be estimated photometrically; this decreases the ductility. The "life" of the supersaturated solution increases with time of heating, mixing and thickness of the preparation, but decreases with the concentration.[1]

Table 126

COMPATIBILITY OF WAXES WITH STAYBELITE RESIN

Wax	Solubility as Estimated by Transparency %	Insolubility as Shown by Hazy to Opaque Wax %	Compatibility Waxy Character Predominant %	Incompatibility Cracks and Checks Develop %
Paraffin Wax	2.5	2.5 to 20	20 to 40	40 to 95
Opalwax	7.5	7.5 to 50	—	50 to 95
Carnauba Wax	2.5	2.5 to 15	15 to 60	60 to 95
Candelilla Wax	5.0	5 to 70	70 to 95	—
Crude Montan Wax	15	15 to 40	40 to 70	70 to 95

CUMAR (CUMARONE–INDENE RESINS) AND WAX COMPOSITIONS

Many useful combinations of Cumar with both natural and synthetic waxes have found application in the electrical industry, rubber manufacture, textileproofing, printing ink and other arts. Cumar has been used, for example, to impart quick setting to wax compounds; increase the melting point of certain waxes; improve the adhesion of wax mixtures to paper or textile fibers; impart gloss; improve alkali resistance; and contribute desirable electrical properties.

[1] G. Jahn, *Kolloid, Beihefte* **13**, 213 (1921).

In very few cases are blends of Cumar with waxes transparent and the homogeneity or stability of a blend must be determined by observing the physical properties of the mix. Homogeneous blends of Cumar and waxes depend very often on the use of optimum proportions which will vary with the nature of the wax used. Where difficulty is experienced in blending certain waxes with Cumar, the mixture is often stabilized by the introduction of a suitable third component.

Noticeable separation of Cumar–wax mixtures from molten blends is more likely to occur in large masses, which cool slowly, than in small quantities, which can be chilled quickly. Stirring during cooling, while the mixture is still mobile, promotes the formation of a homogeneous mixture.

Blends of the (V or W) grades of Cumar were made with the following waxes and homogeneous blends were obtained.

Bayberry Wax	Lanette Wax
Beeswax	Lanolin (Anhydrous)
Carnauba Wax*	Montan Wax
Candelilla Wax	Opalwax
Ceresin Wax	Ozokerite (Refined)
Cetyl Alcohol	Paraffin Wax**
Halowax	Rilan Wax
Fatty Alcohols	Spermaceti
Hydrofol (Hydrogenated Oil)	Stearic Acid
Japan Wax	Stearyl Alcohol

* Small percentages of carnauba wax may be blended with Cumar.

** Blends of Cumar and paraffin can be made in the proportion of 70 (or more) parts of Cumar with 30 (or less) parts of paraffin. To stabilize blends of Cumar with paraffin, rosin, beeswax, metallic resinates, chlorinated diphenyl, or other mutual solvents may be used.

COMPATIBILITY OF VARNISH RESINS WITH WAXES

Phenol-modified resins, as well as ester gum and a fused congo, have been tested with various waxes and have been found to be compatible. However, when this mixture is subjected to continuous heating, as is done in practical applications, there is a separation. It is believed that this is due to oxidation of the resin as the oxidation products are no longer soluble in the wax and precipitate. Therefore, resins have been produced in which the oxidizable portions are reduced to a minimum either by hydrogenating or polymerizing rosin before modification or esterification. This type of resin has been found to be much more compatible and to blend readily with paraffin, carnauba, candelilla, beeswax, montan, ozokerite and other waxy substances.

Table 127

COMPATIBILITY OF NATURAL RESINS WITH WAXES[1]

(Ratio of Resin to Wax, 1:1)

Resin	Beeswax	Carnauba	Japan	Montan	Ozokerite	Paraffin	Petrolatum Wax and Paraffin*	Stearic Acid
Batavia and Singapore Dammar	C–1	C–1	C–1	C–1	C–1	C–2	C–4	C–1,2
Batu	C–2	C–3	C–3	C–3	C–3	C–3	C–4	C–2
Black East India	C–2	C–2	C–2	C–2	C–2	C–2	C–4	C–2
Congo	C–5	C–5	P.C.	C–5	P.C.	P.C.	P.C.	C–3
Elemi	C–1	C–1	C–1	C–1	C–1	C–1	C–1	C–1
Kauri	C–5	C–5	C–5	C–5	P.C.	C–5	C–5	C–2
Pale East India	C–2	C–2	C–2	C–2	C–2	C–3	C–4	C–2
Manila WS	C–5	C–5	P.C.	C–5	P.C.	P.C.	P.C.	C–2
Manila Loba	C–5	C–5	C–5	C–5	P.C.	P.C.	P.C.	C–2

C–1 Compatible at 90–120°C.
C–2 Compatible at 120–150°C.
C–3 Compatible at 150–200°C.
C–4 Compatible at 200–225°C.
C–5 Compatible at 275–325°C.
P.C. Partially compatible.

[1] Booklet of the American Gum Importers' Association.
* 45% Petrolatum wax, 10% paraffin, balance resin.

Table 128

NATURAL–RESIN PARAFFIN–WAX COMPOSITIONS

Resin	Compatibility	M.P. of Mixture, °C.
Batavia Dammar	good	49.0
Pale East India	good	64.0
Black Scraped East India	good	62.0
Batu	good	69.5
Run Boea Medium Dark	not complete	—
Run Congo	not complete*	—
Run Brown Kauri	good	54.0

* Completely compatible if run to a fine melt.

RESIN–WAX COMPOSITIONS

It may be seen from the data in Table 129 that, except in the case of elemi—wax mixtures, addition of resin to the wax usually raises its melting point. Elemi, while lowering the melting point, acts as a plasticizer.[1]

[1] R. W. Allan, *Chem. Industries* **42,** 508 (1938).

Table 129

DATA ON 1:1 RESIN–WAX MIXTURES

Mixture	Combining Tempera-ture, °C.	Loss in Weight, %	Melting Point, °C.
Batavia Dammar B and Beeswax	90–100	0.4	59.5
Batavia Dammar B and Carnauba	90–100	1.0	66.0
Batavia Dammar B and Japan	90–100	0.0	47.0
Batavia Dammar B and Montan	90–100	0.4	76.0
Batavia Dammar B and Ozokerite	90–100	0.0	64.0
Batavia Dammar B and Paraffin	140–150	0.5	49.0
Bold Black Scraped East India and Beeswax	130–140	0.6	65.0
Bold Black Scraped East India and Carnauba	130–140	2.0	82.5
Bold Black Scraped East India and Japan	130–140	1.2	75.0
Bold Black Scraped East India and Montan	130–140	0.4	99.0
Bold Black Scraped East India and Ozokerite	130–140	0.0	73.0
Bold Black Scraped East India and Paraffin	130–140	—	62.0
Pale East India Macassar Bold and Beeswax	120–130	0.4	62.5
Pale East India Macassar Bold and Carnauba	120–130	1.4	75.0
Pale East India Macassar Bold and Japan	120–130	1.0	65.5
Pale East India Macassar Bold and Montan	120–130	0.8	83.0
Pale East India Macassar Bold and Ozokerite	120–130	0.2	77.0
Pale East India Macassar Bold and Paraffin	150–200	—	55.0
Pale East India Singapore Bold and Beeswax	120–130	1.2	62.0
Pale East India Singapore Bold and Carnauba	120–130	1.8	76.0
Pale East India Singapore Bold and Japan	120–130	0.2	80.0
Pale East India Singapore Bold and Montan	120–130	0.6	95.0
Pale East India Singapore Bold and Ozokerite	120–130	0.0	86.0
Pale East India Singapore Bold and Paraffin	150–200	—	64.0
Elemi and Beeswax	98–100	—	56.5
Elemi and Carnauba	100	1.0	57.0

Table 129 (Continued)

Mixture	Combining Tempera-ture, °C.	Loss in Weight, %	Melting Point, °C.
Elemi and Japan	98–100	1.0	45.5
Elemi and Montan	100	1.2	60.5
Elemi and Ozokerite	100	1.0	70.0
Batu Bold Scraped and Beeswax	130–140	0.0	65.0
Batu Bold Scraped and Carnauba	150–200	2.2	85.0
Batu Bold Scraped and Japan	150–200	2.6	75.0
Batu Bold Scraped and Montan	150–200	2.0	127.0
Batu Bold Scraped and Ozokerite	150–200	2.8	74.0
Batu Bold Scraped and Paraffin	150–200	—	69.5
Singapore Dammar No. 2 and Beeswax	100–110	2.6	60.0
Singapore Dammar No. 2 and Carnauba	110–120	0.8	79.5
Singapore Dammar No. 2 and Japan	90–100	0.0	42.0
Singapore Dammar No. 2 and Montan	90–100	0.0	75.0
Singapore Dammar No. 2 and Ozokerite	130–140	0.0	67.0
Singapore Dammar No. 2 and Paraffin	140–150	0.5	50.0
Manila WS and Beeswax	275–325	18.2	69.0
Manila WS and Carnauba	275–325	16.4	80.0
Manila Loba C and Beeswax	300–325	26.3	68.0
Mixture	ture, °C.	%	°C.
Manila Loba C and Carnauba	300–325	17.0	79.0
Congo No. 4 and Beeswax	300–325	18.5	66.0
Congo No. 11 and Beeswax	300–325	17.6	68.0
Congo No. 11 and Carnauba	300–325	21.8	75.5
Kauri Pale No. 1 and Beeswax	300–325	14.3	67.0
Kauri Pale No. 1 and Carnauba	300–325	10.0	81.5
Run Brown Kauri No. 2 and Paraffin	300–325	—	54.0

Natural resins dissolve in waxes with varying degrees of ease. The dammars (Batavia and Singapore dammars, black and pale East India, and Batu), and also elemi, a miscellaneous type, are the only ones that dissolve in the waxes, without being run or at least partially run. The loss of weight is an indication of the relative ease with which solution takes place. Manila resin—Congo—and kauri-wax mixtures lose much more weight in the combination process than do the dammar resins.

A resin, which is soluble with difficulty in wax, may be combined with wax with the aid of a common solvent without marked loss in weight. For example, elemi is easily soluble in all the waxes tested, whereas loba C is not. Loba C is easily soluble in elemi. Elemi may, therefore, be considered a common solvent for the waxes and the resin and will promote their combination. Taking carnauba wax as an example, the Manila (loba C) may be dissolved in the melted elemi and carnauba added to form a ternary mixture in which both resin and waxes are stable.

If liquid resin-wax mixtures are needed, they may be used in the form of aqueous emulsions. When the melting point is sufficiently low, they can be readily emulsified by ordinary methods, using commercially available emulsifying agents. Triethanolamine soaps are effective emulsifiers. The melting point of the mixture may be such, however, that the addition of water, even at the boiling point, will cause the melted compound to solidify, making dispersion impossible. Wax blends containing Manila resins can be emulsified by a different method. The compound is powdered and added slowly to a solution of sodium hydroxide, with stirring and heating. Part of the Manila is saponified. The resulting soap acts as the emulsifying agent. Stirring is continued after all the mixture has been added and the heat removed to insure uniform and complete dispersion.

The natural resins dissolve in various synthetic waxes more or less readily. Dammar and elemi are more compatible. Resins soluble with difficulty are easier to combine by forming a ternary mixture, with elemi as a common solvent for resin and wax, as explained previously.

METHACRYLATE RESIN HOT-MELT BLENDS

Methacrylate polymers do not fuse to a liquid state, when heated above their thermal yield points, but maintain a rubbery consistency over a wide temperature range. For this reason, pure polymethacrylates cannot be used readily as hot melts for coating and adhesives unless dissolved or dispersed. The rubbery consistency can be reduced and the fluidity greatly increased at 150–200°C., by blending with resinous fluxing agents of lower molecular weight, such as rosin, ester gum, dammar and the cumarone-indene resins. By variation of the methacrylate polymer and the nature and concentration of the added fluxing agent, a series of resin blends can be obtained which differ widely in fluidity in the molten state but closely resemble the methacrylate polymers in other properties. The melt viscosity of methacrylate polymers, with a given percentage of fluxing agent, decreases as the molecular weight of the alcohol, esterified with methacrylic acid, is increased. When the blends are used as adhesives, their bond strengths, at elevated temperatures, vary with the nature of the methacrylate polymer and the nature and proportion of the fluxing agent. These blends are heat stable, and do not change considerably on heating for prolonged periods at 150–200°C.

The higher methacrylate polymers, such as butyl or isobutyl, can be dissolved readily in molten waxes, such as paraffin, beeswax and carnauba wax, to produce solutions which, on cooling, form tough, rigid, or flexible masses. The higher-softening methacrylate polymers tend to form harder, more rigid wax blends; the softer, more pliable methacrylates, in general, produce more pliable, elastic blends. The solution temperatures of the wax blends (the temperature at which a clear methacrylate-wax solution is obtained) vary with the methacrylic ester used and with the nature and concentration of the wax. In paraffin wax, for example, the solution temperature decreases as the molecular weight of the alcohol, esterified with methacrylic acid, is increased; and with a given polyester, the solution temperature increases as the percentage of paraffin is increased. The solution temperatures also vary directly with the melting point of the paraffin wax and the molecular weight of the methacrylate polymer. Blending agents of various types can be added to these compositions to reduce the solution temperature and modify the physical properties of the blends at lower temperature. Methacrylate-paraffin blends, with a solution temperature as low as 100°C., can be readily obtained.

These toughened melts are very fluid in the molten state and are suitable for coating by immersion spreading and roller-coating techniques. Paper, coated from these melts, is superior to paraffin-coated paper in water and oil resistance, especially after it is folded, and gives much stronger bonds on heat sealing.

A great many modifications can be made along these lines. In general, increasing the methacrylate content produces tougher, stronger blends being more viscous in the molten state. The use of softer resinous modifying agents, soft waxes, and plasticizers gives greater softness and pliability. Some interesting blending agents, not listed in Table 130, include asphalt, spermaceti, carnauba wax, beeswax, Japan wax, stearic acid and butyl stearate.

Blends such as (1), (2) and (3) listed before (no wax) have

Table 130

COMPOSITION AND PROPERTIES OF SOME MELTS SUITABLE AS ADHESIVES*

Composition	Description at Room Temperature
(1) 67% n-Butyl MA**Polymer 33% Cumar V½	clear, hard, light brown
(2) 60% n-Butyl MA Polymer 40% Dammar	slightly opaque, light color, flexible, slightly tacky
(3) 60% n-Butyl MA Polymer 40% Rosin	clear, light brown, pliable, slightly tacky
(4) 42% n-Butyl MA Polymer 28% Dammar 30% Paraffin Wax	light yellow, not tacky. Good adhesive for moistureproof Cellophane, waxed paper, etc.
(5) 20% Ethyl MA Polymer 80% Rosin or Ester Gum	clear, slightly brown
(6) 20% Ethyl MA Polymer 80% Dammar	light, opaque

Some products, giving good preliminary results in paper-coating tests, which can be applied at about 100°C., are:

Composition	Description at Room Temperature
(7) 37.5% n-Butyl MA Polymer 37.5% Opalwax 25.0% Paraffin Wax (135°F.)	white, fairly tough, not tacky
(8) 34% n-Butyl MA Polymer 34% Opalwax 23% Paraffin Wax (135°F.) 9% Santicizer B-16	tough, not tacky; does not block
(9) 40% n-Butyl MA Polymer 20% Opalwax 20% Paraffin Wax (135°F.) 20% Rosin	tan, tough, slightly tacky

* All are sufficiently fluid for application at below 200°C.
** MA = Methacrylate.

considerable viscosity at elevated temperatures. They are not fluid enough for general dip-coating operations and would require, e.g., roller coating, spreading. The blends are easiest to prepare by the use of an internal mixer of the Werner-Pfleiderer or Banbury type, at an appropriate temperature of 150—220°C.

Blends, such as (4)–(9), are much more fluid since they contain, in general, smaller quantities of methacrylate resin. Also, the Opalwax and paraffin wax, used in many of them, have quite low viscosities at high temperatures. Melts of this class are readily prepared by heating all of the ingredients but the methacrylate to an appropriate solution temperature of 140–180°C. and adding the methacrylate polymer with stirring. The solution temperature and stirring should be maintained until a smooth dispersion is obtained. This generally requires less than 1 hour. A mechanically driven "T"-type stirrer, run at a high enough rate to prevent settling of polymer lumps, is satisfactory.

If melts are used for products which must be odor-free, two points must be kept in mind: 1. Polymers essentially free from

Table 131

COMPATIBILITY OF ETHYL CELLULOSE WITH WAXES, FATTY ACIDS AND ESTERS, TARS, ETC.

Acrawax	C	Glyceryl Esters of Fatty Acids	C
Acrawax B	C	Halowax 1013	C
Acrawax C	C	Japan Wax	C
Anthracene	C	Kollest (Monoglyceride of Hydrogenated Peanut Oil)	C
Armid HT (Fatty Acid Amide)	C	Lauryl Alcohol	C
Aroclor 1260, 1262 (Chlorinated Diphenyl)	C	Linseed Oil Fatty Acids	C
Asphalt (Genasco)	I	Montan Wax (Pure or Crude)	C
Barnsdall Wax M.P. 185°F.*	I	Naphthalene	C
Beeswax (Crude or Refined)	C	Oleic Acid	C
Candelilla Wax	C	Opalwax (Hydrogenated Castor Oil)	C
Carnauba Wax	LC	Ouricury Wax	C
Castor-Oil Fatty Acids	C	Ozokerite	I
Castor Wax (Hydrogenated Castor Oil)	C	Paraffin Wax	I
Ceresin Wax	I	Spermaceti Wax	C
Cetyl Alcohol	C	Stanolind*	I
Chinese Insect Wax	I	Stearic Acid	C
Circasol Waxes*	I	Stearin Pitch	C
Coal Tar	C	Stearyl Alcohol	C
Diethylene Glycol Esters of Fatty Acids	C	Superla Waxes*	I
Diglycol Stearate S	C	Syncera Waxes*	I
Gilsonite	C	Tallow	C
Glycol Esters of Fatty Acids	C	Terphenyl (*o*-, *m*-, *p*-, and Mixed)	C
Glyco Wax S-932	C	Trinidad Asphalt	I
		Tung Oil Fatty Acids	C

* High-melting microcrystalline waxes derived from petroleum.
C–compatible. LC–limited compatibility. I–incompatible.

odor must be employed and 2. direct heating and high temperatures must be avoided to prevent the development of odor. If the nature of the composition will permit, temperature should not be allowed to exceed 150–160°C. for best results.

The best waxes for blending with ethyl cellulose are the aliphatic acids and alcohols. Ethyl cellulose compounded in binary mixtures with beeswax, Japan wax, and coal tar has an excellent toughening effect. Although certain waxy substances in Table 131 are noted as incompatible, appreciable quantities can be added to a solution of ethyl cellulose, dissolved in a compatible wax, before any evidence of incompatiblity is encountered.

Ethyl cellulose and waxes may be blended in solution in a volatile solvent, such as a mixture of toluene and alcohol, or

Table 132

EFFECTS OF INCREASING AMOUNTS OF ETHYL CELLULOSE ON THE SOFTENING POINT OF ETHYL CELLULOSE–WAX MIXTURES

Compositions:					
Wax, %	100	90	75	66.7	50
N-type Ethyl Cellulose	0	10	25	33.3	50
Softening Point Data, in °C., for Mixtures Containing:					
Beeswax, Refined	62.5	103.5	139.0	144.0	182.0
Candelilla Wax	68.0	75.0	130.0	132.0	142.0
Carnauba Wax	82.0	100.5	122.0	135.0	—
Coal Tar (Refined), M.P. 28°C.	27.5	88.0	118.0	132.0	171.0
Gilsonite	158.5	172.5	195.5	208.5	218.0
Japan Wax	41.0	69.0	121.0	151.0	182.0
Linseed Oil Fatty Acids	liquid	liquid	64.0	83.0	110.0
Montan Wax, Crude	83.0	87.5	112.5	125.0	145.0
Montan Wax, Refined	69.5	86.5	115.0	117.5	134.5
Oleic Acid	liquid	liquid	61.0	75.0	106.0
Spermaceti	45.5	91.5	120.0	127.5	131.5
Stearic Acid	55.0	56.0	71.0	87.0	114.0
Stearin Pitch	85.0	106.0	120.0	163.0	182.0
Tallow	35.0	62.0	117.0	151.0	165.0
*Glyco Wax S-932	59.0	—	142.5	—	—
*Glycerol Monostearate (Aldo 28)	56.5	—	125.0	—	—
*Diglycol Stearate S	53.5	—	110.0	—	—
*Acrawax	96.0	—	135.0	—	—
*Acrawax B	82.5	—	120.5	—	—

*N-100 ethyl cellulose was used in these cases.

they may be blended together by dissolving the ethyl cellulose directly in the molten wax heated to a temperature of 150–200°C.

The addition of ethyl cellulose to waxes, in which it is soluble, raises the melting point of the wax, improves its toughness, diminishes surface tackiness, and decreases the tendency of the wax to crystallize. Some data are given in Table 132 on the effect of ethyl cellulose addition on the melting point of waxes. Besides improving waxes alone, ethyl cellulose often improves the gloss, flexibility, and toughness of mixtures of waxes, resins, and plasticizers. These mixtures have a wide variety of uses as coatings for paper and cardboard, adhesives, and plastics. Some of the coal tars are greatly hardened and toughened by ethyl cellulose.

METHODS OF PREPARING MIXTURES OF WAXES WITH ETHYL CELLULOSE*

Mixtures of ethyl cellulose with waxes and related substances are usually compounded by hot-melt methods. The wax is heated to 180–200°C. and, while holding the temperature within this range, all of the ethyl cellulose, previously dried for 1 hour at 100°C., is quickly added with stirring in such a way that it is rapidly wetted by the molten wax. Stirring is continued at the indicated temperature until all ethyl cellulose has completely dissolved. Some foaming usually takes place while the ethyl cellulose is going into solution. Solution may be considered complete when foaming subsides. This seldom requires more than 15–20 minutes after addition of the ethyl cellulose. The temperature of the mixture may then be adjusted for coating or casting. Ethyl cellulose will dissolve in many substances, stearic acid, for example, in much less time than 15 minutes.

For making ethyl cellulose–wax mixtures of the best color, glass-lined blending kettles should be used. Stainless-steel kettles are also satisfactory. Use of iron kettles may result in discoloration. Steam-heated kettles or gas-fired varnish kettles provided with adequate stirring make suitable blending equipment.

In another convenient method of preparation, a heated Ban-

* Hercules Powder Co., Wilmington, Del.

bury mill is used to blend the ethyl cellulose and wax into a homogeneous colloid, after which the mixture is sheeted from a hot two-roll mill. It may be used as strips or may be granulated for remelting.

Some substances may be blended with ethyl cellulose in a volatile solvent, such as a mixture of 80 parts toluene or Solvesso toluene with 20 parts of butanol. Tallow, Japan wax, Aroclor 1260, stearin pitch, and oleic acid are examples of substances which can be blended in this way. In general, however, the solubility of waxes is too limited to make this method practicable.

USE OF INCOMPATIBLE SUBSTANCES

Asphalts, paraffin waxes and high-melting microcrystalline waxes derived from petroleum, all of which are incompatible with ethyl cellulose alone, may be blended into homogeneous mixtures with ethyl cellulose by employing a third mutually compatible substance. This mutual blending agent may be an active solvent-type wax, such as beeswax, stearic acid, or Opalwax; a resin, such as Staybelite Ester 10 or pure phenolic resin; or a plasticizer, such as Hercolyn, raw castor oil, or dibutyl phthalate. Such blends exhibit the beneficial effect imparted by ethyl cellulose, if sufficient solvent-type blending agent has been used to obtain complete compatibility, and are often quite useful mixtures. Resins usually improve the hardness and gloss.

The data in Table 133 plotted in Figure 50, can serve as an example of changes in a ternary mixture, one component of which is ethyl cellulose and two components are cosolvent waxes. Such compositions as these might be used in hot-dip applications for food packages, paper coatings, or as slushing or casting plastics. Five formulas are given, whose composition points are indicated along the line A - - - - A^1 in the ternary chart of Figure 50. The softening points obtained for the four groups of five formulations are plotted at the right-hand side of this figure. Compositions in the four groups differ only in that one of the wax components is different; that is, Opalwax, Kollest, Armid HT, and stearic acid each is used in one set of the five formulas along with the paraffin wax and ethyl cellulose. The latter component was held constant at 25% in all formulations. It should be noted that the softening points of these twenty com-

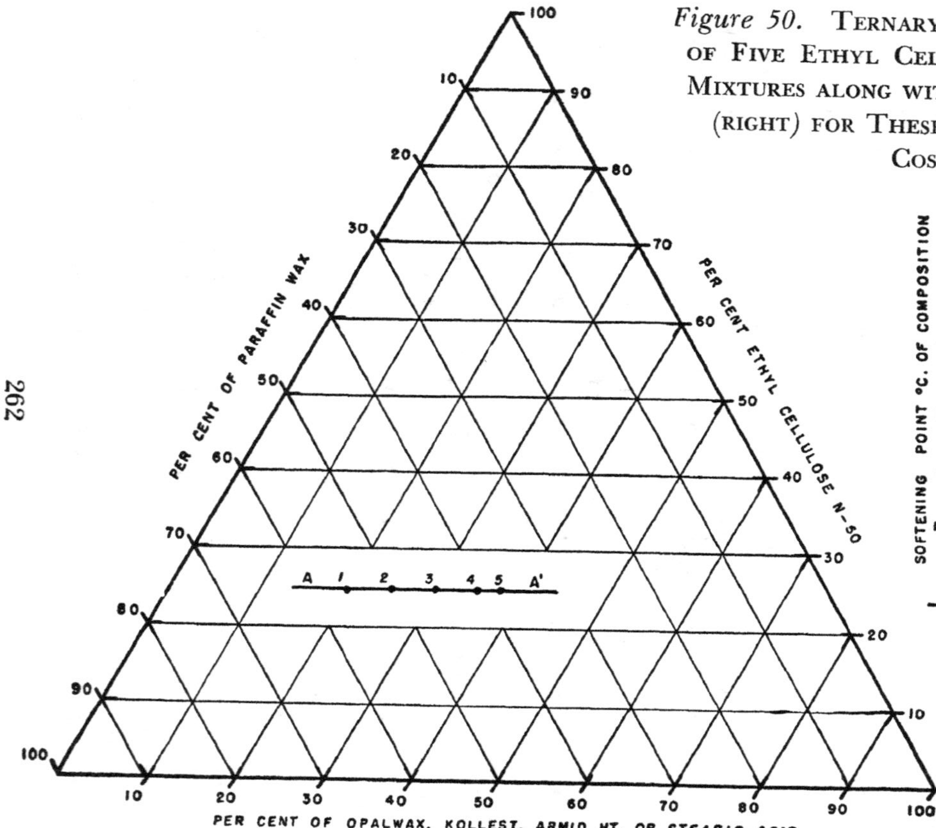

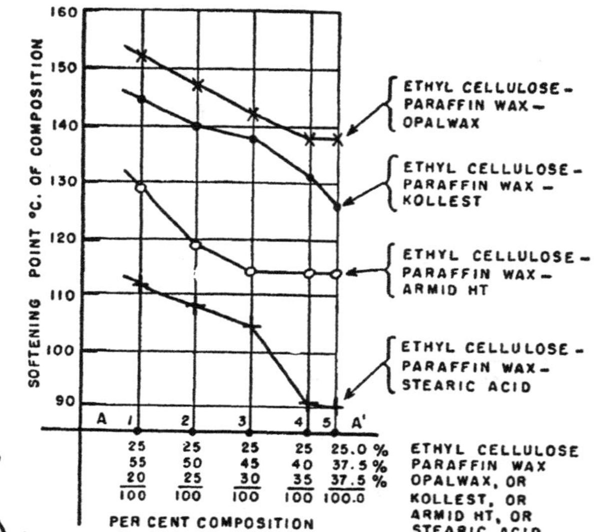

Figure 50. Ternary Chart (left) Showing the Composition of Five Ethyl Cellulose—Paraffin Wax—Cosolvent Wax Mixtures along with Composition—Softening Point Curves (right) for These Five Mixtures when Four Different Cosolvent Waxes Are Used

positions ranged from a low of 90°C. to a high of 152°C. The effects of composition and components on the softening points may be best evaluated through a study of the plotted data at the right-hand side of Figure 50.

Table 133

ETHYL CELLULOSE IN WAX COMPOSITION

	Parts by Weight Formula				
Ingredient	1	2	3	4	5
N-50 type Ethyl Cellulose	25	25	25	25	25
Paraffin Wax, M.P. 60°C.	55	50	45	40	37.5
Cosolvent Wax	20	25	30	35	37.5
Tests	*Test Data*				
Ring-and-Ball Melting Point, Using					
Opalwax as Cosolvent Wax, °C.	152	147	142	138	138
Kollest as Cosolvent Wax, °C.	145	140	138	131	126
Armid HT as Cosolvent Wax, °C.	129	119	114	114	114
Stearic Acid as Cosolvent Wax, °C.	112	109	105	91	90

COMPOUNDING ETHYL CELLULOSE WITH INCOMPATIBLE SUBSTANCES

Normally, incompatible mixtures are of no interest because their nonhomogeneity makes their behavior unpredictable. However, useful incompatible ethyl cellulose mixtures can be prepared, which, if compounded and worked by certain techniques, appear to yield stable dispersions and may be suitable for solving special problems.

Although ethyl cellulose is incompatible with most of the natural and synthetic high polymers, it can be mixed with many of these to obtain homogeneous dispersions which are moldable. For instance, ethyl cellulose and crepe rubber can be homogeneously compounded on hot two-roll mills by placing the ethyl cellulose, either alone or with modifiers, on the mill first and working it into a homogeneous plastic mass. Rubber can be added very slowly, a little at a time, milling each addition thoroughly and homogeneously into the plastic composition before any further addition. By proceeding carefully and keeping the mill temperatures properly adjusted, it is possible to mix ethyl cellulose and rubber in any desired proportions. Chicle, gutta-percha, synthetic rubbers, such as Buna S, vinyl polymers, polyisobutylene, and others can be similarly mixed with ethyl cellulose on the hot two-roll mill; and the mixtures thus obtained can be injection or extrusion molded, or calendered without any evidence of separation. Rubber mixtures can be vulcanized. Some mixtures showed no sign of separation even after several years of aging. However, the microscope clearly reveals that they are dispersions of one substance in the other. Any of these mixtures, if dissolved in a mutual organic solvent, immediately forms an incompatible solvent mixture which, on standing, usually separates into two distinct layers. Films laid down from solutions of such mixtures show striking evidence of incompatibility.

There is a variety of normally incompatible resins, such as high-melting coumarone-indenes, maleic-modified ester gums of high maleate content, pentaerythritol esters of rosin and others which can be homogeneously dispersed in ethyl cellulose either on a hot two-roll mill or in typical hot-melt compositions. High-melting microcrystalline paraffin waxes and asphalts can be incorporated similarly. Normally, however, they are not used with ethyl cellulose alone, but are mixed with it in the presence of one or more compatible modifiers, such as an oil, a solvent-type plasticizer, or resin to aid in obtaining a homogeneous, workable mixture with the incompatible substance. At high compounding temperatures, many such mixtures actually are compatible with ethyl cellulose, but on cooling revert to incompatible mixtures.

Such mixtures are distinct dispersions and, if dissolved in an organic solvent, immediately show unmistakable signs of being incompatible. They are useful because by the techniques used to prepare and apply them there is no opportunity for separation of the incompatible phases. Some interesting hot-melt paper coatings which have good blocking resistance and freedom from tack are based on mixtures of ethyl cellulose with maleic-modified ester gums of high maleic content. Similarly, some very useful hot-melt casting plastics are based on mixtures which contain a high proportion of ingredients incompatible with ethyl cellulose.

Ethyl cellulose is compatible with 42 per cent chlorinated paraffin in ratios up to 1 part ethyl cellulose to 3 parts chlorinated paraffin. Such mixtures are valuable flameproofing impregnating agents. Chlorinated paraffin containing 70% chlorine is a light-colored, high-melting resin which is not compatible with ethyl cellulose unless blended with a mutual solvent-plasticizer such as tricresyl phosphate. Mixtures composed of 5:2:3 and 5:4:3 ethyl cellulose:chlorinated paraffin:tricresyl phosphate are of this type. They are of interest because of their nonflammability.

Ethyl Cellulose and Wax Mixtures

It is to be noted that a small amount of high-viscosity ethyl cellulose (325 cps in 5% solution) raises the softening point of stearin pitch 20°. It appears that smaller amounts of a high-viscosity substance will produce the same general effects as larger amounts of lower-viscosity matter, but the solubility is sometimes impaired.[1]

In some cases, the insoluble waxes may be put into ternary solutions by the use of a mutually miscible substance, for example, paraffin wax 10, montan wax 10, ethyl cellulose 2 give a homogeneous product.

Ethyl Cellulose	10
Stearic Acid, M.P. 54°C.	45
Paraffin, M.P. 44°C.	45

This mixture, fuses at 200°C., gives good solution, and has a drop melting point of 64°C. As noted previously, ethyl cellulose is immiscible in paraffin waxes, but it is soluble in aromatic hydrocarbon waxes, such as naphthalene and anthracene.

The solubility of ethyl cellulose in fatty alcohols, such as cetyl, myristyl and stearyl alcohol, is excellent. In all cases, the melting point of the solid alcohols is much increased by the addition of ethyl cellulose. The hardness and toughness are likewise improved.

---

[1] Hercules Powder Co., Wilmington, Del.

Table 134

Ethyl Cellulose and Wax Mixtures

	Ethyl Cellulose	Fused at °C.	Solubility	Softening Point °C.	Melting Point °C.	Cast
Aroclor, Solid						
100	0	—	—	126.0	129.0	fair
90	10	150	slow	128.0	137.0	fair
75	25	170	slow	125.5	128.0	fair
66⅔	33⅓	200 viscous	slow	127.0	132.0	fair
Asphalt (Genasco)—not miscible.						
Beeswax						
100	0	—	—	61.0	62.5	good
90	10	150 10M (Jells a little.)	good	99.5	103.5	good
75	25	150 10M	good	141.5	144.0	good
66⅔	33⅓	170 10M	good	131.0	139.0	good
50	50	200 darkens	good	178.0	182.0	good
Candelilla						
100	0	—	—	66.0	68.0	poor
90	10	175	fair	69.0	75.0	poor
75	25	175	fair	111.0	130.0	poor
66⅔	33⅓	175	fair	127.0	132.0	good
50	50	200 darkens	fair	135.0	142.0	good

Table 134 (Continued)

	Ethyl Cellulose	Fused at °C.	Solubility	Softening Point °C.	Melting Point °C.	Cast
Carnauba						
100	0	—	—	81.0	82.0	poor
90	10	170	fair	92.0	100.5	quite poor
		1 hour signs of separation				
75	25	200 10M	good	119.5	122.0	better but still poor
66⅔	33⅓	200 viscous	good	130.5	135.0	
Ceresin—not miscible.						
Coal Tar						
100	0	—	—	22.5	27.5	too soft
90	10	170	good	67.5	88.0	too soft
75	25	170	good	95.0	118.0	excellent
66⅔	33⅓	170	good	119.0	132.0	excellent
50	50	200	good	143.0	171.0	excellent
Gilsonite						
100	0	—	—	142.0	158.5	good
95	5	200	good	167.0	174.5	good
90	10	200	good	166.0	172.5	good
75	25	240	good	176.5	195.5	good
66⅔	33⅓	240	good	198.5	208.5	good
50	50	240	good	206.0	218.0	too viscous

Table 134 (Continued)

Halowax—(similar to Aroclor Japan Wax)	Ethyl Cellulose	Fused at °C.	Solubility	Softening Point °C.	Melting Point °C.	Cast
100	0	—	—	40.0	41.0	excellent
90	10	150 darkens	slow	60.0	69.0	excellent
75	25	150 darkens	slow	75.0	121.0	excellent
66⅔	33⅓	170 darkens	slow	127.0	151.0	excellent
50	50	200 darkens	slow	138.0	182.0	excellent
Linseed Oil Fatty Acid						
100	0	liquid	—	—	—	—
90	10	viscous	liquid	good solution	—	—
75	25	150	good	51.0	64.0	semisolid
66⅔	33⅓	150	good	72.0	83.0	semisolid
50	50	160	good	80.0	110.0	solid
Montan, Crude						
100	0	—	—	82.0	83.0	good
90	10	200 15M	good	85.0	87.5	good
75	25	200 15M	good	104.0	112.5	good
66⅔	33⅓	200 15M	good	122.0	125.0	good
50	50	200 15M	good	143.5	145.0	good

Table 134 (Continued)

	Ethyl Cellulose	Fused at °C.	Solubility	Softening Point °C.	Melting Point °C.	Cast
Montan, Pure						
100	0	—	—	68.5	69.5	fair
90	10	150 5M	good	73.0	86.5	excellent
75	25	150–170 10M	good	110.5	115.0	excellent
66⅔	33⅓	180 15M	good	114.5	117.5	excellent
50	50	200 15M darkens	good	129.0	134.5	excellent
Oleic Acid						
100	0	liquid				
90	10	liquid	excellent solution			
75	25	150	excellent	48.0	61.0	semisolid
66⅔	33⅓	170	excellent	55.0	75.0	semisolid
50	50	200	excellent	105.0	106.0	solid
Ozokerite—not miscible.						
Paraffin—not miscible.						
Spermaceti						
100	0	—	—	44.0	45.5	good
90	10	150 10M	good	78.5	91.5	good
75	25	150 10M	good	120.0	127.5	good
66⅔	33⅓	170 10M	good	116.5	120.0	good
50	50	200 10M	good	129.0	131.5	good

Table 134 (Continued)

	Ethyl Cellulose	Fused at °C.	Solubility	Softening Point °C.	Melting Point °C.	Cast
Stearic Acid						
100	0	—	—	54.0	55.0	does not cast
90	10	100		55.0	56.0	fair
75	25	125	excellent	64.0	71.0	good
66⅔	33⅓	140	excellent	75.0	87.0	good
50	50	170 darkens	excellent	91.0	114.0	good
			excellent			
Stearin Pitch						
100	0	—	—	75.0	85.0	too soft
99	1*	200	fair	95.0	106.0	too soft
90	10†	200	fair	100.0	120.0	too soft
75	25†	200	fair	133.0	>250.0	sticky
66⅔	33⅓†	215	fair	148.0	163.0	good
50	50†	250	fair	172.0	182.0	brittle
Tallow						
100	0	—	—	34.0	35.0	excellent
90	10	150	slow	38.0	62.0	excellent
75	25	150	slow	106.0	151.0	excellent
66⅔	33⅓	170 darkens	slow	107.0	117.5	excellent
50	50	200 darkens	slow	112.0	165.0	excellent

* High viscosity.

† Medium viscosity.

Table 135

MELTING POINTS OF COMPOUNDED WAXES

Components and Individual Melting Point	°F.		°F.	Melting Point of Composition Ratio	°F.
Paraffin Wax	150	Opalwax	208	1:1	145
Paraffin Wax	150	Montan	169	1:3	138
Paraffin Wax	150	I.G. Wax OP	221	1:1	145
Paraffin Wax	150	Carnauba	185	1:1	153
Paraffin Wax	125	Albacer	202	1:1	very thick, no flow
Paraffin Wax	150	Albacer	202	1:1	very thick, no flow
Paraffin Wax	150	Albacer	202	3:1	does not melt; slow cooling.
Ozokerite	130	Carnauba	185	4:1	171
Carnauba Wax	174	Acrawax	201–8	99:1	179

MELTING POINT AND CONGEALING POINT OF VARIOUS WAXES

The greater the differences between melting point and congealing point, the greater is the complexity of the wax, for example, blended waxes. Equal parts by weight of carnauba wax and paraffin wax give a melting point of 79.5°C. and a congealing point of 69.0°C., with several intermediate setting points.

Unadulterated vegetable waxes show differences of up to 5°C.

Table 136

MELTING VERSUS CONGEALING POINTS[1]

Wax	Melting Point, °C.	Congealing Point, °C.
Spermaceti	43.5	43.2
Carnauba (Grey)	85.0	79.0
Candelilla	71.0	68.5
Paraffin	60.0	59.0
Bleached Montan	85.0	80.0
Beeswax	63.5	62.5

Table 137

SETTING POINTS OF MIXTURES OF VARIOUS WAXES WITH A DOUBLE-REFINED PARAFFIN WAX HAVING A SETTING POINT OF 50°C. (LUDECKE)

Percentage of Waxes Added to Paraffin Wax Setting Point 50°C.	Crude Carnauba Wax Setting Point 75°C.	Refined Carnauba Wax Setting Point 70°C.	Chinese Insect Wax Setting Point 76°C.	Candelilla Wax Setting Point 67°C.	Natural Yellow Ozokerite Setting Point 70°C.	Crude Montan Wax Setting Point 74°C.	Refined Montan Wax Setting Point 70°C.
3	56.0	52.5	53.4	51.4	50.45	52.0	51.0
5	57.0	53.2	54.4	51.8	51.00	53.0	51.5
8	60.2	53.6	58.0	53.4	51.70	53.0	52.2
10	62.0	54.0	60.0	53.6	52.00	53.2	52.4
15	64.0	56.0	61.4	54.4	53.00	54.0	53.2
20	65.0	59.0	63.6	55.0	54.00	54.0	54.0
25	65.6	62.0	64.0	56.0	54.80	55.0	54.0
30	66.6	62.6	64.4	57.0	55.70	56.0	55.0
35	67.0	63.0	65.0	58.0	57.00	56.4	55.4
40	67.6	63.2	65.4	59.0	58.00	56.4	56.0

MELTING POINTS OF BINARY AND TRINARY WAX MIXTURES[1]

Melting-point curves of the binary wax mixtures are to be found in the three narrow graphs of Figures 51 and 52 surrounding each of the triangular graphs. The triangular graphs represent all possible ternary mixtures of the waxes indicated. Since only compositions can be plotted, a single curve connects all mixtures melting at one specific temperature. The melting point, in °C., for each curve, is indicated on it. In general, a curve is drawn for every 2.5° rise in melting point and the curves then may be compared to the altitude lines on a contour map. Each apex of the triangle represents 100% of the wax indicated, and also 100% of the wax indicated for the binary graph adjoining.

Each side of the triangle represents all binary mixtures of the two waxes at the ends of the line. From the binary melting point data we may select binary mixtures melting at, e.g., 62.5, 65, 67.5, 70, and 72.5°C. By drawing a line on the triangular graph, between two sets of binaries that have the same melting point, we obtain suggested compositions of ternary mixtures for that melting point. Compositions are chosen at uniform intervals over the length of the line and the melting points of the ternary mixtures determined. For a very short line, one ternary mixture is sufficient, while for a long line four or five ternary mixtures are used.

Knowing the qualitative composition of a ternary mixture of waxes, the melting point of this mixture will enable one, by use of the graphs, to determine the approximate quantitative composition of the mixture with a minimum of other data. Likewise, by use of the graphs, mixtures of waxes of any desired melting point can be readily prepared.

It will perhaps seem striking, at first sight, that the isothermals actually are straight lines within the limits of experimental error. This substantiates the idea that wax mixtures are true mixtures and that new chemical compounds are not formed. Because of the comparative chemical inertness of the esters and

[1] J. R. Koch, G. J. Hable, and L. Wrangell, *Ind. Eng. Chem., Anal. Ed.*, **10**, 167 (1938).

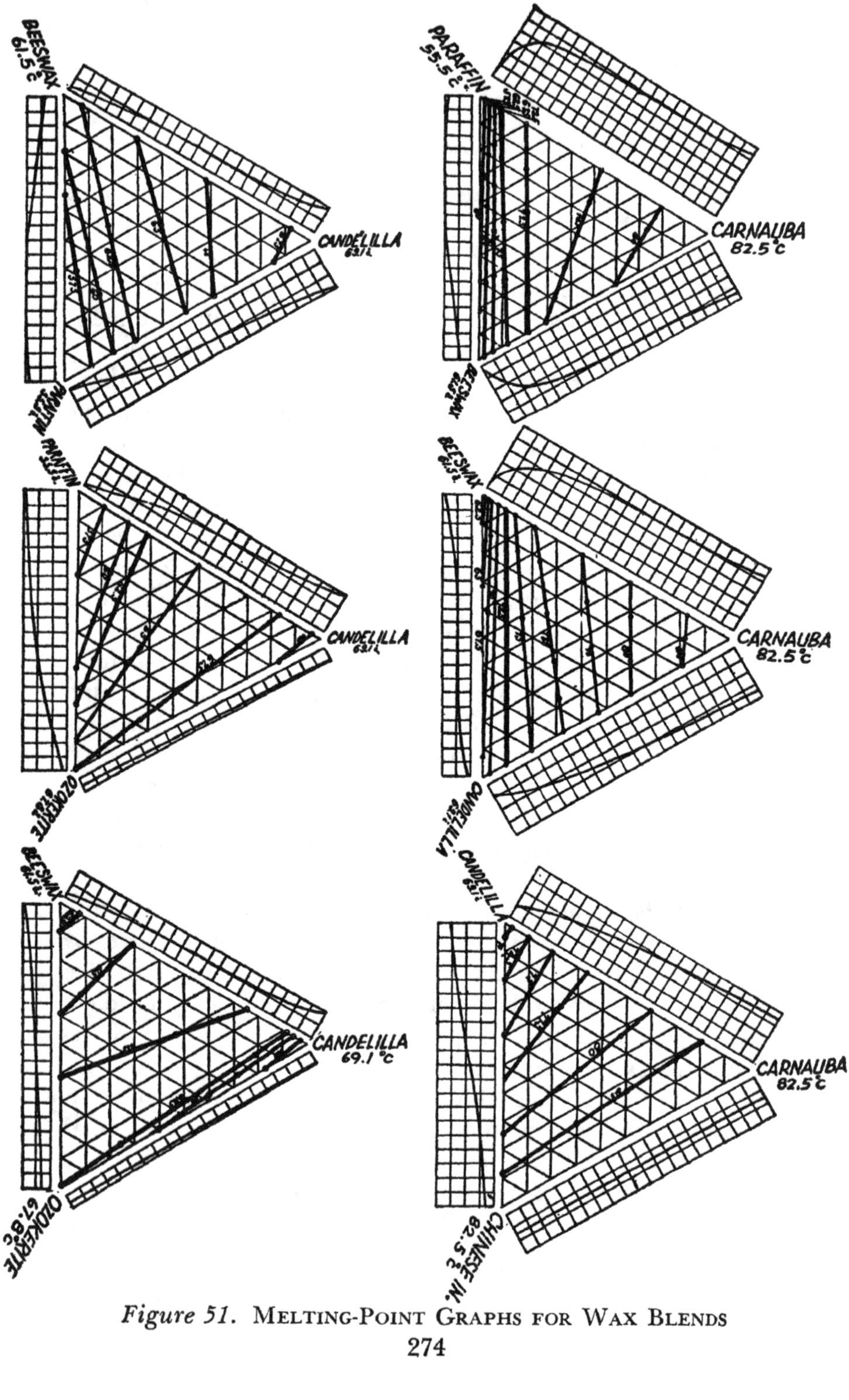

Figure 51. MELTING-POINT GRAPHS FOR WAX BLENDS

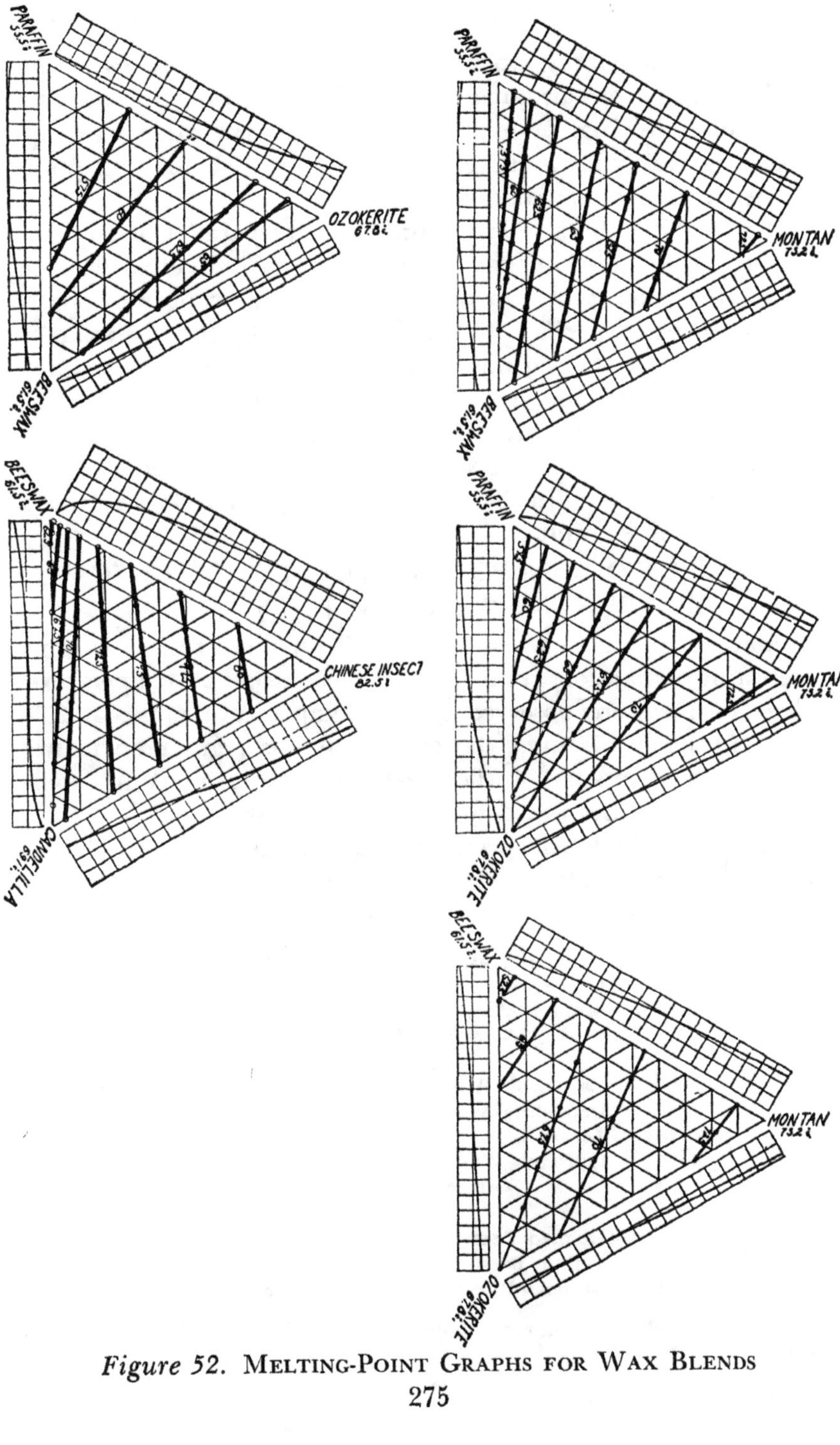

Figure 52. MELTING-POINT GRAPHS FOR WAX BLENDS

hydrocarbons, making up most waxes, this is to be expected. Small amounts of carnauba are effective in raising the melting points of individual or mixed waxes. The only other wax, approaching carnauba in its effectiveness in quick melting point rise, is Chinese insect wax.

The fact that the isothermal lines are not parallel in any case leads to the conclusion that the effect of any individual wax on the melting points of two other waxes is never proportionally the same, but is specific. This is also shown by the dissimilarity in the shape of the various binary mixture curves. A striking example of the effectiveness of each of two waxes in altering the melting point of the other when present in small quantities is shown in the case of beeswax and candelilla. A diametrically opposite effect is shown in the case of candelilla and Chinese insect wax on the same graph.

THE EFFECT OF COMPOSITION ON THE SPECIFIC GRAVITY OF BINARY WAX MIXTURES[1]

Commercial waxes, when used in binary mixtures, show some interesting expansion results with an increase of temperature. These results were observed over a temperature range of 25–80°C. In some mixtures, the expansion was regular no matter what the percentages of the binary composition. With some other mixtures this was definitely not the case.

From the calculated specific gravity values at 25–80°C., the volume-increase values of the waxes were as follows:

Mineral Waxes	%
Paraffin Wax	20.7
Ozokerite	25.9
Montan Wax	11.8
Vegetable Waxes	%
Carnauba Wax	12.0
Candelilla Wax	18.7
Japan Wax	15.1
Ouricury Wax	11.0
Animal Waxes	%
Refined Beeswax	15.8
Yellow Beeswax	15.1
Chinese Insect Wax	8.9

[1] John R. Koch and Sister M. Concetta, *Trans. Kentucky Acad. Sci.* (104-110) 1950.

It will be seen that mineral waxes expand the most and animal waxes the least. The hard, microcrystalline, tough, high-melting waxes show less expansion than those which are soft and have low melting points and Chinese insect wax, which is a fairly hard, crystalline wax, shows the least expansion.

It was found that there is an interesting relation between the melting point and the extent of expansion. Usually, where the melting-point curve follows the normal pattern, the expansion curve will likewise do so.

When the beeswax curves are studied, certain similarities are noticed. The expansion between 25 and 30°C. is very slight. The greatest expansion is between 50 and 60°C. Mixtures of waxes melting below 70°C. show very little expansion between 70 and 80°C.

The most striking graph in this group is presented by the mixture of refined beeswax with yellow beeswax (Figure 53).

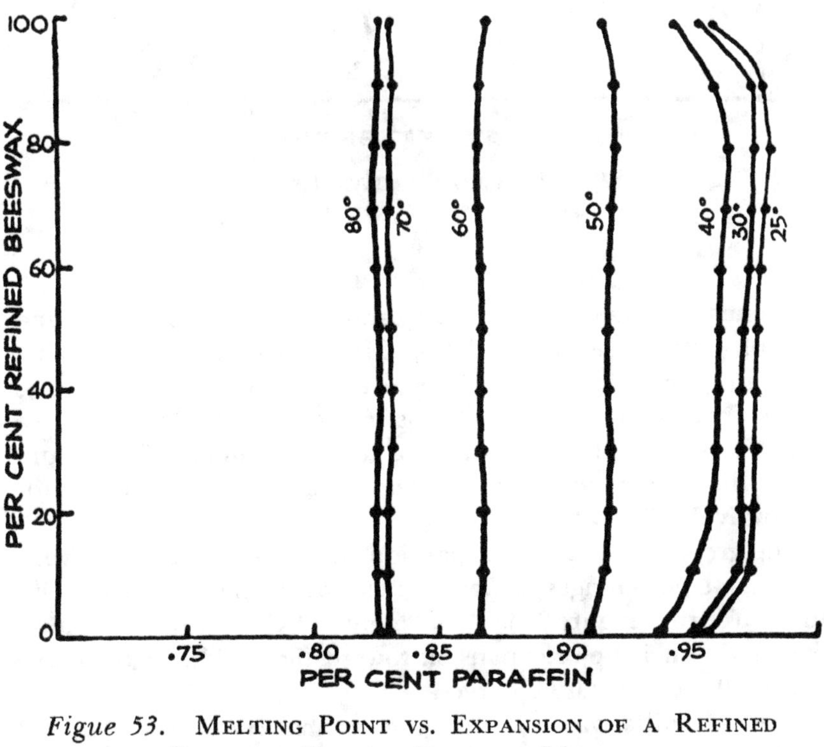

Figue 53. MELTING POINT VS. EXPANSION OF A REFINED BEESWAX—YELLOW BEESWAX MIXTURE

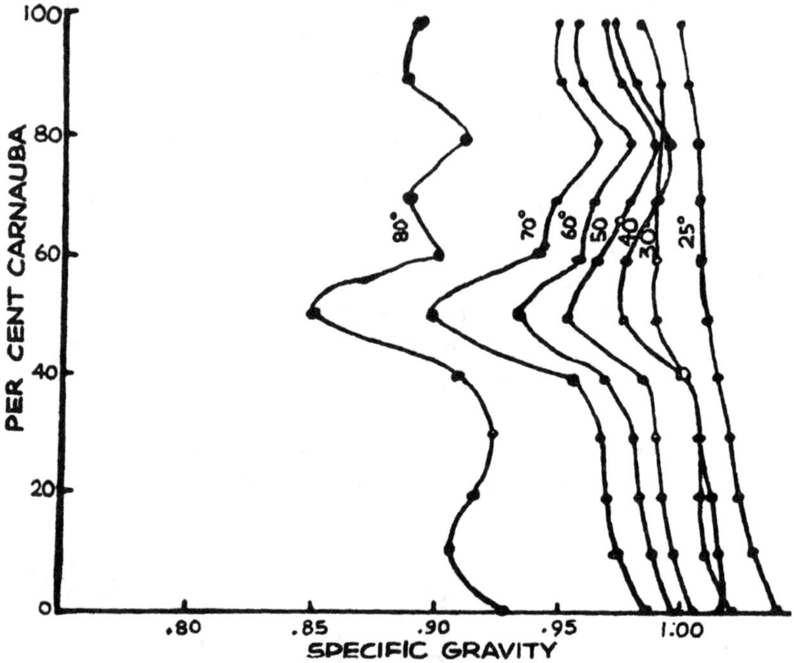

Figure 54. MELTING POINT VS. EXPANSION OF A CARNAUBA WAX–MONTAN WAX MIXTURE

Refined beeswax has a specific gravity of 0.957 at 25°C. and that of yellow beeswax is 0.953. Mixtures of these two waxes show an average value of 0.974 which means that the two waxes shrink in mixtures at 25°C. This change is equally observable at 30°C., less so at 40°C., still less at 50°C., and at 60, 70 and 80°C., the average value is the same for the mixtures as it is for the individual waxes.

The greatest amount of expansion in the beeswax occurs with the largest percentages of beeswax except in the case of paraffin and ozokerite (Figure 56). The 25 and 30°C. curves are regular. We see the increase of expansion toward the middle of the curve in the 40, 50 and 60°C. curves.

The candelilla wax mixtures give normal curves at 25°C. and the individual specific gravity values could be calculated from the values of the original waxes within the limits of experimental error. The mixtures of candelilla wax with vegetable waxes, refined beeswax and ozokerite show a decreased volume at 30°C. In most of the candelilla-wax mixtures, we see that between 30 and 70% of candelilla wax, there is almost no increase in the amount of expansion.

The most interesting sets of curves are obtained with the carnauba-wax mixtures. The greatest expansion with increase of temperature occurs where the percentage of carnauba wax is small. The specific gravity values of the wax mixture at 25°C.

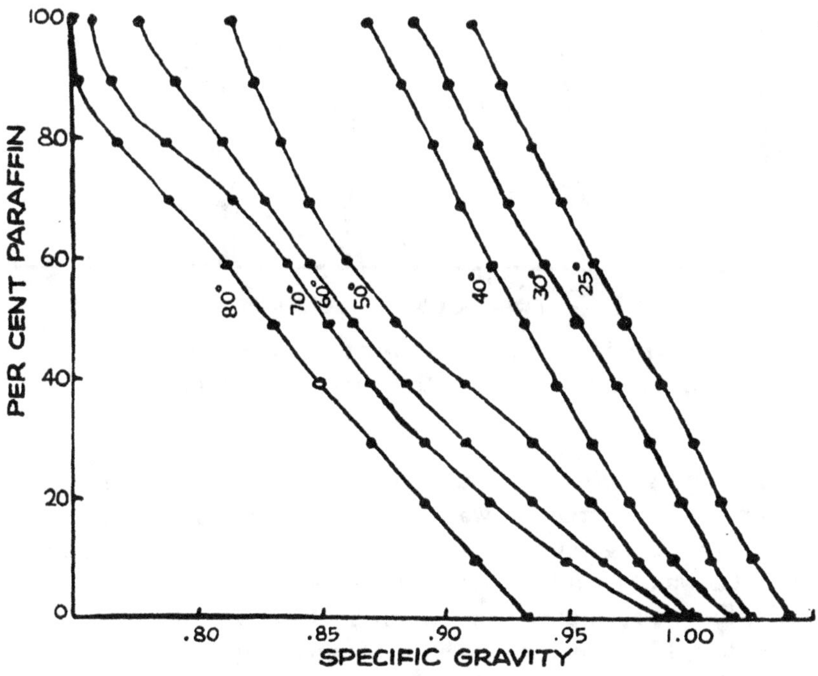

Figure 55. MELTING POINT VS. EXPANSION OF A PARAFFIN WAX—MONTAN WAX MIXTURE

can be calculated within the limits of experimental error from the values of the original waxes. A study of the carnauba wax–montan wax curves will point out this regularity in the 25°C. curve (Figure 54).

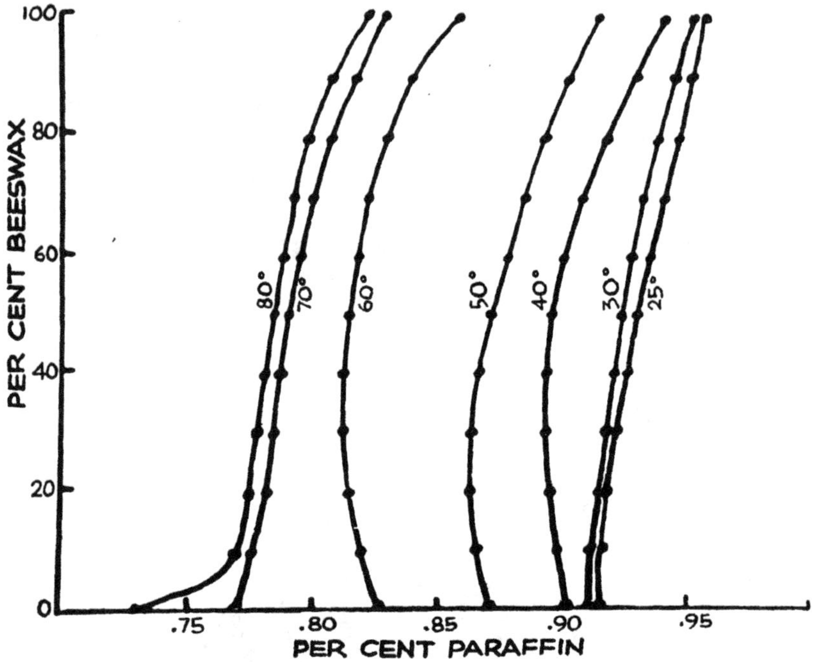

Figure 56. MELTING POINT VS. EXPANSION OF A REFINED BEESWAX–OZOKERITE WAX MIXTURE

The same graph can be used to point out that the 50–50% mixtures with carnauba wax call for special attention. The greatest percentage of expansion is found in these mixtures when carnauba wax is studied with refined beeswax, yellow beeswax, candelilla wax, Japan wax, montan wax, paraffin wax, and ozokerite. The greatest amount of shrinkage is found when equal amounts of carnauba and Chinese insect wax are mixed (Figure 57).

The most regular sets of curves are obtained with paraffin-wax mixtures. The paraffin wax–montan wax curves serve as an example (Figure 55). The 25, 30, and 40°C. curves are very regular. There are only slight deviations in the 50, 60, and 70°C. curves. The 80°C. curve again follows the regular pattern. The experimental specific-gravity values for the mixtures at 25°C.

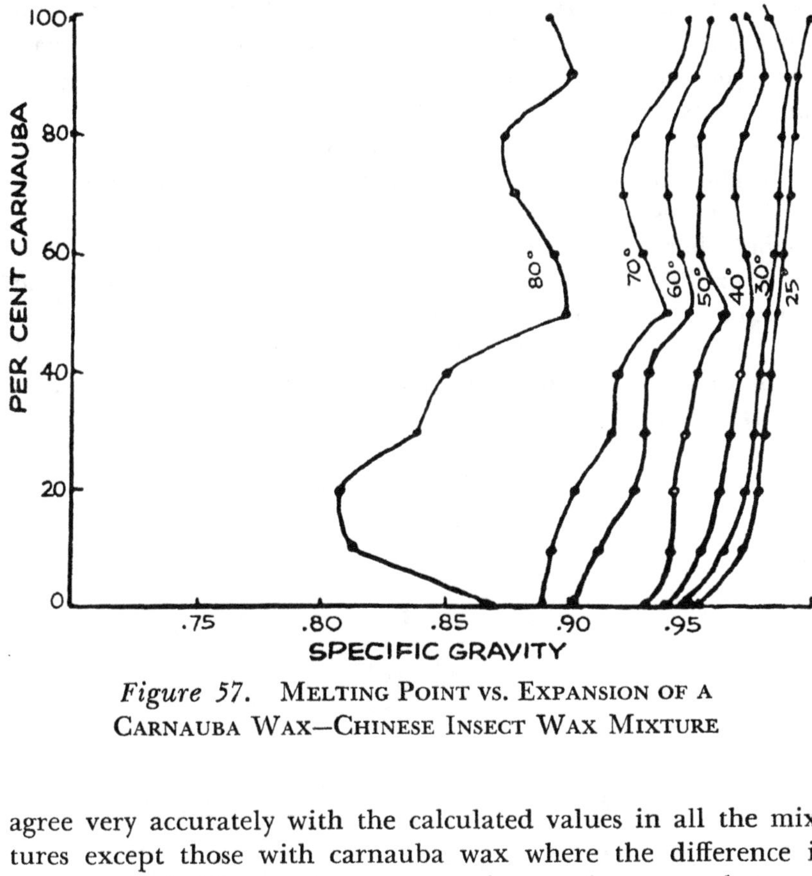

Figure 57. MELTING POINT VS. EXPANSION OF A CARNAUBA WAX—CHINESE INSECT WAX MIXTURE

agree very accurately with the calculated values in all the mixtures except those with carnauba wax where the difference is more marked. The greatest amount of expansion occurs between 40 and 50°C. In most cases, the amount of expansion increases with the increase of the percentage of paraffin.

A careful study of the graphs and a check with mathematical calculations show that in most cases, the values for the specific gravity of the wax mixtures at 25°C. can be calculated from the specific gravities of the components and that the experimental values agree very well with the calculated values, except in the cases pointed out. Most of the calculated values for higher temperatures have been found to agree with the experimental values by 0.01 and rarely have the differences been more than 0.02.

One other set of observations may be made. The mixture of a soft low-melting wax with another soft low-melting wax exhibits a very normal expansion between 30 and 40°C. Past 40°C., the mixtures show more expansion than do the original components.

Mixtures of a soft low-melting wax with a hard microcrystalline high-melting wax or with a hard crystalline wax also give normal curves.

When hard high-melting waxes are mixed together, their curves exhibit marked abnormalities as was pointed out in the carnauba–montan mixtures. When the same waxes are mixed with hard crystalline waxes, marked volume changes are noticed at different temperatures. Carnauba–Chinese insect wax mixtures illustrate this point.

CHAPTER 4

wax technology

ADULTERATION

Waxes, especially when bought at lower than market prices or from unreliable sources, are often adulterated. Flagrant adulterations can sometimes be detected by comparison with a sample of a genuine wax. Other more subtle sophistications require physical, chemical or use tests of varying complexity.

QUALITY VARIATIONS

Natural waxes, like all other natural products, e.g., flour, starch, fossil resins, etc., vary within certain limits because of place of origin, climatic conditions, time and method of gathering, method of handling, storage and shipping, age, exposure, and other factors. Slight variations may not be serious for certain uses. Pronounced variations may make changes in formulation necessary or even rule out a certain lot as unsuitable for a particular use.

SIMPLE TESTS

The following tests are quick and rough in nature. Their purpose is to enable the nonchemist to check on variations in properties and composition of waxes and allied products. Samples for testing should be taken from top, bottom and center of each lot to avoid results that are not average or uniform.

MELTING POINT

Melt a little wax in a dish. Dip into this the bulb of a laboratory thermometer, drawing it out at once and spinning it so that the wax on the bulb forms a fairly uniform layer. Next, work a cork, with a hole of the same size as the thermometer stem, down the top of the thermometer stem to about 4 in. from the bulb. Then insert the cork into a dry, clean test tube which is placed in a beaker of water over a gas burner. Warm the water and watch the wax on the thermometer bulb. When the wax on the bulb becomes transparent or when a small drop of melted wax appears, read the temperature. This temperature may be considered the melting point.

SPECIFIC GRAVITY

(A) For waxes lighter than water: Fill a 100 cc graduate (measuring cylinder) to the 50 cc mark with a mixture of equal volumes of water and alcohol. Take the specific gravity of this with a hydrometer. To this add 20 g of the wax in pieces smaller than the cross section of the graduate so that they will not touch the sides. Read the increase in volume of the water-alcohol mixture. For example, if the increase in volume is 21 cc and the specific gravity of the water-alcohol mixture 0.935, then $20/21 \times 0.935 = 0.89$ is the approximate specific gravity.

(B) For waxes heavier than water: Use the same method, but use plain water, instead of a water-alcohol mixture. For example, if the 20 g of wax raises the volume of the water by 18 cc, then $20/18 \times 1.11$ is the approximate specific gravity. No factor is needed in this calculation, the specific gravity of water being 1.

MOISTURE (WATER) AND INSOLUBLE MATTER

Put a little wax in a test tube standing in a beaker of boiling water and keep the water boiling until the wax has melted. Then turn off the flame and pour benzol into the test tube until it is about half full. Let it warm up and stir or shake until the wax is dissolved. If the solution is turbid, water is present. Then allow to settle. The water will form a layer at the bottom, along with the insoluble impurities present.

ACCURATE DETERMINATION OF SPECIFIC GRAVITY[1]

The accuracy of some of the specific-gravity values given in the literature for natural waxes, such as carnauba and candelilla, and especially some of the very low values reported, is questionable. The temperatures at which these values were determined are often not given. It is possible that they were measured on the molten waxes above their melting point, in which case they would be substantially lower than for the solid waxes, since these waxes shrink markedly on setting. Experience indicates that the lower values are probably in error, due to their having been determined on massive pieces of wax, containing internal voids caused by shrinkage on cooling, to the buoyant effect of adhering air bubbles if determined by liquid displacement, or to some other errors in the determinations.

A method has been developed for determining the specific gravity, which appears to give reasonably accurate and reproducible results.

It is accurate to ±0.001 for the samples examined and variations of more than 0.01 in the specific gravity values for different samples of the same type of natural wax at the same temperature are rather improbable.

The following are average values for the specific gravities, at 25/25°C., of various natural waxes, as determined by this method.

Yellow Carnauba Wax	1.000
Refined Candelilla Wax	0.990
Refined Ouricury Wax	1.060
Yellow, Refined Beeswax	0.960

PREPARATION OF THE SAMPLE

Take a representative sample of the wax to be examined. In the case of natural waxes, such as carnauba, candelilla or ouricury, this may require crushing a substantial amount of wax and quartering down until a representative sample of 10 or 20 g is obtained, or any other suitable method of obtaining a representative sample may be used. Melt this sample, allow any dirt or sand to settle to the bottom, remove any foam or air bubbles

[1] Cartwright, Foster D. Snell Inc.

from the surface and pour into a clean, smooth, flat-bottomed mold, such as a small glass or porcelain dish, to a depth of 2 or 3 mm of the melted wax sample. Allow this to cool, and leave at 25°C. for at least 2 or 3 hours to permit full shrinkage to occur. Then cut a piece of the wax slab, approximately 2 × 4 cm, weighing about 2 g. Scrape the edges smooth and slightly round all corners. This method of preparation yields a representative specimen of the wax, which is substantially free from foreign particles, internal voids due to shrinkage on cooling and air bubbles. It has also suitable size and shape for determination of its specific gravity by the method to be described in the following.

DETERMINATION OF SPECIFIC GRAVITY AT 25/25°C.

Use a Hubbard-type specific-gravity bottle, approximately 70 mm high and 22 mm inside diameter, having a ground-in glass stopper with concave bottom and small hole in the center. Make all weighings accurately to 0.0001 g at 25°C. Weigh the wax specimen. Weigh the bottle, filled with water at 25°C., stoppered carefully and wiped dry, then submerge the wax specimen in the water slowly and carefully so that no air bubbles cling to the surface of the wax, replace the stopper with care. wipe dry, and weigh again. In order to facilitate the wetting of the surface of the wax specimen, it is desirable to add a very small amount of wetting agent to the water in the bottle. Addition of 0.1% Carbowax 4000 is quite effective, has little tendency to cause the water to foam excessively and does not alter the density of the water to an extent as to require consideration within the limits of accuracy of the method.

The most probable sources of error in the method lie in failure to wet the wax specimen perfectly and avoid all air bubbles, failure to seat the ground glass stopper with the same force each time, and failure to maintain accurate temperature control. However, reasonably careful manipulation permits reproducibility of results to within ±0.001 g in the weight of the bottle filled with water, or filled with the wax specimen and water, in successive determinations.

The specific gravity, at 25/25°C., of the wax specimen is calculated as follows:

$$\text{Specific Gravity} = \frac{\text{Weight of Wax}}{\text{A} - \text{B} + \text{Weight of Wax}}$$

Where A = weight of the bottle filled with water; B = weight of the bottle filled with wax and water.

For example, if the wax sample weighs 2.0000 g, the bottle filled with water weighs 50.0000 g and the bottle filled with wax and water weighs 49.9800 g, then the specific gravity = 2.0000/2.0200 = 0.990. An error of ± 0.001 g in either A or B would result in an error of ± 0.0005 in the specific gravity.

IDENTIFICATION OF MIXED WAXES

S. Zweig and A. Taub
Columbia University

In the analysis of commercial wax-containing products, such as cosmetic preparations, polishes, etc., the identification of the wax components has been difficult and complex. With the growth of the wax-polish industry and the increased consumption of waxes for other uses, a need has developed for a definite scheme for the analysis of wax mixtures. In view of the complexity of the problem, research was confined to the identification of the more commonly used waxes such as spermaceti, beeswax, carnauba, candelilla, and montan. Some allied substances, such as cetyl alcohol, rosin (colophony), and stearic acid, which might interfere in the analysis, were also considered.

A comparison of the chemical compositions of the commonly used natural waxes shows that, in general, they consist of similar components; in fact, beeswax, carnauba, candelilla, and montan contain compounds of identical composition, i.e., acids, alcohols, and hydrocarbons with 24 to 34 carbon atoms. The identification is made even more difficult by the similarity of properties and relative inertness of the saturated homologous compounds of high carbon content found in waxes. Separation of the adjacent homologous compounds in the vicinity of C_{24} to C_{34} has been found difficult and often impossible. Therefore, any method based on fractionation of these compounds would not be suitable for analytical purposes. There are, however, outstanding differences in the composite nature of these waxes as a whole which may be used as a basis for analysis. These are: variation in proportion of acids, alcohols, and hydrocarbons;

difference in chain length of the two components of the ester; and the presence of additional distinctive components, such as resinlike substances and ketones.

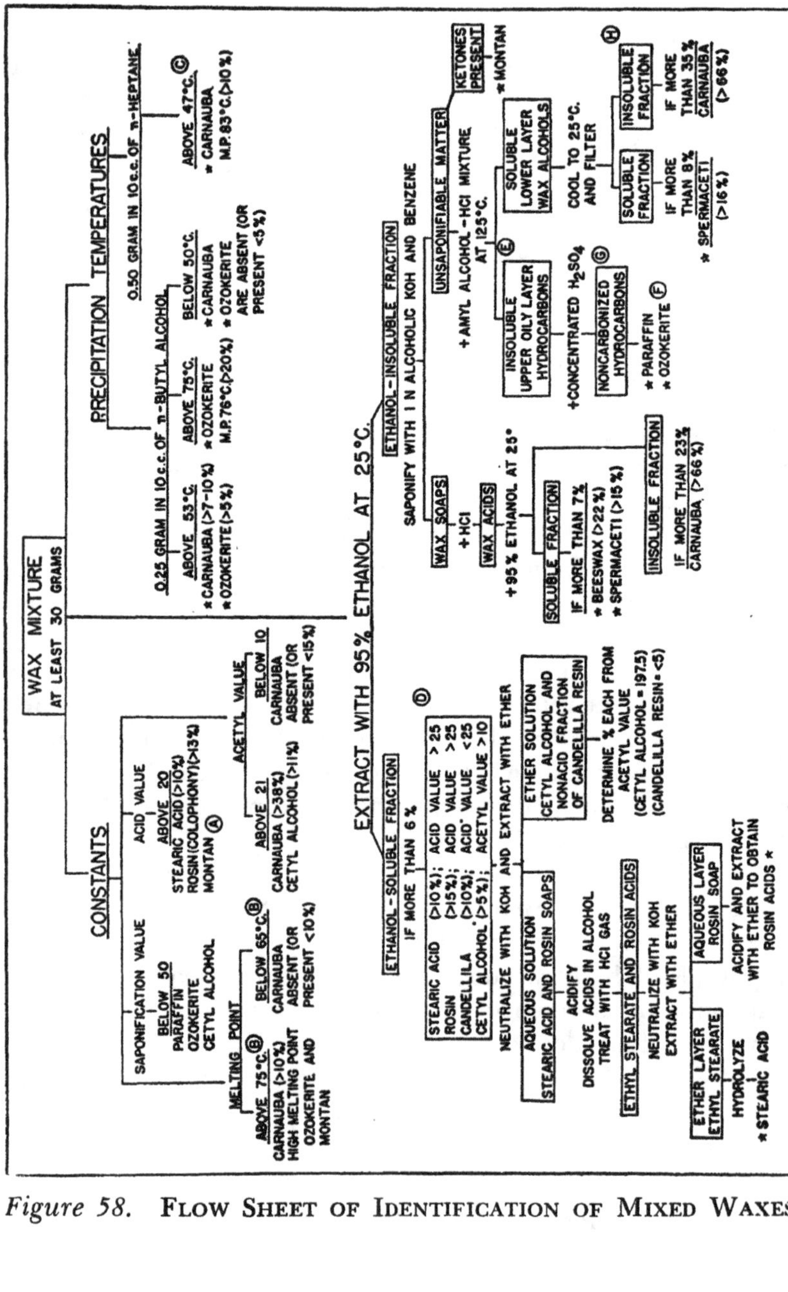

Figure 58. FLOW SHEET OF IDENTIFICATION OF MIXED WAXES

The method outlined as follows utilizes these differences. Thus a practicable system of identification has been evolved, which is based on 1. the determination of pertinent physical and chemical constants, and 2. the quantitative separation of waxes into groups of homologous compounds and the determination of the properties of the separated fractions.

SCHEMATIC OUTLINE FOR IDENTIFICATION (SEE FIGURE 58)

(All percentages are based on the weight of the original sample of wax mixture.)

(A) Variations in the composition of montan wax make a definite statement concerning its quantity impossible.

(B) Based on data from Koch *et al.*[1] and Lewkowitsch.[2]

(C) Candelilla wax forms a cloudy solution in *n*-heptane and, therefore, interferes with determination of precipitation temperatures in this solvent.

(D) Rosin and candelilla resin form cloudy solutions in *n*-heptane.

(E) Absence of a hydocarbon fraction (solid at 25°C.) indicates the absence of not only paraffin wax and ozokerite but also of candelilla wax. (>5%), beeswax (>10%), and montan wax.

(F) 20–30% of raw ozokerite is attacked by concentrated sulfuric acid.

(G) Hydrocarbons from candelilla, beeswax, and montan are carbonized.

(H) Since alcohols of other waxes are also found in this fraction and are closely related to carnauba-wax alcohols, this fraction is of value only when a high percentage of carnauba is present (over 60%).

The percentages in parentheses indicate the quantity that will always be detected by the determination; when preceded by ">" it indicates that the presence of the quantity stated, or more, will be detected. A smaller quantity may or may not be de-

---

[1] Koch, Hable, and Wrangell, *Ind. Eng. Chem., Anal. Ed.* **10,** 166 (1938).

[2] Lewkowitsch, *Chemical Technology and Analysis of Oils, Fats, and Waxes*

tected, depending on a knowledge of the presence or absence of certain other waxes. For example, more than 8% of wax alcohols, soluble in amyl alcohol–hydrochloric acid at 25°C., indicates spermaceti (16%). When the spermaceti, present in the wax mixture, exceeds 16%, the fraction soluble in cold amyl alcohol–hydrochloric acid mixture will always be more than 8%, and spermaceti will always be detected by this determination. However, if the spermaceti in the wax mixture is less than 16%, wax alcohols, soluble in amyl alcohol–hydrochloric acid, may or may not exceed 8%, depending on the nature and quantity of other waxes present in the mixture.

The percentage of the fraction, soluble in amyl alcohol–hydrochloric acid, for all the waxes determined is: spermaceti, 48%; candelilla, 8%; beeswax, 5%; carnauba, 3%. As candelilla contains as much as 8% of substances soluble in amyl alcohol–

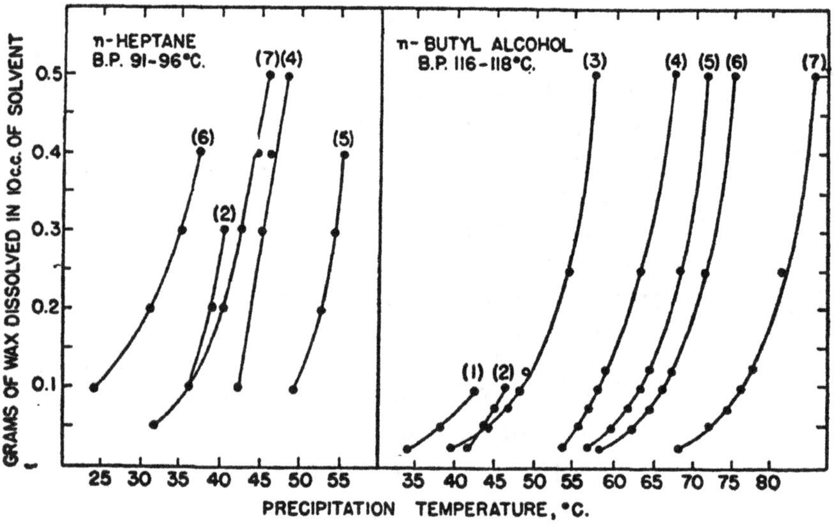

Figure 59. PRECIPITATION TEMPERATURES OF WAXES

1 Beeswax, M.P. 60–62°C.
2 Bleached Montan, M.P. 68°C.
3 Candelilla, M.P. 70.5°C.
4 Refined Carnauba, M.P. 81.5°C.
5 Yellow Carnauba No. 1, M.P. 83°C.
6 Ozokerite, M.P. 65–67°C.
7 Ozokerite, M.P. 76–78°C.

hydrochloric acid, spermaceti cannot be determined with certainty unless this fraction is over 8% of the total wax mixture (16% of spermaceti). However, if the approximate quantity of candelilla and/or beeswax is first ascertained, less than 16% of spermaceti can be detected.

Where the name of a substance is starred, that particular determination is of greatest analytical significance for the identification of the substance.

Procedure

The finely grated wax is weighed and added to 10 cc solvent contained in a jacketed test tube (the inner tube is 1.5 × 13.75 cm; the outer, 2.5 × 15 cm), and then placed in a thermostatically controlled bath which has been previously heated to a temperature at which the wax will dissolve. The temperature of the bath is then allowed to decrease gradually, not more than 1° a minute. The temperature is recorded at the point at which crystals just begin to form. In most cases, this point is fairly sharp. In others, however, because of the gradual precipitation, the exact temperature cannot be determined. The limit of observational error is ± 1°C.

Results

The precipitation temperatures in *n*-heptane and *n*-butyl alcohol are plotted in Figure 59. It is evident that the precipitation temperatures of carnauba and ozokerite are much higher than those of other waxes. It may, therefore, be said that if, for example, 0.125 g of a wax in 10 cc of *n*-butyl alcohol has a precipitation temperature above 50°C., it probably contains ozokerite or carnauba. When the precipitation temperature of 0.100 g of wax in 10 cc of *n*-butyl alcohol is above 64°C., a quantity of high-melting ozokerite is present in the wax mixture. In this case, the presence of carnauba, admixed with ozokerite cannot be detected. However, if the wax mixture is then dissolved in *n*-heptane, the order of solubility of the two waxes is reversed and carnauba precipitates at a higher temperature than ozokerite. A determination of the precipitation temperatures of definite concentrations of wax in both *n*-butyl alcohol and *n*-heptane, therefore, will permit the detection of both ozokerite and carnauba. Their presence is further verified by a quantitative separation of the wax components and by the other constants determined.

Beeswax was omitted from the *n*-heptane determinations because it remains dissolved in the concentrations specified, even at room temperature. Candelilla and rosin (colophony) form cloudy solutions in boiling *n*-heptane; their presence would, therefore, interfere with the determination of the precipitation temperatures in this solvent. However, the precipitation temperatures in *n*-butyl alcohol are not influenced by their presence.

SEPARATION AND COMPARISON OF THE PROPERTIES OF THE WAX COMPONENTS

Alcohol-Soluble Fraction

Alcohol extraction is an established method for the determination of stearic acid and rosin in beeswax. In order to extend this procedure to the estimation of these substances in other waxes, it is necessary to know the quantity as well as the properties of the alcohol-soluble fractions of all the waxes.

Procedure

Fifty grams of wax are melted in a 400 cc beaker and extracted by treating with four 150 cc volumes of boiling 95% alcohol. The alcohol solution is decanted while hot and allowed to cool to 25°C. The precipitate is then collected on a Büchner funnel, washed twice with 50 cc of cold alcohol, and combined with the residue that remains from the original hot extraction. The filtrate is evaporated and the residue weighed.

The alcohol-soluble extract of candelilla is a clear resinous substance of characteristic aromatic odor. It is completely soluble in cold acetone, alcohol, benzene, ether, terpineol and ethylene chlorohydrin and partially soluble in hot acetic anhydride and in boiling *n*-heptane and cyclohexane. In 30 parts of *n*-heptane, 15% of the resinlike substance remains undissolved. A reddish purple color, which changes to dark purple on standing, is obtained in the Liebermann-Storch test.

The alcohol-soluble fraction of carnauba consists of a greenish, gummy, and resinlike matter of pungent, aromatic odor. It shows about the same solubilities as the candelilla resin, and in 30 parts of cold *n*-heptane, 16.5% remains undissolved. The result of the Liebermann-Storch color reaction is similar to that for candelilla resin.

The alcoholic extracts from beeswax, spermaceti, and bleached montan have a soft waxlike consistency and are completely soluble in *n*-heptane and cyclohexane. The corresponding candelilla and carnauba fractions produce cloudy solutions when dissolved in these solvents.

Alcohol-Insoluble Fraction

Ten grams of the fraction insoluble in 95% alcohol at 25°C. is saponified by boiling for 2 hours with a mixture of 50 cc benzene and 25 cc alcoholic potassium hydroxide (45 g of potassium hydroxide in 1,000 cc absolute alcohol). To this, 50 cc hot water is added and the aqueous layer drawn off and washed with three successive volumes of 50, 30, and 30 cc benzene. The combined benzene solution is washed with an alkaline aqueous alcohol mixture (2 to 1). The benzene is evaporated, and the unsaponifiable matter dried at 110°C. To the soap solution, 15 cc of concentrated hydrochloric acid is added and the separated acids are shaken out with hot benzene. The benzene layer, separated from the acid aqueous layer, is washed twice with equal volumes of water, and evaporated. The acids are then dried at 110°C.

By this method, the alcohol-insoluble fraction is divided into: 1. substances soluble in aqueous alkali after saponification (chiefly wax acids); and 2. substances insoluble in aqueous alkali, but soluble in benzene (chiefly alcohols and hydrocarbons, known collectively as the "unsaponifiable matter").

Acids

The acids are dissolved in 20 parts of hot 95% alcohol, and the solution is allowed to cool to 25°C. and filtered. Two fractions result: 1. acids soluble in cold alcohol and 2. acids insoluble in cold alcohol.

Unsaponifiable Matter

The unsaponifiable matter (from 10 g wax) is dissolved in 100 cc hot amyl alcohol and 100 cc concentrated hydrochloric acid (36%) are added. The mixture is stirred and boiled over an asbestos plate for about 5 minutes. The presence of hydrocarbons is evidenced by an oily layer on the surface of the liquid. On cooling, a disk of solidified hydrocarbons can be lifted from the lower amyl alcohol–hydrochloric acid layer. The residual liquid is filtered through a Büchner funnel and the precipitate washed twice with 25 cc amyl alcohol–hydochloric acid mixture. The filter paper with the precipitate is placed in a large evaporating dish and boiling water is added. The oily layer of wax alcohols rises to the top and solidifies on cooling. The filtrate containing the soluble alcohols is washed in a separatory funnel with hot water until free from hydrochloric acid. The amyl alcohol is then evaporated or vacuum distilled and the alcohols are dried and weighed.

This divides the unsaponifiable matter into three fractions: 1. substances insoluble in hot amyl alcohol–hydrochloric acid mixture (chiefly hydrocarbons); 2. substances soluble in hot amyl alcohol–hydrochloric acid mixture, but insoluble in cold amyl alcohol–hydrochloric acid mixture (mainly alcohols of high molecular weight, such as ceryl or myricyl); 3. substances soluble in cold amyl alcohol–hydrochloric acid mixture (mainly low-molecular-weight wax alcohols, such as cetyl alcohol).

Table 138

PROPERTIES OF THE FRACTIONS SEPARATED FROM WAX

	Spermaceti			
Fraction	Entire Wax %	M.P. °C.	n_D 90°C.	Neutral or Acid No.
Entire Wax	—	43.5–44.0	1.4252	—
Soluble in Cold Alcohol	5	32.5–33.0	1.4295	10
Insoluble in Cold Alcohol	95	43.5–44.0	1.4252	—
Total Acids	45	37.8	1.4199	218
Soluble in Cold Alcohol	45	37.8	1.4199	218
Insoluble in Cold Alcohol	None	—	—	—
Unsaponifiable Matter	48	46–50.0	1.4204	—
Hydrocarbons	None	—	—	—
Alcohols (Plus Other Unsaponifiable Matter)				
Insoluble in Cold Amyl Alcohol–HCl	None	—	—	—
Soluble in Cold Amyl Alcohol–HCl	48	44–47.0	1.4200	—
Acetyl Number of Entire Wax	—	2.6	—	—
Saponification Number of Entire Wax	—	128–130	—	—

Table 138 (Continued)

Fraction	Beeswax: Entire Wax %	M.P. °C.	n_D 90°C.	Neutral or Acid No.
Entire Wax	—	60–62.0	1.4327	19.0
Soluble in Cold Alcohol	5.0	60.0	1.4391	88.5
Insoluble in Cold Alcohol	95.0	62–62.5	1.4326	—
Total Acids	40.0	56–59.0	1.4308	170.0
Soluble in Cold Alcohol	30.0	46.6–48.0	—	187.0
Insoluble in Cold Alcohol	9.5	75–78.0	—	105.0
Unsaponifiable Matter	55.0	71–72.0	1.4298	—
Hydrocarbons	20.0	51.7–52.0	1.4264	—
Alcohols (Plus Other Unsaponifiable Matter)				
Insoluble in Cold Amyl Alcohol–HCl	30	79–80.0	1.4310	—
Soluble in Cold Amyl Alcohol–HCl	5	—	—	—
Acetyl Number of Entire Wax	—	15	—	—
Saponification Number of Entire Wax	—	93	—	—
Entire Wax	—	83.0	1.4496	7
Soluble in Cold Alcohol	3	—	1.4916	19
Insoluble in Cold Alcohol	97	83.3	1.4493	—
Total Acids	40	78.0	1.4441	—
Soluble in Cold Alcohol	5	73.8	—	193.5
Insoluble in Cold Alcohol	35	81.6	—	108.5
Unsaponifiable Matter	55	88.3–89.0	1.4375	—
Hydrocarbons	None	—	—	—
Alcohols (Plus Other Unsaponifiable Matter)				
Insoluble in Cold Amyl Alcohol–HCl	52	86–87.0	1.4358	—
Soluble in Cold Amyl Alcohol–HCl	3	—	—	—
Acetyl Number of Entire Wax	—	51–60	—	—
Saponification Number of Entire Wax	—	79–88	—	—
Entire Wax	—	70–70.5	1.4600	22
Soluble in Cold Alcohol	19	—	1.5115	24
Insoluble in Cold Alcohol	81	72–73.0	1.4503	—
Total Acids	29–30	80–80.5	1.4496	115
Soluble in Cold Alcohol	7.5	50–50.3	1.4734	177
Insoluble in Cold Alcohol	22.5	82.5	1.4428	91
Unsaponifiable Matter	50	67–68.0	1.4476	—
Hydrocarbons	30	66.8	1.4325	—
Alcohols (Plus Other Unsaponifiable Matter)				

Table 138 (Continued)

Fraction	Montan (Bleached) Entire Wax %	M.P. °C.	n_D 90°C.	Neutral or Acid No.
Insoluble in Cold Amyl Alcohol–HCl	11	78.2	1.4417	—
Soluble in Cold Amyl Alcohol–HCl	8	—	1.5136	—
Acetyl Number of Entire Wax	—	9–21	—	—
Saponification Number of Entire Wax	—	57	—	—
Entire Wax	—	67.5–68.5	1.4270	45
Soluble in Cold Alcohol	3	52.0	1.4386	29.5
Insoluble in Cold Alcohol	97	68.5–68.7	1.4270	—
Total Acids	38	77–79.0	1.4325	109
Soluble in Cold Alcohol	2.6	—	—	93.3
Insoluble in Cold Alcohol	35.2	78–79.0	—	118
Unsaponifiable Matter	58.2	51.5–51.7	1.4246	—
Hydrocarbons	52.6	52.4	1.4236	—
Alcohols (Plus Other Unsaponifiable Matter)				
Insoluble in Cold Amyl Alcohol–HCl	5	—	—	—
Soluble in Cold Amyl Alcohol–HCl	Trace	—	—	—
Acetyl Number of Entire Wax	—	11.2	—	—
Saponification Number of Entire Wax	—	77	—	—

The properties of the various fractions, separated from each wax by the procedure outlined, are listed in the preceding table.

The results of the preceding determinations will be interpreted as follows. Allied substances, such as stearic acid, cetyl alcohol, and rosin are included, since they are often found admixed to waxes and their presence may affect the interpretation of the results.

From a knowledge of the constants, certain generalities may be inferred, for example, a low saponification number (below 50) indicates the presence of paraffin wax or ozokerite; a very low acid number (below 3 or 4) excludes the presence of more than 10% beeswax, montan, or candelilla, and indicates the absence of rosin or stearic acid. Similarly, a low acetyl number (below 10) excludes large quantities of carnauba or free cetyl alcohol; a low melting point (below 65°C.) shows the absence of more than 10% carnauba wax.

The precipitation temperatures indicate the presence or absence of ozokerite and carnauba.

An examination of the alcohol-soluble fraction yields valuable information concerning a wax mixture. For example, when the quantity of alcohol extract is below 5%, the absence of candelilla, rosin, stearic acid and cetyl alcohol may be definitely inferred. Candelilla contains 20% of an alcohol-soluble, resinous substance, of low acid number. Consequently, the presence of more than 5% of an alcohol-soluble substance of high acid number (above 25) indicates the presence of, e.g., stearic acid or rosin.

The separation of the ingredients of the alcohol-insoluble fraction and the determination of their properties yield more specific information concerning the wax mixture. The wax acids are divisible into two groups, on the basis of solubility. The acids, soluble in cold alcohol, are compounds of low molecular weight, chiefly myristic, palmitic, and adjacent homologous acids, melting below 51°C. Those insoluble are compounds of high molecular weight, with 24–34 carbon atoms and melting point above 75°C. The entire acid fraction of spermaceti and 75% of the acids of beeswax are composed of lower-melting compounds. Therefore, if the alcohol-soluble acid fraction is absent, or present in small quantities (less than 7.5% of the entire wax), spermaceti and beeswax are either absent or present in small quantities only. A further verification can be obtained by an examination of the wax alcohols.

The wax alcohols have similarly been divided into a fraction soluble in cold amyl alcohol–hydrochloric acid mixture and a fraction insoluble in this solvent. The entire unsaponifiable fraction from spermaceti is composed of alcohols soluble in cold amyl alcohol–hydrochloric acid, whereas all the other waxes, including beeswax, yield comparatively small percentages of substances soluble in this solvent at 25°C. Therefore, the presence or absence of spermaceti would be indicated by the determination of the quantity of matter soluble in cold amyl alcohol–hydrochloric acid mixture. Since spermaceti is essentially the cetyl ester of myristic and lauric acids, the ratio of soluble acids to soluble alcohols is approximately 1 to 1. A higher ratio, such as 1.5 parts or more of soluble acids to 1 part of soluble alcohols, would, therefore, indicate the presence of beeswax.

The hydrocarbon fraction offers a means for detecting the presence of ozokerite or paraffin. The wax hydrocarbons are carbonized when heated to 110°C. with concentrated sulfuric acid,[1, 2] while the paraffin hydrocarbons are unattacked and can be extracted from the diluted acid mixture with a volatile solvent, such as ether or petroleum ether. The quantitative determination of ozokerite and paraffin, when both are present, is difficult. Graefe's method,[3] often employed for this purpose, is based on the fact that 50–60% of ceresin hydrocarbons is precipitated from their solution in carbon disulfide by the addition of a mixture of ether and alcohol (1 to 1), whereas paraffins remain in solution. Other methods, based on differences in solubility, melting point, and refractive index, are described by Holde.[4]

The following description of the important distinguishing characteristics of each wax is presented as a supplement to this discussion.

SPERMACETI

The unsaponifiable matter, totaling 48%, is found in the fraction soluble in cold amyl alcohol–hydrochloric acid. All the acids are soluble in cold ethyl alcohol. When added to other waxes, spermaceti causes a lowering of the acid number, melting point, and acetyl number. The ethanol-soluble fraction is small, and hydrocarbons are absent.

BEESWAX

Cold alcohol dissolves 75% of the acids. If spermaceti is present, as indicated by the quantity of cold amyl alcohol–hydro-

[1] Better and Davidsohn, *Taschenbuch für die Wachsindustrie* (Stuttgart, Wissenschaftliche Verlagsgesellschaft, 1932).

[2] Lüdecke, *Die Wachse und Wachskörper* (Stuttgart, Wissenschaftliche Verlagsgesellschaft, 1926).

[3] Graefe, *Chem. Z.* **27**, 248 (1903); *Analyst*, **28**, 190 (1903).

[4] Holde, *Examination of Hydrocarbon Oils and of Saponifiable Fats and Waxes* (New York, John Wiley, 1922).

chloric acid fraction, the amount of acids in excess of the combining ratio of alcohols and acids, 1 to 1, which form the ester of spermaceti permits the calculation of the approximate quantity of beeswax.

CARNAUBA WAX

The precipitation temperatures of carnauba wax in *n*-butyl alcohol and in *n*-heptane are much higher than those of other waxes (ozokerite excepted). Hydrocarbons are present in but very small amounts (about 1%). The acetyl number is much higher than that of any other wax studied. The melting point of a wax mixture is considerably increased by the addition of small quantities of carnauba wax.[1, 2] The acids and alcohols are mostly of high molecular weight; consequently, both the cold alcohol-soluble acid fraction and the fraction soluble in cold amyl alcohol–hydrochloric acid are very small.

CANDELILLA WAX

The alcohol-soluble fraction consists of resinous matter and constitutes approximately 20 % of the wax. This fraction has a low acid number as compared to rosin (colophony) which has an acid number of about 160. The resinous matter may be separated from rosin and stearic acid by neutralizing the alcohol-soluble fraction with potassium hydroxide, adding water, and extracting with ether or benzene.

The acid fraction of candelilla is dark brown and more than 75% is insoluble in cold alcohol. The hydrocarbon fraction is relatively large and differs from paraffin or refined ozokerite hydrocarbons in being readily carbonized by concentrated sulfuric acid

MONTAN WAX

There are considerable differences in the composition of the three varieties of montan wax, but all invariably contain montanic acid ($C_{28}H_{56}O_2$), free or combined as an ester. The propor-

---

[1]Koch, Hable, and Wrangell, *Ind. Eng. Chem., Anal. Ed.,* **10,** 166 (1938).

[2] Lewkowitsch, *Chemical Technology and Analysis of Oils, Fats, and Waxes* (London, Macmillan Co., 1922), Vol. II, p. 889.

tion of hydrocarbons is large, and ketones have been found in the raw and the distilled products. The presence of resinous substances and sulfur-containing compounds differentiates raw montan wax from the distilled and bleached products.

Because of the absence of alcohols, the acetyl number of the unsaponifiable matter of distilled wax is for all practical purposes zero, whereas the raw wax and refined wax contain varying amounts of alcohols of high molecular weight, such as ceryl or myricyl. The presence of ketones in the unsaponifiable matter is an important clue to the identification of montan. Ketones are estimated by reducing with sodium in alcohol to secondary alcohols and determining the difference in hydroxyl number before and after reduction. They may also be estimated by the method of Mangouri,[1] which is based on the insolubility in boiling methyl alcohol of the barium salt of a hydrazone formed when the ketones are treated with *p*-hydrazinebenzoic acid.

PARAFFIN

The hydrocarbons are not attacked by concentrated sulfuric acid, whereas most other wax hydrocarbons are carbonized by this acid. The addition of paraffin to a wax mixture appreciably lowers the saponification number.

OZOKERITE

The precipitation temperatures in *n*-butyl alcohol and *n*-heptane are much higher than those of all the other waxes (carnauba excluded). The hydrocarbons of the refined ozokerite are not attacked by concentrated sulfuric acid.

The precipitation temperatures of the unattacked hydrocarbons can be used to differentiate paraffin qualitatively from ozokerite. Graefe's method* may be applied to determine whether both are present.

ROSIN (COLOPHONY)

Rosin, which is found in the cold alcohol-soluble fraction, has a very high acid number (about 160), and can be differen-

[1] Mangouri, *Biochem. J.*, **31**, 1978 (1937).

*See reference 3 on page 298.

tiated from stearic acid in that rosin acids are not converted into ethyl esters when treated in alcohol solution with hydrogen chloride gas.[1] Addition of rosin to a wax mixture raises the acid number considerably (10% of rosin increases the acid number by 16). Color reactions for rosin, such as the Libermann-Storch reaction, are not very reliable since candelilla wax and montan wax produce similar colors.

STEARIC ACID

This acid, like rosin, is found in the alcohol-soluble fraction, and is differentiated from rosin as indicated under Rosin. The acid value of a wax mixture is increased by 20 when 10% stearic acid is added.

CETYL ALCOHOL

This alcohol is identified by the high acetyl value of the alcohol-soluble fraction. The acetyl value of the entire wax mixture is increased by 20 for each 10% of added cetyl alcohol.

RATIO NUMBERS OF WAXES

The ratio number often serves as a good guide in detecting adulteration of beeswax.[2]

$$\text{Ratio Number} = \frac{\text{Saponification No.} - \text{Acid No.}}{\text{Acid No.}}$$

Ratio Numbers	
Ceresin	0
Paraffin Wax	0
Ozokerite	0
Stearic Acid	0
Japan Wax	11
Insect Wax	29
Carnauba Wax	39
Tallow	48
Myrtle Wax	68

[1] Lewkowitsch, *Chemical Technology and Analysis of Oils, Fats, and Waxes* (London, Macmillan Co., 1921), Vol. I, p. 639.

[2] Fryer and Weston, *Technical Handbook of Oils, Fats and Waxes* New York University Press, 1917.

DETECTING WAX ADDITIVES[1]

Of the number of possible methods for detecting additives in vegetable waxes, the technique of solvent leaching followed by specific tests on the extracted residue holds most promise.

LEACHING TECHNIQUE

An accurately weighed sample of about 10 g vegetable wax is finely pulverized with mortar and pestle and placed in a 250 ml Erlenmeyer flask with 100 ml chloroform. The sample is agitated with an electric stirrer for 2 hours at 25°C. Then the flask is immersed in an ice bath (optional step) for 1 hour, with occasional manual stirring. The sample is then filtered and the solvent evaporated from the filtrate. A vacuum oven can be used near the end of the evaporation to remove last traces of solvent and prevent charring. The residue is accurately weighed. After obtaining an additive-enriched residue, specific tests may be developed to detect individual components.

The best solvent to use is one which has maximum solubility for a wide range of possible additives and minimum solubility for the vegetable waxes. Of a large number of solvents tested, chloroform, carbon tetrachloride, and toluene were the best.

Tables 139 and 140 indicate the effect of varying time on the extraction and cooling steps in the procedure.

Table 139

VARIATION OF TIME IN EXTRACTION OF SAMPLES (PURE CARNAUBA–FATTY GRAY)

	Sample grams	Residue from $CHCl_3$ Extraction, grams	Acid No. of Residue	Acid Equivalent*
Extraction 2 Hours Ice Bath 2 Hours	10.015	0.810	14.2	11.5
Extraction 2 Hours Ice Bath ½ Hour	10.011	0.865	13.4	11.6
Extraction 3 Hours Ice Bath 1 Hour	10.000	1.036	10.4	10.8
Extraction 1 Hour Ice Bath 1 Hour	10.005	0.798	13.0	10.4

* Obtained by multiplying acid number of residue times weight of residue.

[1] Marsel, Treacy and Godino, *Soap and Sanitary Chemicals,* July 7951.

Table 140

VARIATION OF TIME IN EXTRACTION OF SAMPLES
(5% DUREZ WITH CARNAUBA WAX)

	Sample grams	Residue from $CHCl_3$ Extraction grams	Acid No. of Residue	Acid Equivalent *	% Durez Extracted as Calculated
Extraction 3 Hours Ice Bath 1 Hour	10.006	1.369	31.7	43.4	93
Extraction 1 Hour Ice Bath 1 Hour	10.017	1.271	32.1	39.2	82

* Obtained by multiplying acid number of residue times weight of residue.

One of the essential facts to be established is the natural variation in the weight of residue from extraction of a series of representative samples of a natural wax. Based on eight samples of carnauba wax, there is a maximum 20% variation from the average for the weight of extracted residue.

The extracted residue may then be examined qualitatively and quantitatively. This was done by: 1. weight of residue compared to normal weights of extract; 2. acid number; 3. saponification number; 4. special tests, such as concentrated sulfuric acid extraction to indicate paraffins.

A group of control samples was prepared from a basic sample of No. 3 pure crude carnauba wax, whose acid number was 3.8 and saponification number, 84. A series of possible wax additives or extenders was selected and individual samples, containing 5% of additive added to the wax, were prepared.

Table 141 indicates results of solubility studies on these wax compounds. As a result of these studies, chloroform was selected as the solvent for further work.

Table 142 lists some chemical constants for the extenders. The constants were determined by methods developed earlier. For example, the solvent mixture: isopropanol, *n*-amyl alcohol, and toluene (8:1:1) was used in determining the saponification number of all samples, except Damar Singapore, Piccolyte S-115 and Batu East India. The solvent used for the last substances was a 50-50 mixture of ethanol and toluene.

Table 141

SOLUBILITY OF EXTENDERS IN SOME COMMON SOLVENTS

Components	$CHCl_3$ 25°C.	$CHCl_3$ 0°C.	Toluene 25°C.	Toluene 0°C.	Acetone 25°C.	Acetone 0°C.	C_6H_6 25°C.	C_6H_6* 0°C.	CCl_4 25°C.	CCl_4 0°C.	Ethyl Acetate 25°C.	Ethyl Acetate 0°C.
Shanco No. 300	1	2	1	2	2	2	1	—	1	1	1	1
Damar Singapore	1	1	1	1	2	3	1	—	1	1	2	3
Lewisol	1	1	1	1	1	2	1	—	1	1	1	1
Piccolyte S–115	1	2	1	1	2	2	1	—	1	1	2	3
Batu East India	1	1	1	1	2	2	1	—	2	2	2	2
1047 Fine-Melt Pale Congo	1	2	1	2	2	2	1	—	1	1	2	2
Hydrogenated Castor Oil	1	1	3	3	3	3	2	—	2	2	2	2
Durez No. 219	1	2	1	1	1	1	1	—	1	1	1	1
Candelilla	2	3	3	3	2	2	3	—	3	3	2	3
Pure Carnauba	3	3	3	3	4	4	3	—	3	3	2	3

* Benzene freezes at this temperature.

1 = Soluble.

2 = Nearly all soluble.

3 = Slightly soluble.

4 = Not soluble.

Tests made with 5 g extender to 25 cc solvent.

Table 142

CHEMICAL CONSTANTS OF THE PURE SUBSTANCE

Pure Components	Acid No.	Saponification No.	Iodine No.
Shanco No. 300	54.60	110.0	11.9
Lewisol	43.10	182.0	16.4
Durez No. 219	70.00	110.0	—
Damar Singapore	32.00	28.2	17.0
Batu East India	15.00	83.2	—
Piccolyte S–115	4.00	4.0	13.2
1047 Fine Melt (Pale Congo)	77.00	108.0	28.5
Hydrogenated Castor Oil	2.85	193.0	—
Candelilla Wax	12.00	70.0	11.7
Pure Carnauba Wax	3.76	84.0	—
Bleached Carnauba Wax	8.10	87.6	—

Table 143

CHEMICAL TESTS ON CHLOROFORM EXTRACTS OF EXTENDED WAX SAMPLES

Pure Carnauba with 5% of the Following	Sample g	Residue from $CHCl_3$ Extract g	Acid No. of Extract	Acid Equivalent	% Resin in Sample as Calculated from Acid Equivalents
Shanco No. 300	10.021	1.184	32.5	39.5	4.8
Lewisol	10.005	1.139	23.8	27.1	3.3
Durez No. 219	10.042	1.150	36.9	42.4	4.2
Damar Singapore	10.013	1.233	21.2	36.1	7.2
Batu East India	10.011	1.030	12.7	13.1	—
Piccolyte S–115	10.016	1.200	11.3	13.5	—
1047 Fine Melt (Pale Congo)	10.011	1.191	37.8	45.0	4.2
Hydrogenated Castor Oil	10.006	1.233	12.5	15.4	—
Candelilla Wax	10.004	0.933	15.2	15.1	—
Paraffin Wax	10.032	1.100	12.0	13.2	—
Pure Carnauba Wax	10.029	0.796	17.1	13.6	—
Pure Carnauba Wax	10.109	0.929	13.4	12.5	—
Bleached Carnauba Wax	10.051	0.989	28.6	27.3	—

Many of the compounds have an acid number very much higher than carnauba wax; if these extenders are extracted, this can be used to detect them. In the case of hydrogenated castor oil, for example, the high saponification number of 193 can be so used. Similarly other tests may be devised.

Table 143 lists the results of determining acid numbers of chloroform extracts of the series of extended carnauba–wax samples mentioned earlier. In a number of cases, both the greater weight and the acid number of the residue obtained, as compared to the values for normal carnauba wax, indicate the presence of resin. For example, the residue from the Durez-extended sample contained 42.4 acid equivalents compared with 13 for pure carnauba wax. Similarly, the residue from the Piccolyte S-115 extended sample was 1.2 g compared to 0.86 g for pure carnauba, clearly indicating the presence of the Piccolyte, although not shown by the acid number. In the last column are the percentages of resin in the sample as calculated, based on values from Table 140. It will be noted that bleaching of

Table 144

CHEMICAL TESTS ON CHLOROFORM EXTRACTS OF EXTENDED WAX SAMPLES

Pure Carnauba with 5% of the Following	Sample g	Residue $CHCl_3$ Extract g	Saponification No.	Basic Equivalents	% Resin in Sample as Calculated from Basic Equivalents
Shanco No. 300	10.022	1.383	91.7	127.0	5.4
Lewisol	10.056	1.267	112.0	142.0	4.1
Durez No. 219	10.004	1.139	93.0	106.0	3.5
Damar Singapore	10.080	1.269	71.7	91.0	—
Batu East India	10.002	1.193	77.8	93.0	—
Piccolyte S–115	10.040	1.230	65.0	80.0	—
1047 Fine Melt (Pale Congo)	10.033	1.202	87.4	105.0	3.4
Hydrogenated Castor Oil	10.047	1.224	122.5	150.0	4.3
Candelilla Wax	10.024	0.885	72.3	64.0	—
Pure Carnauba Wax	10.067	0.747	93.7	70.1	—
Bleached Carnauba Wax	10.056	0.739	88.0	65.0	—

the sample would interfere somewhat with the test unless this fact was known.

Table 144 shows similar results for saponification numbers. It will be noted that glycerides, such as of hydrogenated castor oil, could be easily detected by the high saponification number of the residue from chloroform extraction.

Table 145 indicates that there is not enough variation in iodine numbers to be of any benefit. Any unsaturated compound would show up here.

Table 145

CHEMICAL TESTS ON CHLOROFORM EXTRACTS OF EXTENDED WAX SAMPLES

Pure Carnauba with 5% of the Following	Sample g	Residue $CHCl_3$ Extract g	Iodine No. of Extract*
Shanco No. 300	10.142	1.295	20.7
Lewisol	10.138	1.240	22.6
Durez No. 219	10.002	1.306	19.5
Damar Singapore	10.040	1.300	22.1
Batu East India	10.050	1.157	23.6
Piccolyte S–115	10.175	1.315	21.1
1047 Fine Melt (Pale Congo)	10.000	1.266	21.5
Hydrogenated Castor Oil	10.000	1.230	19.7
Candelilla Wax	10.032	1.023	24.4
Pure Carnauba Wax	10.002	0.869	29.6
Pure Carnauba Wax	10.105	0.840	27.1

* Hanus Method.

It would be of great value, in working on a residue by these methods, to know beforehand what components are present. Tests are being conducted to ascertain the value of infrared absorption for preliminary identification of the chloroform extracts.

Other tests are in the process of development for special components.

ODOR AND TASTE

To detect odorous or flavoring components, a thin layer of the wax is brought into contact with diluted 50 and/or 10%

alcohol for about 24 hours. The alcohol must not have a different taste and smell from pure alcohol of the same strength after having been in contact with the wax.

An odor which is not noticeable or only faint may develop on aging, heating or interaction with other components. Samples of the wax should be put into clean jars and stoppered or closed to be airtight. Two jars should be kept in a dark place, one at room temperature and the other near a radiator or on a hot-water line (not hot enough to melt the wax). Another jar should be placed in direct sunlight. After a few days, each jar should be opened and smelled. Time and temperature develop and intensify most odors.

THE MELTING OF WAXES

Commercial waxes are sold as slabs, cakes, lumps, flakes or powders. They are best melted in steam-jacketed kettles, double boilers (water baths) or electrically heated kettles whose temperatures can be readily controlled. Direct-flame heating is to be avoided, because local overheating leads to discoloration and decomposition and the fire hazard is very great.

For waxes that are not sensitive to oxidation, open kettles may be used. For waxes sensitive to oxidation, closed kettles, with or without a blanketing gas (nitrogen or carbon dioxide), are required. Since many natural waxes are contaminated with foreign matter, such as insects, leaves, dirt, fibers, etc., it is customary to separate the impurities from the molten waxes by settling or infiltration. Where the circumstances permit it, this step may be postponed until the wax is dissolved or emulsified.

Waxes should not be heated any higher than it is necessary. The heating process, at any given temperature, should be of the shortest duration. These precautions are necessary to minimize discoloration and chemical or physical changes. Good stirring, by a mechanical mixer, helps to melt waxes quickly and uniformly. Care should be taken to avoid beating in too much air which may oxidize the wax. Regulating the speed of the mixer and using suitable propellers or arms on the mixing shaft tend to minimize the amount of air that is beaten in. The number, size, shape, pitch, and location of the propellers or arms are influential in this connection.

Certain waxes contain varying amounts of water. When they are melted, the water tends to volatilize and then boil off. This results in foam formation which may cause the batch to boil over. To avoid this, the heat must be judiciously applied and, in some cases, a whipping stirrer must be used to break up foam at the surface of the molten wax. Wax-melting kettles should never be filled so high that the expansion of the wax (caused by heat) and the foam produced will come too near the top of the kettle. Carnauba wax, containing 5% water, will expand five times as much as the dry wax does when heated.

Pigments, lakes, clays, abrasives and other substances that are mixed with waxes often contain appreciable amounts of water. They should be dried before addition to a wax to be melted.

Foam is especially troublesome in the hot emulsification of waxes, where large quantities of foam develop when soap or other foam-producing emulsifying agents are used. In some cases, it is even necessary to use small amounts of defoaming agents, such as oils, higher alcohols or Foamex.

To avoid discoloration, chemical reaction or contamination, it is usual to melt waxes in glass-lined, aluminum, stainless steel, Inconel or other special-alloy kettles. Iron and copper are not recommended as they are attacked by the fatty acids present or liberated in molten waxes.

When melting together waxes and other fusible ingredients, it is best to melt that substance first which has the highest melting point. To this is then added the next highest-melting component and any other ingredients are added one at a time in the descending order of their melting points. Slow stirring should be resorted to at the start and the rate of stirring can be gradually increased as the harder components are melted. This routine saves time and heat and also prevents overheating, discoloration and decomposition. Of course, these conditions may be changed after a number of batches of a special composition have been made and their characteristics learned.

COMMERCIAL WAX SOLUTIONS

SOLVENTS FOR WAXES OR WAXLIKE SUBSTANCES

The physical properties of solvents are an important consideration in the formulation of wax or waxlike mixtures. The properties of most importance to be considered are:

Solubility and compatibility
Drying time and evaporation rate (distillation range)
Flash point
Odor
Toxicity
Clean final dry
Possible tarnishing or corrosive effects on metals

A solvent may be defined as a substance by means of which a solid can be brought into a fluid state. This definition, although inadequate, expresses the general idea of the use of solvents. The prime purpose of wax solvents is to dissolve or disperse the wax, so as to obtain suitable application properties.

In many processes, the object of employing a solvent is merely to provide a means of transferring a solid from one place to another in a convenient and desirable manner; when the transference has been accomplished, the solvent is of no further use and has to be removed as rapidly and as completely as possible, but in a manner compatible with concomitant factors. These factors make it desirable that certain characteristics of a solvent should be known and that a solvent suitable for the purpose should be used.

Liquid wax mixtures generally require moderately fast-drying solvents. For this purpose, petroleum solvents are widely used. These are available in various grades to suit regular and special requirements.

Paste–wax mixtures require a solvent drying moderately slowly. For this purpose, a petroleum solvent, like No. 10 mineral spirits or Apco 125, is often used. Apco 125 is preferred where it is desirable to improve aging qualities and reduce shrinkage. It facilitates application, particularly under unusual conditions, such as outdoor application, hot weather and rough surfaces.

Troluoil, Apcothinner, No. 10 mineral spirits, and Apco 125 are petroleum solvents which are considered to be nontoxic.

SOLVENT–WAX COMPOSITIONS

Since most waxes are not soluble in the usual commercial solvents, it is necessary to form a uniform stable paste or gel. To attain this end, while retaining the properties that make the finished product useful, it is necessary to balance carefully the waxes and solvents both as to kind and quantity. Thus, although ozokerite will give a stable gel with naphtha, it is not used alone in polishes, because of insufficient luster and because of its relatively high price. The addition of carnauba wax improves its hardness and luster and the addition of paraffin wax reduces the unit cost while improving luster.

Carnauba and paraffin waxes are crystalline and only slightly soluble. They do not have good binding or gelling properties with solvents. Ozokerite and other amorphous waxes have excellent binding or gelling properties with many solvents and can carry along and prevent the crystallization of certain amounts of crystalline waxes. The consistency of waxes in solvents is commonly regulated with paraffin wax. Such compositions usually contain 2–6% ozokerite (in addition to the other waxes) for the best consistency and stability.

In solvent–wax compositions, the high-melting wax content is usually not higher than the hydrocarbon–wax content (ceresin, ozokerite and paraffin). Only for wax pastes that are subjected to higher temperatures, e.g., tropical goods, should the amount of high-melting waxes be increased.

When adding a solvent to a melted wax composition, the solvent should be heated to a temperature above the setting point of the wax to avoid crystallization. It should be run in slowly, while mixing thoroughly by a mixing device that does not splash the contents or beat in air.

SOLVENT RETENTION OF WAXES[1]

A solution of 22 parts of 75°C. paraffin-free, Russian, refined ozokerite and 78 parts of oil of turpentine (sp. gr. 0.9800) is used as a standard for determining the amount of solvent necessary to produce pastes of like consistencies from various waxes. The parts of solvent per 100 parts of wax that will give the same consistencies are: 75°C. ozokerite, 354, 65°C. petrolatum, 233, beeswax, 294, 52–54°C. (solidification point) paraffin wax, 426, crude carnauba wax, 900, crude montan wax, 525.

In general, solvent retention decreases with increase in the

[1] H. Lux, *Allgem. Oel-u. Fett-Ztg.* **32,** 153 (1935).

setting point of waxes. This is the case with common varieties of ozokerite and paraffin wax. Beeswax possesses a markedly lower and bleached montan wax a markedly higher retention than would be expected merely from their setting points. As a measure of retention, the *retention number* (Rn) is used.

$$\text{Rn} = 100\,\frac{\text{S} - \text{Sv}}{\text{S}}$$

where S = original solvent content and Sv = loss of weight of paste. For the power of a wax (ozokerite) to increase the retention of another wax, the term *retention effect* (Re) is used. The retention effect of an ozokerite is the greater, the smaller its own retention and the higher its setting point. The same applies to paraffin waxes. As a first approximation, the retention effect can be calculated by means of the equation $\text{Re} = R_{max}/C_{Rmax}$, where R_{max} is the highest value of the retention that can be imparted to a given wax, having originally a retention of R_0, by the addition of another wax (ozokerite). C_{Rmax} is the amount of such added wax expressed as a percentage of the mixture (ozokerite and paraffin wax).[1]

Ozokerite, in the form of a solvent paste, is transparent and exhibits uniform contraction of the entire mass on evaporation of the solvent; paraffin wax is opaque and gives off solvent only at the surface.

It is impossible to pass from dilute solutions to pastes by solvent evaporation even though the threshold concentration for paste formation is reached.[2]

FAULTY SOLVENT–WAX COMPOSITIONS (SHOE CREAMS AND POLISHING WAXES)[3]

Trials, on a small scale, from the beginning of the process to the storage of the filled shoe cream, are very important and even indispensable in avoiding the following expensive faults:

Folds or wrinkles with matt (dull) and often patchy surfaces

---

1 L. Ivanovszky, *Oil Colour Trades J.* **96**, 365 (1939).

2 L. Ivanovszky, *Petroleum* **4**, 100 (1941).

3 J. and A. Davidsohn, *Shoe Creams and Polishing Waxes* (New York, Chemical Publishing Co., 1940).

owing to harsh cooling with cold air currents, especially with a high content in montan wax.

Bad mirror surface because of unsuitable filling temperature, overcooling, overheating of the wax melt, too high content in hard wax, too much heavy benzine or too volatile benzine.

Veiled bloom on the surface, with black creams, caused by too little or too much nigrosine base, vapory filling room, too much hard wax.

Incrustation on the surface due to too much vegetable wax, too volatile turpentine substitute, poor binding, which disturbs the uniform internal crystallization of the cream, so that after the cream has set, a further crystallization takes place on the surface, and the wax components dissolved in the solvent remain, after evaporation, as a mouldy incrustation. It may also be caused by absorption of the easily soluble particles of the cream, by sweat, to the point of saturation so that, on evaporation, the same process take place.

Sweating in the tin owing to putting highly chilled cream into too warm a store, insufficient internal binding, too much paraffin, too short storage of the fresh cream or too warm storage. The highly dispersed wax molecules absorb more thinning agent at cooler external temperatures, which, when increasing the temperature, separate in the form of drops on the edges and surface of the tin.

Granular condition may be caused by lack of homogeneity through too rapid or too cold addition of the solvent, too much montan wax, too much paraffin, too high melting or thinning temperature, too prolonged setting process. Sometimes granulation can be avoided by including ¼–2% Aerosol OT (100%).

Rattling in the tin is due to shrinkage of the cream owing to too much hard wax, use of paraffins of too high a melting point, bringing still warm cream, which has not yet fully bound itself with the solvent, into too cold a storeroom, too hot filling, badly fitting tin lids, too warm storage.

Filling Containers with Wax Composition

When filling containers with a wax composition, care must be taken to avoid bubbles which mar the surface and cause excessive shrinkage or collapse on cooling. One method of avoiding this is to allow the melted wax to stand until most of the air, which may have been introduced by mixing, pumping or flowing, is dissipated. The filling should be through a funnel or tube that reaches as far down as possible into the container. Sometimes, especially in the case of large containers, these are filled nearly to the top and allowed to "set." Then the rest of the space is carefully filled. During the process of setting or congealing, the containers should not be disturbed. The lids should be applied only after complete setting and cooling. Even after this, the wax composition should be allowed to age for a few days.

Where the appearance of the surface layer of a wax compound is of importance, certain techniques in formulation and handling are necessary. The characteristic surface rings developed in some carnauba wax–solvent compositions are due to fractional crystallization of one or more components. This ring formation and high gloss are enhanced by rapid evaporation or cooling. Artificially chilled air is used by some, which is led over the open containers. The containers may also be conveyed through a cooled compartment, on a moving belt. Too rapid cooling must be avoided as it may yield nonhomogeneous products and cause separation.

The filling operation is best done in a room having an exhaust system to remove fumes and with air that has been passed through a filter and drier to remove dust and excessive moisture. A temperature of 55—65°F. is best for most purposes. For every wax composition, there are optimum filling conditions of temperature, speed, etc. These can be determined only by experiment. When air-conditioning equipment is not available or is out of order, the temperature of the wax and speed of filling must be adjusted to meet the changes in temperature and humidity of the room.

Solvent-containing wax compositions are usually packed in tins or jars rather than in tubes or bottles, because, if the solvent separates, leaving a wax mass beneath, it is difficult to remove or to stir together. Water-containing (emulsified) waxes, when properly made, do not harden and so are usually packed in tubes or bottles.

Table 146

SOLVENT–BINDING AND LUSTER–PRODUCING PROPERTIES OF WAXES

Wax	Solvent–Binding Power	Luster Value
Acrawax	good	good
Albacer	slight	good
Bayberry Wax	good	medium
Beeswax	good	slight
Carnauba Wax (Bleached)	medium	medium
Carnauba Wax (Crude)	medium to slight	good
Carnauba–Wax Residues	medium	medium
Candelilla Wax	medium	good
Ceraflux	medium	good
Ceresin	good	good
Cetyl Alcohol	good	medium
Chinese Insect Wax	medium	good
Cotton Wax	medium	good
Fiber Wax	medium	good
I-G Waxes	except I–G Wax E, which is suitable for saponified products, good. I–G Wax E, medium	except I–G Wax BJ, bleached and unbleached, which only shows a slight polish effect, good
Japan Wax	good	good
Lanette Wax	good	medium
Microcrystalline Paraffin Wax	good	medium
Montan Wax	good	good
Opalwax	slight	medium
Ouricury Wax	medium to slight	good
Ozokerite	excellent	good
Ozowax	good	good
Paraffin Wax	medium	medium
Petrolatum Wax (Protowax)	excellent	medium
Rilan Wax	good	good
Shellac Wax	good	good
Spermaceti	good	good
Sugar-Cane Wax	medium	good

UNITED STATES GOVERNMENT SPECIFICATIONS FOR SOLVENT–WAX COMPOSITIONS

P-W-134. Wax, Floor; Solvent-Type, Liquid (with Resins). October 29, 1942. Intended for use on all floors except those containing asphalt or rubber. Shall consist of waxes and resins dispersed in turpentine or petroleum distillate. The flash point of the solvent shall be not less than 28°C. The wax shall dry in not over 45 minutes. There shall be not less than 11% of *non-volatile matter* having the following properties: *softening point* not less than 72°C.; *acid value* not over 18; *saponification value* 40—85; *Wijs iodine value* 15—22; *ash* not over 0.5%.

P-W-158. Wax, General-Purpose; Solvent-Type, Liquid and Paste (for Floors, Furniture, etc.), September 17, 1942, superseding PW-141, February 12, 1936. Shall consist of waxes dispersed in turpentine or petroleum distillate. The flash point of the solvent shall be not less than 28°C. The wax shall contain no rosin. The wax shall *dry* in not over 45 minutes. Type I (liquid) shall contain not less than 11%; Type II (paste) not less than 20% *nonvolatile matter.* In both types the nonvolatile matter must have a *softening point* of not less than 71°C.; *acid value* of not over 8; *saponification value* 20—70; *ash* not over 0.5%.

COLORING WAXES

The color of waxes and wax compositions may be altered by means of dyes, pigments, bleaching or other chemical treatment. A temporary (until the wax is melted) lightening in color is often produced by beating air, or varying amounts of water, into a melted wax.

When dyes are used for coloring waxes, as in shoe polishes, a stock solution of the dye is usually made with oleic acid (red oil), montan wax or stearic acid. For example, heat 100 lb of oleic or stearic acid in a kettle (half full) to 90—95°C. Then stir in slowly, a little at a time, the dye (e.g., nigrosine base) to the extent of 30—50 lb and, while stirring, raise the temperature to 101—105°C. Continue stirring until all the dye is dissolved. If it is desired to dilute the dye solution, more oleic or stearic acid may be mixed in or a solvent, such as turpentine or naphtha, may be used. Where it is not desired to introduce oleic or stearic acid or solvents in making up the dye solution, certain waxes like montan, beeswax, Japan wax or carnauba wax may be used instead. They are not so good solvents for basic dyes and thus will yield solutions less concentrated than those made with oleic or stearic acid.

Since dyes are often affected by heat, it is best to dissolve them at the lowest possible temperature in the shortest period of time. Thus, where they are to be dissolved in melted wax, they should be added only after all the components have been introduced and melted and at the lowest possible temperature before completing the batch.

Proper mixing is necessary to avoid undissolved dye, which is not only an economic loss of a high-priced ingredient, but also mars the appearance because of specks or lumps and produces different shades of color in different parts of the composition.

Dyes are standardized by their makers or distributors with sodium chloride, Glauber's salt, dextrin or other fillers. These are not soluble in fatty acids, waxes or wax solvents and sometimes may be harmful. If such is the case, their solution in the wax or solvent should be filtered before use. Where dyes are used to color wax emulsions, sodium chloride, Glauber's salt or other electrolyte present may disturb the emulsion or cause corrosion of the metallic container. Accordingly, only dyes standardized with dextrin or other harmless fillers should be used.

The dye used should be tested for permanence or the supplier should warrant the quality. Thus, if it is to be used in an alkaline, acid, oxidizing or reducing medium, it should not be affected by these. Suitable dyes are now available for all specific uses.

Wax compositions or solutions are usually colored with oil-soluble dyes with or without insoluble pigments. Wax emulsions may be colored either with oil-soluble dyes, dissolved in the wax or solvent that may be present, or with water-soluble dyes, dissolved in the water.

EMULSIONS

An emulsion may be considered as a homogeneous suspension of tiny droplets of oil in water or of water in oil. The oil-in-water type may be represented by the usual furniture polish (milky) and the water-in-oil type by butter. The term "oil" is used here in a broader sense to include oils (mineral, vegetable, animal or essential), fats, greases, waxes, hydrocarbons (benzol, naphtha, turpentine, etc.), synthetics (ethylene dichloride, nitrobenzol, etc.), that is, substances which do not mix with water.

Emulsification formulae and methods have been evolved chiefly through practice, by actually making a great many emulsions. Because of the vargaries and eccentricities of emulsions, practical workers have made greater technical advances in this field than the research chemists. Too often, the trained chemist does not achieve as good emulsions as the lay worker, because the chemist rebels instinctively against empirical formulae and does not follow instructions so implicitly as the man "who knows he doesn't know." Moreover, each new emulsion represents a new problem having numerous variables. These should not be underestimated if a good stable emulsion is desired. The technique of preparation of any particular formula should first be mastered before any variations are attempted.

FORMULATION

The first factor in formulation is the ratio of the wax to the water phase. Usually, an increase in amount of the water phase will result in a lowered viscosity of the emulsion. Often, the decreasing of the amount of the water phase may cause inversion. Low percentages of water, in many emulsions, produce pasty or even solid emulsions. Since increased viscosity usually increases stability, viscous (concentrated) emulsions are frequently made so that aging and mechanical shock will not break them. These are diluted as required before use.

The amount of the emulsifier used is dependent on many factors, such as stability, viscosity, type, color, costs, etc. It is usual to start out with an amount equal to that used in similar emulsions and work up or down, depending on desired results. It is best to work with parts by weight for all ingredients in order to avoid errors due to differences in densities.

Since a test of stability may require "shelf testing" for some length of time, it is best to make up a number of experimental emulsions, varying the components within certain limits and allowing them to age.

METHODS OF EMULSIFICATION

Just as one man's food may be another's poison, so one method, which will give a perfect emulsion in one case, may produce a perfect failure in another. Thus no one method or emulsifying agent will serve universally.

When an emulsion of a solid, melting above 100°C., is desired, it should first be melted with sufficient solvent or oil to reduce the combined melting point below 100°C. For example, naphthalene with naphtha or other hydrocarbons; synthetic resins with hydrocarbons or vegetable oils.

It must be borne in mind that perfect results cannot be obtained until a few experimental emulsions are made in order to become familiar with working conditions. This is why experience shows that one of the given formulae should be mastered before attempting any variations.

Variations in raw materials, procedure, errors in proportions, etc., produce poor results. The formulae given in this book have been repeated many times and will work if they are strictly adhered to.

Of course these formulae cannot fill every individual requirement. Variations are necessary. In order to work out successful formulae, patience is essential. That which is worth while getting is worth while striving for. It is suggested to vary only one ingredient or proportion at a time. This enables one to know exactly what produces the change in the finished product.

Emulsifying with Ammonium Linoleate

This is an excellent agent for emulsifying vegetable and fish oils, waxes, fat, resins, hydrocarbons and many other water-insoluble products. When emulsifying a water-insoluble product, having a melting point of over 100°C., it should be first dissolved in naphtha, ethylene dichloride, turpentine, or a similar solvent. Alcohol, as a rule, should not be used as it breaks most emulsions. Similarly, acids, esters and salts must be avoided.

Procedure

Using proportions given in Table 147, first dissolve the indicated amount of water in the ammonium linoleate by covering it with the required amount of water and allowing to soak overnight. Work in slowly the next day until dissolved completely. Do not attempt to dissolve in any other way or lumps will result.

To this, add slowly, with vigorous agitation, the indicated amount of oil and continue stirring until homogeneous.

When a wax is to be emulsified, it is melted and considered as an oil. In this case, the water must be heated above the melting point of the wax. Most trouble is encountered in making wax emulsions when the solution of ammonium linoleate in water and the melted wax are not heated sufficiently. To play safe, keep each of these solutions between 95 and 100°C., not allowing the temperature to drop below the melting point of the wax while adding one to the other. These emulsions have been made many times with uniformly good results. If your emulsion is grainy or forms a film of wax on the surface, then the fault is in manipulation. Good wax emulsions cannot be made by hand or with a slow-moving paddle. The vigorous agitation of a fast electric stirrer is essential.

Emulsions of the various, inflammable hydrocarbons give products of high cleansing power and of a much higher flash point.

Table 147

TYPICAL EMULSION FORMULAE

Substance Emulsified	Parts	Water Parts	Ammonium Linoleate Parts
Kerosene	90	90	8
Naphtha	90	100	7
Benzol	90	100	7
Gasoline	90	100	7
Pine Oil	90	90	10
Carnauba Wax	90	620	12
Beeswax	90	500	12
Ozokerite	90	400	14
Turpentine	90	100	8
Nitrobenzol	90	100	8
Orthodichlorbenzol	90	100	8
Methyl Salicylate	90	100	8

These formulae can be lessened in cost by reducing the amount of emulsifier used. The minimum can be determined by experiment. Increasing the amount of water will give thin emulsions. Certain emulsions are improved by the addition of 1% or so of ammonia dissolved in water when making the emulsion.

Some emulsions, particularly those containing saponified

candelilla, carnauba or Japan wax or stearic acid, may thicken on standing.[1] New emulsions, therefore, should be observed for about a week to note any physical changes.

In certain wax emulsions, particularly for shoe creams, emulsification is brought about by partially saponifying one or more of the saponifiable waxes present. It is not necessary to effect complete saponification, but only to an extent that will produce a good emulsion with the water added. Most manufacturers of wax emulsions of this type use solutions of alkali carbonates in place of caustic alkalies. Carbonates release carbon dioxide gas during saponification. The turbulence, produced by the gas, seems to aid in the dispersal of the wax and yields a better emulsion. In order to avoid boiling over of the batch, the wax mixture should be kept below 70°C., and the hot alkali solution added, a little at a time, with slow stirring.[1] Sodium soaps give firmer creams than potassium soaps, but the second are preferred because the finished cream shrinks less. Saponification of waxes proceeds with increasing difficulty in the following order: Japan wax, beeswax, montan, carnauba, candelilla, fiber and shellac waxes. Paraffin, ozokerite, ceresin, petrolatum, microcrystalline and other hydrocarbon waxes are not saponifiable and thus require an emulsifier.

In making emulsions of this type, it is usual to melt the waxes as described under Melting of Waxes. If a solvent is to be incorporated, it is warmed to a temperature a little above the melting point of the wax composition and then run in slowly, while mixing. If the emulsifier is soluble in the melted wax, it is dissolved in it before the water is introduced. The water is warmed to a temperature above the melting point of the wax composition and run in slowly, with good mixing. In some cases, it is desirable to add just enough water to form a gel of the wax and emulsifier and then mix thoroughly until all of the water added has been absorbed. Then the balance of the water is added slowly, while mixing well. If a sample is put on a glass plate, it should not slide easily or show any water separation when pressed with a finger.

Wax emulsions corrode tins and, therefore, the inside of the container should be lacquered, enameled or otherwise coated.

---

[1] Bennett, *Practical Emulsions* (New York, Chemical Publishing Co., 1943).

CHAPTER 5

waxes in industry

DENTAL WAXES

Frederick Gruenwald

Dentists and dental laboratories use quantities of various wax compounds as: base-plate wax, inlay-casting wax, impression compound, set-up wax, bite wax, sticky wax and carving wax. In the hands of the dentist and of the technician, the dental waxes have to work and serve as ideal plastic materials. They must have high plasticity at normal temperatures, tensile strength, fusibility, hardness, adhesiveness and noncrystallinity. The requirements for base-plate wax, for example, include plasticity, fusibility, hardness and light color. Inlay-casting wax must soften without becoming flaky. It must not show appreciable chipping or flaking when trimmed to a fine margin at room temperature. It should become very plastic when heated and harden very quickly when cooled. The melted wax, when vaporized at 500°C., must leave no solid residue other than carbon. It must have a certain thermal expansion and flow.

Carving wax should be hard, of very fine grain, and easy to carve. Sticky wax must have good adhesion, but should not stick to the fingers.

The following formulae are representative of these compounds:

Base-Plate Wax

Formula No. 1		No. 2	
Paraffin Wax	70	Paraffin Wax	80
Beeswax	20	Acrawax C	5
Carnauba Wax	4	Rosin	5
Gum Dammar	6		

INLAY WAX

Paraffin Wax	18.0	Candelilla Wax	1.0
Carnauba Wax*	3.5	Resin**	2.5

STICKY WAX

Flexowax C	11	Ozokerite	13
Beeswax	11	Venice Turpentine	2

CARVING WAX

Paraffin Wax	30	Montan Wax	20
Ozokerite	30	Carnauba Wax*	20

* Carnauba wax can be replaced by a synthetic wax, such as Acrawax C.
** The resin may be a phenol modified resin or a petroleum resin.

In all these formulae the resins and waxes are being replaced more and more by synthetic products, because these are free from impurities, are of uniform composition, and often cheaper than the natural products.

METHOD OF PRODUCING BASE-PLATE WAX

The resin, which must be very light, is melted; then the waxes are added and heated till homogeneous. The hot waxes are filtered through a fabric and again heated. An oil-soluble dyestuff, previously dissolved in a portion of the paraffin wax, is added for discoloration and the mixture stirred till completely uniform, filtered again, and poured in flat pans. After a few days, the cakes are heated in warm water and milled on a two-roll mill, the rolls having the same speed and being made of granite or stainless steel. After cooling in cold water, the milled sheets of wax are stamped.

LIPSTICKS[1]

The outstandingly important property of lipstick waxes is their melting point, which means that synthetic substitutes must always be selected with this factor well in mind. Melting points of the more common waxes, normally used in lipsticks, are approximately as follows:

	°C.
Beeswax	63–65
Ceresin	74–80
Carnauba Wax	84–86

[1] S. P. Jannaway, *Soap, Perfumery, Cosmetics* (1942).

Bleached Ozokerite	70–82
Bleached Montan Wax	70–85
Japan Wax	52–56
Paraffin Waxes	42–65

Hydrogenated fats and synthetic waxes cover this range quite effectively. Mixtures of different waxes do not have the melting point that one might expect from the individual melting points and ratio of the constituent waxes.

In addition to these waxes and their synthetic counterparts, including hardened oils and diglycol stearate, etc., other substances, commonly used in lipstick manufacture, include mineral oil, castor oil; sesame, peanut and other vegetable oils; petroleum jelly (short fiber), lanolin absorption base, deodorized cocoa butter, cetyl and stearyl alcohols, wool wax, spermaceti, stearic acid, benzoinated lard; perfume, color, antioxidants, and preservatives.

Beeswax is the most frequently found substance in a lipstick base. It gives a good consistency to the base and, together with the lanolin, aids in binding the castor oil, for beeswax is almost as soluble in a lanolin-castor mixture as it is in a hydrocarbon combination, such as paraffin oil. Beeswax also, because of its shrinkage properties, makes the expelling of the finished stick from the mold a much simpler process. Chemically, the wax stabilizes the eosin (color) in the stick by the presence of the free melissic and cerotic acids contained in the beeswax itself. However, to produce a shiny, brilliant lipstick, a low beeswax content is necessary. For ordinary glossy sticks, 10% of beeswax should never be exceeded.

Ceresin resembles beeswax in many of its properties and may be used along with beeswax. This earth wax is a fair binder, with about the same consistency as beeswax, but it does not have the same shrinkage property. The crystalline structure of ceresin does not tend to change with time, for once this ingredient has crystallized out in the mixture, no further change will be noted. Beeswax and ceresin can be used together in order to secure a fine-grained mixture, as each product complements the properties of the other.

Bleached ozokerite, including genuine ceresin, is especially valuable because of its property of retaining oils uniformly and permanently throughout the wax structure of the lipstick.

Carnauba wax, when present in about 2–8% of the total mass, helps to promote rigidity and good finish.

Cetyl alcohol is a good solvent for the eosin in the stick and helps in the permanence of the stain. However, one must use this product with discretion, never exceeding 5% of the total base, or else, the desired "cetyl effect" will be replaced by a tendency to dryness.

Paraffin wax and mineral and vegetable oils are also important constituents of many lipstick bases.

The lipstick must have a high melting point (about 55–65°C.) so that it will not soften unduly in the summer; it must be hard enough to withstand pressure without breaking. It must spread easily and smoothly, but not leave too thick and greasy a deposit. These properties are by no means easy to balance, especially when utilizing a more or less "scratch" lot of ingredients.

Although the pigments used do not have a great tendency to settle, heating should not be unduly prolonged. The melt is poured into molds which are then rapidly cooled. Slow cooling is not only a less satisfactory production job, but it also tends to promote the growth of wax crystals, with consequent weakening of the stick. Sticks are removed from the mold, inserted in the case and then "flamed" to remove finger marks. "Flaming" consists of rapidly turning the lipstick through a small Bunsen flame in order to melt its surface. This must be done uniformly and rapidly.

PHARMACEUTICAL PREPARATIONS

Frederick R. Greenbaum

Ointments, in pharmacy and medicine, serve the purpose of bringing the medication in intimate contact with the skin. They are used to treat various skin conditions, burns, wounds and injuries with specific medications. Each particular medication offers an individual problem as to the compounding of the ointment.

There was a time in pharmacy and medicine, when almost every medication was placed in a grease base, such as lanolin, petrolatum, ceresin or combinations of these.

This type of ointment, lotion, etc., prepared from benzoinated lard, petrolatum, lanolin, or combinations of such vehicles is distasteful to the patient.

In addition to this, physicians, particularly dermatologists, have pointed out that a grease-base ointment will greatly delay the action of the drug, particularly if the drug is soluble in water, and should be made available in a form which allows the drug to go into solution in the excretin of the skin or the exudates of the wounds, whichever the case may be. This is why greaseless bases have recently come to the foreground and more and more greaseless ointments, salves and lotions are formulated.

A greaseless base is one in which stearic acid is used as a wax. Some of the stearic acid is combined with potassium hydroxide, or better with triethanolamine, which forms a soap with the equivalent portions of the stearic acid. This soap acts as an emulsifying agent for the excess stearic acid, forming a white ointment of the vanishing-cream type. The medication, if soluble in hot or cold water, is dissolved in water and added to the molten stearic acid, with the triethanolamine in the water phase. A uniform and smooth emulsification occurs, yielding a greaseless-base ointment, which is very effective clinically.

Instead of stearic acid, Glycostearin, or a mono- or a distearate may be used. Of course, depending on conditions, other waxes may be added, for instance, carnauba wax to impart hardness, or beeswax. Very often, in order to avoid the crystallization of stearic acid, a small amount of lanum, or an absorption base of the type of Aquaphor or Parachol is added.

Recently, sulfanilamide, sulfathiazole, and sulfadiazine have been recommended for topical application in wounds, war injuries, burns, etc., and these substances have been put up in an ointment base. Usually this ointment is a combination of a greaseless base made from stearic acid, with another base made from lanolin or a hydrogenated vegetable oil.

Usually, ointments made from stearic acid as a base cannot be milled and, of course, it is very desirable to mill a sulfanilamide or a sulfathiazole ointment in order to break down the crystals of the sulfoneamides. Other waxes are added to make milling possible.

It must also be kept in mind that a greaseless base is more effective and, very often, the concentration of the drug must be lowered if it is used in a greaseless base. This is the case with ammoniated mercury. The 10% ammoniated mercury ointment, U. S. P. XII, has wool fat and white ointment as a base. If ammoniated mercury ointment were made with a stearic acid greaseless base, a 10% ammoniated mercury would produce severe burns. In a greaseless base, the highest concentration of ammoniated mercury ointment that can be used with safety, is 5%. This is an example to show that each drug represents a specific and individual problem and no generalized rules can be laid down.

According to work done by the American Pharmaceutical Association,[1] the requirements of an ideal ointment base are as follows:

1. It should wash off with water
2. It should be stable
3. Therapeutically efficient, penetrating to the pathological site
4. Nongreasy
5. Free from objectionable odors
6. Nonirritating
7. Nondehydrating
8. Neutral in reaction
9. Easily compounded by a pharmacist
10. Nondegreasing in action
11. Nonhygroscopic
12. Compatible with all medications
13. Nonstaining
14. Efficient on dry, oily, or moist skin
15. Capable of serving as a medium for oil-soluble as well as water-soluble medicaments
16. Capable of holding at least 50% of water.

Many washable ointments have been made, but all have been found to fall short of the ideal. Some bases are incompatible with many of the medicaments commonly prescribed by dermatologists. Others are apparently reversed by certain medicaments and become nonwashable. Still others are un-

---

[1] *J. Am. Pharm. Assoc.* **2,** 467 (1941).

necessarily involved or call for the use of substances which are very difficult to obtain.

Modifications of suggested bases have been studied and an attempt has been made to develop new bases which would more closely approach the ideal. The base, which we shall discuss in the following, is the result of these studies and, although complete evidence as to the compatibility is not as yet available, it appears to be more satisfactory than any other base thus far suggested in the literature. It is apparently compatible with every medicinal substance with which it has been tested and until a better formula is developed it is suggested for use by the pharmacist and pharmaceutical manufacturer.

This formula uses cetyl alcohol and Wetanol (sodium lauryl sulfate) as a base.

Cetyl alcohol is a white, waxlike solid, melting at 49°C., insoluble in water, soluble in mineral oil and vegetable oils, fats, alcohols, ether, carbon disulfide, glycol and glycol ethers. It is stable in the presence of acids, alkalies, light and air. It is used as a hardening agent, emulsion stabilizer, and emollient for creams. It may be used as additive in 1–5% concentration or 5–20% as the principal wax base.

Cetyl alcohol is greaseless, a good emollient and makes the skin velvety. Its penetrating power in combination with a wetting agent, such as sodium lauryl sulfate (Wetanol), causes medicaments to be readily absorbed by the skin and, in some cases, they may be used in lower concentrations than in older types of bases, such as petrolatum and lanolin.

Wetanol and similar wetting agents are obtained from alkyl derivatives of aryl sulfonates, esters of higher fatty alcohols and dibasic acids, such as sulfosuccinates and many others which are of value in preparing more efficient and finer ointment bases. Their effectiveness is based on the power of reducing surface tension and lowering interfacial tension between solid and solvent, thus permitting rapid penetration and dispersion of the solid. Therefore, they are useful as detergents and penetrants and some of them are also efficient emulsifiers.

An important factor in determining the desirability of an ointment base is its pH value. The pH value of an ointment is difficult to determine and the pH value of skin is as yet not known. The pH of this new ointment base is 7–9, that is, decidedly on the alkaline side.

Temperature changes between 0 and 50°C. do not affect the consistency or promote separation of this base, so that extreme temperatures encountered in transportation will have little effect on its value as a washable ointment base.

No apparent incompatibilities have been found for the base during the experimental work.

As an illustration of an ointment, using some of the newer wax bases, the following examples may be given:

Wetanol	0.5	White Petrolatum	20
Cetyl Alcohol	8.0	Water, to Make	100
Cocoa Butter	6.5		

This base is compatible with the following medications: phenol 2%, Whitfield combination, Ichthammol 10%, ammoniated mercury 1%, rectified oil of birch tar, sulfur 10%, Burrows solution 5%.

Another formula is as follows:

Cetyl Alcohol	6.4	White Petrolatum	14.3
Octadecyl Alcohol (Stenol)	6.4	Liquid Petrolatum	21.4
Wetanol	1.5	Water, to Make	100.0

Modifications of these formulae are obtained by adding various waxes, like beeswax, paraffin wax, spermaceti, or one of the newer synthetic waxes, such as Carbowax. Another base, suggested as a possibility for a greaseless base, was made in accordance with the following formula by N. Sorg and J. W. Jones.[1]

Glyceryl Monostearate	15	Glycerin	35
Cetyl Alcohol	15	Diethyleneglycol	35

In the treatment of athlete's foot, a very frequent fungus infection, various ointments are used, such as Whitfield Ointment, N. F. VII, diluted to ¼–½ strength with petrolatum or Ung. Aquae Rosae. Frequently, ¼–1% thymol is added.

Another ointment for this purpose is Ung. Hydrarg. Ammoniati, U. S. P. XII, Ung. Acidi Borici and Ung. Aquae Rosae, mixed in equal portions.

One of the more modern treatments of athlete's foot is with a

[1] *J. Am. Pharm. Assoc. Practical Pharm. Ed.* **2**, 400 (1941).

phenyl mercuric nitrate 1:1500 ointment incorporated in a greaseless base, usually of the stearic acid triethanolamine type.

Whitfield Ointment, also known as compound ointment of benzoic acid, N. F. VII, consists of:

Benzoic Acid	12	Wool Fat	5
Salicylic Acid	6	Petrolatum, to Make	100

In recent work, Francis J. O'Brien[1] found that the addition of thymol and the use of a cold-cream base increase the fungistatic action of this popular ointment. The following formula is proposed:

Benzoic Acid	6.0	Mineral Oil	2.5
Salicylic Acid	3.0	Cold Cream, to Make	100.0
Thymol	2.5		

The cold cream, which is used for this purpose, has the following formula:

White Wax	13.0	Borax	0.6
Ceresin	6.4	Water, to Make	100.0
Mineral Oil	55.0		

In the treatment of psoriasis, a common skin condition, different ointments are used, some supposedly specific for this condition, for example:

Salicylic Acid	1	Petrolatum, to Make	30
Anthralin Ointment	0.5–6.0		

An ointment containing chrysarobin may also be used, for example:

Salicylic Acid	1.0	Petrolatum, to Make	30
Chrysarobin	0.6		

Coal–tar ointment is also suitable for this purpose:

Crude Coal Tar	4	Corn Starch	15
Zinc Oxide	2	Petrolatum	15

An effective sulfur ointment is as follows:

Precipitated Sulfur	1–2	Petrolatum	30
Salicylic Acid	1–2		

The function of beeswax in ointment bases is not only stiffening, but also to increase the water number or water value

[1] *J. Am. Pharm. Assoc. Practical Pharm. Ed.* **2,** 376 (1941).

of the ointment to a considerable extent. Other waxes have been found to have properties similar to beeswax, one of these being ceresin, which has the ability to stiffen ointment bases and at the same time serves to minimize or eliminate oil leakage, or separation of oils from the ointment when water is incorporated.

Recently, sulfated hydrogenated castor oil and an emulsified base consisting of 10% diglycol stearate, 20% sulfonated hydrogenated castor oil, 30% petrolatum, and 40% water were substituted for the official ointment bases in several antiseptic ointments. Bacteriological tests indicated that the antiseptic value of these ointments is equal or superior to that of the official ointments.

Stearic acid and cetyl alcohol are also used as waxes for a base of very satisfactory hand lotions, which may or may not be medicated. The addition of certain substances, such as pectin, tragacanth, sodium alginate, Methocel, etc., may be desirable.

The use of pectin as addition to an ointment base has recently been recommended. The following formula will serve as an illustration:

Boric Acid	5	Wool Fat	7
Glycerin	8	White Petrolatum	25
Pectin	3	Ringer's Solution, to Make	100

This formula was recommended by Goldner.[1] Tannic acid, salicylic acid, and alkalies cannot be incorporated in an ointment containing pectin.

Suppositories are made of waxes or wax combinations. Cocoa butter is one of the old substances used for this purpose. Very often, other waxes or wax combinations are used, but, of course, it must be remembered that in suppositories the wax combinations must melt at body temperature, so that the medicaments will get into intimate contact with the mucous membrane and exert their beneficial action.

Another interesting use of waxes in medicine is in connection with silver nitrate solution, which is placed, as a routine measure, in the eyes of new-born babies. This silver nitrate solution is put up in ampules made of wax. A special paraffin wax mixture is

[1] *J. Am. Pharm. Assoc. Practical Pharm. Ed.*, **3**, 24 (1942).

poured into molds and the silver nitrate is added to it and then the ampules are sealed. On cooling the silver nitrate solution is enclosed in the wax ampules. Before use, the doctor or nurse snips the ampule open and drops the silver nitrate solution into the eye.

Another use of waxes in pharmacy and medicine is for coating pills and tablets.

Tablets and pills, containing various medicaments, are sugar coated in special rotary drums. Several layers of sugar coating are applied, often color is added and then, in order to impart a nice shiny appearance, a very fine layer or coating of carnauba wax or other waxes is applied. This is carried out by tumbling the carnauba wax or other wax combinations in the rotary drum with the sugar-coated pills. A very fine layer of wax is taken up by the pills which assume the familiar shiny appearance.

EMBALMING PREPARATIONS

Hilton Ira Jones

The chief uses of waxes in embalming are in the production of wound fillers, feature modeling waxes and lip wax, and in making the corpse appear more natural.

WOUND FILLERS

No. 1

a	Yellow Beeswax	25	*b*	Soap Flakes	10
	Paraffin Wax	50		Water	25
	White Petrolatum	50			

Heat each separately to a clear solution, then add *a* to *b*, with stirring, and color to suit.

No. 2

Protowax No. 33448	50
Ceresin	50
Color	to suit

No. 3

a	Paraffin Wax	25	*b*	Stiff Paste of Bentonite in Water
	White Petrolatum	50		
	Yellow Beeswax	25		

Melt *a* and add to *b*, with stirring, and color to suit.

No. 4

Protowax	20	Starch	40
White Petrolatum	10		

Melt the waxes together, but the melt should not be hot, and beat in the starch. Add color to suit.

No. 5

Protowax	40	Oil Red Solution (10% in Kerosene)	2
White Petrolatum	20	Oil Brown Y (10% in Kerosene)	1
Starch	80		

These two waxes are widely used as modeling waxes for making noses, ears, etc., in reconstruction of accident cases. If this formula is to be used as a wound filler, it is better to have it softer as:

No. 6

Protowax	20	Starch	40
White Petrolatum	20		

Lip Wax

No. 1

Wound Filler	75	White Petrolatum	100
White Beeswax	75		

Add color to suit with oil red.

No. 2

White Lead Paint	100	Solution of Oil Red (10%)	1

PYROTECHNICS

George W. Weingart

The different colors of the flames of burning pyrotechnic compounds are due to the fact that certain colored lines predominate in the spectra of the elements used in their production. For example, yellow lines predominate in the spectrum of sodium and thus sodium salts are used to make yellow fire. The strontium spectrum shows the predominance of red lines, therefore, strontium gives a very decided red light.

To produce these color effects, it is necessary to volatilize a portion of the element. This is accomplished by combining an oxidizing agent with a source of carbon, which, when ignited, develops a high temperature, sufficient to vaporize a portion of the color-producing element.

The oxidizing agent is usually potassium chlorate and many substances have been utilized for supplying the required carbon. Among these are the waxes which have been found to work very satisfactorily with potassium chlorate and, in some cases, e.g., in the production of blue fire, stearin and paraffin wax gives a smooth burning fire of deep color unobtainable with any other carbon source. Kauri and carnauba waxes are more suitable for red, green and yellow fires. The following mixtures will illustrate the use of waxes:

BLUE

Potassium Chlorate	7	Stearin	1
Copper Oxalate	2	Calomel	3

RED

Strontium Nitrate	16	Kauri Wax	3
Potassium Chlorate	8		

GREEN

Barium Nitrate	60	Carnauba Wax	10
Barium Chlorate	25	Paraffin Wax	1
Potassium Chlorate	5		

YELLOW

Potassium Chlorate	7	Barium Nitrate	1
Sodium Oxalate	2	Kauri Wax	1

CANDLES[1]

A candle consists of a cylinder of solid, fusible, combustible matter surrounding and saturating a combustible wick. When lighted the surrounding matter is absorbed as it liquefies by combustion. The gases formed are raised to the temperature of ignition and as the carbon particles burn they provide illumination.

Candles developed from pine torches, rope strands covered with resin or pitch and pitch coated with beeswax. Strands of yarn were coated in the early dip process by repeated dipping and cooling in molten tallow or wax. Introduction of molds in

[1] U. S. Dept. of Commerce, O.T.S., 1947.

the fifteenth century by Sieur de Brez marks the beginning of candles as we see them today.

Practical considerations of candle making include the nature of the combustible, its melting point, viscosity, and burning power. Also essential is the blending of suitable substances from several sources, selection of the wick that is adapted to the combustible chosen in number of threads, substance and structure. The equipment, methods of mixing, coloring and production by hand or molding machines must also be determined.

WICKS

Wicking is made from the best quality of cotton and spun carefully so that there will be no loose fibers or threads. After bleaching, the cotton is spun into yarn and, as wicking, is cut into uniform lengths. Pickling of wicking in a dilute solution of mineral salts for 24 hours counteracts the tendency to smoke. The mineral salts used comprise borax, potassium chloride or nitrate, ammonium chloride, or chloride sulfate, or phosphate. Pickling causes the wick to curve in burning and to vitrify the ash, as well as prevent burning too rapidly, and obviates the necessity for snuffers. The hanks, after steeping, are placed in a centrifuge for expelling the solution. Further drying may be necessary for obtaining a characteristic crisp feel. Enough mineral matter is retained in the wicking to affect materially the burning of the candle.

WINDING

The wicking is wound on spools or bobbins by use of a spool winder, which is also used for gathering threads from any number of spools according to the size of the wick. A thirty-thread wick, for example, may be made of single threads from ten spools wound together with three of these united in the wick braider. Spools of the finished wick are inserted on the spindle of the candle machine.

The size of the wick must be adjusted to the diameter of the candle and its fusibility. The capillarity of the wick must be such that the exact quantity of the fluid combustible shall rise through it and be burned with a flame sufficiently large and hot to consume as much combustible as it melts.

A candle of very fusible and combustible substance should have a looser plait than one made of material of higher melting point and less ready combustion, with the number of threads depending on the diameter of the candle.

CANDLE SIZES

Sizes of candles are indicated by the number required to weigh a pound. Diameter and length are other variables.

Equipment necessary, besides the wick-plaiting and wick-winding machines, is a boiler to supply steam to melt the waxes and to heat the molds; water under pressure to cool the molds; jacketed kettles in which to melt the waxes; candle-makers' tools, such as a tin swimmer, tin pouring bucket, steel cutter or scraper, steel wick knife and thermometer. A foot- or steam-power butt saw and a steam stamping machine, or a combination sawing-stamping-polishing machine, may be added.

MACHINERY

Candle machines vary from the hand mold to the modern machine in which 24—360 candles and more are produced according to their size and the speed of operation. They consist essentially of a series of vertical cylinders with the top terminating in the bottom of a shallow tray. The heads of hollow pistons within the cylinders, or molds, form the tops or tips of the candles. The cylinders and the movable piston heads form the candle molds. Wicks are inserted through the movable pistons from the wicks wound on spools below the pistons.

WAX CANDLES

Wax candles are difficult to mold satisfactorily because of their tendency to stick and to shrink during solidification, with consequent cracking.

PARAFFIN CANDLES

Paraffin and stearic acid are the principal raw materials of candle making. Paraffin is a mixture of the solid hydrocarbons of the paraffin series and varies within relatively wide limits in hardness, melting point, and color, according to the source from which it has been derived and to the degree coloring matter and softer bodies of similar composition have been separated. Below its melting point, it becomes plastic. This disadvantage is overcome by the addition of (5% and more) stearic acid, according to the hardness of the paraffin.

Petroleum paraffin is available in five grades, melting at 120, 125, 130, 135, and 140°F. Paraffin obtained by distillation of shales melts at about 140°F., while that from ozokerite melts at 140–158°F. If paraffin from ozokerite melts at a temperature lower than 136°F., it may be assumed that hydrocarbons of lower melting point have been added to cheapen the product.

Refined ozokerite, known as ceresin, or ozokerite paraffin, is admirably adapted to candle manufacture because of its high melting point.

The paraffin candle has grown greatly in favor because of its greater illuminating power and the absence of the pungent and disagreeable odor that characterizes the old stearin candle when blown out. A high degree of luster, combined with translucency, and ready adaptability to ornament in shape and color are other desirable qualities of paraffin. Addition of stearic acid or other substances minimizes the chief objection to paraffin candles, their inability to support their own weight in warm weather.

Ceresin candles have a high melting point. Their illuminating power is greater than that of the best paraffin candles from other sources. Besides being free from smell and not greasy to the touch, they burn with a dry cup, and are not so liable to gutter as ordinary transparent candles.

TEMPERATURE CONTROL

Proper temperature control of the molds during and after casting is important to avoid streaking, shrinking and sticking. When molding is properly done with smooth molds, the paraffin candles obtained have high luster. In the case of stearin and wax candles, and sometimes with other varieties also, extra gloss and polish are given to the surface by rubbing and rolling them by hand or between cloth-covered rollers. Several machines have been constructed for this purpose.

COLORS, SIZES AND SHAPES

Candles are produced in various colors, sizes and shapes. They may be made with tapering butts to fit any holder, either by use of a tapering machine or of additional molds inserted in the fluted or regular candle mold. The most common shape is that of a plain cylinder of varying diameter and length. A great variety of ornamental forms for religious and decorative purposes are molded in numerous prismatic cross sections, such as square, triangular, octagonal, etc. They may be molded longitudinally or spirally fluted. Decorated candles are of every color, shape and design. The decorations are made of wax and put on the candles by hand. Long tapering candles are made by dipping and rolling on a marble slab.

STANDARD CANDLE

The standard candle of United States and Great Britain is the practical unit of illuminating power. Its weight is one sixth of a pound and should burn 120 g of spermaceti per hour. The wicks shall be made of three strands of cotton plaited together, each strand consisting of eighteen threads. There are other specifications for the wick including its weight. For a 12 in. length, the weight should be 6–6.5 gr. The weight of the ash from ten wicks shall be not more than 0.025 gr.

ADDITIVES FOR RAISING THE MELTING POINT OF CANDLES

Stearic Acid is a good hardening agent for paraffin-wax candles. It raises the melting point and can be used in any quantity, without impairing the burning quality of the candle. For example, 5–10% of stearic acid, which is used for candles that are to be burned in hot climates, is mixed with paraffin wax, to counteract the tendency of the paraffin wax to bend when hot. The brittleness of spermaceti is corrected by the addition of beeswax, stearin, paraffin, or ceresin (obtained from the mineral wax ozokerite).

Palmitic and stearic acid mixtures solidify at temperatures varying between 130–155°F., according to the percentage of each present.

Hydrogenated oils and fats also serve as hardening agents for paraffin wax and are used chiefly for candles which are consumed in glasses. Synthetic or natural resins can be applied in the form of coatings only and are used chiefly for decorative candles.

Crystallization is prevented and removal from molds is facilitated by adding 5% or more paraffin wax to stearic acid. Opacity equal to that produced by using 40% stearic acid is obtained by adding 2% Santowax M to paraffin candles. Addition of 1% "Hertolan" (B-naphthol benzoate) produces a whiter and more opaque appearance than when at least 40% stearin is added.

Paraffin obtained from brown coal has much less bending action than the paraffin obtained from mineral oil. Paraffin candles, composed merely of hydrocarbons, emit more light per unit than either pure stearic acid candles or composition candles which are mixtures of the two. The disadvantage of using paraffin alone (even if it is of high melting point) for candle making is that its softening point is below its melting point. Instead of one paraffin wax, e.g., of a melting point of 125°F., it is possible to get a paraffin of the same melting point by mixing two paraffins, one of 96°F. and another of 140°F.

The chilling time of paraffin wax is a better indication of value for candle making than the melting point. The melting point of wax, in itself, is no measure of quality or chemical composition, since it is derived from the melting points of the constituents which influence one another in various ways. Other characteristics in determining quality are oil content, light stability and tensile strength.

The melting point is, however, the chief difference between various grades of paraffin wax, each of which usually covers a range of 2°F.

Crude scale waxes contain more oil than refined waxes and consequently are of poorer quality for comparable melting points. Those with higher oil content have lower tensile strength.

Candles of a better grade than those made with a standard formula are made by increasing the amount of stearic acid and decreasing the amount of paraffin, or vice versa.

Waxes commonly used in candle making include: beeswax,

candelilla and carnauba wax, ceresin, insect or Chinese wax, montan wax, ozokerite, spermaceti, ouricury wax and opalwax (hydrogenated castor oil). Abril wax is a new synthetic wax which is said to be more uniform than natural products. Its melting point is reported as high as 250°C. (482°F.), with possible use as an additive in the higher-melting-point types of paraffin compositions.

Beeswax has a melting point of 145–158°F. Beeswax produced by wild bees is considered better than wax produced by domesticated bees. Generally speaking, the light-yellow crudes are of better quality since they can be bleached by natural processes. Beeswax is often adulterated with other waxes which are not always good for candle making.

Candelilla Wax is not quite so hard as carnauba wax and does not emulsify or saponify so readily. It sets slowly in melted compositions and often does not reach its maximum hardness for a few days. Mineral waxes and Acrawax C reduce its stickiness, but increase the aging and the setting time. Additions of oleic or similar acids slow down crystallization and rapidly increase softness. Its melting point is 149–156°F.

Carnauba Wax is the hardest, highest-melting, commercial natural wax, except for some crude grades of ouricury wax. It is added to other waxes to increase their melting point, hardness, toughness and luster and to decrease stickiness, plasticity and crystallizing tendencies. Addition of 1% Acrawax to carnauba wax raises the melting point 3°C. (5°F.). Its melting point is 184–196°F.

Acrawax is compatible with carnauba wax and other vegetable waxes, rosin, shellac, stearic acid and other high fatty acids, fats and ester gum. It is not compatible with paraffin and other mineral waxes. Approximately 1% aluminum stearate, stirred into a mixture of melted waxes until the entire mixture is uniform, will prevent separation of some waxes. Thus 10% Acrawax and 89% paraffin wax are held together by this treatment. Considerable variation is permissible. The viscosity of the melt is increased and its clarity is not perfect, but the compatibility is maintained, apparently indefinitely.

Acrawax B has a good luster and is very similar in properties to Acrawax. It differs from Acrawax in that it is compatible with paraffin wax and forms gels with mineral spirits and kerosene.

Acrawax C blends with vegetable waxes, paraffin and other mineral waxes, including carnauba wax and rosin. It is of particular interest where a high-melting wax, which is not brittle, is desired. Both Acrawax A and B have various commercial uses which may encourage their use for candle making.

Ceresin was originally a purified ozokerite, but for economic reasons commercial grades usually contain 50–80% paraffin wax. Such grades are yellow to white and melt at 130–170°F.

Chinese Wax (insect wax) is yellowish white, quite hard, shiny, fibrous and crysalline. It melts at 149–176°F.

Montan Wax is the hardest of the natural nonvegetable waxes, with a melting point of 161–197°F. It is brittle, lustrous and sets more slowly than carnauba wax solutions. For this reason, it should be filled into containers at a lower temperature. The bleached or highly refined wax loses some of its original characteristics, becomes softer and is less easily emulsified. It is used in place of carnauba wax and beeswax in many compositions because of its high melting point, hardness, toughness and emulsifiability.

Ozokerite (imported) is a natural wax which sets more rapidly and shrinks more than beeswax. To offset these properties Venice turpentine, or another plasticizer, is often added. It has excellent absorbing properties for oil, grease and certain solvents. In this respect, it is superior to paraffin wax which tends to crystallize from mixtures and allows fluid components, present in the composition, to "sweat out." Low temperatures do not make it as brittle as paraffin wax. It becomes harder on aging.

Ozokerite is often adulterated with paraffin wax to reduce its cost. It is also sophisticated with carnauba wax, resins and asphaltum to increase its melting point or hardness. It has a melting point of 149–176°F.

American Ozokerite melts at 135–150°F. By centrifuging, a very plastic grade, melting up to 165°F., is obtained. By vacuum distillation, harder grades, melting up to 220°C., are produced. The American product is now made in large quantities and is considered equal in quality to the imported.

White Ozokerite, melting at 75.7°C. (168°F.), mixed with paraffin wax melting at 58.3°C. (137°F.) in the following proportions, has the new melting points as indicated.

Proportions				New Melting Points °C.	°F.
4	Ozokerite	1	Paraffin Wax	74.4	166.0
4	Ozokerite	2	Paraffin Wax	73.2	163.7
4	Ozokerite	3	Paraffin Wax	72.5	162.5
4	Ozokerite	4	Paraffin Wax	72.0	161.6

Commercial Spermaceti is translucent, shiny, somewhat greasy, brittle and crystalline. It is partially saponified by aqueous alkalies and forms emulsions. It has a melting point of 107–120°F.

Ouricury Wax (uricury wax) has chemical characteristics similar to those of carnauba wax, but has a higher resin content, is more soluble in common solvents, its fusion point is lower, and its iodine and saponification numbers are higher than those of carnauba wax for which it is substituted in making candles and other products. Its melting point is 177–185°F.

Opalwax (hydrogenated castor oil) is only slightly compatible with paraffin wax, but more so with ceresin, ozokerite, montan and carnauba waxes and various natural and synthetic resins. The addition of 20% naphthalene tetrachloride, behenone or stearone makes opalwax approach carnauba wax in hardness, fracture and solvent retention. It has a melting or solidification point of 187–190°F.

IG Wax Z is a chemically inert product developed in Germany. It is colorless, odorless and without taste, very hard without being brittle, and of excellent luster. The melting point is above 212°F., consequently, a small amount added to paraffin wax, stearin or similar substances produces considerable hardening effect. Among its uses is increasing the stability to heat of paraffin candles.

MELTING POINT OF WAX MIXTURES

It is advisable to understand something about the properties of the mixed products obtained either by adding stearic acid to paraffin or paraffin to stearic acid. The disadvantage of using paraffin only (even if it is of high melting point) for candle manufacture is its great drawback of having a softening point below its melting point. As a result, such candles either get stuck up or get bent or both and cause annoyance to the consumer. It should be noted in this connection that instead of one paraffin wax melting, e.g., at 52°C., it is possible to get a paraffin of the same melting point by mixing two paraffins, one of 36°C. and another of 60°C. The melting point and the proportion of the mixtures can be worked out by the following formula:

$$\frac{fa + f'b}{a + b}$$

where f and f' stand for the melting points of the two varieties and a and b for the quantities used: as, for example, 4 parts of paraffin melting at 55°C. and 2 parts of paraffin melting at 49°C. will give according to the formula:

$$\frac{4 \times 55 + 2 \times 49}{4 + 2} = \frac{220 + 98}{6} = 53°\text{C.}$$

Even in the case of two separate paraffins (not mixtures) having the same melting point, the paraffin obtained from brown coal has much less bending action than the paraffin obtained from mineral oil in what is known as the bending test. Candles of different material but of the same length and diameter are suspended horizontally, with one end fixed, in rooms heated to definite temperatures for at least 3 hours. In this test, in the case of mixtures of paraffins, the effect of softening point on the low-melting paraffins is to produce a greater bending than in the case of unmixed paraffins of the same melting point. Again, the candles made out of brown-coal paraffins are less liable to bend than those of paraffins obtained from mineral oils.

So far as light-giving properties are concerned, paraffin candles which are composed of hydrocarbons emit more light per unit than either pure stearic acid candles or composition candles which are mixtures of the two. Stearic acid is already an oxidized product, that is, it is made up of carbon, hydrogen and oxygen, and it will naturally emit less light and heat, but this difference in the burning of candle is hardly of any disadvantage, although it may be of interest from an academic point of view.

Congealing Point of Candles

When molten candle material is allowed to cool, it begins to congeal at a point much lower than the softening or melting point. This is known as the congealing point and is a characteristic property of oils and fats. This property is of special importance in candle manufacture, because the candle material, when it is poured into the machine, must congeal and solidify before it can be pushed out of the mold. Even when the melting point of the candle material is over 50°C., the congealing point is as low as 30°C. Consequently, the water used in the candle machine must be at least 20°C. to effect rapid cooling.

The chilling time of a paraffin wax is a better indication of its value for candle making than the melting point.

Under specified conditions the chilling time is as follows:

M.P. °C.	Chilling Time Minutes
54–56	50–60
52–54	60–70
50–52	70–80

Table 148

Melting Points of Candle Compositions

Mixed Ceresin and Paraffin Wax

Ceresin %	Paraffin Wax %	Melting Point °C.	Solidifying Point °C.
100	0	70–73	69.5
95	5	69–73	68.5
90	10	68–72	66.5
80	20	66–71.5	65.0
70	30	64.5–70	63.0
60	40	62–69	62.0
50	50	58.5–67	60.0
40	60	56.5–65	59.0
30	70	54.5–62	57.0
20	80	52.5–58.5	54.0
10	90	49.5–54.5	49.0
0	100	47–52	47.0

Table 148 (Continued)
Mixed Paraffin Wax and Stearic Acid

Paraffin Wax %	Paraffin Wax Melting Point °C.	Stearic Acid %	Stearic Acid Melting Point °C.	Mixture Melting Point °C.
90	38.9	10	49.3	37.7
80	38.9	20	49.3	37.0
70	38.9	30	49.3	37.7
60	38.9	40	49.3	40.3
50	38.9	50	49.3	43.6
40	38.9	60	49.3	43.9
30	38.9	70	49.3	45.3
20	38.9	80	49.3	47.5
10	38.9	90	49.3	48.3
90	57.7	10	49.3	50.5
80	57.7	20	49.3	49.3
70	57.7	30	49.3	48.3
60	57.7	40	49.3	47.5
50	57.7	50	49.3	45.5
40	57.7	60	49.3	43.9
30	57.7	70	49.3	41.7
20	57.7	80	49.3	45.5
10	57.7	90	49.3	47.3
90	54.5	10	49.3	53.3
80	54.5	20	49.3	52.0
70	54.5	30	49.3	50.5
60	54.5	40	49.3	49.3
50	54.5	50	49.3	48.0
40	54.5	60	49.3	45.5
30	54.5	70	49.3	42.8
20	54.5	80	49.3	46.4
10	54.5	90	49.3	47.7
90	55.8	10	54.3	54.75
80	55.8	20	54.3	53.60
70	55.8	30	54.3	52.60
60	55.8	40	54.3	51.25
50	55.8	50	54.3	49.30
40	55.8	60	54.3	47.55
30	55.8	70	54.3	48.60
20	55.8	80	54.3	51.85
10	55.8	90	54.3	53.00
90	43.3	10	50.5	47.7
80	43.3	20	50.5	46.9
70	43.3	30	50.5	45.5
60	43.3	40	50.5	44.4

Table 148 (Continued)

Mixed Paraffin Wax and Stearic Acid

Paraffin Wax %	Paraffin Wax Melting Point °C.	Stearic Acid %	Stearic Acid Melting Point °C.	Mixture Melting Point °C.
50.0	43.30	50.0	50.5	43.30
40.0	43.30	60.0	50.5	42.70
30.0	43.30	70.0	50.5	44.90
20.0	43.30	80.0	50.5	48.00
10.0	43.30	90.0	50.5	48.60
90.0	43.45	10.0	54.3	48.00
80.0	43.45	20.0	54.3	47.00
70.0	43.45	30.0	54.3	45.50
60.0	43.45	40.0	50.5	43.75
50.0	43.45	50.0	50.5	45.00
40.0	43.45	60.0	50.5	48.15
30.0	43.45	70.0	50.5	49.90
20.0	43.45	80.0	50.5	51.40
10.0	43.45	90.0	50.5	55.00
90.0	36.50	10.0	54.0	36.50
66.6	36.50	33.3	54.0	39.00
33.3	36.50	66.6	54.0	45.75
10.0	36.50	90.0	54.0	51.75
90.0	37.50	10.0	54.0	36.5
66.6	37.50	33.3	54.0	35.5
33.3	37.50	66.6	54.0	47.0
10.0	37.50	90.0	54.0	52.0
90.0	40.75	10.0	54.0	39.75
66.6	40.75	33.3	54.0	40.50
33.3	40.75	66.6	54.0	47.50
10.0	40.75	90.0	54.0	52.00
90.0	45.00	10.0	54.0	44.00
66.6	45.00	33.3	54.0	40.75
33.3	45.00	66.6	54.0	40.00
10.0	45.00	90.0	54.0	52.50
90.0	48.50	10.0	54.0	47.50
66.6	48.50	33.3	54.0	45.00
33.3	48.50	66.6	54.0	47.75
10.0	48.50	90.0	54.0	52.50
90.0	50.00	10.0	54.0	49.00
66.6	37.50	33.3	54.0	35.50
33.3	54.00	66.6	54.0	47.00
10.0	54.00	90.0	54.0	52.50

Table 148 (Continued)

Mixed Paraffin Wax and Stearic Acid

Paraffin Wax %	Paraffin Wax Melting Point °C.	Stearic Acid %	Stearic Acid Melting Point °C.	Mixture Melting Point °C.
90.0	56.5	10.0	54.0	55.5
66.6	56.5	33.3	54.0	52.0
33.3	56.5	66.6	54.0	47.5
10.0	56.5	90.0	54.0	52.5

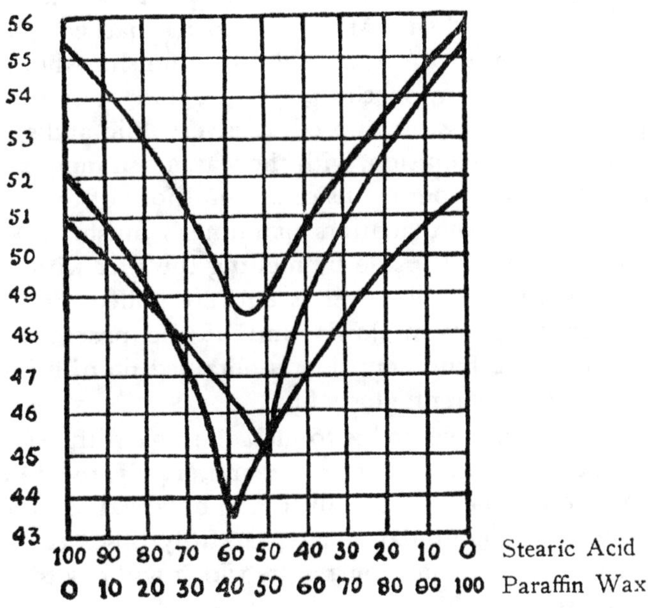

Figure 60. MELTING–POINT DIAGRAM OF A STEARIC ACID–PARAFFIN WAX MIXTURE

LEATHER FINISHES

W. A. Kingman

In the finishing of leather and also of products made from leather, wax of some description is one of the important constituents of the finish mixture. In the case of tanned hides or skins, all leather, except sole leather, splits or suède-finished leather, is coated with a clear or pigmented finish mixture. This finish is composed primarily of wax, casein and shellac, with the possible addition of other ingredients, such as albumen, latex, synthetic-resin emulsions, or water-soluble colloids. The wax most commonly used is carnauba wax and the grade selected depends on the desired color of the finished products, although Number 1 or Number 2 grades are most commonly used. Formerly, this wax was emulsified by long boiling with a castile soap solution, the finished emulsion containing about 2 lb of wax and 4–5 oz of castile soap to the gallon. The film of wax formed on drying such an emulsion was dull and opaque and had a matt surface, so that the finished leather had to be "jacked" or polished by friction in order to produce a glossy finish. About 1925, it was discovered that by special procedure and by special emulsifying agents, it was possible to produce a very fluid and semitranslucent carnauba wax emulsion, with the wax much more finely dispersed than in the previous emulsions. Emulsions of this type were soon used by the manufacturers of tanner's finishes. Many of these emulsions are still made by melting the wax with 15–20% soap and cooking till the water is entirely boiled off. Boiling water is later added very slowly and, if the procedure is properly carried out, a fluid emulsion is obtained on dilution, which will dry to a transparent, glossy film.

The modern method is to melt the wax with a fatty acid, then add a base to form a soap *in situ* and, finally, to add boiling water slowly until the desired dilution is obtained. If the proper precautions of manipulation are carefully observed, a thin, translucent or milky emulsion results which will dry to a transparent, glossy film.

By varying the fatty acid and the base used, as well as the proportions of the two, countless varieties of emulsions can be made, which may be used for some particular method of application or to give some particular effect on the finished leather. By the selection of the proper fatty acid and base, an emulsion of very low surface tension may be produced, which will spread very readily, with a minimum tendency to "crawl." A different combination will yield an emulsion of a much higher surface tension which will give less foam when spread on the leather with a sponge or brush.

In addition to the use of wax in finishing skins, wax emulsions are very largely used in shoe factories to give a final dressing to finished shoes. These finishes are made up of shellac, wax and sometimes casein and dyestuff. The shellac is dissolved in water with the aid of an alkali, and the wax is emulsified to produce a thin, bright-drying emulsion. These two solutions are mixed in varying proportions, according to the finish desired and the type or nature of leather being finished. If the finish is applied by a sponge, a foam-retarding agent is necessary, but if applied by a spray, the foam is of less importance; the finish must level off quickly and show no tendency to crawl. In both the tanner's finish and the shoe factory finish, the same type of wax emulsion is used.

A few typical formulae will show the general method of preparation of such emulsions.

Formula No. 1

Carnauba Wax	11.2	Morpholine	2.2
Oleic Acid	2.4	Water	67.0

Melt the wax with the oleic acid and bring the temperature to 90°C. Add the morpholine and stir till clear. Add the water, which should be boiling, very slowly with vigorous and constant stirring, so that the mixture forms a clear but viscous mass of petrolatumlike consistency. As further water is added, the mixture gradually thins and becomes more or less translucent.

No. 2

Carnauba Wax No. 1	60	S489 (Emulsifier)	14
Water-White Distilled Linseed-Oil Fatty Acids	16	Water, to Make	375

Use the same procedure as described in the previous formula.

No. 3

Carnauba Wax No. 1	120	Caustic Soda (50%)	8
Oleic Acid	20	Water	32
Trigamine	20	Water	800

Melt the wax in the oleic acid, add the Trigamine and heat to 90°C. Then add the caustic soda and the first water at 95°C. Stir till a gel is formed. Then add the balance of the water, which must be boiling, and stir till cool.

Although carnauba wax, on account of its hardness, color and emulsifiability, has been very largely used as most suitable for emulsions for tanner's finish and shoe dressing, other natural waxes, such as ouricury wax and vegetable fiber wax, have been substituted for carnauba wax with very good results during its scarcity. One new synthetic wax, S688, is a good substitute for carnauba wax for certain applications, as it can be emulsified to a thin emulsion, drying to a transparent film. The film produced is slightly tacky but this "tack" or "drag" can be eliminated by the addition of a small percentage of an organic colloid, such as casein; 15–20% of casein with the S688 wax forms a mixture which dries to a transparent film with no objectionable tack. This can be used to replace carnauba wax in a number of applications to leather or leather goods.

PAPER FINISHES

W. A. Kingman

In the finishing of paper, it has been the practice for many years to use wax in the finishing or coating mixture for certain effects in the finished paper. In flinted and in friction-calendered papers, wax has to be used in the coating to act as a lubricant during the mechanical polishing and to aid in producing the desired gloss.

At first, beeswax was emulsified and the emulsion added to the coating mixture, but, on the introduction of carnauba wax, this was adopted by the flinted-paper trade and largely by the friction-calendered paper finishers. For such purposes, the wax is emulsified as follows:

Carnauba Wax	50 lb	Castile Soap	12 lb
Water	50 gal		

The three ingredients are boiled together, until the wax is emulsified and sufficient water is added to bring the volume to 50 gal.

The following formula will illustrate the use of such an emulsion in a coating mixture:

Clay	300 lb	Wax Emulsion	12 gal
Water	20 gal	Casein Glue	25 gal
Talc	18 lb		

The casein glue may be made up as follows:

Casein	100 lb	Borax	17 lb
Water	50 gal	Ammonia (26° Bé.)	1 qt

Another formula, for making a wax emulsion for flinted papers, is as follows:

Laundry Soap	7 lb	Water	12½ gal
Carnauba Wax	50 lb		

Boil with live steam till thoroughly emulsified (generally 4—5 hours). Cool to 35°C., and add 2 lb. of 26° ammonia. Make up to 50 gal with cold water. The emulsion should be allowed to stand before using, as it seems to improve with age.

A beeswax emulsion, suitable for friction-calendered papers, is made as follows:

Yellow Beeswax	150 lb	Water	25 gal
Castile Soap	28 lb		

Melt the soap in the water and add the wax. Melt and stir till emulsified and smooth. Add water to make 150 gal.

In the manufacture of coated boxboard, a wax emulsion is sometimes added to the casein-coating mixture to make the coating more flexible and to improve the bending and folding properties of the coated board. Such an emulsion is usually made with Japan wax.

Japan Wax	75 lb	Water	25 gal
Stearic Acid	17 lb		

Boil till the waxes are melted. Cool slightly and add 10 lb borax dissolved in 10 gal water. Boil till a smooth emulsion is formed, cool, add 1 qt of 26° ammonia and make up to 75 gal.

In the manufacture of special grades of paper and of boxboards, a recent development in sizing is to use an emulsion of paraffin wax with the usual rosin size and, when this is thoroughly mixed with the paper stock in the beater, the wax and rosin are simultaneously precipitated on the fiber by the addition of alum. This combination wax-and-rosin size is claimed to produce a sheet of paper or boxboard with superior waterproofness and folding properties.

WATERPROOFING KRAFT PAPERS[1]

Water repellency is characterized by a resistance to wetting by water. A water-repellent paper should shed drops of water or show a "ducks-back" effect yet may, under slight pressure, allow free passage of water and may offer little or no resistance to the transmission of moisture or water vapor. Waterproofness indicates an ability to resist the passage of water. A paper, heavily impregnated with wax, may resist the passage of water under normal conditions, yet allow some passage of water vapor at the same time. Moistureproofness indicates an ability to resist the passage of water vapor. A paper, coated with a heavy, continuous film of wax, will probably be moistureproof to all intents and purposes under normal conditions. The normal conditions is the qualifying statement in these definitions, because true water- and moistureproofness are extremely difficult to obtain under all conditions.

Paper consists of countless cellulose fibers felted together and bonded with an adhesive or size. As a result, there is a continuous contact of cellulose to cellulose throughout the sheet, with minute openings between the fibers. Since water, in passing through the sheet, does so primarily through the interstices rather than through the fibers, the paper can be made water tight by filling these openings with wax and the degree of water tightness will depend on how completely and uniformly this is accomplished. Paper, waxed in this manner, i.e., where the wax is confined entirely to the interior of the paper, is called "dry-waxed."

Moisture (water vapor) is transmitted both by cellulose fibers and through the interstices. Dry waxing, even though filling the openings with wax, does not completely cover the cellulose fibers and many of them protrude above the surface of the paper. These surface fibers serve as wicks which absorb moisture from a humid atmosphere at one side of the sheet and carry it through to the other side, where it is given off to a less humid atmosphere. Dry waxing is, therefore, incapable of producing a moistureproof sheet. This can be accomplished only by completely covering the surface fibers with a continuous film. If wax is used, this paper is called "wet-waxed" paper.

---

[1] Killingsworth, *Paper Ind.* 8, 526 (1943).

A paper, having the advantages of both wet- and dry-waxed papers, is obtained by laminating with wax. Here, two layers of paper are bonded together with a layer of wax, forming a continuous barrier resistant to the passage of water vapor. Since the wax film is covered by paper on both sides, the continuity and effectiveness of the film are protected against injury by careless handling or scratching. The most important, fundamental precaution, to observe in lamination, is to prevent contact of surface fibers through the bonding medium.

The most common type of dry waxing, for imparting water repellency to paper, is by beater sizing. Here, a wax emulsion, or a dispersion of wax in water with the aid of emulsifying agents, is required because of the necessity of depositing an accurately controlled, small amount of wax on the fibers. The emulsion must be compatible with the pulp and rosin size during the beating and capable of being broken by alum for complete deposition and retention on the fibers.

Another form of sizing with wax emulsion for water repellency is known as top sizing. Here, the proper concentration of emulsion is applied to the paper, usually in the water box or size press. The emulsion must not wet the paper too easily and should be resistant to alum which is leached from the paper as it passes through the bath.

Where higher degrees of water repellency or fair waterproofness is desired, paper is often passed over a roll partially immersed in molten wax. The roll picks up wax and transfers it to the paper. The amount and type of transfer depends on the speed of the roll, the density of the paper, the temperature of the wax, the type of wax, the time allowed for penetration while the wax remains molten, the pressure on the squeeze rolls, etc. With modern machines, it is possible to obtain almost any wax content desired from 10% up to saturation. Below 10%, it is extremely difficult to obtain uniform coverage by hot waxing methods. To aid in the distribution of molten wax through the sheet, it is common practice to use high temperatures in the wax bath and to pass the coated sheet over the hot rolls or even to wind the paper while still hot.

Wet waxing for water- and moistureproofing is usually conducted on the same machines as used for hot dry waxing. By keeping the temperatures low, the speed of the machine high, and chilling the coated paper, the wax film may be solidified on the surface without much penetration. Where both penetration and surface wax are desired, the paper is actually submersed in molten wax. Frequently, paper coated with molten wax is immediately plunged into cold water or passed over cold rolls, which produces a glossy, shiny surface wax film, a characteristic desirable in certain applications. In this operation, the thickness of the film is adjusted by means of scraper bars or by the pressure of a set of squeeze rolls.

Lamination of two sheets of paper is easiest carried out on equipment designed especially for this purpose. However, quite often, existing equipment may be used if care is given to the essential details. It is necessary to apply molten wax to one of the papers to be laminated and effect the lamination by the squeeze rolls, before appreciable penetration has occurred. The laminant should set almost at the time of contact with the squeeze rolls. Where two sheets of porous paper are to be laminated, extremely close control of temperatures and speed, or the use of very viscous laminating mass, is required. The alignment of perfectly machined combining rolls is also essential.

Petroleum waxes have been extensively used for waterproofing in the paper industry and, from the viewpoint of price, supply and ease of handling, they probably represent the most satisfactory substances available at this time. They are divided into two broad classes, paraffin-type waxes and microcrystalline-type waxes. The paraffin-type waxes include fully refined waxes and crude scale waxes. Both are relatively hard and brittle, white, crystalline solids and are sold in various melting grades, ranging from about 120 to 140°F. The microcrystalline-type waxes, also called amorphous or petrolatum waxes, differ from paraffin-type waxes in the following important respects:

1. Considerably smaller crystals
2. Higher melting points (145–190°F.)
3. Greater ductility at all temperatures
4. Greater tackiness
5. Generally darker colors (lemon yellow to dark brown)
6. Higher viscosities in the molten state.

It is important to keep these differences in mind when a wax is chosen for a particular application. Each type has its drawbacks and often blends of the two are required to meet service conditions.

From a practical standpoint, a heavy, uniform film of a crystalline wax may be as water- or moisture-resistant as an equal film of microcrystalline wax, but because of its crystalline and brittle nature, this film cannot stand rough handling or flexing and retain its water or moisture resistance. It is primarily the tough, ductile character of the microcrystalline waxes over wide ranges of temperature that makes these waxes more effective water- and moistureproofing agents.

The crystalline waxes are employed for coating paper where a hard, nontacky, glossy film is desired. Also because of their low viscosity when molten, they are usually preferred for hot impregnation of paper. The microcrystalline waxes and their blends are useful for wet waxing where the wax film may be subject to adverse conditions. Also because of their tackiness, rather sharp solidification point, and fairly light color, they are superior laminating compounds. The operations are clean, the bond good, and the aging qualities of the laminated paper excellent. The microcrystalline waxes are more grease resistant than the crystalline types and are useful where this characteristic is desired.

EVALUATION OF WAXES FOR PAPERBOARD

In order to understand the problems of wax application more completely, practical tests have been developed for microcrystalline and other waxes. The impregnation and wax coating of paperboard involve so many variables that the resultant data can be very confusing. This is particularly true when it is necessary to communicate with the manufacturer. The standard melting-point and penetration data do not seem adequate to explain the problems of wax protecting paperboard.

The following tests have acted as a guide in predicting certain performance characteristics:

1. Stain and flow at 130°F.
2. Softening and melting point
3. Flexibility and strength
4. Viscosity

5. Surface tension
6. Permeability
7. Degree or percentage of impregnation
8. Effect of wax on the strength of board
9. Electrolyte-penetration test
10. Water absorption
    a. Humidity
    b. Submersion

These data permit the classification of waxes by quality ratings. It is possible to identify from a series of blended waxes those which are unfit for protective purposes.

STAIN AND FLOW TEST AT 130°F.

Equipment

Casting molds (½ in. cube); galvanized 16-gage steel sheet, mounted at a 30° angle to horizontal; 0.025 in. white blotting paper; No. 330 red paraffin-soluble dye (Glidden Company, Cleveland, Ohio).

Test Method

Cast six ½ in. cubes of wax.

Stain Test

Dust red paraffin-soluble dye on the bottom of the cube of wax. Place three cubes of wax each on a 4 in. square of blotting paper. Place the blotting paper in an oven at 130°F., with circulating air. Test for 24 hours. Waxes which melt below 130°F. or oil will diffuse into the blotting paper and form a red stain. The extension of this stain beyond the square of wax is reported as the degree of staining.

Flow Test

Heat seal three cubes onto a 16-gage galvanized sheet. Arrange the cubes in a line at the top of the sheet, mounted at a 30° angle. Place the mounted sheet in a 130°F. oven for 24 hours. The amount of flow in inches from the starting line is the desired result.

SOFTENING AND MELTING POINT

The purpose of the melting-point determination is to know accurately at what point the wax will melt and at what point the wax will soften. The melting point indicates the type of the wax as well as its thermal handling characteristics. The softening point indicates not only the flow point but also the temperature below which the wax is not soft and cheesy and has lost its strength. This technique is a rigid method for controlling uniformity of the wax.

Equipment

Silver-plated copper bar $\frac{1}{4} \times 1\frac{1}{4} \times 15$ in.; Bunsen burner on one end of the bar and a basket of dry ice on the other end; suitable mounting for the bar supported from the ends; iron-constantan 30-gage thermocouple with flattened high-temperature solder or No. 8 braising rod tip, 1/16 in. in diameter; wooden thermocouple pressure spindle; Leeds and Northrup temperature compensating potentiometer.

Test Method

Brush a thin layer of wax, approximately 0.010 in. thick, the entire length of the bar. Apply heat to one end and cooling to the other end of the bar. When the line of melted wax has progressed near the center of the bar, apply the thermocouple so that it straddles the line of demarkation between melted and unmelted wax. Check this point twice at 5-minute intervals. If the temperature checks within 0.50, this may be considered the melting point. Apply 0.002 in. antitarnish paper by means of a squeegee photographic roller across the face of the wax. Where the wax is melted, it will be absorbed into the paper. The pressure of the roller on the paper will make an impression on the softened portion of the wax. The first softening point can thus be determined. The thermocouple is pressed against the bar by means of a wooden spindle to avoid conducting heat away from the thermocouple. This equipment is the Parr melting apparatus which may be obtained from the Central Scientific Company.

FLEXIBILITY AND STRENGTH

The purpose of these two tests is to indicate the behavior of the wax under physical shock and assembly. It is important that the wax is flexible enough to withstand a sudden blow or fall. In addition, the wax must be strong enough to resist shock without permanent deformation. According to the arrangements made, these factors are known as:

E = Modulus of elasticity
Sr. = Modulus of rupture

The lower the number, the more flexible the wax will be in the first case and the higher the number, the stronger the wax will be in the second case. Both of these factors have a bearing on the handling of the impregnated paperboard stock. If a wax is flexible enough to withstand 180° of bending, it will be probably flexible enough to be bent and impregnated in a paper carton. However, the wax, although soft and flexible, may be so weak that beyond a certain limit of flexing, it will soften to a mush. This means that the wax will no longer support the fiber as it is being bent and the paper will tear, not because the wax is not flexible enough, but because the wax is too weak.

Equipment

Casting molds, $\frac{1}{4} \times \frac{3}{4} \times 3\frac{7}{8}$ in.; device for clamping the wax sample; device for loading the wax sample uniformly; scale for measuring the degree of deflection of the bar of wax as the load is increased.

Test Method

The wax sample is cast into $\frac{1}{4} \times \frac{3}{4} \times 3\frac{7}{8}$ in. These should be chilled and held at 21°C. at least 2 hours. No more than two bars at a time should be tested in the testing machine to avoid thermal change. Reasonably close checks on four bars will be considered satisfactory tests. A suitable support should be made for one end of the bar extending no more than ½ in. to provide a nonslipping attachment. The innermost edge of this clamp will be considered the indicator for the length of the tested sample. From that point, the bars should extend 1½ in. from the supporting clamp. The bar should be clamped in a horizontal plane and a uniform weight should load it. The deflection of the bar at 20° intervals should be recorded with the weight required to bend the bar. The weight should be added at as uniform a rate as possible and the weight necessary to break the bar ultimately should be determined.

Calculations should be made as if the bar were a simple beam loaded at one end and the modulus of elasticity and modulus of rupture should be calculated in the usual manner.

Table 149A lists the angular deflection of a horizontal beam in degrees, with the corresponding calculated deflection "F" given for a radius of 1½ in.:

Table 149A

DEFLECTION OR F

Angular Deflection, θ	Sine θ	Calculated Deflection, F at 1½ in.
0	0	0
1	0.0175	0.0250
10	0.1736	0.2605
20	0.3420	0.5135
30	0.5000	0.7500
40	0.6428	0.9645
50	0.7660	1.1500
60	0.8660	1.2950
70	0.9397	1.4080
80	0.9848	1.4770
90	1.0000	1.5000

From theoretical and applied mechanics, the maximum deflection is:

$$F = \frac{W}{EI} \times \frac{l^3}{3}$$

W = Load in pounds
E = Elastic modulus to be found

$$I = \frac{bh^3}{12}$$

b = Width of beam in inches
h = Height of beam in inches
l = Tested length in inches

By taking the ultimate breaking point, the modulus of rupture may be determined by the following formula:

$$Sr = \frac{M^c}{I}$$

$\frac{I}{C}$ = Moment of Inertia—Considered as the distance from the axis of bending to the outermost portion of the beam

M = The internal moment beyond the elastic limit

By calculation, Sr = 957 W

VISCOSITY

The viscosity test will determine the cooling cycle so that the dipping temperature may be adjusted to an optimum. The flow and dipping characteristics of the wax will depend on the viscosity.

Test Equipment

Stormer viscosimeter.

Test Method

The standard Stormer viscosity curve is run. The wax is held at the measuring temperature for at least 5 minutes before making the first test. Successive tests are made at 5-minute intervals until a check is obtained, within 1/10 second, the temperature being held constant. A curve is drawn as the test proceeds from the chilled, mushy wax to the highly fluid wax at around 300°F. A check is made at two temperatures, such as 190 and 230°F. after the curve is completed. If these checks come within 1/10 second of the original determined value, the curve is considered to be valid.

SURFACE TENSION

Surface tension measures the wetting ability and is a definite indication of the impregnation characteristics. It indicates the spread of the compound when used for dipping.

Equipment

Standard capillary tube.

Test Method

The surface tension is determined by a standard instrument which can be obtained from Fisher Scientific. It consists of an outer glass container in which a capillary tube is inserted. The capillary tube has a measuring scale in millimeters. Wax is poured into the outer container and the outer container is submerged in an oil bath. The temperature of measurement is arbitrarily set at 100°C. Since this is close to the impregnation and dipping temperatures involved, sufficient time must elapse to permit equilibrium temperatures to be established throughout. The capillary tube is heated by a 250-watt infrared bulb 6 in. away from the capillary tube in order to prevent the wax from freezing the capillary. The capillary tube, after being heated, is inserted into the wax. The measuring scale is set at zero at the wax level. The capillary rise in the tube is observed. A reading is taken after ½ hour at constant temperature. The wax is sucked out of the capillary by means of a vacuum and successive readings are made at 10-minute intervals until a constant reading within 0.2 mm. is obtained. Under such conditions thermal equilibrium has been established and a surface tension measurement can be made which is comparative. A better system of heating would be the insertion of the entire measuring unit in a 100°C. oven for 1 hour. The samples could be brought out and measured within a reasonably short period of time and far greater accuracy would result due to less thermal difference along the capillary tube.

PERMEABILITY

The permeability is a mechanism which will determine the ultimate protective value of the wax. Paper fibers will absorb whatever water is brought to them. Therefore, the permeability of the wax determines the amount of water per hour which is brought to the paper fiber. Unless this is an extremely low figure, the paper will swell and likely cause the wax film to crack off.

Equipment

Standard 3-in. aluminum measuring cup with supporting ring; desiccator.

Test Method

Permeability is given in terms of milligrams of water loss per hour per square centimeter per millimeter thickness. The standard aluminum 3-in. cups are used. Water is poured into the cup. The wax sample is prepared by dipping 0.002 in. Superior Manifold paper into relatively cold wax and withdrawing slowly until a uniform 0.004 in. of wax is deposited on the paper. The total thickness at that point is 0.006 in. The paper thickness is deleted because of the wicking action of paper fibers. It is considered that the paper is a supporting medium and the wax is not a protective membrane across the paper fibers.

The wax-paper sample is cemented in place by pouring black Barnsdall 160°F. into the sealing ring provided and insuring positive wetting of the aluminum dish by means of a hot soldering iron.

The sample is placed in the desiccator with sulfuric acid and the test is conducted at 45°C.

The sulfuric acid is changed each time a reading is taken. The first reading is taken at the end of the first 24-hour period. The water loss is determined, and if duplicate samples do not check closely with each other, they are examined for possible leaks. Having eliminated the leak, the test will proceed for 5 days. The calculation is made as follows:

$$\frac{\text{Loss in grams} \times 1000 \times (\text{mils thickness})}{\text{Area in square centimeters} \times \text{hours on test}}$$

DEGREE OR % IMPREGNATION

The permeability of the wax determines the amount of water reaching the paper per hour of testing period. The per cent impregnation determines how well the paper fiber has been covered with wax and protected or prevented from swelling. Thus, if the impregnation is sufficient, a poor dipping wax may be used without detrimental effects. However, with the proper impregnation and the proper external coating, a good waterproof sample should result. The final test as to whether the dipped sample will stand up under high humidity and water submersion conditions is dependent almost entirely upon the performance of the impregnated board. It is for this reason that the per cent impregnation is important. It is obvious that if on most boards the per cent impregnation maximum is 70% of the initial weight, under these circumstances both the fiber is covered and the voids filled. Under most conditions, such a board cannot be bent without breaking the fibers. The ideal impregnation is the amount necessary to cover through the paper fibers and leave most of the voids virtually open so that there is freedom of motion of the fibers. A satisfactory per cent impregnation seems to be 35–45% of the initial paper weight, which is a little over one-half of the maximum weight which can be added filling the voids and covering the paper.

Equipment

Standard oil extraction apparatus, or weight balance; scale for determining area of paper.

Test Method

The per cent impregnation is obtained by weighing a sample of paper before and after impregnating. The sample must be impregnated in a centrifugal spinner to assure minimum adsorption of wax on the outside of the paper. A check may be by extraction in toluol in a Soxhlet thimble.

EFFECT OF WAX ON THE STRENGTH OF BOARD

In order to relate the physical tests which have been made to the behavior of the board, a strength test is made on the impregnated board.

Equipment

Dipping tank; testing unit to apply uniform load.

Test Method

The board which has a standard score in the center of it is impregnated in the usual fashion. One inch wide strips are taken from this impregnated board. The board will be bent once in both directions at 180°, then fastened in a clamp, and the load applied until the board breaks. The number of pounds per inch of board is the measure of the strength; the control is measured without wax impregnation. By comparing the effect of various impregnating waxes, a relative evaluation can be obtained. This also checks the quality of board from time to time.

ELECTROLYTE PENETRATION TEST

The purpose of this test is to determine how the board will withstand zinc chloride or electrolyte.

Equipment

Standard megger; class cylinders approximately 1 in. in diameter and 2¼ in. tall; Barnsdall 160°F. wax or equivalent.

Test Method

The glass cylinder is cemented or fastened to the impregnated board by means of the Barnsdall wax. The tube is filled with 33°Bé zinc chloride solution saturated with sal ammoniac. The board is tested after exposure for 24 or 48 hours at room temperature. One contact with the megger is placed in the solution. The other contact of the megger is made onto a stainless steel plate. The test sample is pressed against the stainless steel plate and a reading is made on the megger. A fairly accurate evaluation may be made as to the degree of protection the board has afforded to electrolyte transfer.

WATER ABSORPTION

The purpose of this test is to determine the amount of water which will be picked up by the board due to its hygroscopic action, and the amount of water which will be picked up in the voids due to its capillary action. This allows comparison of various processes of impregnation and the relative effectiveness of various waxes as a moisture barrier around the individual paper fibers.

Equipment

Test cabinet of 45°C.; closed container to give 100% humidity; closed container in which samples may be submerged in water.

Test Method

Humidity

The impregnated board is placed in the 45°C. 100% humidity cabinet and allowed to remain undisturbed for 48 hours. The difference in weight is the percent absorption representing the hygroscopic action of the board. It is a measure of the quality of impregnation.

Submersion Test

Samples of the board are weighed and submerged in water at 45°C. in a closed container. The sample is weighed at the end of 48 hours.

The increase of water after submersion represents:

The water which would have been absorbed due to exposure to high humidity.

The water which has filled the voids.

By combining the humidity and submersion tests, it is possible to estimate the amount of water absorbed by the fiber and the amount filling the voids in the board. The quality of the impregnation is determined by the water absorbed by the fiber, as the impregnation is designed to coat the fibers and thus provide a water-vapor-inhibiting barrier. The voids cannot be filled with wax and maintain a flexible board which can be handled in the manufacture of the battery. For this reason, the outer dip is provided which prevents water from penetrating the voids. If impregnation has not been properly carried out, any imperfection or break in the outer dip, or ingress of moisture vapor through the socket can cause the fibers to swell due to the absorption of the moisture.

ADHESIVES

The familiar sealing wax, that is sold in sticks and used for sealing letters and for taking impressions of seals, etc., on important documents, is a misnomer. Such sealing waxes rarely contain any wax but are composed of resins, plasticizers, fillers and pigments.

Heat-sealing agents, containing wax, are, however, made and used commercially in large quantities. They are thermoplastic substances which are applied hot (150–300°F.) to each of two surfaces. On bringing these surfaces in contact with each other and cooling, the two layers will combine to form a single layer which unites the surfaces. Similarly, surfaces may be coated with such compositions and allowed to cool, without contact. Such surfaces, at ordinary temperatures, are nonsticky and may be stored, under proper conditions, for indefinite periods. To unite such treated surfaces, they are pressed together and heated to fuse the coatings. On cooling, the surfaces are firmly united. Heat-sealing agents are also made as foils or sheets that are interposed between two surfaces to be united.

There are two types of heat-sealing agents that are applied to foils or sheets, one that really melts to a liquid and the other that changes to a sticky mass.

Paraffin wax is typical of the agents which are coated on a film or sheet and which truly melt. Waxed paper, still the most widely used heat-sealing material, was the earliest product to bring the advantages of heat sealing to packaging. Paraffin wax is not particularly sticky at ordinary temperatures, but melts at 130–140°F., depending on the grade of wax used. If two pieces of waxed paper are pressed together and are sufficiently heated through to their interface, the paraffin on their surfaces will melt. While molten, the wax is not sticky and thus will not hold the two sheets together, but if they are held in position until the fused wax, in the interface, chills, the wax will have sealed the two sheets together. It is characteristic of most true melting agents that their seals must be held in position while the sealing compound solidifies.

The force required to separate the two pieces of sealed waxed paper, that is, the tightness of the seal will depend on the cohesion of the wax and the adhesion of the wax to the paper.[1] It is common experience that neither the cohesion of paraffin nor its adhesion to paper is very great, therefore, the tightness of the seal, or, the seal strength of waxed papers, is not very great either, although, it is obviously adequate for many purposes. It should be noted that within limits, the seal strength of waxed paper increases with an increase of the thickness of the waxed coating. It should also be noted that until a certain

[1] Leinbach, *Modern Packaging* **16,** 85 (1943).

thickness of wax is coated on the surface of paper, it does not become heat sealing. If, under the heat of the seal, too great a proportion of the wax is absorbed into the paper or among the irregularities of its surface, too little remains on the surface to effect a seal. Thus, it takes less wax on smooth-surfaced, non-absorbent glassine to make it heat seal than it does on rougher, more absorbent sulfite papers. However, given enough wax on a sulfite paper, the rougher surface has the effect of better adhesion of wax to paper and the seal of the waxed sulfite paper will be stronger than that of waxed glassine, whatever the other attendant, functional qualities of either paper.

The relationship between the absorbency of the base sheet and the minimum coating weight, for effective seal, is important and applies to any type of heat-sealing agent and to any type of base sheet.

In order to improve the seal strength of waxed papers, it is necessary to cause both the cohesion of the wax and its adhesion to its base sheet to be strengthened. This is done by adding rubber, rubber derivatives, resins, or related compounds. Suitable substances not only raise the melting point of the wax, but also make it become quite sticky when hot. In fact, blends of this type do not have a true melting point, but as the temperature increases, they become sticky ("tacky") and softer until they finally do approach what might be termed a molten state, though never so fluid as molten wax. The relation between temperature increase and progressive softness is known as the "melting" or "softening" range of the blend. As indicated, such blends can be made to be quite tacky when hot and thus yield quick and firm heat seals with a minimum of chilling required before the seal "sets." In some cases, at sealing temperatures, the coating is so tacky that no chill is required, for the tackiness is sufficient to hold the seal together. Blends of this type usually have excellent adhesion to many materials and papers or films, coated with fairly heavy coatings, have found widespread application, such as in bags for wet-pack frozen foods, in cheese wrappers and in "heat-seal" labels, where the tackiness of the hot coating on the back of the label is useful in adhering the label to almost anything including glass, the sealed ends of waxed bread paper, individual frankfurters, etc. As may have been noted, such modification of wax tends to direct it into the class of the second type of agents which on heating do not melt, but become very sticky.

FOIL COATINGS

Harold A. Levey

Since food products have been packaged, it has always been a desideratum to prepare a wrapping material of such a nature as to retain, as long as possible, the original moisture content of the food product. In certain types of foods, such as most confections and candies, in particular, the objective is to prevent the taking up of moisture, and this is equally true of crackers. Certain types of cakes, meats, tobacco goods, etc., lose their moisture, even if the surrounding atmosphere is of very low relative humidity, e.g., as low as 15%. In order to maintain the proper conditions, a wrapping membrane which offers unusually high resistance to the transmission of moisture or water vapor is desired.

If complete transparency of the wrapping is not required, a wrapping sheet of relatively low moisture transmissibility offers no serious problem. Wrapping sheets of this type are used in the manufacture of large paper bags for such food products as granulated sugar and other hygroscopic commodities. Such sheets are obtained by coating two sheets of kraft paper, each on one face, with an asphaltic solution in some suitable, volatile, organic solvent, or with an asphaltic water emulsion, or a combination of the two in what is known as a "cut back." After the emulsion is broken, which occurs quickly in view of the fact that the water phase is readily drawn into the paper stock, leaving the minute globules of asphalt, together with the stabilizing agent (which may be bentonite, or casein) on the surface of the paper stock. The coated faces are then allowed to dry and placed in contact with each other. The two sheets are then passed through steam-heated calender rolls and, under the action of heat and pressure, quickly and permanently bonded together. This type of packaging sheet of low moisture permeability is extensively used as a wrapping for war equipment shipped in crates and it relieves the more costly complete covering, with a film of grease or heavy oil, of the exposed metal parts which would otherwise rust.

When transparency or translucency is required, glassines, vegetable parchments, tissue papers, etc., are used as wrappings. These are made moisture impermeable, to a high degree, by impregnation and/or coating with waxes or waxlike bodies. The animal and vegetable waxes used for this purpose include carnauba, bayberry, and candelilla, which are the most effective ones, though beeswax and other animal waxes may also be used. Among the waxlike bodies are paraffin and ceresin, as well as the higher fatty acids, such as palmitic and stearic, and certain synthetic products, such as the chlorinated naphthalenes (Halowaxes), and other chlorinated hydrocarbons of very high molecular weight which are usually solid at room temperature. Frequently, these waxes are blended with resins to improve their working characteristics. The harder waxes are often compounded with certain softer waxes which function as plasticizing agents, providing flexibility. When papers are both impregnated and coated with wax to such an extent that a film can be scratched off from their surface with the fingernail the ultimate product of high moisture resistance is designated as "wet-waxed."

All of these sheets are translucent to varying degrees, none, however, is completely transparent as they all diffuse light to a greater or lesser degree. The cellulose or cellulose derivative sheets, such as those of the viscose type or cellophane, and the cellulose acetate and ethyl cellulose sheets are the chief types of completely transparent wrappings on the market. Other types of transparent sheets on the market in a limited quantity are gelatin sheets, rubber sheets made either from rubber chloride or rubber hydrochloride, and vinyl sheets made from the copolymer of vinyl acetate and vinyl chloride, formal vinyl acetal, butyl vinyl acetal, and starch hydrate. These sheets are moisture transmissible to varying degrees, but none of them has a low enough value to be suitable for a food wrapping when the food product must maintain a constant moisture content. In order to improve the moisture resistance of these sheets, they are frequently coated with a composition containing a wax or waxlike substance. The wax alone would substantially reduce the moisture transmission of such sheets, but their transparency would be affected. Accordingly, a thin lacquerlike coating is used which contains relatively small amounts of waxes and yields a surface which is not only flexible, but also completely transparent. A representative formulation of such a coating, which is used for making moisture-impermeable viscose sheets marketed as moistureproof cellophane, is as follows:

Solids	Weight %	Solids	Weight %
Cellulose Nitrate	52.0	Dibutyl Phthalate	8.7
Ester Gum	35.0	Paraffin Wax	4.3

It is interesting to note that a cellophane sheet coated with a film of this composition will transmit about one hundredth of the moisture transmitted by the same sheet uncoated. The solids are dissolved in a suitable solvent mixture, such as is used for most cellulose nitrate solutions, e.g.:

Solvents	Volume %	Solvents	Volume %
Ethyl Alcohol	25	Toluene	18
Ethyl Ether	43	Butyl Acetate	14

Approximately 4 parts by weight of solvent are used to 1 part of the solids. The sheet is passed through this solution, and then, in vertical direction, through a tower which is heated and which operates at reduced pressure, carrying off the solvent vapors which are subsequently condensed for reuse. After evaporation of the solvents, the paraffin exudes to the surface, and is deposited as a continuous film. The total thickness of the moisture-proofing film approximates 1/10,000 in., and it has a remarkable moisture resistance, i.e., it actually transmits less than 0.5 g of water as vapor in 24 hours, per 100 sq in., with a relative humidity pressure difference of more than 75% on each side of the sheet.

This solution can, of course, also be applied to other types of transparent sheets to make them moisture impermeable to about the same degree. In addition, it may be applied to paper stock including glassines, vegetable parchments, and tissue papers, which will then be more translucent, approaching transparency, with the same high moisture impermeability.

There are obviously three commercial methods for applying wax finishes to paper and similar forms of films or sheets:

1. Dissolving the waxes in suitable solvents
2. Liquefying the waxes by application of heat
3. Emulsifying the waxes in water

The solvents or solvent mixtures for the waxes may be found in the technical literature, or by experimentation. It may be necessary, in some cases, to use the solutions at elevated temperatures, in view of the fact that the waxes, or certain types, separate out as suspensions or dispersions at room temperature.

Emulsions of the waxes may be readily prepared by known means, and usually contain 10–20% by weight of the wax or wax mixture, and 1–2% of a suitable emulsifying agent, of which the triethanolamine fatty acid soaps are representative. Other emulsifying agents are also used, and those containing morpholine are particularly desirable where it is necessary to avoid reemulsification of the wax coating. Other emulsifying agents, which can be readily destroyed by heat, are equally effective.

It is interesting to note that many of the animal and vegetable waxes are compatible with certain varieties of ethyl cellulose, particularly those with a combined ethoxy content of 44–50%. In addition, a limited amount of paraffin wax may be introduced into the mixture with the waxes which are compatible with ethyl cellulose, provided the paraffin wax content is kept under one third of the total weight of waxes used. The addition of ethyl cellulose makes the coating free from smear, and more permanent, also as maintaining a degree of flexibility, particularly at sub-zero temperatures, not obtainable with the usual plasticizing agents. Only very thin coatings are required to produce the desired degree of moisture impermeability. In using ethyl cellulose wax mixtures, it is important to maintain a high content of aromatic hydrocarbons, approximating 80% by volume of the solvents, and to avoid alcohols and ketones which would precipitate the paraffin wax.

PARAFFIN WAX-POLYETHYLENE PACKAGING[1]

The importance of protecting perishable and corrodible items from the deteriorating effects of moisture is well recognized.

[1] *Technical Memorandum,* Socony-Vacuum Oil Co., Inc.

Petroleum waxes are the basis of most protective packaging; laminating strength and flexibility are their prime requisites for use in fabricating protective barriers.

The effect of polyethylene on the use of paraffin wax in an important segment of paper converting illustrates the extent of the changes in packaging materials. The advent of polyethylene made gloss achievement relatively simple, bringing it within the reach of all converters. By using polyethylene, in some cases, it is necessary to modify the composition of the coating. As a result, in the hard-coat papers, paraffin wax is no longer used alone, but a three-component system is applied; while the formulation remains essentially wax, the paraffin-wax portion may be as little as one half of the blend, and averages perhaps two thirds of such compositions. The balance of the wax is microcrystalline.

Such high percentages of microcrystalline wax could not be used in most coatings before the adoption of polyethylene because of its detrimental effect on gloss and blocking. Nor could polyethylene be used were it not for microcrystalline wax, which lends the necessary toughness and flexibility to the three-component system. The properties which enable the microcrystalline wax to contribute protection at –40 to +165°F. are related to the present demands, but are not necessarily the same. "Zero" flexibility has been supplanted by "maximum" versatility, and it is far more necessary for the microcrystalline wax to "stay put" than to "stick."

WAXED PAPER[1]

The wax composition used consists essentially of a moistureproofing wax or waxy substance, a hard wax (either vegetable or animal) and a blending agent which is generally a resin. As the moistureproofing wax, a white crystalline high-melting paraffin wax, melting over 50°C., e.g., 52, 55, or 60°C., or even higher is preferred. Amorphous paraffins, particularly those having a high melting point, may be used to give increased flexibility of the composition. The low-melting noncrystalline waxes, petrolatums, etc., if used, will only be in minor quantities, since they

[1] T. G. Finzel and D. E. Drew, U. S. Patent 2,108,809.

tend to decrease the moistureproofness and also to give a greasy and smeary surface.

The hard wax imparts toughness. It also contributes greatly to the production of hard, nontacky, nonsmeary surfaces and heat-sealing properties. The hard wax may be of vegetable or animal origin and preferably candelilla or carnauba wax. These waxes are generally esters of higher fatty acids and high-molecular-weight, aliphatic alcohols. Other examples are montan wax and palm wax, as well as certain synthetic waxes of a similar nature, such as may be made by esterifying the acids of montan wax with higher alcohols.

The composition of paraffin and hard waxes, although miscible well above their melting points, produces cloudy liquids or semisolids in the region of their melting points. Even rapid chilling will not remedy this condition. The blending agent serves to prevent this separation and to inhibit haziness in the chilled composition. Thus, the primary function of the blending agent is to give the rapidly chilled composition its glass-clear appearance. It may also, of course, be chosen to assist in increasing the flexibility, gloss, nontackiness, heat-sealing, etc.

In general, the blending agents are resins. Illustrative examples of blending agents are the common varnish or lacquer resins, such as ester gum, elemi, dammar residue or the resins from gutta-percha, or some of the coal-tar or anthracenelike resins or some of the synthetic resins, e.g., properly modified alkyd resins and hydrorosinates, including diethylene glycol hydrorosinate.

Other substances which serve some special purpose, such as imparting greater flexibility to the product, greater slip between adjacent surfaces, preventing blushing, increasing the heat-sealing, etc., may also be incorporated in the composition. Rubber, gutta-percha, balata, etc., when used in small quantities, impart a greater flexibility and toughness, as well as greater heat-seal joint strength, when the composition is coated in thin films. Other toughening agents may be used, such as a suitable cellulose derivative (ethyl cellulose) or small quantities of film-forming resin, as vinyl acetate resins. For decreasing the tendency to blush, small quantities of substance, such as hydrogenated castor oil, paraffin oils, hydroxy fatty acids, β-naphthol, hydronaphthalene resins, etc., may be used. Greater smoothness or slippage between adjacent surfaces is secured when small quantities of talclike substances, such as zinc stearate, are added to the composition.

To increase the penetrating power of the compositions, particularly when they contain substances which give them increased viscosity in the molten condition, a small quantity of a volatile solvent, such as toluol or solvent naphtha, may be added. These solvents are used only in limited amount, such as is necessary to give the composition the desired fluidity and melting point, for good penetration and ready application. The amount of solvent will generally be approximately equal to the amount of the composition or less. It is preferable to use the minimum amount of solvent to save solvent expense and time required to eliminate the solvents from the impregnated paper.

The practical limits of the ingredients are as follows, the percentages being by weight:

	%		%
Paraffin Wax	2–30	Blending Agents	
Hard Wax	20–80	(Resins)	20–80

It is, of course, impossible to make every combination of components within these ranges. Some of the combinations will not tolerate, e.g., 30% paraffin. Some of the hard waxes require a much greater amount of resin than does candelilla to produce glass-clear films, but at some point within these limits a coating composition may be formulated which will, when properly processed, be transparent, glass clear and have other desirable properties.

One of the characteristic features of these compositions is that they have an index of refraction close to that of cellulose. In general, the refractive index of the composition can be readily controlled by the kind and quantity of resins used. For example, the refractive index of the composition may be increased by additions of highly refractive resins, of which coumaron and chlorinated diphenyl resins are examples. The compositions, in the molten state, have a viscosity higher than melted paraffin. This characteristic is highly desirable when the composition is used for coating, as it makes possible to obtain a more moistureproof product than with paraffin wax, the consumption of impregnating media being substantially equal or somewhat less.

Formula No. 1

a. A thin, porous sulfite tissue, basis weight 13–18 lb (24 × 36 — 500) is treated with the following solution:

Ammonium Oleate	1.9	Denatured Alcohol	77.1
Water	21.0		

The excess soap solution is removed by means of squeeze rolls and the wet tissue is passed over dryer rolls, drying the tissue until it feels distinctly moist (10–15% moisture). The damp tissue is passed through a supercalender stack, the rolls of which are cooled. This is followed by a second supercalendering in heated rolls. It is possible to combine these operations using a stack in which the first half of the rolls is cooled and the remainder heated. As the tissue emerges from the hot calender rolls, it is dry and has a fine finish. Although its solid fraction has been increased, the calendered paper still retains a good proportion of its absorptiveness.

The soap-treated, calendered paper is passed into a molten bath of the following composition:

Refined Carnauba Wax	34.5	Wax B–430	17.2
Glyco Wax C–991	34.5	Paraffin Wax (M.P. 62°C.)	13.8

The temperature of the melt should be about 90°C. After several seconds' travel through the melt, the paper is drawn up through a set of doctor knives, suitably heated to prevent crystallization of the waxes on them, one on each side of the paper, which scrape off all but about 15 lb of the melt per 100 lb of paper. The paper then passes between hot rods of small diameter which smooth out the wax surface. The impregnated paper is allowed to cool to room temperature and then wound up. At this stage of the process, the paper is thoroughly impregnated, flexible and nontacky, but, due to the incompatibility of the slowly cooled melt components, is so badly blushed as to be practically opaque.

The blushed paper is then unwound and passed over a head roll and down through an electrically heated chamber in which the impregnated composition is thoroughly melted, but is not heated so hot as to harm the paper or to vaporize the components of the wax mixture. While still in the molten state, the hot film is passed into water at 10–25°C. and then immediately out of the water. The contact of the film with water need be no more than 1 second. The paper is then passed between scraper rods which remove the greater portion of the water that adheres to the surface of the film. If the paper is running at high speed, a sheet of water is carried for some distance vertically out of the water bath on the film surface. It is desirable to break this sheet of water soon and allow the greater portion of the water to run back into the bath, so that the small amount of water remaining on the film can form into easily removable globules. After leaving the scraper rods, a portion of the remaining drops of water is blown or shaken from the film. The paper then passes between felt pads or cloth-wrapped rolls which remove the remainder of the water from the film.

The product obtained is transparent, very clear and free from any haziness due to blushing of the composition or to incomplete impregnation. It is glossy, difficult to smear or finger mark, and nontacky. It does not stick when stored at ordinary weather conditions in roll form or in sheets. It is free from odors which would be objectionable for a food wrapper, sealed by heat to a strong permanent joint, moistureproof, and can be printed with inks, such as are used on lacquered moistureproof, transparent, regenerated cellulose sheeting.

b. The same calendered tissue as in Formula 1 (*a*) is treated in the same fashion as in Formula 1 (*a*), but using the following transparentizing composition:

Candelilla Wax	55.5	Paraffin Wax (M.P. 62°C.)	16.7
Low-Melting Ester Gum	27.8		

If desired, the impregnated paper, after being scraped, is passed directly into a bath of cold water without the intervening cooling step, as described in Formula 1 (*a*).

The transparent product obtained is in every way similar to that obtained in Formula 1 (*a*), except that it is softened at a slightly lower temperature. However, it is more easily clarified

c. The same calendered tissue as in Formula 1 (*a*) is treated by the cooling step.

with the following composition in the same manner as in Formula 1 (*a*).

Refined Carnauba Wax	43.4	Wax B–430	21.7
Low-Melting Ester Gum	21.7	Paraffin Wax (M.P. 62°C.)	13.2

Instead of cooling the molten composition absorbed by the tissue, the paper is passed over polished cold iron rolls which are hollow for circulating cold water or brine.

A product is obtained that is transparent, clear, glossy, moistureproof and otherwise similar to that of Formula 1 (*a*).

d. The same calendered and soap-treated tissue as in Formula 1 (*a*) is impregnated and cooled in the same manner as in Formula 1 (*a*), using, however, the following transparentizing composition:

Low-Melting Ester Gum	37.0	Dewaxed Shellac	37.0
Refined Carnauba Wax	18.5	Paraffin Wax (M.P. 62°C.)	7.5

The product obtained is very transparent, free from haze, nontacky, nonsmeary and very flexible.

No. 2*a*

Wax B–430	30.0	Paraffin Wax (M.P. 62°C.)	15.0
I. G. Wax E	30.0	Crepe Rubber	10.0
Ester Gum	15.0		

Solvent: Toluol
Solids content: 50%

Paper is impregnated with this solution and scraped as in Formula 1 (*a*), except that such an amount of solution is left on the paper when removing the excess that there will be about 20 lb of solid composition to 100 lb of paper. In this case, the scraping knives need not be heated, since the components of the composition do not readily crystallize out of solution during the scraping operation. The temperature of the impregnating solution should be 40–60°C. After leaving the scraping knives, the paper is passed through a drying chamber at 60–90°C. in which the toluol is evaporated. The impregnated paper is cooled and wound up, then momentarily exposed to an elevated temperature and chilled as in Formula 1 (*a*).

The product obtained will be transparent, very flexible and tough, and will have excellent heat-sealing properties and good slip. It will, however, tend to blush unless chilling is carried out when the composition is somewhat hotter than required in the previous example.

b. A good 20 lb glassine (24 × 36 — 500) is coated with a composition of the same solid components as in Formula 2 (*a*), but the solvent is omitted. Solvents are not necessary in this case, since the glassine is not impregnated and as the composition without a solvent is not too viscous for good coating.

The glassine is passed through the molten composition and the excess is scraped off with suitable heated knives. No time is allowed for impregnation as was the case with the calendered tissues. The coated glassine is then led over smoothing rolls and, while the composition is still molten, is passed into cold water. The adhering water is removed and the film wound up.

The product obtained is moistureproof, glossy and somewhat improved in transparency. It is, however, slightly hazy due to imperfections in the glassine base. The wax coating is readily chilled to be free from haze, since the thickness of the composition is substantially less than that of Formula 2 (*a*). The glassine base is increased in thickness about 0.00005–0.0001 in. by the coating. The coated glassine is nontacky and nonsmeary. It can be heat sealed to give a very strong joint, even with its thin coating.

No. 3

A tissue of basis weight 13–18 lb (24 × 36 — 500) is moistened with water to about 10–15% moisture content by any suitable means, such as on a glassine dampener, in a mist chamber, or by wetting the paper and drying it down to the desired moisture content. The dampened paper may be rolled up and allowed to stand until the moisture is equally distributed. The uniformly moistened paper is calendered on hot calender rolls, during which operation it is compacted, given a smooth surface and dried down to normal moisture content. The calendered paper is then transparentized by the same composition and process as described under Formula 1 (*a*).

The product obtained is moistureproof, glossy, nonsmeary, of good slip, and heat sealing. Its clarity and transparency will be somewhat inferior to that of Formula 1 (*a*). It will have a slightly hazy appearance, that shows up as a gray cast over a black background due to somewhat poorer impregnation.

No. 4

a. A good 20 lb glassine is coated with the following composition according to the procedure outlined in Formula 2 (*b*):

Refined Carnauba Wax	66.6	Paraffin Wax (M.P. 62°C.)	6.7
Diethylene Glycol Abietate	26.7		

About 5–10 lb of this composition are applied to 100 lbs glassine.

The product obtained is moistureproof, glossy, nonsmeary and nontacky and slightly improved in transparency. It is somewhat hazy, since the glassine base is not impregnated, but the wax coating applied is glass clear. It is essential to apply a thin film of the composition, since a thick film, such as would be obtained by impregnating a calendered tissue as described in Formula 1 (*a*), would have a slight whitish haze even with the best rapid cooling.

b. An about 0.001 in. thick film of transparent regenerated cellulose sheeting, or other transparent sheeting, such as cellulose acetate, ethyl cellulose, gelatin, etc., is coated as in Formula 4(*a*). On 100 lb film, 5–10 lb of wax composition are applied.

The product obtained is moistureproof, glossy, nonsmeary, nontacky and glass clear. The wax coating is tough and flexible. The film can be crumpled without chipping off portions of the coating or leaving objectionable marks or cracks in the coating.

HOT-MELT COATINGS

W. S. Lawrence

Coating, in industry, is done mostly with a liquid solution of the ingredients in a volatile solvent which plays no part in the finished film but merely serves as an agent to deposit the nonvolatile substances over the surface to be coated. Unless the solvent is recovered, which is a somewhat troublesome and expensive operation if no large quantities are involved, it is a total loss and disappears into the atmosphere. Some substances used in coatings tend to hold the solvent tenaciously and it is very difficult to free the coating from the last traces of volatile ingredients.

With these problems in mind, it is easy to understand why, for certain purposes, it is desirable and economical to apply the coating in liquid form by the use of heat and nonvolatile ingredients. There is no expense involved for lost solvent and the coating is deposited in its final form as far as ingredients are concerned.

There is no hazard of explosive vapors in the air and solidification is accomplished by cooling to or near room temperature.

The disadvantage is that, since there is no liquefying agent to aid in the spreading of the solids, there are serious limitations imposed as to the viscosity and type of ingredients that can be used in a hot-melt coating spread by knife or roller. Some substances, commonly used in solvent coatings, which are susceptible to change or decomposition by heat, cannot be used at all or only in very limited amounts. Nitrocellulose and vinyl copolymers may be mentioned in this connection. The requirement that the fused composition must have a suitable viscosity and spreading behavior constitutes one of the serious limitations of hot-melt coating by knife and roller.

Calendering, where it is used to apply a coating to fabric, etc., is, in a sense, a type of hot-melt coating, but a specialized form of application, and its use greatly extends the field and the classes of substances that can be applied hot. It is not intended to discuss here the use of waxes in compounds that can be calendered on, since the problems of application in this field are in a class by themselves.

It is broadly true that film-forming elastomer-type substances have a relatively high viscosity, under heat, as compared with most other resins, and it is in this respect that the limitations of coating are most keenly felt. Viscosity, in the melted state, generally decreases with increasing temperature and this offers some help, but increasing temperature introduces problems of stability and discoloration that have to be considered in relation to the properties of the finished coating.

The ingredients of hot-melt coatings, to be spread by knife and roller, are quite similar to those used in solution coatings, but the proportions are different.

Resins are widely used in hot-melt coatings to promote adhesion and to give hardness, toughness and tensile strength to the film. Film formers are particularly valuable achieving the last three properties, but their proportion is seriously limited because of high viscosity. Non-film-formers are not too viscous, when used in higher percentages and, in general, are valuable ingredients of hot melts for certain purposes.

Both oil and chemical plasticizers, are widely used to adjust tack, melting point and viscosity, but they have to be carefully chosen for their effect on the other ingredients of the coating and on the properties of the coating itself.

Pigments, dyes and fillers may be used for colored or opaque coatings or to reduce cost and their use is similar to that in solvent coatings.

Stabilizers and special chemicals may be employed in small quantities for special functions.

Waxes have been widely used in hot melts for numerous reasons. In fact, coatings of wax alone are very common on paper, producing the familiar waxed paper used in the home and elsewhere. Many, if not most, coatings made by the use of a hot melt contain a large proportion of wax.

There are many different types of waxes, including natural waxes of animal, vegetable and mineral origin, synthetic waxes and fatty acids and esters that are not true waxes but function the same way.

In contrast to their use, in a restricted measure, in solvent coatings, waxes in hot melts contribute highly desirable properties, both in facilitating the application of the melt and in modifying the properties of the finished product. Since the term wax covers such a wide scope of products, both natural and synthetic, with a correspondingly wide range of properties, any wax employed has to be first considered from the viewpoint of what its presence will contribute to, or how its presence will modify, the finished product, independently of its performance in the coating job. Waxes, with few exceptions, aid the actual coating job since their viscosity, when hot, is low and they make for ready flow and smooth spreading at lower temperatures. Thus they permit the use of more highly viscous substances which contribute valuable properties to the finished film.

The utility of finished products is wide and varied, as would be expected from a class of compounds some of whose members are completely water soluble, others water insoluble, some soft, flexible and tacky, others hard, brittle and dry.

Waxes have one unique property, their limited solubility in the other ingredients of the melt in the cold. Although this is not always the case and is a variable characteristic, depending in great measure on the nature of the other components of the system, it is a fairly common property among waxes which makes possible the production of coatings that have good slip and are free from blocking tendencies. The value of this should not be underestimated, especially when it is realized that hot-melt coatings are most often wound in a roll, under pressure, and often not completely cooled.

Coated papers, produced by use of hot melts, have been expanding because of metal-foil shortage. Since moisture plays such an important part in some products, either by the necessity to retain it or by the necessity to exclude it, the moisture-vapor transmission of a film is often highly important. Waxes, especially those of an amorphous nature, are very valuable in this respect.

Heat-sealing glue is often used on foils and paper to seal containers; in a hot-melt-coated paper or cloth, heat sealing is inherent and is accomplished by running a hot wheel or bar over the seam which cements the surfaces together.

A hot-melt coating is always heat sealing to some degree, but this property can be greatly increased by a proper choice of ingredients. In some cases, sealing is accomplished by pressure alone, especially if a temporary or not too strong seal is desired.

Waxes enable the formulator of hot melts to produce a coating for package and container liners that will seal readily, be soft, flexible and free from blocking tendencies even when the product made in rolls is sheeted and stacked. In a broad sense, waxes are rightly the principal ingredient of most hot melts since, as a class, they make an important contribution to easy working characteristics at lower temperatures and, at the same time, to the properties of the finished film.

Some coatings will increase water and oil resistance, some will decrease it. They will decrease or increase moisture-vapor transmission, as well as produce slip, tack, freedom from tack, hardness, flexibility, brittleness and softness. The range of properties is very wide and since many of the waxes are solvents, in the hot state, for film formers and resins, an almost endless variety of melts can be produced by a careful choice of ingredients.

As in so many other fields, the development of synthetic waxes has extended the range of application and there is a wide range of desirable waxes with which to tailor-make a melt to meet the specific problem at hand.

TROPICAL WAXED PAPER

Leonard S. Levitt

When candies and food products are wrapped hot in waxed paper, wax-transfer has to be prevented to the food product during the wrapping operation and in storage during hot weather or under abnormal heat.

In order to overcome this difficulty, some manufacturers have specified to their converter that the waxed-paper coating should comprise 7–10% of a higher-melting wax, e.g., carnauba. Satisfactory results were obtained within the temperature limits determined by the carnauba–wax addition. However, where especially hot wrapping was desirable, temperatures during or after storage were abnormally high, the problem arose again. To increase the percentage of carnauba meant to sacrifice flexibility of the sheet and increase the cost beyond reasonable limits. Thus began the search for a "synthetic" which would not only satisfy the physical requirements, but also bring the paper cost down. The wax needed had to have unique properties, which may be classified into two categories: 1. properties of the wax itself and 2. properties of the blend.

PHYSICAL PROPERTIES OF THE WAX

1. Color—light, preferably white
2. Form—as hard as carnauba
3. Luster—preferably very high
4. Melting point—85°C., or higher
5. Compatibility—give a homogeneous blend with paraffin wax
6. Toxicity—none
7. Reactivity—noncorrosive to metal

8. Solubility—water insoluble

In addition, the cost should be much less than that of carnauba.

PROPERTIES OF BLEND

Since volume production could be obtained best with the water-waxing system, the following properties are requisites of the wax blend:

1. Melt to a thin liquid below 250°F. (scorching point of paper and paraffin)
2. Compatibility with paraffin in following ways:
   Give a homogeneous melt
   On quick cooling (down to 40°F. in a water system), the higher-melting wax should crystallize partially or bloom to the film surface
3. Have quick-setting properties
4. To give a higher-melting coating 7% high-melting wax should be sufficient to limit cost
5. Give good sheet flexibility for automatic wrapping and not detract from heat-sealing properties should they be desired.

A blend of a synthetic wax (Albacer) and paraffin wax was found suitable for this purpose. This is a white, hard wax having a very high luster and a melting point of 95–97°C. It is insoluble in water, but completely soluble hot in hydrocarbons. It has been used for its high luster as a coating for linoleum and paper products as well as in various types of polishes.

Although there is little or no increase in the EMP (English melting point) of the synthetic paraffin blend, the wax film of paper, coated with this combination by the water-waxing system, has a relatively high melting point. This suggests the probability of a partial "blooming" or orientation of the synthetic wax to the paraffin surface, which results in a high melting point of the film, approaching that of the synthetic wax itself. The fact that Albacer has a melting point about 14°C. higher than carnauba is another distinct advantage. Also, the property of "bloom" seems more pronounced with the new blend, which gives a somewhat harder surface film without sacrificing flexibility of the sheet. Still another desirable feature of the use of the synthetic is the much quicker setting of the film, which allows for speedier production.

In the use of this type of wax, two important factors must be borne in mind: 1. A water-waxing system is required; and 2. a fairly critical temperature of the melt must be maintained. This temperature, 240°F., has been determined by several commercial runs. It seems that below this point poor results are generally obtained and when the temperature rises as much as 8–10° above this optimum, scorching of the paper or souring of the paraffin occurs. Water-waxing is necessary, since the action of the new wax is dependent on a partial crystallization from the blend, obtainable only by a quick drop in temperature. Luster, too, which is highly desirable, is dependent on this lowering of temperature that only the water-waxing unit can give. Because temperature plays such an important part, a thermostatically controlled system is suggested for consistently good results.

Excellent results are now being obtained by the addition of this synthetic blend to 138–140° EMP paraffin. The waxes are melted together, with slow agitation, to insure complete blending up to 240°F. From this point on, the converter needs only to apply the regular procedure of his waxing system.

Paper made with this new wax is being used for wrapping caramels, hard candies, chocolates, etc. Advantages are: lowering of cost by as much as one half, increased temperature limits of wrapping and storing, higher luster of the wrapper and good sheet flexibility for automatic wrapping.

NONSTICKING (ANTIBLOCKING) COATINGS

Certain materials, or coated surfaces, have a tendency to adhere to each other in hot weather and when subjected to even slight pressure as when they are superimposed on each other. This is especially true during the summer in our climate and all year round in the tropics. Such materials are natural and synthetic rubber, resins, thermoplastics (e.g., polyvinyl butyral), hygroscopic materials (e.g., glue), paraffin-waxed paper, etc.

Carnauba wax, because of its high melting point, has been used effectively to prevent sticking or blocking of coated paper or cloth, rubber, etc. Its disadvantages are that it is brittle and not too adhesive. Synthetic waxes, such as Albacer, Ozowax and Acrawax C are gradually displacing Carnauba wax in this field. Synthetic waxes are usually incorporated in the coating or in the composition (as in the case of rubber or resins). They bloom to the surface and form a nonsticky film which does not crack off and thus gives permanent protection.

TEXTILE PROCESSING

Waxes and waxlike substances are among the most valuable assistants in the textile processing industries. Their use in sizing, finishing, printing, waterproofing, polishing, etc., is as old as the finishing industry. Almost every formula for sizing, finishing, waterproofing, etc., contains, in one form or other, a wax or waxlike substance.

Most of the natural fibers contain some natural wax or waxlike substance as a protective agent against atmospheric influences. These natural waxy substances are eliminated, during the manufacturing process, by scouring or boiling, to assure uniform dyeing and bleaching of the fibers. By removing the waxes, we change some of the most valuable properties of the fibers, such as softness, pliability and elasticity. It is often possible to regain these properties by treating the fibers or finished yarns with waxes or waxlike substances.

In textile processing, waxes and waxlike substances can be used in two different ways:

1. To produce a nonpermanent effect, that is, one which will be removed from the yarn or goods in a following operation, but which is absolutely necessary to assure a better workability. It is quite natural to use waxes for this purpose which are easy to remove without affecting the fibers. For instance, it would be a mistake to use for wool a wax which requires strong alkalies for saponifying.

2. To produce a permanent effect, which is desired to give the finished goods or yarns a specific appearance, e.g., high luster, softness or water repellency. In this case, the application of the wax is always combined with a mechanical aftertreatment, like calendering, polishing, rubbing, etc.

SIZING

The purpose of yarn sizing is to provide:

1. Better weavability

2. Specific finish
3. Increased weight

The basic agents, used in the sizing process, are starch, glue, gums, etc. They increase the tensile strength of the yarns, but diminish the elasticity and softness. The softeners compensate for this loss, maintaining the higher tensile strength of the yarns. The softeners are as important as the starches themselves. They make the starch film pliable and flexible. They also reduce the harshness, lubricate the surface of the yarns, increase elasticity, and improve hand. A very important role of the waxes in sizings is that they improve the adhesion of the size to the fiber, diminishing the formation of dust during weaving.

The most frequently used waxes are Japan wax, beeswax, paraffin wax, spermaceti and tallow. Under trade names, there are many softeners, on the market, containing emulsified waxes and other ingredients.

In sizings, only waxes should be used that can be easily removed later. An unsaponified wax very often causes unlevel dyeing and difficulties in bleaching.

The very same precaution should also be taken in the use of polishes. The polishing of yarns is useful where softness and smoothness are more important than higher tensile strength, such as for knitting. Polishing or waxing is made simply by passing the yarn over wax rolls. The use of paraffin wax, which is very difficult to eliminate, often causes troubles in dyeing and bleaching.

PRINTING

The reserve printing processes used large amounts of waxes. From the great variety of these processes, only the batik printing process is still in use, all the others having been replaced by more modern methods.

The batik technique originates from Java, where it is still in use as a home industry, producing highly artistic designs. The fundamental difference between batik and ordinary printing is that in the batik, the design is dyed completely through. It appears on both sides of the cloth in the same strength. The batik is a dyeing rather than a printing process. In the original Javanese batik, the dyeing is always combined with some sort of printing process, like block printing, to produce various effects. There is a great variety of designs which can be produced by batik, depending on the mode of application of the wax reserve. It can be applied to the whole surface of the cloth, or only in spots, before dyeing, between two or more dyeings, or after dyeing. The waxes used are paraffin wax, ozokerite, vegetable waxes and other ingredients, such as rosin.

The beauty of the designs achieved by batik led to a great variety of imitations produced on a large scale. It is natural that for large-scale manufacture, special processes were developed none of which could ever produce anything as artistic as the original Javanese batik.

The bleached cotton goods are impregnated with the molten wax, by passing through squeezer rolls. After the wax hardens, the goods are irregularly broken, followed immediately by the dyeing. The dyeing is done mostly in cold or at low temperatures, using vat dyes. The dyestuff penetrates the fiber only where the wax resist is broken, leaving the covered spaces undyed. After drying, the wax is removed, e.g., by organic solvents. The effect is a one-colored marblelike design. By applying the wax reserve again and dyeing in another color, or by repeating the whole process several times with different colors, beautiful, multicolor effects can be achieved. As in the Javanese batik, the process can be combined with block, screen or roller printing effects.

Regarding the waxes used in batik printing, almost always the cheap varieties are used. Ozokerite is more in use than paraffin wax.

FINISHING

To produce goods, with a high and possibly lasting luster, has always been the desire of the finishing industry. To imitate the soft, silky luster on cheap materials, like cotton goods, has always been the dream of the finishers. Mercerizing was an important step forward in this direction.

Waxes can produce a high luster, when applied to textiles and properly aftertreated. There are thousands of types of goods with specific effects obtained by wax finishes and a proper, mechanical aftertreatment. "Chintz," the well-known furniture covering, a typical high-luster material, is finished as follows: The dyed and printed goods are humidified, rolled, and kept in humid surroundings for over 24 hours, to distribute the humidity evenly all over the goods. Before humidifying, a finish, containing starch or dextrose, or methyl cellulose and a large amount of a hard wax, is applied. The humidified goods are then calendered by a three-roll, friction calender.

A similar process is used to produce rayons and acetates, with a chintz effect, for ribbons and other goods for the millinery industries. It is evident that all these effects are not lasting. The first washing or dry cleaning eliminates them.

YARN FINISHES

The ability of waxes to produce smoothness and luster on textile materials, is used also in the finishing of special yarns as sewing thread, shoemaker's thread, fishing lines, thread for shoe laces and many others. The wax, usually along with other ingredients, such as starch or gum, is applied to the yarns. After drying, the yarns are polished by brushing and rubbing. Another method consists in brushing over the finished yarns wax as an emulsion or dissolved in an organic solvent. This treatment gives high luster, a deeper, more brilliant color and higher elasticity and tensile strength.

WATERPROOFING

The purpose of waterproofing is to prevent the penetration of water through a textile material. There are two different types of water-resistant finishes:

1. Waterproofing
2. Water-repellent impregnation

The waterproofing of a textile material consists in forming a highly hydrophobic layer on one or both sides of the fabric, which fills the pores of the goods and is sufficiently resistant to prevent the penetration of water through the cloth. A great variety of substances is used to attain this goal. Waxes, in combination with rubber, drying oils, metallic soaps and fillers, are commonly used. The waterproofing composition can be applied by a doctor blade or by passing through the liquid and squeezing out the excess. After drying, a suitable aftertreatment, like calendering, follows. Articles, treated in this manner, are used for raincoats, heavy ducks, osnabrucks, etc.

The water-repellent impregnation is a more recent development in the finishing industry. Applied chiefly to clothing, it does not fill the pores of the fabric, but makes its surface highly repellent by the application of hydrophobic substances. The most widely used wax for this purpose is paraffin wax in the form of a stable emulsion. The water repellency produced in this manner does not resist repeated washing or laundering or dry cleaning.

The newest development in the water-repellent impregnation embraces modifying the surface of the cloth, by chemically binding the waxy substance to the fiber. A cloth so treated will resist repeated washing and dry cleaning.

Water repellency is sometimes combined with other properties, like softness. Ladies' hose is made smooth and water resistant in one single operation by the use of waxy finishes. Sometimes, smoothness is desired without water repellency. Certain wax emulsions are frequently used for giving smoothness and hand.

The development of synthetic waxes opened many new fields for the application of waxes in the textile industry. The finisher now has waxes which are very easy to handle and which are chemically pure and uniform. By an appropriate choice, combining different types, almost any kind of finish can be produced. Waxes, which are compatible with synthetic resin finishes, have a big future. The modern trend in the finishing industry is to produce lasting finishes, which withstand repeated washing and dry cleaning, maintaining the original finish and appearance of the textile material. The task of the synthetic–wax industry is to investigate such waxes and possibly develop waxes which combine chemically with the fiber, giving a more lasting bond.

TEXTILE APPLICATIONS OF SYNTHETIC WAXES

Jack H. Dollinger*

Waxes are used in many phases of textile processing by application either to various fibers during processing or to the finished fabric. Most natural fibers contain some wax or waxy substance which gives resiliency and protection against atmospheric influence, but since during the manufacturing process these waxy elements are removed by boiling or scouring, the loss of workability and elasticity of the fiber is restored by treatment with waxlike substances.

Basically, waxes are used to give either a temporary effect during processing and are removed before the yarn or fabric is marketed, or a permanent effect which enhances the utility or appearance of the finished product. The nonpermanent application of waxes is for the purpose of improving the workability of the fiber by lubrication and increasing elasticity and reducing fuzziness. Such waxes must be easily removable by boiling and scouring with mild alkalies and are generally readily saponified. The waxes used to give more or less permanent effects are relatively high-melting hard types which are not easily removed by scouring or solvents. They give softness, luster, weight or water repellency to the finished yarn or fabric.

The use of synthetic waxes in textile processing is of comparatively recent origin and their diversity and volume of application are steadily growing. The natural waxes, which are often replaced or modified by synthetics, are of animal, vegetable or mineral origin. Beeswax, candelilla, carnauba, paraffin and microcrystalline waxes are most commonly used, although Japan wax, ozokerite and montan wax also find special application. In most sizing and finishing operations the waxes are used with starches, gums, glues and resins which give added strength, weight and finish.

Many synthetic waxes are commercially pure organic products manufactured to close specifications under strict laboratory control. Such waxes are free from foreign matter and are not subject to variations found in natural waxes which are caused by variations in atmospheric and growing conditions from year

* Glyco Products Co.

to year. As they are produced from domestic raw materials, they are not subject to wide price fluctuations as are imported natural waxes because of climatic or political conditions. Some so-called synthetic waxes are not truly synthetic, but are blends of natural waxes and natural and synthetic resins.

As natural waxes, synthetic waxes may also be grouped, according to their use for temporary or for permanent effect, as readily saponifiable or difficultly saponifiable. The saponifiable group consists of the fatty acid esters of polyhydric alcohols and as such are related to the vegetable waxes, fatty acid esters of monohydric alcohols with long carbon chains, and animal fats and oils, fatty acid esters of glycerin. The alkyl amides with waxlike physical properties and certain long-chain ethers are members of the difficultly saponifiable group.

Synthetic waxes offer a very wide range of properties from which to choose and many special "hand-tailored" blends and compounds can be made with natural and synthetic waxes and resins. The range of melting points is also very wide and one unique product, Acrawax C, has a melting point of 285°F. while the highest-melting natural wax is carnauba wax, which melts at about 185°F. Many are free-flowing fluids of low viscosity just above their melting point. Most of them are hard waxes but plastic waxes are also available. Some are quite light in color and are stable to color change during processing. The solubility of the various synthetic waxes in water and solvents is an important factor in their over-all usefulness. Some are water soluble or dispersible while others are insoluble in, and unaffected by, boiling water. However, most of them can be emulsified or dispersed in water, which is important in many textile operations, and their solubility in solvents is as varied as their other properties. Some of the synthetic waxes, listed in Table 149B, will give an indication of the wide variations.

In textile processes, waxes are applied from water emulsions and dispersions, solvent solutions and hot melts. Water emulsions are the easiest to use and are, of course, least expensive. Solvent solutions are the most costly because of the solvent and the equipment needed for precautions against fire and toxicity and, where possible, for solvent recovery. Although the most direct method, hot-melt application is limited.

The production of waterproof or water-repellent fabrics is an important field for the application of synthetic waxes. Because of its exceptionally high melting point, 285°F., and insolubility in water, even at boiling, as well as its insolubility in cold solvents of all types, Acrawax C has found effective application. Generally, it is applied to fabrics from aqueous dispersion, Acrawax C Dispersion S933, of 33.3% solids content, which can be further diluted with water to any desired concentration. The fabric is impregnated by dipping into the dispersion and is then calendered at about 300°F. to set the wax and give a continuous coating on the fibers. An interesting application is for impregnating the cotton-duck liners used as mold material in sponge-rubber curing, where it must withstand repeated use under high heat and humidity.

Water-repellent coatings made with blends of synthetic waxes may have ingredients which will give mold and fungus protection to the fabric. The following formula yields coated fabrics having a water repellency of over 90 in accordance with government specification and good fungicidal properties.

Flexo Wax C*	10
Glyco Wax S932*	4
Cunilate No. 2137**	10
Aroclor No. 1254†	2
Xylol	74

Synthetic waxes are also used in special finishes and coatings to give luster and flexibility. Cotton and rayon braid can be run through a hot-melt blend of Flexo Wax C and Glyco Wax S932 and then over the brushes on polishing rollers. A lustrous finish and good flexibility are imparted to the braid and, if desired, the addition of oil-soluble colors to the wax blend will make the coating suitable for different colored braids.

Ceramid is an alkyl amide which has very interesting properties. It is a hard, yellow wax with a sharp melting point of 176°F. Ceramid can be readily emulsified in water, is soluble in most hot solvents and to the extent of a few per cent in many solvents (alcohol, aliphatic and aromatic hydrocarbons, etc.)

* Glyco Products Co.; ** Scientific Oil Compounding Co.; † Monsanto Chemical Co.

after cooling from a hot solution. It is compatible with most waxes and resins and it hardens paraffin wax and raises its melting point. It also tends to improve its water resistance since this wax is insoluble in water. The hardness and luster on polishing of Ceramid indicate its suitability to replace or to blend with carnauba wax with which it is compatible and its low price, uniformity, hardness and light color make it very attractive for this purpose. It may also be used on textiles for semi-permanent finishes.

In textile delustrants, diglycol stearate S and the stearates of the Carbowax‡ are used not only for their waxlike characteristics but also for their suspending action on pigments, such as titanium dioxide and also their thickening effect on water dispersions. Since they are emulsifying agents, other waxes, such as paraffin, may also be incorporated without the addition of emulsifying agents. In some cases, their thickening properties permit the elimination of thickeners, such as starch or vegetable gums.

Plastic- or resin-coated fabrics have attained a substantial position on textile markets. Special problems of reducing tackiness and improving water resistance of these surface coatings were solved by use of synthetic waxes. Outstanding in this regard are the Acrawaxes, particularly Acrawax C. Addition of small percentages of this wax to plastic coating compositions, organosols, plastisols and water dispersions or laticies gives very effective control of tackiness and increases water and salt-spray resistance. Depending on the rubber or resin used as the coating, the method and temperature of application, etc., one or another of the synthetic waxes has generally proved an efficient detackifying agent.

The textile screen-printing industry has found increasing use for a special adhesive wax as a "table wax." Prior to the use of table waxes, a fabric was pinned down on the table and due to slippage of the fabric during printing, a poor register was not uncommon. An almost colorless, plastic, adhesive wax, Flexo Wax C Light, is now extensively used as a table wax by applying from a hot melt, levelling and allowing it to solidify. The fabric to be printed is stretched taut and pressed to the wax surface which holds it firmly during the printing process. There is

‡ Carbide and Carbon Chemicals Co.

very little danger that the material will slip and good register is readily obtained. After printing, the fabric is lifted clean without the wax adhering to it and the wax surface can be reused a number of times before renewal.

Table 149B

SYNTHETIC WAXES

Wax	Color	Approx. Melting Point °F.	Sp. Gr. at 70°F.	Solubility Water	Solubility Hydrocarbons
Acrawax	tan	203	1.04	I	SH
Acrawax C	tan	286	0.97	I	SH
Ceramid	yellow	175	0.98	I	SH
Carbowax 4000 (Mono) Stearate	yellow	134	1.10	SH	SH
Diglycol Stearate S	white	135	0.96	DH	SH
Flexo Wax C	tan	153	0.82	I	SH
Glyco Wax S932	white	140	0.96	I	SH
Stroba Wax	tan	211	1.04	I	SH

I = Insoluble S = Soluble
D = Dispersible H = Hot

TEXTILE PRINTING

W. D. Baird

Waxes are used to a limited extent in the pigment type of printing. For pliability and "hand" of the printed goods, a wax, such as beeswax, is incorporated into the varnish or other types of fixing vehicle.

Essentially, batik printing consists of applying molten waxes or resins, in any desired pattern or design, to the carefully calendered and stretched fabric. The work is generally carried out on a table or bench which has been covered with several layers of cheap cotton cloth. A thin layer of sand or china clay separates the fabric under treatment from the layers of cotton. A batik pencil, brush or mold is used to apply the molten waxes or resins which will form the required resist after solidification.

The wax resist printed goods, after hardening for at least 24 hours, are placed in cold water where the resist is cracked in order to produce the desired "veiny" design.

The goods are immediately transferred to the dye bath where

the desired depth and shade are obtained. The wax is then removed in a hot or boiling water bath. Multicolor effects are produced by repeating the entire procedure with different dye baths.

Cold-dyeing Amanthrene, Indigosol, Sulfur and Amanil Naphthol AS dyes are used when good fastness is required. Where speed of dyeing and shade brilliance are of greater importance than fastness to color-destroying agents, basic dyestuffs are applied. Fast-dyeing acid colors, such as the Amacid Milling colors, which draw well from a weakly acid bath, are also used.

The kind or composition of wax resist varies with the type of goods treated and the nature of the dyes to be used. For example, a dye bath containing caustic soda, as used in vat or naphthol work, would require a large proportion of an unsaponifiable wax, such as paraffin wax, in order to protect the resist against the strong alkali.

A knowledge of melting points is essential because a moderate elevation of dye-bath temperature is sometimes necessary to increase the fastness, affinity of dye to fiber, etc.

A listing of some of the usable waxes and resins, together with their resistance to saponification, follows:

Wax	Melting Point °F.	Resistance to Saponification
Paraffin Wax	90–120	excellent
Carnauba Wax	190	very good
Raw Palm Wax	220	very good
Japan Wax	110	poor
Chinese Wax	180	very good
Stearic Acid	150	poor
Spermaceti	110	poor
Rosin	210	poor
Petroleum Jelly	150	excellent
Beeswax	170	poor

It may be seen that where cold dyeing from a strongly alkaline bath is used, a firm resist which will remain hard below 75°F. and is highly resistant to saponification may be obtained with paraffin wax as its base. Should a slight rise in temperature be necessary, such as for certain vat colors, retaining the resistance to saponification, carnauba or palm wax or both may be combined with the paraffin in amounts depending on the temperature increase.

Brittleness is sometimes imparted to the wax resist to produce a sharper break or crack. In the absence of a strongly alkaline dye bath, rosin serves well.

PROTECTIVE AND DECORATIVE COATINGS

Austin O. Allen

Natural, synthetic and chlorinated waxes are widely used in protective and decorative coatings. In many cases, the quantities used are small, but they impart very important properties for certain applications.

The principal uses of waxes in protective and decorative coatings are as follows:

Coatings containing wax have improved water repellency and moisture-vapor resistance. Such coatings are applied to fabrics and paper. Waxes are also widely used in coatings for waterproofing masonry surfaces.

Coatings containing chlorinated waxes, when properly pigmented, confer fire resistance as well as protective and decorative effects on such burnable surfaces as wood, composition board, fabric, etc.

Waxes are widely used in protective and decorative coatings, particularly in baking finishes, to impart nontackiness. This is very important in lithographic finishes for metal plate, where the metal sheets are stacked directly after lithographing. The waxes used for this purpose are usually natural or synthetic waxes of the low to medium melting-point range, such as paraffin wax, ozokerite, etc. The waxes are sometimes incorporated in the lithographic coating but are more often added to a solvent and stored in a hot room to insure fluidity for addition to the finish at the time of use.

Waxes are very important in making marproof a wide range of protective and decorative coatings, such as baked furniture finishes, lithographic finish and other specialty coatings where marproofness is essential. These finishes can be applied by spray, dip or roller coating. They may be incorporated in the coatings during the manufacturing process or may be added in solution form at the time of use.

Certain waxes incorporated in coatings, particularly baking

finishes, will increase salt-spray resistance. Acrawax C is particularly efficient for this purpose.

Certain waxes added to pigmented coatings during the mixing operation help disperse the pigments during the grinding operation.

Waxes are often used in both clear and pigmented film to reduce gloss and give a satin or rubbed-effect finish.

Some types of paint and varnish removers contain waxes to prevent rapid evaporation.

Waxes are often used in protective and decorative coatings as plasticizers to improve flexibility.

ANTIDIM (NONFOGGING) COATINGS

The glass surfaces of binoculars and telescopes, the eyepieces of divers' masks, gas masks, spectacles and goggles, the surfaces of windshields, and similar surfaces have a tendency to condense moisture. As a result of such condensation, the vision of the observer is diminished and often totally obscured.

Of the many compounds used for their antidimming properties, the most frequently employed have been those based on soaps, sulfonated oils, saponified Turkey-red oil and some hygroscopic agent, such as glycerol. However, compounds prepared from these substances have not proved entirely satisfactory. They have a tendency to form films which pass too readily into solution in the water condensed on their surfaces. The compounds prepared from soap bases are alkaline, while those prepared from sulfonated oils are either acid or alkaline, or tend to develop acid characteristics. Either of these deviations from chemical neutrality is undesirable, because the compound is then likely to damage the treated surfaces or adjacent materials. In the case of fine, delicate, optical equipment and instruments, these potentially harmful characteristics are matters of considerable concern.

Furthermore, the sulfonated oils, used as surface-tension depressants in antidimming compounds, when exposed to moisture-laden atmospheres, have a marked tendency toward liquefaction due to their inherent solubility in the water condensed on the lenses or other surfaces. Under such conditions, the antifogging film is readily washed away. When the atmosphere is dry, the sulfonated oil compounds tend to undergo spontaneous desiccation. Under these conditions, the antidimming film, spread over the treated surface, tends to dry out and develop perceptible opacity. The desiccated film shrinks and cracks, exposing parts of the underlying surface which condense moisture.

Other deficiencies or inadequacies of the older antidimming preparations will be recognized on consideration of the following properties which should be present in an ideal agent for such application.

- It should greatly lower the surface tension of water
- It should readily form a uniform, adherent, cohesive, and transparent film of good optical properties
- If soluble in the water condensed on the filmed composition, the solution should not run off the treated surface easily.
- It should be chemically inert to the filmed surface and to the adjacent materials
- It should not lower the dielectric strength of the materials with which it must come into contact
- It should be nonirritating to tissues and eyes
- It should be nontoxic on direct contact or on inhalation in vapor phase
- It should not contain gritty substances or have a tendency to develop them so as to cause damage through scratching on filming or defilming the optical surface
- Its consistency should not vary greatly due to changes in temperature or atmospheric conditions
- It should not dry out on standing and become too hard or brittle
- It should not absorb moisture too rapidly so as to become too soft for storage, handling, or application
- It should not undergo spontaneous chemical or physical changes during storage that will impair its utility in any respect.

Preparations markedly superior to those mentioned before can be produced from certain polyhydric alcohols that have been partially esterified with higher fatty acids. They give antidimming agents which can readily be spread out in thin layers, approaching monomolecular dimensions, over optical surfaces to form films of excellent transparency and great durability, resulting in clearer vision and lengthened life of the film. The films formed from these antidimming preparations reduce the glare of sunlight or other bright lights, thus producing better vision. Camera lenses so treated minimize the ill effects of glare in photographic operations.

These partial esters also lower surface tensions to a decided degree. Although they absorb moisture, the hydrated products thus formed do not thin out and run off. On the contrary, when these esters absorb moisture, they actually thicken. They spread evenly over the entire lens or other surface to which they are applied. Since their fluidity is not dependent on the presence of water, as are sulfonated oils, they will not dry out either in storage or in film. These characteristics eliminate the need for incorporating hygroscopic agents, such as glycerol, or nondrying mineral oils in order to maintain the degree of fluidity demanded of an antidimming composition. They are not injurious to the surfaces over which they are filmed and do not damage textile fabrics on which they may be spilled.

They are nontoxic, nonirritating and have a bland odor. Since they are neither acid nor alkaline, they do not etch or spoil the surfaces of glass, cellulose nitrate, cellulose acetate, acrylic or vinyl resins, ethyl cellulose, cellophane, Pliofilm, Koroseal or other synthetic resins or plastics. Their low surface tension causes them to spread well over all these surfaces.

ANTIDIM (NONFOGGING) COATING FORMULATION

Formula No. 1

Nonaethylene Glycol Monostearate	50
Ethylene Glycol Stearate	50

No. 2

Nonaethylene Glycol Stearate	50
Ethylene Glycol Palmitate	50

No. 3

Polymerized Glycol Stearate	60
Polymerized Glycol Oleate	40

RUBBER

W. E. Leistner

The use of waxes in rubber is almost as old as the history of rubber compounding. It dates back to 1881, when it was first added to retard the deterioration of rubber products, which was so prevalent at that time. Strangely enough, this still constitutes its principal use in the rubber industry.

Hydrocarbon waxes are added to rubber in proportions of 0.25–3%, based on the rubber content, for protection against deterioration by the action of light, oxygen, and ozone. This phenomenon is commonly known as sun checking or atmospheric cracking. Waxes are a very effective age resister for this type of aging, but are of no value in combating the other factors which also cause the deterioration of rubber.

These waxes, which are usually paraffin, ceresin, and ozokerite, probably function by blooming to the surface of the rubber article and thus forming a thin, flexible protective film. In actual compounding practice, these waxes are added together with an antioxidant, which is usually some kind of organic amine. This greatly intensifies the protective action of the wax and constitutes the best-known protection for sun checking or atmospheric cracking.

Waxes are sometimes added to rubber as softeners. Due to their slight swelling action, they cut the time required for mastication and aid in the general processing which follows the compounding of the rubber. In the calendering of highly-pigmented stocks, waxes are of definite aid.

A large amount of wax is used by the electrical wire industry in insulation compounds. Here, waxes aid in the tubing of the wire and improve the moistureproofness and insulating characteristics.

Waxes are also used in molded goods, where they aid in processing and impart smoothness to the finished article. Waxes are generally used with rubber in all cases where moistureproofing is very important, such as in marine cables and water-impermeable diaphragms.

PLASTIC MOLDING

The use of waxes in plastics is entirely for lubrication and this may be broken down into two phases: 1. The wax lubricates the surface and reduces the tendency of the plastic to stick to the mold; 2. it lubricates the particles of the compound, thus reducing the friction and increasing the effective plasticity of the molding powder during its plastic stage in the mold. It is important that the wax should be at least partly compatible with the resin in order to avoid undesirable "blooming" on the surface of the molded article.

It is necessary that the wax should melt above 70°C. in order that it should not become soft during grinding. Mostly vegetable waxes have been found satisfactory for use in plastics.

More specific information as to tests for choosing the wax is not available and it appears that a more or less foolproof method of cut and try must be used in establishing whether or not a wax is suitable as a lubricant in plastics.

ELECTROPLATING

C. B. F. Young

In the field of electroplating, waxes are being used for coating insulating racks so that the supporting mechanism of the plating work is not covered with the metal being deposited, as a stop-off for multicolored electrodeposits and/or bright and dull polished areas, or for polishing certain areas, and as insulation on anode racks used for electrolytic polishing.

In rack insulation, the unit supporting and leading the current to the work will receive appreciable amounts of the metal being deposited if it is not insulated. This acts as a thief and upsets plans in specification work and can be quite expensive in plating with noble metals.

Many designs are produced by the combination of colors of different metals or alloys. For instance, an object is plated with nickel and then certain areas are coated with a relief of some type and design. The unit is again plated with copper and the coating removed. Thus, a red design on a white background is produced. In this procedure, which is becoming more and more important, the rack holding the work is made the anode in a special solution which contains, in general, the following acid radicals: perchlorate, acetate, phosphate, sulfate, nitrate, chromate and fluorides. Being the anode, the rack material will soon dissolve, unless the metal or alloy used is insoluble in the solution (which seldom occurs) or is insulated from the solution by an inert substance such as wax.

The properties of an ideal wax for plating should be as follows:

In the first group mentioned before, the wax should have good adhering qualities, be applied easily, not chip if used at room temperature, be firm if used at boiling temperature in alkali solutions, be insoluble in most acid and alkali solutions and be insoluble in boiling trichloroethylene. In some cases, if possible, the wax should be insoluble in most lacquer thinners or liquids which are used to strip lacquers. This is not essential in most cases, however. The wax should have a long life and be easy to repair.

In the second group, the wax must have good adhering qualities in alkali cleaning baths, acid, alkali, and cyanide plating baths. It should not chip or flake easily and be firm at 75—80°C. It should be easy to apply and remove. If the wax is used to produce multifinished buffed surfaces, it must have good resistance to abrasion.

The last group requires that the wax should produce a rugged surface which will stand temperatures up to 80°C., without flaking or chipping when the rack is handled roughly. It should be possible to reapply the wax over old coats and it must not be attacked by chemical or electrochemical oxidation at temperatures up to 80°C.

ELECTROLYTIC CONDENSERS

G. F. Coggins

Waxes are used extensively in the condenser industry. Large quantities are used for potting the capacitor elements in their cases; for moistureproofing paper containers; and for seals. Smaller quantities of miscellaneous types of waxes are also used in other categories.

Each application will be discussed separately. Condensers vary considerably in type as well as application and since modern electrical equipment may be required to operate under widely varying and sometimes extreme conditions of temperature, humidity, voltage and vibration, severe specifications and highly specialized test methods were developed for waxes.

The paper tubular condenser, used so widely in radio circuits, contains much wax and its life is dependent on it. This type of condenser is used in almost all electronic equipment and must operate in swampy tropical climates, such as are encountered around the Gulf of Mexico, the Dutch East Indies, Ecuador, etc., where the humidity remains at a high level month after month and the temperature varies over wide limits during each day. The cyclic temperature changes cause the tubular condenser to "breathe" and, unless the protection is very good, due to the lowering of the insulation resistance, failure will result.

This type of condenser consists of a condenser section encased in a paper tube, having the ends plugged with wax through which the wire terminals of the section protrude. The section is composed of high-grade paper impregnated with wax or another dielectric, such as oil. This paper is separated by alternate sheets of foil, the whole being rolled into a small round mass. Most of the sections are about the thickness of a fountain-pen barrel and about 1 in. long. These sections are fitted into a paper tube, allowing about ¼ in. on either end for filling with end-fill wax. The paper tube is wax impregnated before assembly and the whole assembly is given an over-all dip in moistureproofing wax.

The end-fill wax extensively used has the following characteristics:

It will withstand a cold-flow test at 185°F. for 24 hours. This means no flow from a full can measuring 2 × 1¾ × 1 in. stood on end.

It will pour easily from a pour pot having a hole of ⅛ in. diameter when the temperature is about 400°F. When small-diameter tubes are being filled, the temperature used is 375–425°F., and when large-diameter tubes are used, the temperature is lowered to 275–340°F. The temperature limits are determined by experience and are fixed so as to allow the wax to flow well into the end of the tube and to adhere to the walls and wire terminal. If the adherence is not strong enough, moisture leaks easily from the exterior to the interior of the unit and a weak mechanical assembly will also result.

It will have a plastic range such that air bubbles, forced out of the assembly because of expansion by the heat of the end fill, will not freeze on the surface of the seal so as to mar the appearance.

It will not be attacked excessively by mineral oil or Halowax, the two impregnating agents used in the section.

It will contain no soluble chlorides or sulfates.

This wax is inexpensive, is manufactured by several companies and is composed largely of rosin and a filler. Use of a synthetic resin is preferred, for then better control is possible during the wax preparation. Careful control of polymerization is a critical requirement for control of the cold-flow point and color (tan preferred).

The over-all dip wax is usually the same as that used for moistureproofing the paper tube. This is not necessarily so, however. This wax must have a melting point of no lower than 160°F.; (200°F. is better) and afford a high degree of moisture seal. Paraffin waxes are not suitable. Compounded waxes, containing petroleum bases and antiwetting compounds, such as aluminum stearate, are the rule. Ceresin is too brittle and a little low in melting point, but is good for moistureproofing. Microcrystalline waxes (both yellow and white) are commonly used. They have the advantages of being clear enough so that printing on the tube can be seen distinctly and being high melting and nontacky. The moistureproofing of an over-all dip wax is always determined by subjecting complete condensers to life tests at 95–100% humidity and noting the time of breakdown when on voltage.

The wax used to impregnate the paper of the paper tubular condenser section is generally Halowax No. 1001, but any wax having a dielectric constant above 6, a low power factor, high insulation resistance and chemical stability can be used. Hydrogenated vegetable oil has been used extensively. Opalwax is also in use.

The electrolytic condenser uses wax for potting, sealing and moistureproofing.

The potting waxes are usually pitches and many suitable compounds of this class are available. The common use is to pot the condenser element in the can. Adhesion is necessary and ability to withstand 185°F. without flow. Freedom from soluble chlorides, sulfates, and chemicals corrosive to aluminum is required because any corrosion of the delicate aluminum connections would cause faulty operation or failure of the unit in service.

The sealing waxes are the same as those used in the paper tubular condenser, but the range of choice is larger, because larger tubes of ½–1⅜ in. diameter are common.

Moistureproofing of paper tubes is done the same way as for paper tubular condensers, although on some cheaper assemblies, there is one notably different impregnant—paraffin wax.

Electrolytic-condenser manufacture also consumes large amounts of waxed paper, usually waxed with paraffin. This paper is used for covering benches, trays, etc., and, in general, for the handling of merchandise to prevent the contamination of component parts, which might cause corrosion in the final assembly.

Any waxes, designed for use in connection with electrolytic condensers, should be free from soluble chlorides, sulfates, and any chemicals that will corrode aluminum.

The mica condenser uses waxes for potting, moistureproofing, and also during processing. The potting "waxes," usually pitches, must have low power factor (electrical energy loss) or high "Q" and are usually not acceptable unless they can withstand 200°F. without flow and have a "Q" of 400 or better 550 kc.

Moistureproofing waxes are used to impregnate the Bakelite molded cases of the smaller mica condensers. These waxes are similar to those used for paper-tube moistureproofing. Vacuum impregnation secures penetration into the molded casing. Supula Wax No. 1254 and ceresin have been used extensively. This wax has very good electrical characteristics.

Low-melting, highly-refined waxes of the petroleum series are used, during mica condenser assembly, to hold stacks of foil and mica elements together. After the heating of the element, the wax flows between the mica and foil sheets providing a solid mass that can be handled easily. This wax must have high insulation resistance and "Q."

High-voltage condensers use wax for filling creepage space between parts of porcelain insulators. Such compounds must have high-voltage breakdown characteristics and be moisture resistant. The Aroclors have been used widely, particularly No. 1262.

The trend in the condenser industry is toward the use of higher temperatures in the neighborhood of 220°F. and low temperatures of the order of 50°F. Waxes will have to stand up to these temperatures or else the condenser industry will have to eliminate them.

WAX COATINGS FOR FRUITS, SHRUBS AND PLANTS

M. P. Hofmann

The use of various waxes to produce water resistance or water repellency on fabrics, paper, etc., is so firmly fixed in the mind of the average person that the use of wax in connection with fruit, growing plants and shrubs is rarely thought of. Although apples, pears and various vegetables have been coated with wax to enhance their keeping qualities and appearance, the use of wax as a coating for citrus fruits has been developed to such an extent as to use a considerable tonnage yearly.

The appearance of oranges, lemons, limes and grapefruit is very much improved when wax coated and polished and these coatings aid materially in preventing the drying out of these fruits.

One of the best grades of Florida juicy oranges ripens to a russet or greenish color. Although the fruit is superior to other varieties in both juice and vitamin content it sells poorly because of its lack of eye appeal. To remedy this defect, these oranges are coated with a wax to which red or orange color has been added. When this coated fruit is polished, its appearance is so much improved that it compares quite favorably with California fruit. These coating waxes are prepared in emulsion and their formulation is quite similar to the Dry-Brite types of waxes used for floor and linoleum finishes. As hardness of the coatings is not essential, addition of such waxes as paraffin, montan, etc., is permissible. They must, however, be used with discretion, as when dry and polished, they must not show any signs of tack or stickiness. Taste or flavor of the wax is not of great importance as the skins of these fruits are almost always discarded. However, waxes having marked odors or flavors should be avoided.

For the commercial coating of oranges, a long shallow tank is provided. A continuous belt or chain runs through it, which picks up the fruit and brings it out of the bath. The fruit then moves along an inclined chute for drying. From this it is fed between rapidly revolving, long-haired brushes which impart the final polish just before the fruit is packed.

These waxes are sold in concentrated form, containing 50% or more wax. These are diluted before use to a wax content of 6–10%. Color is usually added to the wax before emulsifying, but it can also be added to the bath. If added to the wax, oil-soluble colors are used, whereas certified food colors are used as additions to the bath.

A typical formulation is as follows:[1]

		oz
1.	Carnauba Wax	121
	Oleic Acid	37
2.	Triethanolamine	22
	Water	56
3.	Montan or Paraffin Wax	48

Both 1 and 2 are heated to 90°C. and combined with stirring. A dark, clear jellylike mass results which must be kept at 90°C. and 3 is added. The mass is stirred until the montan or paraffin is all incorporated into the mix; then water, which has been heated to 90°C., is added slowly with constant stirring until the desired dilution is reached. When made up to 10 gal the formula contains approximately 14% solids.

Some of the newer microcrystalline waxes are used as fruit coatings, as well as some of the newer self-emulsifying carnauba-wax substitutes.

The addition of color depends largely on the ripeness of the fruit, its color, the color desired, amount of wax used on the fruit, etc., and is determined by experiments. Small additions of ammonia-dissolved shellac, loba or pontianac gums may be made to the formulations, as in floor waxes. They will be found to increase luster and diminish tackiness.

[1] U.S. Patents 1,943,468 and 2,057,413.

The use of various combinations of shellac, beeswax, etc., by nurserymen to seal grafts and other wounds in shrubs and trees is, of course, well known. The volume of wax used for this purpose, however, is very small.

TREE, SHRUB AND PLANT COATINGS

A standard procedure formerly used by nurserymen for coating roses, trees, shrubs, etc., to prevent drying out during winter dormancy and shipment was to dip them into hot paraffin wax. The procedure was very unsatisfactory because if the wax was not hot enough, a very thick coating was formed, while if it was too hot, the plants were killed.

This practice has largely been supplanted by the use of emulsified wax coatings. In the treatment of citrus fruit with wax coatings, the object was to seal the pores of the fruit tightly, so that all moisture was retained. In the coating of growing plants, trees and shrubs, this is not the case, as a sufficiently porous coating must be applied to permit the plant to "breathe," yet not permit the escape of too much moisture.

Much experimental and research work was done in this field in the last few years and among others, the Agricultural Experiment Station of Michigan State College has conducted a great amount of field and laboratory research on this subject. Among other interesting discoveries, it has been demonstrated that on apple and cherry stock, the fruit yield has been appreciably increased when the trees where sprayed with wax.

In the development of these wax emulsions for use on nursery stock, toxicity was found to be highly important. After much research, it was found[1] that the damage to stock was caused principally by the emulsifying agents used. Of these, ammonium linoleate was found to be the least injurious, and a critical concentration of ammonium linoleate was found below which no injury was noted. A formula having the desired characteristics and being nontoxic to plants was developed as follows:[2]

---

[1] Technical Bulletin 218, *Studies on the Development, Preparation, Properties and Applications of Wax Emulsions for Coating Nursery Stock and Other Plant Materials,* Michigan State College, Agricultural Experiment Station, Department of Agricultural Chemistry, East Lansing, Mich.

[2] U.S. Patent 2,013,063, Dow Chemical Co.

Formula No. 1

	%
Paraffin Wax	8.6
Ammonium Linoleate	2.6
Bentonite	2.6
Water	86.0

No. 2

	%
Paraffin Wax	10.0
Trihydroxyethylamine Stearate	0.7
Triethanolamine	0.7
Bentonite	2.7
Water	85.9

Addition of a small amount of a drying oil to this formula is found to improve moisture resistance.

No. 3

	%
Water	48.8
Paraffin Wax	32.3
Linseed, Corn or Soy Oil	10.8
Trihydroxyethylamine Stearate	8.1

Heat to 90°C. and agitate until an emulsion forms. Stop agitation when temperature drops to 49–52°C.

In use, 1 part is added to 9 parts water in which enough bentonite has been dispersed to give a bentonite concentration of 1–2.5%. The finished emulsion is not very stable and should be used as prepared.

PHOTOMECHANICS

J. S. Mertle

Beeswax, paraffin and certain mineral waxes are extensively used in various photomechanical processes.

An early employment of wax in this field was for the production of so-called artificial (factitious) negatives, or those made by engraving line designs in an opaque or nonactinic wax coating on a transparent glass plate. The image so produced can then be used as a negative for exposure on sensitized paper, glass or metal.

For protection of photographic equipment, particularly darkroom benches and plateholders used in the wet collodion process, coatings of wax or paraffin offer a surface impervious to the corrosive silver nitrate solution employed with collodion plates.

Beeswax and paraffin wax are also used for transparentizing by making translucent the support of paper negatives, the added transparency reducing the time of exposure of line and half-tone negatives when used for contact printing in photolithography and related processes. For this purpose, the wax can be applied to the paper base with a hot iron, though a more convenient procedure is to apply a solution, consisting of 1 oz white wax dissolved in 5 oz hot turpentine.

For production of photographic prints of high gloss and for drying rotogravure carbon tissue, after sensitization in bichromate baths, it is usual to apply the wet surface to glass or special ferrotype plates of japanned iron or chromium-plated steel. To prevent sticking of the print or tissue to these surfaces after drying, the plates are sometimes coated with paraffin wax, using a solution of 10 gr paraffin in 1 oz benzol of carbon tetrachloride.

The use of waxes in printing inks is well known, but they are also employed in special inks for photomechanical plate making, either because of their acid resistance in photoetching methods or because of their greasy nature, which is a requisite in photolithography. An etching or starting ink for rolling up bichromated albumen (ink top) exposures on zinc and copper for relief halftone and line etching (photoengraving) may consist of this simple formula:

Black Letterpress Ink	4	Thin Litho Varnish	1
Beeswax	4		

The ingredients are melted together and then thoroughly ground.

A soft etching ink for photoengraving comprises:

Black Litho Ink	8	Thin Litho Varnish	8
Tallow	3	Beeswax	3
Asphaltum	1		

First melt the solid ingredients, then add the litho varnish. When cool, work well with a muller on an ink slab.

A finishing ink for rolling up the plate prior to the final bite or etching period is composed of:

Black Litho Ink	8	Rosin	2½
Beeswax	4		

Add sufficient turpentine to give the desired working consistency.

Special acid-resisting inks are employed for *rolup* in halftone zink etching, an example being as follows:

Black Letterpress Ink	8	Beeswax	1½
Asphaltum	2	Turpentine	5
Rosin	1½		

The etched plate must be warmed before applying the ink, the heat then causing the applied ink to melt and run down the sides of the partially etched dots.

For staging halftone plates during relief etching, acid-resistant fluid inks are locally applied, so that the remainder of the plate may be further etched to obtain correct tone values. An ink for this purpose may contain:

Powdered Asphaltum	4	Beeswax	2
Turpentine	6	Dragon's Blood	2
Benzol	6		

Processes of mechanical staging have been devised for relief halftone etching on zinc and copper, which aim at dispensing with staging by applying a waxy composition to the partially etched plate. A representative composition of this type is:

Petrolatum	8	Beeswax	8

The ingredients are melted together and enough printing ink added to impart a black color to the mixture. The final composition is rubbed into the plate, after the first flat etch. The plate is then rubbed with fine sawdust, leaving the middletones and shadows of the image more or less protected by a greasy layer.

In etching powders, mineral waxes have been proposed as a partial substitute for natural resins and at least one such "synthetic" powder has been offered to the American photoengraving trade.

Opaquing of images is necessary in photomechanics, not only for the elimination of pinholes and other blemishes in the image, but also for partial obliteration of parts of the image, such as is sometimes required in color reproduction. A popular substance for this purpose is known as "turpentine opaque," containing the following ingredients:

Turpentine	2½ oz	Beeswax	40 gr
Asphaltum	¼ oz	Carbon Black	20 gr

The wax is dissolved in the turpentine by heating, after which, the asphaltum is added, finally stirring in the carbon black.

For correction of photolithographic halftone images in color reproduction, dot etching is used, in which the halftone negative or positive is staged during etching, following the principle of halftone relief etching. Several solutions have been proposed for staging on plates and films, one entailing the use of wax:

Powdered Asphaltum	50	Yellow Beeswax	15
Venetian Turpentine	20	Turpentine	150

Dissolve by heating the mixture, but do not boil off the turpentine.

In photolithography, a fluid developing ink is used for imparting a greasy foundation to bichromated albumen images on grained zinc and aluminum. Such inks are articles of trade, though a suitable substitute can be had from the following formula, all ingredients of which contain wax in some form:

Transfer Ink	4	Etching Ink	8
Black Letterpress Ink	4		

The ingredients are ground in 16 oz of oil of turpentine, to which 20 drops of oil of lavender have been added to give an agreeable odor to the final ink. For use, the mixture may be thinned with rectified turpentine to the consistency of heavy cream.

For inking up images on photolitho-transfer paper, a stiffer ink is required, such as the following:

Palm Oil	1	Burgundy Pitch	2
Beeswax	1	Black Litho Ink	8
Asphaltum	2	Litho Varnish	2½

If desired, the ink can be thinned by working up with litho varnish.

Work can be introduced on litho plates by drawing or painting with litho crayons or tusche. A simple formula for the grease crayon is:

Lampblack	2	Wax	4
Soap	4	Spermaceti	4

The ingredients are melted together and the mass poured into paper molds in pencil form.

Like crayons, litho tusche is an article of trade, the following formula giving an idea of its composition:

Yellow Wax	2	Shellac	3
Mutton Tallow	2	Lampblack	1½
Marseilles Soap	6		

The tusche should dissolve fairly easily in distilled water and the resulting liquid should flow freely from brush or pen. Liquid tusches are supplied commercially; they differ in principle from the listed formula merely by having the ingredients ready, dissolved in water or other solvent.

Washout solutions are preferred by some lithographers for strengthening the image or design prior to its use on the litho press. An old formula for a washout solution is:

Powdered Asphaltum	6	Yellow Wax	1
Mutton Tallow	3	Turpentine	40

The ingredients are dissolved in the turpentine with the aid of heat and a thin film of the cool solution is allowed to dry on the litho printing surface.

Both beeswax and paraffin wax have been suggested as damp protectors for engraved steel plates, their presence preventing the formation of rust. Although either substance can be used separately for the purpose, some engravers prefer a coating containing equal parts of beeswax and paraffin wax, as paraffin does not absorb oxygen and thus retards oxidation of the stored plates.

POLISHES

Wax polishes are of two types: those made with water (emulsions) and those made with organic solvents. In both types, the water or solvent evaporates after application to leave a waxy film which is rubbed to give a polish or luster. Since one specific wax rarely gives the result desired, it is usually blended with other waxes and sometimes resins or other substances to obtain certain effects. Price often is a factor that prompts the introduction of excessive amounts of low-priced waxes (e.g., paraffin) in many polish compositions.

FLOOR POLISHES[1]

The cost per pound of a wax means very little because a pound of one may give one and a half to two times as much coverage as another and look just as well. Then, experience has shown that some waxes take several times as long to apply and polish as others. Again, there is a wide variation in the wearing qualities of waxes and in the appearance of the finished job. Some of the more subtle properties, such as odor, appearance, and color, weigh very heavily in selling and may or may not add a great deal to the cost of the wax. The red and yellow oil dyes most commonly used cost, as a rule, very little per pound of wax and the perfumes, such as alpha terpineol, oil of cassia, benzaldehyde, etc., usually add but slightly to the manufacturer's cost. Some plants have expensive machinery to obtain the right texture and appearance of the paste, whereas others seem to accomplish good results with hand apparatus. Waxes, such as carnauba, candelilla, or I. G., decrease cost by allowing the addition of cheaper solvents, whereas beeswax, paraffin, and ozokerite work in the opposite direction.

Nearly all buyers of waxes desire a paste with a smooth or glossy surface, free from oozing solvents or cracks. Waxes of the carnauba type usually give a flat, rough, paste surface when melted in solvents and allowed to cool in cans. The adding of beeswax or ozokerite, as well as high-melting paraffin and ceresin, will improve the smoothness. I. G. wax is reported to produce mirror surfaces in certain combinations. Rapid cooling aids greatly in developing smooth surfaces on paste wax, and it eliminates cracking in some cases.

Yellow and orange colors seem to hold the field, no doubt, on account of the stimulating effect on the eyes of the buyer.

Evaporation from the can depends on the solvents used (usually benzine and turpentine) and the solvent retention of the waxes. Ceresin, paraffin, and ozokerite are quite poor in this respect, whereas carnauba and candellila are usually very good.

Almost any paste wax will lose solvents if it gets hot enough

[1] Faucett, *Drugs, Oils and Paints,* 64 (1935).

and many a good wax will become soft when standing in the sun. It would be an excellent idea for wax makers to use labels which have great ability to radiate light and heat. White seems to be the best. Pastel colors could be used without increasing heat absorption materially and yet lending sales appeal to the product. The use of too much very volatile petroleum thinner will cause wax to form deposits around the lid of the can. Allowing the melted waxes and their solvents to cool too slowly will often cause a granular texture which bleeds out solvents badly. Too large a percentage of volatile solvent is also a source of this defect, whereas the noncrystalline waxes, as a rule, tend to prevent it.

The odor of a wax is very important in retail sales. The perfume chosen should arouse ideas of cleanliness, luxury, etc. A mixed perfume will, no doubt, please the largest number of people. One should avoid perfumes which are used in cheap soaps, etc., on account of the suggestion of cheapness which may become associated with one's product in the retail buyer's mind. The use of carefully selected petroleum solvents is necessary because there is a great difference in their odor. Some distillates and benzines smell like gasoline or kerosene and give a suggestion of cheapness by their disagreeable odor.

The consistency of a paste wax should be such as to be easy to remove from the container with the hand and to apply with a cloth. The consistency is easily controlled by increasing or decreasing the percentage of liquid ingredients.

Ease of application or spreading is improved by slow-evaporating solvents and nonsticky waxes: carnauba, montan (candelilla is fair), ceresin, shellac wax, paraffin wax, etc.

A paste wax will not spread uniformly if the solvent evaporates too rapidly, or if it contains too much of a resin-forming liquid, such as turpentine. The selection of proper waxes for the combination is vital. Beeswax and ozokerite should not be used liberally because of their gumminess. Carnauba, montan, ceresin, paraffin, glyceryl stearate, etc., will aid uniform spreading,

Some of the best-selling waxes dry in about 5 minutes after application. This may be partially accomplished by the use of 10–20% petroleum benzine with No. 1 distillate. Turpentine may be substituted for part of the distillate, but its expense precludes the use of large quantities. It has also a tendency to reduce paste consistency. Paraffin wax, beeswax, and ozokerite also inhibit rapid drying.

Carnauba and candelilla give gloss, but should be accompanied by montan or ceresin for uniform spreading and beeswax or ozokerite to give plasticity to the almost dry waxes, so that uneven smears of dried paste may be spread out with a little rubbing. Some paraffin wax also gives gloss but large amounts produce a hazy or slightly dulled surface.

For light floors, and especially linoleum, light-colored waxes should be used, because repeated applications may darken the floor. Overheating the waxes, to combine them with the solvent, will darken the paste, especially carnauba or beeswax.

A smearproof combination will require the sparing use of low-melting paraffin wax, Japan wax, and low-grade ceresin. Slow-evaporating solvents, such as pine oil, No. 2 or 3 distillate, etc., should be very sparingly used or avoided except for special waxes. Carnauba wax, I. G. wax, candelilla, and the other hard waxes are usually quite smearproof.

A wax should not dry to a brittle surface which will powder off when used. The use of beeswax increases toughness. Adding calcium stearate to a wax will give an apparent toughness or scratch resistance to a wax finish.

The coverage of a wax finish is improved by moderately slowly evaporating solvents, high nonvolatile content, and sparing use of sticky waxes. Using lots of high-boiling solvents may make a wax spread farther, but it is likely to give such a thin film that gloss and utility will be impaired.

If a wax is free from high-boiling solvents and low-melting paraffin, too much beeswax or ozokerite and other soft waxes, it will greatly reduce the labor of household operation.

Some agents which reduce the slip on waxed floors are beeswax, ozokerite and, best of all, calcium stearate.

Water-spotting will be noticed in waxes containing too much beeswax. Calcium stearate will greatly improve nonspotting.

There are waxes on the market which dry without polishing, but some of them wash off with water. This may be caused by the use of some soluble soap of caustic soda or borax. If ammonia soaps are used, the ammonia will evaporate and leave the resin or fatty acid in its original, waterproof condition. Triple-pressed stearic acid is a good emulsifying acid for use with ammonia.

The formation of a white or opaque light substance in cracks and nail holes is avoided by the use of moderately slowly-evaporating solvents; but more especially by a moderately increased use of beeswax or, better still, about half as much ozokerite. These waxes give a hornlike texture to dried finishes. A solution of carnauba, in light petroleum thinner, will dry to a soft, porous, crumbly cake which, when broken up, forms a product similar to a powder polish sold for use on floors.

Some lacquers and a few other finishes will turn white under a wax finish which contains too much nonvolatile solvent. The use of powerful solvents such as toluol or xylol may wrinkle fresh long-oil varnishes or lacquers. Petroleum benzines and distillates are best for general use on all surfaces.

There are two well-known methods for preventing paste waxes from solidifying in layers, with the heavy or most insoluble waxes at or near the bottom: One is quick cooling and the other, rapidly agitating the wax while cooling in a water-jacketed kettle until the first solidification point is reached. A few seconds after this, the wax is quickly filled into cans. This transferred wax resolidifies the second and last time. The first solidification point is sometimes visible or it may be detected by falling of the mercury in a thermometer followed by a slight rise before the final solidfication.

One should select a wax or a combination of waxes which will not partially melt when the sunlight strikes the surface on which it is placed. Surprisingly high temperatures are sometimes attained by dark surfaces when exposed to sunlight.

A very cheap wax can be made by melting 42 lb of candelilla wax and adding 22 gal of No. 1 distillate.

A better-grade wax is composed of: 36 lb of ceresin wax, 28 lb carnauba, 26 lb beeswax, 28 gal of No. 1 distillate, 4 gal of petroleum benzine, ½ oz of alpha terpineol and yellow and red dye solutions to suit.

A wax which is especially good for old floors is as follows: 31 lb carnauba, 33 lb bleached montan, 32 lb ozokerite, 5 gal pure-gum turpentine and 39 gal No. 1 distillate.

A good nonslipping wax may be made by melting: 6 lb carnauba, 6 lb beeswax, 8 lb bleached montan, 20 lb calcium stearate and thinning with 12½ gal No. 1 distillate.

A liquid polish may be made by diluting any of these to about 11% by weight of nonvolatile matter.

A noninflammable polish may be made by heating 32 lb carnauba, 33 lb bleached montan and 22 lb ozokerite in a drum until just melted; then adding 8.7 lb stearic acid and just melting. In a separate vessel, 25 gal boiling water is mixed with 4 lb triethanolamine and the whole poured, with stirring, into the wax and acid melt. A third vessel contains 25 gal boiling water which may be added as samples are removed and tested on a cold sheet of metal until the desired paste consistency is obtained.

SAPONIFIED SHOE CREAM[1]

Saponified creams offer an opportunity for cutting down manufacturing costs and reducing fire hazards. From other aspects, however, they have serious drawbacks. They do not keep particularly well in very cold weather, or in metallic containers; their viscosity is relatively high and the applied film is not so resistant as that of ordinary creams to dampness and rubbing; in course of time they are liable to make shoe leather hard and brittle. This at first sight formidable list of drawbacks can be greatly reduced. It must also be considered that these creams are more suitable than the water-free colored types for the treatment of colored leather.

Waxes can be thinned with water after partial saponification with an alkaline emulsifying agent. Owing to the variations in acid value and other properties, each wax requires separate consideration as regards its behavior on emulsification.

Carnauba wax (saponification number: 83) is readily emulsified by soap, preferably rosin curd or soft soap and also by Japan wax or rosin with potassium carbonate. In itself it is only employed in special cases, its usual function being to improve the quality of cheaper waxes. Bleached carnauba wax with the much lower acid number of 11 is more resistant to emulsification

---

1 Kirchdorfer, *Mfg. Chemist,* 328 (October, 1934).

and can be used only in saponified creams together with other more readily emulsifiable waxes. It is occasionally used in the manufacture of high-grade, snow-white creams.

Beeswax (saponification number: 92) is best emulsified with potassium carbonate in very little water at boiling temperature. Excessive water in this case causes granular precipitation of the myristic acid, while soaps do not yield good emulsions. When properly carried out, saponification of unadulterated beeswax leads to perfectly homogeneous, smooth creams which do not smear during polishing, are free from tackiness, and leave behind a durable, elastic, highly glossy film.

Crude montan wax (saponification number: 80) is just as popular for black saponified creams as for the solvent creams. As it is not readily saponified by alkali carbonates, it is usual to incorporate 20–30% rosin and up to 50% paraffin wax (to reduce tackiness). In addition to, or in place of, the rosin, beeswax, carnauba wax residues, or Japan wax may be used. Pure montan wax (acid number: 77) is even more difficult to saponify than the crude wax and the creams produced are less glossy or homogeneous. For this reason, the addition of the before-mentioned waxes is again advisable.

Rosin (saponification number: 160) promotes saponification in addition to improving stability and smoothness. In fact, it is almost indispensable in shoe-cream compositions containing ingredients which are hard to emulsify.

Paraffin wax (saponification number: nil) increases the body and makes the cream thinner and more uniform when rubbed; it improves water resistance, reduces the greasy, tacky nature of the soft waxes, and prevents bubble formation when the cream is being filled off. The best grade is the 50°/52° scale or block paraffin with a low oil content.

Potassium carbonate is a saponifying agent far superior to the cheaper sodium carbonate. For example, when the manufactured creams are stored at low temperatures, sodium carbonate is apt to crystallize in fine needles, but potassium carbonate does not have this disadvantage.

Soaps act not only as emulsifiers but also as stabilizers (protective colloids).

Calculation of theoretical quantity of alkali required for saponification is a very simple matter. The saponification number of the wax (expressed in grams) is multiplied by 1.5, the result being the number of grams of technically pure 96–98% potassium carbonate required for each kilogram of that particular wax.

The most suitable dyes for black water creams are the alkali-resistant nigrosines.

For preservation of water creams, of the older agents by far the best is formaldehyde (formalin), while modern milder-smelling substances are represented by the hydroxybenzoic esters, such as Moldex.

MANUFACTURE OF SAPONIFIED CREAMS

In the most customary procedure, the so-called "two-pot process," the wax is first cautiously melted in a polished iron or enameled pot and separate solutions of saponifying agent and dyestuff, each in about one third of the total water required by the formula, are kept in readiness. The dye solution is filtered before use if at all cloudy. When the temperature of the molten wax has fallen to 70°C., cautious additions are made of: 1. saponifying solution, 2. dye solution, 3. one half of the remaining quantity of water. Violent foaming of the wax can be arrested by more rapid addition of the solutions, but slow introduction should then be resumed. Good stirring is essential from the very start and the stirrer must be rotated in the same direction throughout the operation. If the wax is too cold before mixing begins, this will be revealed by "curdling," which can be rectified by rewarming until the mass is perfectly fluid. Perfect incorporation of all the ingredients is indicated by the absence of skin formation when the stirrer is stopped and by no change in the homogeneity of the cream.

To sample the batch, a little is placed on a clean, dry glass plate and allowed to cool to room temperature, when it should be uniformly firm and have a smooth, delicate, and glossy surface. No moisture should appear under finger pressure. A deficiency in any of these respects can be corrected by one or more of the following measures: brief boiling or warming; successive additions of potassium carbonate or soap solution (to improve smoothness); a little ammonia; some Japan wax, semi-saponified rosin, olein, or stearin (to promote the binding power and uniformity). Stirring should be avoided as far as possible during these corrective measures in order to reduce foaming. Bubbles can be eliminated either by incorporating 1% of spindle oil (although it reduces surface gloss somewhat) or by reducing the height through which the warm mass falls into the jars during filling off. The temperature for the last operation is much less critical than in the case of unsaponified creams. In general, the lowest practicable temperature is desirable, and the tins should be sealed only after complete cooling, which may take 24–48 hours.

RUBLESS EMULSION POLISHES*

Aqueous wax emulsion polishes have become standard items for home and building maintenance. Ease of application, low cost, and low flammability of these emulsions are advantages which have led to their steadily growing use. The so-called dry-bright or rubless polish industry originated with the discovery that carnauba wax could be emulsified to give a system, which, when spread on a surface such as linoleum, would dry to a lustrous film. Since that time, many improvements have been made in the properties of such emulsions and of the films deposited from them.

Waxes other than carnauba may be emulsified and various other vegetable and mineral waxes or combinations of waxes can be utilized to give products of improved quality or some desired property. New emulsifying agents have also been found and the use of synthetic resins in place of a portion of the wax has been introduced. All of these changes have been directed toward the development of an improved and lower-cost commodity.

The art of emulsion manufacture has developed faster than our understanding of the scientific principles involved, with the result that many fundamental questions on the origin of stability of the emulsion and the relationships between its com-

* Petrolite Corp. Ltd., New York City.

position and the properties of the polish film cannot be answered logically.

Rubless polishes are not emulsions in the strict sense of the word, since an emulsion is one immiscible liquid dispersed in another. Rubless polishes are dispersions of a solid waxy phase in an aqueous phase, flocculation of the particles being prevented by soap and electrolytes in the aqueous phase. It is believed that soap molecules are strongly adsorbed on the surfaces of the solid particles and, through partial ionization, form an electrical double layer which prevents close approach or contact of particles.

Another tenable theory to account for the stability of rubless polishes is that the wax is dissolved in soap micelles to form a kind of sandwich micelle, the outer surfaces of which consist of the carboxyl groups of the soap molecules, the whole having stability characteristics similar to an ordinary soap micelle.

Electron-microscope studies of typical dry-bright emulsions[1] have revealed that the wax particles present are extremely small, varying in diameter from 0.005–0.12 microns. Such particles are not resolvable in an optical microscope.

As a result of the small particle size, these emulsions are translucent. The particles, having diameters much less than the wave length of visible light, are not efficient reflectors of light. This property is undoubtedly related to the dry-bright character of the films formed from the emulsion. When a typical wax emulsion is spread on a surface, such as a floor, and the water allowed to evaporate, the wax particles are deposited as a layer roughly 2 microns thick. Since this layer is composed of particles originally about 0.1 micron or less in diameter, it seems likely that the surface of the wax film will have relatively few irregularities of much more than 0.1 micron in depth. Thus the film reflects light like a mirror, rather than scattering or absorbing light as a rough surface would. When the emulsion contains large particles, the dry-bright quality is lost, since the film then contains large irregularities which scatter light and give the surface an opaque or dull appearance.

Besides being shiny, the wax film must have other properties,

---

1 D. Schoenholz and C. S. Kimball, *Soap and Sanitary Chemicals* (August 1947).

such as good wear, low tack and resistance to water spotting. In addition, the emulsion itself must remain stable even on long storage and must spread and level properly when applied to a surface.

All of these qualities are interrelated in their demands on the nature of wax and emulsifying agent and thus it is difficult to obtain the optimum in all qualities simultaneously. Furthermore, not all manufacturers desire the same properties in their emulsions. Some may want a polish which can be easily removed by mild soap and water. Others may want a highly water-resistant film. The application to which the polish is to be put will often determine the special properties required.

For these reasons, no one particular polish formula can be said to be the best. The manufacturer must consider the application and the market for his product and use his judgment in selecting a suitable formula.

Dry-bright polishes consist of suspensions of wax or waxlike solids in an aqueous medium, the whole being stabilized by soaps and electrolytes. In addition, these emulsions usually contain a resin solution, such as ammoniacal solution of shellac, Manila resin, etc., which will facilitate spreading and leveling of the emulsion and will improve smoothness and gloss of the surface film.

A wide variety of emulsifying agents has been proposed for stabilization of emulsion polishes. In actual practice, however, oleic acid soaps have been found to be most efficient and reliable.

The emulsifying agents employed in polish formulations do not contribute to gloss, hardness and water repellency of the final dry polish film and, in fact, are detrimental to these properties. For this reason, it is highly desirable to keep the amount of emulsifying agents as low as possible without loss of the dry-bright property or stability.

Commercial dry-bright polishes usually contain 12% or more nonvolatile solids, including waxes, resins, electrolytes, leveling agent and nonvolatile soap constituents. The formulas listed are for polishes containing about 12% nonvolatiles. The formulae given are intended only to be illustrative of the types of polish emulsions obtainable. Many variations of any one formula are generally possible.

EMULSION–POLISH

Formula No. 1

No. 3 Refined Carnauba Wax N.C.	50.0
Crown Wax 23	25.0
Durez 14140	32.1
Oleic Acid	6.9
AMP*	6.4
Borax	4.0
Water	858.0

Wax-to-Water Method

Blend the first four ingredients in a melt at temperatures up to but not exceeding 270°F. Cool to 210°F. with agitation. Add the AMP with stirring at 200–210°F., followed immediately by the borax in a saturated solution. Cook at 200–210°F. for 5 minutes, then pour the wax melt slowly into the water at 200°F. with rapid agitation. When the emulsion has smoothed out, add a cold solution of a leveling agent during agitation and cooling to room temperature.

Water-to-Wax Method

In a kettle of sufficient capacity for the completed emulsion, blend the first four ingredients in a melt at temperatures up to but not exceeding 270°F. Cool to 210°F. and add the AMP while stirring at 200–210°F., followed immediately by the borax in a saturated solution. Cook at 200–210°F. for 5 minutes, then with no agitation rapidly add the total required amount of boiling water. Agitate slowly until the emulsion has smoothed out, add a cold solution of a leveling agent during agitation and cooling to room temperature.

Leveling Agent

For this formulation we recommend the use of 10–20% maximum of the following leveling agents. Increasing the amount of leveling agent decreases tack, increases film hardness, and increases slip.

1. Shellac Solution	(12%)
2. Amberol Solution	(12%)

* 2-Amino-2-methyl-1-propanol.

No. 2

No. 3 Refined Carnauba N.C.	25.0
Crown Wax 23	50.0
Durez 14140	32.1
Oleic Acid	6.0
AMP*	7.2
Borax	4.0
Water	851.0

This is a very satisfactory product for the manufacturer who is concerned with costs as well as quality. The gloss is good, but not quite so good as formula No. 1. The film will not be so hard as formula No. 1 but the slip resistance is better.

No. 3

Crown Wax 23	75.0
Durez 14140	32.1
Oleic Acid	6.0
AMP*	7.6
Borax	4.0
Water	851.0

For preparing formulations 2 and 3, use the same methods and temperatures as for formula No. 1.

This product has high gloss, slip resistance, and resistance to water spotting. The film is slightly softer than formula No. 2, but the slight tack should not be objectionable. This formula represents good quality with considerable saving in manufacturing cost over the formulas containing carnauba.

No. 4

Crown Wax 23	50
Durez 219	50
Oleic Acid	10
AMP*	6
Borax	4
Water	830

Wax-to-Water Method

Blend the first three ingredients in a melt at temperatures up to but not exceeding 350°F. Cool to 210°F., with agitation. Add the AMP with stirring at 200–210°F., followed immediately by the borax in a saturated solution. Cook at 200–210°F. for 5

* 2-Amino-2-methyl-1-propanol.

minutes then pour the wax melt slowly into the water at 200°F., with rapid agitation. When the emulsion has smoothed out, add a cold solution of a leveling agent during agitation and cooling to room temperature.

Water-to-Wax Method

In a kettle of sufficient capacity for the completed emulsion, blend the first three ingredients in a melt at temperatures up to but not exceeding 350°F. Cool to 210°F. and add the AMP,* while stirring, at 200—210°F., followed immediately by the borax in a saturated solution. Cook at 200—210°F. for 5 minutes, then with no agitation rapidly add the total required amount of boiling water. Agitate slowly until the emulsion has smoothed out. Add a cold solution of a leveling agent during agitation and cooling to room temperature.

Leveling Agents

This formula is a good base emulsion for the addition of large amounts of leveling agents. The high resin content of the original mix allows the use of up to 50% of the usual 12% leveling agents, such as shellac, Amberol, etc. Some reports indicate that even up to 80% leveling agent may be added.

With increased addition of resin, either to the original mix or in the leveling agent addition, the water resistance and gloss are excellent, but the ability to wash off with soap solutions is lost and the film becomes subject to scratching which cannot be buffed out. The film is very hard and bright when first laid down. This emulsion will pass three cycles of the freezing tests and perhaps more.

No. 5

Crown Wax 23	50.0
Durez 13560	61.2
AMP*	6.0
Borax	4.0
Water	839.0

Wax-to-Water Method

Blend the first two ingredients in a melt at temperatures up to but not exceeding 270°F. Cool to 210°F., with agitation. Add

* 2-Amino-2-methyl-1-propanol.

the AMP,* with stirring, at 200–210°F., followed immediately by the borax in a saturated solution. Cook at 200–210°F. for 5 minutes, then pour the wax melt slowly into the water at 200°F., with rapid agitation. When the emulsion has smoothed out, add a cold solution of a leveling agent during agitation and cooling to room temperature.

Water-to-Wax Method

In a kettle of sufficient capacity for the completed emulsion, blend the first two ingredients in a melt at temperatures up to but not exceeding 270°F. Cool to 210°F. and add the AMP,* while stirring, at 200–210°F., followed immediately by the borax in a saturated solution. Cook at 200–210°F. for 5 minutes, then with no agitation rapidly add the total required amount of boiling water. Agitate slowly until the emulsion has smoothed out. Add a cold solution of a leveling agent during agitation and cooling to room temperature.

Leveling Agents

Large quantities of leveling agents may be added, up to 50% of the usual 12% leveling agents, such as shellac, Amberol, etc., because of the high resin content of the original mix.

This is a high-resin-content formula comparable to formula No. 4, but allows a lower temperature level in formulation due to the use of Durez 13560, which is an 18.2% oleic acid blend with Durez 219.

With increased addition of resin, either to the original mix or with the leveling agent, the water resistance and gloss are excellent, but the ability to wash off with soap solutions is lost and the film is subject to scratching which cannot be buffed out. The film is very hard and bright when first laid down.

No. 6

No. 3 Refined Carnauba N.C.	50.0
Crown Wax 36	25.0
Crown Wax 1035 or 700	25.0
Oleic Acid	12.0
AMP*	5.2
Morpholine (Alternate)	7.6
Water	816.0

* 2-Amino-2-methyl-1-propanol,

This emulsion will pass three cycles of the freezing tests and perhaps more.

This formula compares favorably with formula No. 1, but contains no resinous substances. The gloss on drying is not so high as that of formula No. 1, but the finished film is easier to buff and the gloss after buffing is equal or superior to that of formula No. 1. This is an excellent product for use where buffing is desirable.

No. 7

Refined Ouricoury	30
Crown Wax 36	30
Crown Wax 1035 or 700	40
Oleic Acid	8
AMP*	6
Morpholine (Alternate)	10
Water	785

The following recommendations refer to both formulas No. 6 and No. 7.

The wax-to-water method of formulation is not suitable for formulations which contain inert microcrystalline wax, such as Crown 1035 or Crown 700. It has been found that best results are obtained by the following procedure: Melt the waxes and oleic acid together with agitation and bring the temperature to 200–205°F. Add the AMP* or morpholine with very close temperature control to avoid volatilizing. Add the water at 200–210°F. very slowly, with agitation, until a heavy gel forms and inverts to the oil-in-water type emulsion. This change can be detected by a change in color and viscosity of the gel. With the inversion, the rate of water addition can be increased, but the temperature of the added water should be kept at 200–210°F. until approximately half of the total required water has been added. The remainder of the water and the leveling solution may be added cold and very rapidly with agitation and cooling to room temperature.

Leveling Agents

We do not recommend the addition of more than 10% of the 12% leveling solutions, such as shellac, Amberol, etc.

* 2-Amino-2-methyl-1-propanol.

This formula yields a good-quality, resin-free polish at a considerable saving in manufacturing cost over formula No. 6. Gloss of the finished film is improved by buffing to which this polish is more amenable than are the resin-containing formulas.

No. 8

	a	*b*
Crown Wax 15	50	50
Durez 225 or Shanco 320	50	50
Oleic Acid	12	12
AMP*	6	8
Borax	4	—
Water	844	813

Wax-to-Water Method

Blend the first three ingredients in a melt at temperatures up to but not exceeding 400°F. Cool to 210°F., with agitation. Add the AMP*, with stirring, at 200–210°F., followed immediately by the borax in a saturated solution. Cook at 200°–210°F. for 5 minutes, then pour the wax melt slowly into the water at 200°F. with rapid agitation. When the emulsion has smoothed out, add a cold solution of a leveling agent during agitation and cooling to room temperature.

Water-to-Wax Method

In a kettle of sufficient capacity for the completed emulsion, blend the first three ingredients in a melt at temperatures up to but not exceeding 400°F. Cool to 210°F. and add the AMP* while stirring at 200–210°F., followed immediately by the borax in a saturated solution. Cook at 200–210°F. for 5 minutes, then with no agitation rapidly add the total required amount of boiling water. Agitate slowly until the emulsion has smoothed out. Add a cold solution of a leveling agent during agitation and cooling to room temperature.

Leveling Agents

To this high-resin-content emulsion as much as 80% leveling agent (12% solution of shellac, Amberol 750, etc.) may be added.

Crown Wax 15 was developed specially for this type of high-

* 2-Amino-2-methyl-1-propanol.

resin-content emulsion. It is equal in quality to formula No. 4, but allows the use of the harder high-melting resins, such as Durez 225 and Shanco 320, as well as the addition of more leveling agent.

The high temperature of the original melt is necessary because of the use of Durez 225 or Shanco 320. Durez 14140 or Shanco 380 may be substituted, but in this case, the base will be softer than is necessary due to the presence of an excess of oleic acid. The optimum amount of oleic acid for Crown Wax 15 is 19.3%, which is less than the amount contained in Durez 14140 and Shanco 380.

As has been pointed out under formula No. 4, increased addition of resin, either to the original mix or in the leveling agent, gives excellent gloss and water resistance, but wash off with soap solution is lost and the film is subject to scratching which cannot be buffed out. The film is very hard and bright when first laid down. This emulsion will pass three cycles of the freezing tests, and perhaps more.

No. 9

	a	*b*
Crown Wax 15	50.0	50.0
Durez 225	50.0	50.0
Oleic Acid	12.0	16.0
Morpholine	8.8	15.2
Borax	4.0	—
Water	841.0	835.0

Use the same methods and the same quantities of leveling agents as described under formula 8, but add the morpholine instead of the AMP.*

SILICONE–WAX POLISH FORMULAS

No-Rub Automobile Polish

Crown Wax 23	34
Silicone (5000 centipoise viscosity)	20
Mineral Spirits**	1300

Mix the wax and silicone with 20% of the solvent by heating to 190–200°F. When solution is complete, the remaining solvent is added at about 120°F., with agitation. The mixture is

* 2-Amino-2-methyl-1-propanol.
** Solvent boiling range about 275–325°F.

cooled with agitation and the agitation continued during packaging to prevent settling.

No-Rub Furniture Polish

Crown Wax 23	3.0
Silicone (5000 centipoise viscosity)	3.6
Oleic Acid	3.6
Morpholine	1.7
Mineral Spirits*	67.5
Water	40.0

Dissolve the wax, silicone and oleic acid in the mineral spirits. Form the emulsion either by the wax-to-water technique or the water-to-wax technique as outlined under formula 8.

These formulas make good high-gloss polishes of the no-rub, quick-drying type. Most silicone polishes require thorough cleaning of the surface of application for best results. With clean surfaces, the polish is spread smoothly, allowed to dry to a powdery film and the film wiped off leaving a high gloss with no rubbing.

The ratio of silicone to wax should be maintained while the variation in the total solids (wax and silicone) is 4—10%.

LEVELING AGENTS

12% Shellac Solution

Refined Dewaxed Shellac	102
26°Bé Ammonia (28%)	26
Water	722

The water is heated to 175°F. and the ammonia added with stirring followed by the ground shellac. The temperature is gradually raised to 200°F. until the shellac is dissolved. The solution is then cooled and stored for addition to emulsions.

Advantages of shellac as a leveling agent are that it has somewhat better water resistance than the synthetic resins used for leveling and it is more compatible (in larger amounts) than synthetic resins.

Disadvantages of shellac are high cost and necessity of short storage. Old shellac will not dissolve to form clear solutions.

In making shellac leveling agent, morpholine is frequently

preferred to ammonia because of smoothness of the wax film as compared to some granular roughness when ammonia is used. The amount of morpholine for this formula should be about 17 parts, but may need slight adjustment to give a clear solution.

12% AMBEROL 750 SOLUTION

Formula No. 1

Amberol 750	22.0
Aqueous Ammonia (26%)	3.2
Water	158.0

No. 2

Amberol 750	25.0
Morpholine	7.0
Water	176.0

Heat and stir the mixture at about 130°F. (up to 195°F. in the case of morpholine) until the resin is completely dissolved. The solution is clean and filtration is not needed.

Compared to shellac, Amberol 750 has better color, greater hardness and higher melting point. Compared to Manila loba, Amberol 750 has higher melting point, greater hardness and approximately equal flexibility.

However, Amberol 750 is not so tough and not so water resistant as shellac or manila loba.

Morpholine instead of ammonia, in the amount indicated, improves the leveling effect.

12% MANILA–LOBA SOLUTION

Manila Resin	100
26°Bé Ammonia (28%)	27
Morpholine as Alternate	40
Water	706

Add the ammonia to the powdered Manila resin, followed by about one third of the water at 140–160°F. Then allow the mixture to stand overnight. Then add another third of the water at the same temperature and continue stirring to a clear solution. If the solution does not clear at this point, add more ammonia before the addition of the final one third of the water. After cooling, filter to remove undissolved resins.

The advantage of this cooling agent is its low cost. Its disadvantages are that it is nonuniform, does not have as good leveling characteristics as shellac or synthetic resins and usually requires more alkaline agents.

Morpholine instead of ammonia, in the amount indicated, improves the leveling effect.

PRINTING INKS

C. R. Bragdon

Waxes have long been used in printing inks for controlling their working and drying properties and to improve scratch resistance. On account of their tendency to build up a flocculated structure within the vehicle, they promote "shortness," and aid the production of sharp, clear-cut impressions from the type or other matter, without fuzziness such as may result when the ink is too "long" and stringy, or has too much penetration. The effects produced are out of all proportion to the minute amounts required, ranging from fractions of 1% up to a maximum of about 3%.

Animal and vegetable waxes were first used for this purpose, and beeswax, candelilla, carnauba and Japan wax are still in use. Beeswax improves the lifting properties of process inks, preventing crawling. Carnauba, on account of its hard, brittle nature (its melting point being higher than that of any other common natural wax), its solvent power for organic dyes, and its ability to absorb mineral oil almost as a sponge takes up water, has a unique place among the ink maker's raw materials which very few other ingredients, natural or synthetic, can quite fill.

When the cheaper waxes derived from petroleum became available, they were found to have other useful characteristics which adapted them for an even wider market and paraffin wax now enters into a greater variety of formulations than any other wax. Both the refined and microcrystalline types have been used. The crystalline paraffins are almost insoluble in litho-varnish vehicles of the usual typographic and litho inks, etc., and a sort of granular layer forms on the surface of such inks on standing for a short time in the can. This is less noticeable when microcrystalline wax is used, but these waxes are often stickier and do not impart so much scratch resistance.

Another use of paraffin wax, not so common now as it was before the advent of the modern heat-set inks, is an antioffset spray. Linseed-base magazine inks require hours to dry thoroughly, especially where several layers are built up, as in four-color process work, and to permit the paper to be rolled up or stacked quickly without making an impression on the back of the facing sheet. Here, a mist of molten wax is sprayed over the print as it emerges from the impression rolls. This wax solidifies before it reaches the paper, so that the droplets do not coalesce but remain as a cushion of fine solid spheres to hold the surfaces apart by an infinitesimal distance, preventing offset and permitting oxygen to reach the ink to complete its drying. Such an expedient is unnecessary with the newer resin-solution-base inks, which dry perfectly in a fraction of a second by evaporation of the solvents under intense heat. These are now used in the typographic printing of almost all magazines of national circulation.

Wool-grease or lanolin, which according to chemical composition should be classed as a wax rather than a fat, is also used in inks partly to regulate shortness and partly as a lubricant to improve the "slip" characteristics. It has more length or string than most other waxes and, when used with paraffin, for which it has some solvent power, it tends to prevent crystallization and graininess. It is an ingredient of many "ink compounds," sold for use by the printer in adjusting the working properties of his ink to the temperature, humidity, type of paper stock, etc., for a particular job.

A patent[1] issued to Bradley in 1932 discloses the use of 0.5 to 2.0% of paraffin or ceresin wax in nitrocellulose printing compositions to retard the evaporation of volatile solvents and thus improve the printing qualities by forming a thin solid crust at the surface. This is similar to the action of wax in paint and varnish removers, but the claim is made that much smaller ratios of wax are effective.

[1] T. C. Bradley, U. S. Patent 1,884,553 (1932), *Composition Intended for Coating, Decorating and Printing Wood, Metal, Paper, Cloth and Other Surfaces.*

Radical changes in the use of waxes are reflected in patents[2] issued since 1938. These refer to compositions containing major proportions of wax, with resins, plasticizers and pigments, which are solid at ordinary temperatures and are applied in the molten condition, using heated fountains, distributing rollers and forms. Such inks have been developed for both intaglio and typographic printing. One metropolitan daily tabloid newspaper, of considerable circulation, uses the second process. The ink solidifies instantaneously on contact with the cold web of paper, and prints sharply and relatively smudgeproof. Waxes are particularly well adapted for such applications because of their sharp freezing point and quick transition on cooling from liquids of comparatively low viscosity to rigid but usually slightly flexible solids. This is in contrast to the higher viscosity of most molten resins and their gradual thickening through a stage of extremely high viscosity to final rigidity and even brittleness.

Wax-set inks, such as are covered in a particular modification of the Mock patent,[3] need not, and usually do not, contain any wax, but they are instantaneously dried by absorption of their solvents in the excess of molten wax when the printed web is passed through a wax bath, as in waxing bread wrappers. The ink binder, which may be a cellulose ester, alkyd resin or other binding agent insoluble in wax, is precipitated along with the pigment in intimate contact with the fibers of the paper. In addition to the speed gained by combining the printing and waxing operations in a single machine, a further great advantage is the freedom from odor and taste.

As new synthetic waxes are developed which combine more and more qualities necessary for successful use in inks, they will undoubtedly find a much larger market in this field. One important property present to an unusual degree in some of these products, particularly those which are made by esterifying various fatty acids with polyhydric alcohols and those of the amide type, is that of acting as wetting agents, thus aiding the grinding and improving the degree of dispersion of the pigments.

---

[2] W. L. Jones and E. H. McLeod, U. S. Patent 2,147,651 (1939), *Method of Multicolor Intaglio Printing.*

[3] W. W. Mock, U. S. Patent 2,081,949, *Method of Printing.*

FLUORESCENCE OF WAXES

Jack De Ment

When various waxes and waxlike substances, either natural or artificial, are placed under filtered ultraviolet light, they are often seen to fluoresce or glow strongly and characteristically. This light emission is frequently characteristic of the substance, its chemical and physical nature, origin, manufacture, age or previous history. In a pure wax, the fluorescence may be due to specific chemical bondings or to radicals which act as light emitters or fluorophors.

Fluorescent impure waxes often have a phosphorlike constitution in which the wax itself acts as a solvent or host to smaller amounts of organic, or even inorganic, impurities which act as centers of emission. Although these organophosphors, as they are called, have not been investigated extensively, the development of the science of fluorochemistry has come to depend on them. Anthracene, for example, fluoresces a medium blue color when pure, but the commercial product contains small amounts of chrysene which activates luminescence and causes a spectacular green glow. This appears to hold for many commercial waxes and related substances which are of varying purity.

There are two important uses of the fluorescence of waxes: 1. in fluorochemical analysis, for testing waxes and waxlike substances for contamination and adulteration and 2. in formulation of fluorescent and phosphorescent paints, pigments, inks and other strongly glowing liquids or solids.

In the fluorochemical analysis of waxes, long-wave ultraviolet light, i.e., wave lengths approximating 3650Å, are most reliable, although at times short-wave radiation, i.e., wave lengths approximating 2537Å, has been found valuable. The ultraviolet absorption by waxes and similar organic substances is greater, in general, in the longer-wave regions. However, in the excitation of luminescent substances which contain waxes and for special studies, shorter wave lengths make it possible to introduce almost twice as much energy into the system.

In testing the newly developed synthetic waxes and waxlike substances, the examination under ultraviolet light is a very efficient and reliable method. It is simpler and requires much less time than chemical analysis.

It is beyond the scope of this discussion to list the fluorescence of all known waxes and synthetic waxy products, but a short table will give an idea of the results which may be obtained by simple inspection of the unchanged sample, or preferably its freshly broken surface, under ultraviolet light. Some of the more important natural and synthetic waxes are listed in table 150. The reader is referred to books which cover this aspect of waxes, and their fluorescence under various types of radiation.[1]

Carnauba wax fluoresces a bright yellow color, with a faint brown mottling. In organic solvents, this fluorescence changes to milky blue or violet. In chloroform solution, it is possible by the Radley method, to detect as little as 1% carnauba mixed with other waxy substances. Chinese wax usually luminesces a bright white, and it has been found that coconut oil (white-colored solid) glows with a bright white color also, although a tinge of blue may be present according to the purity of the specimen.

Beeswax fluoresces variably. The English variety exhibits a brownish-yellow fluorescence with a green tinge; foreign beeswax glows a brown color which frequently shifts into the orange. White beeswax fluoresces white with blue. Chloroform solutions of most of these waxes are practically nonfluorescent.

Solution of the specimen is the best approach to reliable fluorochemical assay of waxlike substances. In liquid glyceryl compounds, dissolved waxes are often easier to characterize by their fluorescence. The addition of paraffin, which fluoresces a brilliant blue, to ceresin or beeswax may intensify the luminescence, but the method has been found to be uncertain.

Diglycol phthalate is an interesting substance from the standpoint of its fluorescence. The solid phthalate luminesces a strong blue color but pulverization changes the glow to a deep violet. Many other solid polyhydric alcohol esters behave in a similar

[1] Dake and De Ment, *Fluorescent Light and Its Applications;* De Ment, *Fluorescent Chemicals and Their Applications,* and Dake and De Ment, *Ultraviolet Light and Its Applications* (New York, Chemical Publishing Company, 1942).

Table 150

FLUORESENCE OF WAXES AND WAXLIKE SUBSTANCES

Substance	Fluorescence
Ambergris	bluish tinged
Beeswax	bluish (variable)
Carnauba	strong yellow
Ceresin	strong white
Diethylene Glycol Monooleate	strong blue
Diethylene Glycol Phthalate	bright blue-green
Diethylene Glycol Monoethyl Ether Citrate	bright blue
Diethylene Glycol Monoethyl Ether Phthalate	bright blue-green
Ethylene Glycol Monolaurate	very bright blue
Ethylene Glycol Monolaurate	bluish
Ethylene Glycol Monoricinoleate	green-blue
Glyceryl Formal	pale blue
Glycerin	bluish
Glyceryl Tristearate	bright violet-blue
Glyceryl Monostearate	bright yellow-blue
Glyceryl Monooleate	bright yellow-blue
Glyceryl Ricinoleate	blue-green
Mannitol Glyceryl Monolaurate	deep green-brown
Mannitol Glyceryl Monooleate	bright brown-blue
Mannitol Glyceryl Monostearate	blue-gray
Myricyl Alcohol	blue-yellow
Myristic Aldehyde	bluish
Myristone	yellowish
Palmitic Aldehyde	blue-white
Palmitone	yellowish
Polymerized Glycol Monolaurate	strong blue
Polymerized Glycol Dilaurate	yellow-blue
Polymerized Glycol Phthalate	strong pale blue
Polymerized Glycol Maleate	blue-yellow
Polymerized Glycol Monooleate	strong blue-brown
Polymerized Glycol Monostearate	brown-yellow
Stearone	blue-green
Stearic Aldehyde	bluish
Sorbitol Triricinoleate	green
Sorbitol Lactate	deep blue-green
Sorbitol Tartrate	strong yellow-blue
Zinc Palmitate	blue-white
Zinc Sebacate	white
Zinc Stearate	strong blue

fashion. Butyl carbitol palmitate fluoresces a brown-blue color. Substances, such as Albacer, Yumidol, Abopon, and trihydroxymethylamine stearate, fluoresce strong lavender, blue, pale blue, and coral-brown.

The fatty acid esters of mannitol often fluoresce with brown or yellow hues and the oleates, stearates and palmitates frequently emit in the blue and green regions of the spectrum. Ricinoleates nearly always fluoresce some shade of green, although compounds which contain organic substances of a high degree of fluorescence may modify this to blue-green or yellow-green. In general, most glycol and glyceryl esters of fatty acids containing six or more carbon atoms are characteristically luminescent. Fatty acid esters of polyhydric alcohols having a high molecular weight are also fluorescent.

Remarkable fluorescence was noted in ambergris, the waxlike substance formed in the intestinal tract of *Physeter macrocephalus,* the sperm whale, which is so valuable for fixing the odor of delicate and expensive perfumes. Genuine specimens of ambergris examined exhibited a strong blue fluorescence, overshadowing the original brown or tan color and in ethanol solution, this emission is more readily observed.

The use of waxes in fluorescent paints and similar products is not so widespread as their properties would indicate. However, some pure waxes and also those of a phosphorlike constitution, i.e., with a brilliant glow under ultraviolet light, find limited application in the arts, sciences and industries.

For example, polymerized glycols have a strong characteristic fluorescence. This fluorescence is similar to that displayed by natural waxes. This means that these waxlike synthetics might be used as substitutes for some of the natural products.

In fluorescent drafting inks and invisible hand-stamping inks used in theatrical and police work, these fluorescent polymerized glycols should serve the purpose well. It is realized that fluorescent inks and paper treated with luminescent substances can be of utmost importance in both peace and wartime for reading and writing in the dark.

Carbon is rarely depended on as the activator in luminescent pigments, but the boric acid phosphors are a series of phosphorescent substances in which carbon acts to enhance light absorption and/or emission. The introduction of easily reducible waxes or solid glycol derivatives is suggested in lieu of tartaric acid and other chemicals.

Waxes can play an important role in making phosphorescent

or fluorescent substances into paint or ink. They may serve as a vehicle or solvent when comparatively nonfluorescent or, if strongly luminescent, they may be added to alter the final hue. They may serve as media in which to grind, store, and handle inorganic phosphors which are prone to lose their luminescent qualities on comminution. They may serve as plasticizers or drying agents where the phosphor must not be acted on over a period of months or years.

The glycerol esters of fatty acids, for example, are almost always brightly fluorescent and if these substances are harmless to living tissue, they may find their way into the preparation of an important branch of cosmetics, protective creams and lotions, whose effectiveness depends on their strong fluorescence.

In protection from sunburn, especially important in tropical and semitropical areas, these fluorescent waxes and related substances may be incorporated into a cream or preferably a lotion, which absorbs the erythemic or burning (ultraviolet) rays from sunlight and emits the energy as harmless visible light through degradation by fluorescence. Petrolatum, lanolin and similar compounds already play this role in protective creams, but the use of brightly luminescent substances, whose efficiency is relatively high, may well supplant them. These waxes may be used singly or in combination. The selection of luminescent, waxlike chemicals should depend wholly on data obtained by spectrophotometric study of the absorption characteristics in the abiotic ultraviolet regions.

In general, many of the fluorescent waxes, particularly those which are solid and have a high melting point and good stability in air, can be employed where a bright and characteristic emission is required. In addition to the previous examples, a few others might be cited from the many which have already developed.

Waxes can go into solutions of fluorescent dyes, e.g., uranine, dispersed over the water from life rafts in signaling to aircraft during emergencies. They may be used in the patented processes of detecting hair flaws in metals because of their ease of handling, inexpensiveness and the wide choice possible in wetting power, viscosity, homogeneity, and stability to heat. These substances may also be employed in the patented processes of fluorographic printing and engraving. In criminology, they might serve for tagging objects and catching thieves. In the theater, they can be used in fluorescent cosmetics for a wide variety of comic and weird effects.

The military applications of fluorescent waxes are very promising. They may find their way into cartography for preparing luminescent maps and in intelligence work for preparing secret messages (butter and animal fats, and natural waxes have been used on occasion), for inter-tank signaling by use of ultraviolet light, in marking food, narcotics, explosives and other objects, and for fingerprinting by the methods of fluorophotography. The solid and brilliantly luminescent, waxlike substances may serve in testing gas masks and the gastightness of food boxes, medicine containers, suits, etc., because of their flourescence.

For tracing sewage, drinking water and streams, liquid waxlike substances may be used, to a limited extent, when the expensive dyes are not available. The detection of mercury in the laboratory is possible by use of fluorescent wax screens and short-wave ultraviolet light. Waxes are suggested as binders and cementing agents in delicate screens used in X-ray work, where retention of the crystalline structure of the inorganic phosphor is important. For vitamin and drug standards in fluorometric assay, pure liquid waxes and waxlike substances, either alone or in solution, might be employed because of their stability under continued irradiation with ultraviolet light.

For the study of minute skin structure, diseased tissue and other surfaces, where intricate and detailed investigation is significant, fluorescent waxes used with ultraviolet light and photography may be of considerable value.

To simulate the lifelike appearance of natural tissue, some of the waxes can be incorporated into supporting media for dentures and serve as artificial replacements for tissue burned or injured beyond repair and of unsightly appearance. The low order of brilliance presented by the fluorescence of many substances is being employed to give a lifelike luster to artificial teeth.

Petrolatum, paraffin and a few other brightly glowing waxes and similar substances can be used, because of their fluorescence, to provide illumination which will not violate the restricted conditions of a blackout. Examination of documents, paintings and art objects in laboratory practice, etc., may entail the use of fluorescent waxes or waxlike substances.

COLORING WAXES

by Dr. Jesse Werner*

Wax products must often be dyed or colored in order to make them more attractive for marketing or because the very existence of certain products depends almost entirely on the fact that they are colored. Cases in point which belong to the former category are decorative candles, toys and textile dressings. Examples of the second type are shoe and other polishes, china-marking pencils and copying papers.

Dyes applicable to waxes fall into two groups: 1. those that are soluble in molten waxes and 2. those that are insoluble. In the first case, the dye is usually dissolved in the molten wax at as low a temperature as possible, or first mixed with stearic acid or polychlorinated biphenyl[1] before dissolving in the molten wax. In the second case, the dye or pigment is dispersed in the molten wax or first dispersed in stearic acid and this dispersion added to the molten wax. This type of wax-insoluble dye is usually more expensive than the soluble type, because it must be manufactured in a very finely divided, easily dispersible form, or else has to be incorporated into the wax melt by a special procedure to give completely uniform dispersions. These pigments are, therefore, not used unless very special effects are desired. Spirit-soluble dyes are sometimes also employed. They are dissolved in a minimum amount of alcohol and the solution is mixed with an alcoholic solution of stearin. The mixture is left to set hard and is then added to molten wax in the amount necessary to impart the desired shade.

The wax of wax emulsions is colored in the melt or in solution with wax-soluble dyes, while the aqueous phase, before being combined with the wax melt or solution, is colored with water-soluble dyes.

---

* General Aniline and Film Corporation.

1 British Patent 459,797.

Decorative candles and similar articles may be prepared from colored stock or from colorless stock. If prepared from colorless stock, in a final step, it is coated by dipping the object in molten colored wax for a moment and removing. For coloring candles, pigments containing volatile poisonous compounds should not be employed. Among the pigments which should not be used for candles or for anything that will be handled, such as toys, textile dressings, etc., are orpiment, verdigris, vermilion, red lead, red chrome, zinc yellow and chrome yellow. Insoluble earth colors and lampblack are sometimes used for the production of very cheap articles. They are dispersed in a very finely divided state in the molten wax, but it is extremely difficult to get completely uniform dispersions.

The dyes and pigments useful for coloring waxes belong to several different chemical classes. Azo dyes containing no strongly acidic groups may be used and also the oil-soluble nigrosines and indulines. Complex metal salts of the mordant colors and triphenyl methane colors are sometimes employed. Various simple anthraquinone compounds are also used and if it is desired to make the wax fluorescent, complex polycyclic compounds of known and unknown structure are applied.[2]

Among the suitable dyes and pigments the following Sudans are soluble in waxes and have good fastness to light and sublimation: Sudan Yellow GRD, Sudan Red BBA, Sudan Red 4BA, Sudan Brown SBA, Sudan Corinth B and Sudan Black BR. A general formula for making candle coatings using the Sudan colors follows.

Dissolve:

0.05 g Sudan Color in a melt of
25.00 g Stearic Acid and add a melt of
75.00 g Paraffin Wax

However, where very excellent fastness to light, sublimation, and aging is required, the following Fast Candle colors are suitable: Fast Candle Yellow A, Orange A, Orange B, Red A, Blue B, Green A, Green B and Green C. These dyes are very easily dispersible. The formula for their application is similar to that of the Sudans, except that a melt of 2 g Fast Candle Color with 25 g stearic acid is used for 75 g paraffin wax.

---

[2] U. S. Patents 2,123,245; 2,123,246; 2,135,259; 2,138,381; 2,158,296; 2,163,950; 2,178,521; 2,183,629; and 2,187,819.

If very intense coloring effects are desired and fastness is not important, the use of the following so-called bases is recommended: Auramine Base, Chrysoidine Y Base, Bismarck Brown TSS Base, Rhodamine B Base, Victoria Blue BA Base, Victoria Green Base, Methyl Violet Base A, etc. These colors require the use of at least three to five times as much organic acid, such as stearic acid or oleic acid, in order to develop full tinctorial strength. The nigrosine and induline bases are in the same class and are of good tinctorial strength and good fastness to light.

When it is desired to produce more opaque or so-called "whiter" effects in wax products, such as candles, Hertolan N Powder or very finely divided titanium dioxide may be used for this purpose.

The Sudans are suitable as dyes for various types of polishes, such as floor and shoe polishes. In addition, Autol Brilliant Red BND, Sudan Blue GA, Alizarine Irisol N Powder (soluble in oil), Alizarine Sky Blue B (soluble in oil), Sudan Green 4B, and Sudan Violet BR may also be used. The last five products have good fastness to light.

For copying papers, the before-mentioned bases may be used as well as various basic dyes, such as Magentas, Methylene Blues, Methyl Violets, Victoria Blues, Victoria Greens, etc. For spirit-hectograph paper Crystal Violet AO Concentrated Pure and AON Supra are suitable. They require special milling in the oil-wax composition in order to obtain proper dispersion.

CARBON-PAPER INK COMPOSITIONS

B. J. Staneslow

The following wax mixtures have been studied: carnauba-montan; carnauba-paraffin; esparto-montan; esparto-paraffin; montan-paraffin.

Wax mixtures with maximum desirable properties for use in carbon-paper manufacture range from a mixture of 50% esparto and 50% montan to 100% esparto and include a mixture of 25% esparto with 75% paraffin.

Carnauba wax contracts greatly on cooling. This is not desirable. Mixtures of carnauba with montan or paraffin show undesirable contraction and rough surfaces. The surface is grainy and the mixtures are not homogeneous throughout.

Montan and paraffin waxes do not form satisfactory mixtures.

The surfaces are not smooth or flat. Although montan wax itself is very hard, it is not dominant in mixtures with paraffin which is soft.

Esparto wax forms good mixtures with either montan or paraffin wax. In this respect, it is superior to carnauba. Alone, it contracts very little, is less brittle than carnauba, has a scratch hardness nearly equal to carnauba, but a much lower melting point.

Mixtures of various percentages of "ink oil" No. 512 (a mineral oil) with carnauba, esparto, montan and paraffin waxes have also been tested.

The vegetable waxes (carnauba and esparto) form better mixtures with ink oil than do the mineral waxes (montan and paraffin). The first can combine with more ink oil than the second. Of the mineral waxes, montan seems to be somewhat soluble in ink oil. Paraffin is also soluble but to a lesser degree.

The only beneficial quality of ink oil in mixtures with these waxes is its decreasing the brittleness of the waxes. The resulting mixtures are very tough.

Montan-wax mixtures with ink oil show the greatest change in scratch hardness. Although ink oil tends to decrease the scratch hardness progressively with increasing amounts of the oil, yet this change is most pronounced with montan wax-ink oil mixtures.

Flatness of surface is not appreciably affected by ink oil, although 40–60% ink oil with carnauba decreases the flatness slightly, while 20% ink oil with paraffin wax increases the flatness slightly.

Ink oil decreases the smoothness of surface of waxes. With carnauba and esparto waxes, greasy surfaces are formed due to the better combination of the wax with the oil, while with montan and paraffin waxes, oily surfaces are produced. This checks with the previous statement that the vegetable waxes can combine with larger amounts of the oil than the mineral waxes.

Ink oil, unless 60% or more is used, does not affect the graininess or pittiness of the wax surface.

Ink oil seems to decrease the contraction of wax with increasing amounts of the oil. The exception is esparto wax, where the contraction is appreciably increased with 20–60% ink oil.

The preceding discussion indicates that the use of ink oil must be carefully regulated, especially with montan and paraffin waxes present. Excessive amounts of ink oil will be brought to the surface by temperature rise or by pressure. This may explain the transfer of oil from the carbon-coated sheet to the printed form.

Esparto wax holds 40–60% of ink oil satisfactorily according to a rough test made. This is also true of carnauba wax. Montan wax will hold 20% ink oil while 40% is in excess. Paraffin will not hold 20% satisfactorily. These results were obtained by pressure tests.

EFFECT OF VARIOUS PERCENTAGES OF PARAFFIN WAX ON ESPARTO–MONTAN MIXTURES

Different percentages of paraffin wax were added to a 50% montan–50% esparto wax mixture and a 25% montan–75% esparto wax mixture. The following observations were made:

The best all-around mixtures are as follows:

With 50% montan–50% esparto wax, satisfactory mixtures were obtained with 100, 125, and 150% paraffin wax. The percentage of paraffin wax is based on the combined weights of montan and esparto waxes.

With 25% montan–75% esparto wax, satisfactory mixtures were obtained with 75, 100, 125, and 150% paraffin wax.

A minimum ratio for satisfactory mixtures exists between the paraffin wax and the montan wax. In a 50:50 mixture, the minimum satisfactory ratio is 2 parts paraffin to 1 part montan. In a 25:75 mixture, the minimum satisfactory ratio is 3 parts paraffin to 1 part montan. The ratio of 2 parts paraffin to 1 part montan may also be satisfactory.

The mixtures become softer with increasing amounts of paraffin wax. Montan wax, being hard, retards somewhat the effect of paraffin. This is true only with smaller amounts of paraffin.

Small amounts of paraffin wax increase the toughness of the mixture, especially with higher montan wax content. Beyond a certain point, the changes are not so pronounced. Paraffin wax does not exert any undesirable effects on the flatness of the mixture.

Small ratios of paraffin to montan wax decrease the smoothness of the mixtures.

The cross section or homogeneity of the mixture is but slightly affected by paraffin wax. In any case, it is never greater than that of carnauba wax alone.

Small ratios of paraffin wax to montan wax greatly affect the graininess of the mixture. With 50% montan–50% esparto wax, graininess is most apparent in the ratio of 1–1½ parts of paraffin to 1 part montan wax. With 25% of montan–75% esparto wax, graininess is most apparent in the ratio of 1–2 parts paraffin to 1 part montan wax.

Paraffin wax decreases the contraction.

These observations show that the ratio of paraffin wax to montan wax must be controlled for best results.

Study of mixtures of dark-green petrolatum with carnauba wax, esparto wax, crude montan wax and paraffin wax led to the following conclusions:

Increasing amounts of petrolatum gradually decrease the starch hardness of carnauba and esparto wax. The scratch hardness of montan wax is appreciably decreased with 20% petrolatum. Change in scratch hardness is more gradual with paraffin wax.

Petrolatum noticeably decreases the brittleness of waxes, especially montan wax.

Large amounts of petrolatum with carnauba wax decrease the surface flatness. With other waxes, flatness is improved.

Increasing amounts of petrolatum tend to produce a more homogeneous mass, but, at the same time, to increase tackiness, with gumlike characteristics.

Small amounts of petrolatum (about 20%) tend to increase the graininess of the surface while larger amounts have very little effect. Exceptions are carnauba wax where large amounts of petrolatum increase the graininess considerably and paraffin wax, where graininess is decreased noticeably by a large percentage of petrolatum.

Petrolatum decreases the contraction of carnauba wax to a small extent, but increases that of esparto wax to the same extent. No change is shown with montan wax except in large amounts. There is no change with paraffin wax.

A mixture of 20% esparto wax and 80% dark-green petrolatum has high gloss.

Additions of ink oil to mixtures of 42.8% esparto wax with 14.3% montan wax and 42.8% paraffin wax, and 37.5% esparto wax with 12.5% montan wax and 50% paraffin wax have been studied.

Ink oil noticeably reduces scratch hardness of these mixtures.

Ink oil decreases brittleness producing a tough material.

Ink oil has no effect on surface flatness.

Ink oil slightly reduces surface smoothness by forming a greasy surface. This greasiness increases with increasing amounts of ink oil and is more apparent in the second mixtures.

Ink oil increases homogeneity in small amounts, but decreases it in amounts 30% and over.

Ink oil has very little effect on graininess of surface except in large amounts, which decrease graininess slightly.

Ink oil has no effect on contraction in small amounts, but increases it in moderate amounts.

STENCILS

The most common stencils are of cellulose or of protein base. Waxes play an important part in protein-type stencils.

The manufacture of stencils is a very exacting and delicate operation not to be attempted without much experimental work. A general procedure of operation can be broken down into the following steps: compounding the "dope" or solution; coating each sheet individually; conditioning and drying; setting or hardening with a second coating; drying; trimming; and aging.

The principle involved in any stencil is to coat a loose web of paper, like Yoshino, with a continuous film. This film must be absolutely free from pinholes. The film must also be even, uniform and nonsticky, while the sheet itself remains flexible. The texture must be firm enough to withstand a cutting type blow without cutting the letters out.

On examining a cut stencil, ready prepared for use, it is seen that the type, on striking, has squeezed the coating aside leaving the letter held in place by a mesh of paper fibers. Since no stain is made on the backing sheet as a result of striking the stencil, the material has not been disposed of in the manner of ink from a carbon paper but has been broken down and pressed aside.

In discussing the relation of wax to the function of the stencil solution, it is advisable to review a general formula. This formula cannot be claimed as reliable since the very quality of the component parts may even necessitate a reformulation.

Mineral Oil	130	IG Wax S	10
Sulfonated Oil	60	Colored Pigment	25
Clay Filler	4	Gelatin	10
Stearic Acid	9	Water	300

Many of the hard waxes can be substituted in this formula; other IG waxes, crude montan, and carnauba have proved suitable. The wax gives body to the coating. However, in order to obtain a fluid, workable mixture, the wax must be readily emulsified by sulfonated oil.

Wax is also important in producing the proper firmness and in waterproofing the dried film.

As, in most cases, where wax is employed as an ingredient, advantage is taken of its very special physical properties, especially of the fact that waxes are solid at room temperatures but can be liquefied by heat. Waxes have many other special advantages, such as ease in blending and good wettability.

In stencils, as in other wax formulae, physical advantages are utilized without a thorough understanding as to what has taken place chemically.

CARBON-PAPER INKS

Alan B. Olsen

There are two classes of waxes employed in carbon-paper inks. The amorphous waxes, such as ozokerite, ceresin and petroleum waxes, body the ink or absorb and hold the oil vehicle. The hard crystalline waxes are the essential ingredient in carbon paper inks. They hold the ink in a solid form and permit the handling of the coated papers with a reasonable amount of cleanliness.

In the manufacture of carbon-paper inks and typewriter-ribbon inks, it is essential that all the ingredients are completely non-drying. Most waxes conform to this specification. Carnauba wax is particularly desirable because it gives a product with a good shine and hard finish that is easy to handle. A carnauba ink is usually free from stick and smudging. Some of the other hard waxes do not set so hard as carnauba when mixed with an oil vehicle.

Other hard waxes that can be used in carbon inks are montan, ouricury, candelilla and numerous compounded waxes sold as carnauba substitutes. There is, however, no carnauba substitute that can be used in a carnauba formula without proper adjustments.

The compounding of a carbon-paper ink is very specialized and requires experience and familiarity with the usual ingredients. Essentially, a printing ink made with nondrying animal, vegetable or mineral oil containing a sufficiently high percentage of wax could be called a carbon-paper ink. The ink, which must remain hard at room temperature, necessarily has to be mixed and handled in the melted state. This is best carried out in steam-jacketed equipment to prevent burning.

A simple workable carbon ink can be made by mixing 2 parts of hard wax, 2 parts of nondrying oil, and 1 part of carbon black. The requirements for a good carbon-paper ink are: It must be liquid with good flow to smooth out in a thin even coating on paper. If the formula listed fails to do this, it may be that the color chosen has too high an oil absorption and thickens the ink, or it may be that the wax lacks the proper wetting and flow qualities. If we let the ink cool, it should harden into a solid mass that will fracture. If it is soft when cool or pasty, the carbon paper will be unsatisfactory.

In better carbon-paper inks, the black is made to appear blacker through the incorporation of blue or purple pigments or dyestuffs. If a dye is used, a nonevaporating and nondrying solvent must be used with it.

The amorphous waxes are used in carbon-paper inks as secondary waxes, in relatively small quantities, to prevent the oil from sweating out of the ink. Other purposes of these waxes are to remove the greasy feel and to prevent excessive brittleness.

The ideal carbon-paper ink would be a mixture of wax and color, but this is not practicable. This ink would not adhere to the paper and would flake off when the paper is bent or folded. Further, the ink would be too hard and would not write when struck by the type. Therefore, usually oil is added to produce a softer and better-adhering wax.

Each wax has characteristics that make for specialized use in formulation. There is perhaps no other use for wax that demands so much or depends so completely on the physical and chemical properties of a wax as does carbon-paper ink.

SOLUBILITY OF COMMON WAXES USED

Carnauba Wax

This wax will mix in all proportions with all oils whether of animal, vegetable or mineral origin. Whenever the amount of oil exceeds 70% of the oil-wax mixture the oil is usually not retained.

This wax mixes with most other waxes both of vegetable and mineral types in all proportions. Fatty acids act like the oils.

Resins and ethyl and nitro cellulose are miscible to some extent, but are not soluble in the wax.

Candelilla Wax

This wax behaves much like carnauba in its blending and solubility qualities, but will carry a smaller percentage of the oils.

Ouricury Wax

It has similar qualities to carnauba and holds about the same amount of the oils, but retains a somewhat harder consistency due to its resin content.

Montan Wax

This wax has an unusually smooth curve when the oil proportions are plotted and will retain at least as high a proportion of oils as carnauba.

Ozokerite

It blends in all proportions with all types of oils. In carbon papers, its main purpose is to increase the absorption of the oils into the harder waxes.

Esparto Wax

It has about the same blending qualities as candelilla wax, perhaps holding as much as 60% of the combined oil-wax mixture in oil. (40% esparto to 60% oil.)

WETTING QUALITIES OF NATURAL WAXES

The wetting qualities of natural waxes seem to be inversely proportional to their acid numbers when tested for the wetting of pigments. To explain, the lowest acid number wax will wet the most pigment with the best flow to the ink. The highest acid content will give the thickest mixture. This holds for an acid number of about 4, but when the acid number is below this, the rule does not hold.

Nypene and Piccolite resins are compatible with waxes. Mixes in carbon-paper ink formulas of 4 parts of wax to 1 part of resin, with the usual mineral-oil diluent, give waxlike behavior with little if any resinlike qualities. In other words, these resins are cheap partial substitutes for the more expensive natural waxes, without sacrificing the necessary waxlike qualities. If the resin is first dissolved in the mineral oil at temperatures above the melting point of the resins and then cooled down to temperatures tolerated by the waxes, formulation is quite simple and easy due to the oil solubility of the resin.

CHAPTER 6

glossary, special tables, and substitutes

GLOSSARY

Absorption. Taking up of a gas or vapor by a fluid in which it dissolves.

Acetyl number. Number indicating the extent to which a substance may be acetylated; in fats, oils, and waxes, measure of the number of hydroxyl groups in the molecule.

Acid number. Number indicating the amount of free acids in a substance; in fats, oils, and waxes, measure of the degree of hydrolysis.

Adsorption. Taking up (concentration) of a substance on the surface of another substance in a very thin (often monomolecular) layer.

Amorphous. Noncrystalline.

Aniline point. Lowest temperature at which equal volumes of freshly distilled aniline and the substance being tested are completely miscible.

Ash. Solid residue, remaining after a substance is completely burned.

Autooxidation. Oxidation that takes place only when another oxidation occurs in the same system.

Boiling point. Temperature at which the vapor pressure of a liquid equals the atmospheric pressure.

Coefficient. Factor by which a value must be multiplied to make the calculation correct.

Cohesion. Property, due to attraction between the molecules of a substance, which resists separation of the substance into parts when acted on by external forces.

Color, Lovibond. See Lovibond color.

Color, Saybolt. See Saybolt color.

Conductivity. The relative ability of a material to carry an electric current; the reciprocal of the resistance of a circuit.

Conductivity, effective A-C. Effective conductance between the opposite, parallel faces of an object having unit length and unit cross-section.

Density. Weight per unit volume, e.g., pounds per cubic foot or grams per cubic centimeter.

Dielectric constant. Reciprocal of the force of attraction between two electric charges of unit value separated by a distance of 1 cm.

Dispersion. Fairly permanent suspension of finely divided (undissolved) particles in a fluid.

Ductility. Property that permits the drawing out of a body in the direction of its length.

Edibility. Food value and not being harmful when eaten in normal amounts.

Elasticity. Property of recovering original shape and dimensions on removal of stress.

Emulsifiability. Ability to form an emulsion with water or other fluid.

Emulsion. Mechanical mixture of two immiscible liquids, usually milky in appearance.

Ester. Organic compound, formed by the union of alcohol and acid with elimination of water, e.g., glycol stearate.

Ester number. Difference between the saponification and acid number; measure of neutral fats.

Estolide. Polymer of hydroxy acids.

Eutectic. The property of a mixture or solution that all of its components solidify or liquefy simultaneously.

Explosiveness. Tendency to enter into violent chemical reaction, evolving heat and high-speed gaseous reaction products, thus producing enormous pressure.

Fire point. Temperature at which a substance, when ignited, begins to burn freely after withdrawal of the ignition agent.

Flash point. Temperature at which a substance begins to give off inflammable vapors, measured under specified conditions.

Flexibility. Ability to be bent repeatedly, within limits, without cracking or breaking.

Free fatty acid. The amount of free fatty acids present in a fat, oil, or wax, measured by the acid number.

Freezing point. Temperature at which a liquid solidifies or begins to form crystals; for pure substances, the same as the melting point.

Fusion, heat of. See Latent heat of fusion.

Gelling. Formation of a gel or jellylike substance.

Glyceride. An ester of glycerol and a higher fatty acid, e.g., glyceryl tristearate.

Hardness. Resistance to external forces, e.g., crushing, abrasion, indentation, or scratching. Moh's scale of hardness indicates hardness with 1 as softest and 10 as hardest:

1. Talc
2. Gypsum
3. Calcite
4. Fluorite
5. Apatite
6. Orthoclase
7. Quartz
8. Topaz
9. Corundum
10. Diamond

Heat of Combustion. Amount of heat evolved when 1 gram molecule of a substance is burned in oxygen at constant volume.

Hydrocarbon. Compound containing carbon and hydrogen only, e.g., methane, benzol.

Hydroxyl number. See Acetyl number.

Hygroscopic. Tending to absorb moisture from air or other gases.

Iodine number. Number indicating the amount of iodine absorbed by a substance, expressed as a percentage of the molecular weight; in fats, oils, and waxes, measure of the double bonds present.

Kreis test. Test for rancidity (autoxidation) of oils, fats and waxes, with an ether solution of phloroglucin and hydrochloric acid; rancid products give red or pale-red color.

Lactone. Cyclic anhydride formed by splitting off one molecule of water between carboxyl and hydroxyl groups of the same molecule.

Latent heat of fusion. Quantity of heat absorbed when the unit mass of a solid is changed to a liquid at constant temperature.

Length. Stringiness of a fluid when poured or when a rod is dipped into it and pulled out; proportion of oil in a varnish.

Linear expansion coefficient. Ratio of change in length per degree to length at 0°C.

Lovibond color. Color value given as the ratio of the three basic colors: yellow, red and blue.

Luster. Natural or artificial brilliancy or sheen.

Malcrystalline. Having a nonrecognizable crystal form.

Melting point determination.

Capillary method. A small quantity of a substance is heated, in a capillary tube, to the temperature at which it reaches its fluid form.

Ball-and-ring method. The aperture of a metal ring is filled with the substance, and a metal ball is placed on it. It is heated to the temperature at which the substance softens and is forced out of the ring by the weight of the ball.

The second method is used for substances having no definite melting point, but softening before melting. Values obtained by the two methods differ from each other.

Molecular weight. The sum of the weights of the atoms of a molecule.

Monoclinic. Crystal type having two of its three unequal axes perpendicular to the third and oblique to each other.

pH. Logarithm or the reciprocal of hydrogen-ion concentration; a measure of alkalinity or acidity of a solution or dispersion. A pH of 7 is neutral. Values above 7 are alkaline; values below 7 are acid.

Penetration. Method of expressing the hardness of a solid by the depth to which a loaded needle penetrates under standard conditions of load, time, temperature, etc.

Plasticity. Property that permits permanent alteration of shape and/or size by slight stress and without rupture.

Poisson's ratio. Ratio of the transverse contraction, per unit of a material forming a bar of uniform cross-section, to its elongation per unit length when acted on by tensile stress.

Polenske number. Number of cubic centimeters of 0.1 N alkali, used in titrating water-insoluble, volatile acids from 5 g of fat, wax or oil.

Refractive index. Ratio of the velocity of light in vacuum to that in the substance being tested.

Reichert-Meissl number. Number of cubic centimeters of 0.1 N potassium hydroxide used to neutralize the water-soluble, volatile acids liberated from 5 g of oil, fat or wax under specified conditions.

Resin. Semisolid or solid mixture of complex organic compounds having no definite melting point; insoluble in water, but soluble in certain organic solvents. Resins may be of natural origin, e.g., rosin or ester gum; or synthetic, such as phenol-formaldehyde or alkyd resins.

Rhomboid. Parallelogram whose opposite sides and angles are equal, but the adjoining sides and adjacent angles are different.

Saponification number. Milligrams of potassium hydroxide necessary to completely saponify 1 gram of a fat, oil or wax.

Saybolt color. Color value measured by the Saybolt colorimeter.

Saybolt viscosity. Viscosity measured by the number of seconds required for a sample to pass through a standard orifice in a Saybolt viscosimeter under certain conditions.

Setting point. Temperature at which a melted wax stops flowing when acted on by a small, definite pressure under certain conditions; temperature of solidification.

Solubility. Degree to which a substance will dissolve in a particular solvent, usually expressed as grams dissolved in 100 grams of solvent.

Specific gravity. Ratio of the weight of any volume of a substance to the weight of an equal volume of water at 4°C. The specific gravity of water at 4°C. is 1.

Splitting. Decomposition of glycerides by hydrolysis.

Stability. Property of a substance to remain substantially unaltered over a certain period of time.

Structure. Relative arrangement and organic union of particles in a material; crystalline or amorphous.

Surface tension. Tendency of the surface of a liquid to contract to the smallest possible area.

Sward hardness. Hardness measured by the number of oscillations of the Sward hardness rocker on a polished plate-glass surface coated with the substance to be tested.

Tack. Stickiness or tendency to adhere.

Tenacity. Resistance to breakage under the action of stress.

Tensile strength. Longitudinal stress per unit cross section required to break a specimen; measured, e.g., in pounds per square inch.

Thermoplastic. Capable of becoming plastic on heating.

Thermosetting plastic. One that irreversibly hardens under the action of heat.

Titer (titre). Solidification temperature of a mixture of fatty acids and unsaponifiable matter, obtained by saponifying, acidulating, and washing the water-insoluble fraction of fats and oils.

Toxicity. Harmful physiological effects of a substance.

Transition point. Temperature at which two crystalline forms of a substance are in equilibrium with each other.

Triclinic crystal. One having three unequal axes intersecting obliquely.

Unsaponifiables. Residue extracted from the alkaline solution, after complete saponification of a fat, oil or wax.

Vapor pressure. Pressure of any vapor above its liquid or solid form, at the temperature at which equilibrium is established.

Viscosity. Resistance offered by a fluid to flow, resulting from the combined effects of adhesion and cohesion.

Volume resistivity. Reciprocal of conductivity, measured by the resistance of a body of unit cross section and of unit length at 0°C.

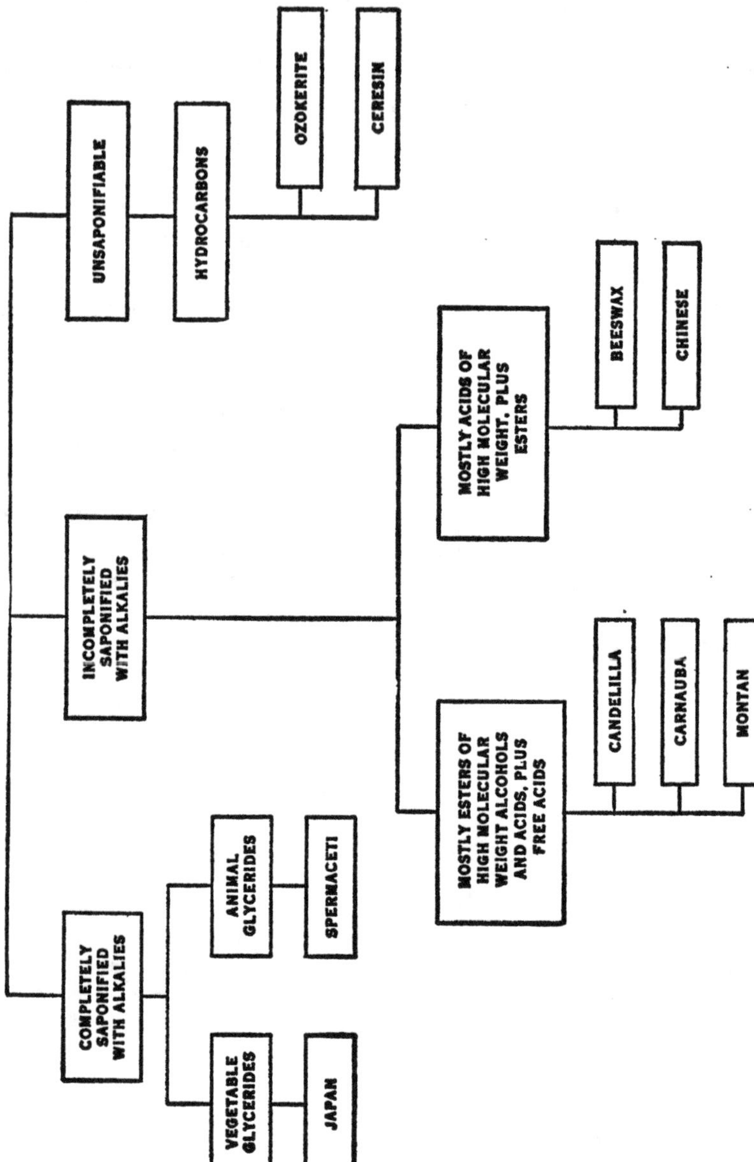

Figure 61. Action of Alkali on Waxes

Table 151

Origin and Use of Waxes

Wax	Type	Source	Locality	Original Color	Hardness	Commercial Grades
Beeswax	insect	*Apis mellifica*	worldwide	light yellow to dark greenish brown	fairly brittle, but plastic when warm	crude; yellow refined: bleached white
Candelilla	vegetable	*Euphorbia antisyphilitica*	Mexico and southern Texas	dark brown, or brownish to greenish	harder than beeswax, not so hard as carnauba	technical
Carnauba	vegetable	*Corypha cerifera*	Brazil and South American countries	sulphur-yellow or yellowish green	very hard and brittle	No. 1 yellow; No. 2 yellow; No. 2 north country; No. 3 north country; No. 3 north country refined; No. 3 chalky
Ceresin	mineral	purified ozokerite from lignite		yellow or white		white; yellow; orange, sold according to melting-point range, 130–133°, 135–138°, 140–145°, 150–160°, 160–170°F.
Insect Wax or Chinese Wax	insect	*Coccus ceriferus*	Western China	white, opaque, fibrous		
Japan	vegetable	*Rhus succedanea, Rhus vernicifera*	China, western provinces of Japan	greenish, pale yellow or light brown	hard and brittle	kitagumi and chickusan
Montan	vegetable	lignite	Saxony, Thuringia	dark brown, brownish black	hard and brittle; conchoidal fracture	crude and bleached
Ozokerite	mineral	lignite	chiefly Galicia, Tcheleken Islands; also in Utah, Austria, Egypt, Yugoslavia, Orange Free State	pure yellow to dark brown, greenish brown, or dark amber	varies from quite soft to as hard as gypsum	crude; natural yellow; white
Spermaceti	animal	*Physeter macrocephalus*		white, crystalline and glistening	very brittle, readily powdered	blocks, cakes, technical, U.S.P.

Table 151 (Continued)

Wax	Containers and Weights in pounds	Special Features	Uses	other names
Beeswax	barrels 300, kegs 100, cases and bags	distinctive odor	manufacture of wax polishes, furniture polishes, wood polishes and finishes; candles; transparent papers; pomades, toilet preparations, cosmetics; food products, chewing gums, sweetmeats; adhesive compositions; dressing and polishing leather; modeling fruits and flowers; base for plasters, etc., engraving, lithography; sizing and finishing textiles; in medicine	
Candelilla	bags 150–200	distinctive odor	manufacture of leather dressing, furniture polish, shoe polish; candles; cements; varnishes; sealing wax; electrical insulating compositions; phonograph records; paper size; celluloid, rubber; waterproofing and insectproofing containers; in paint removers; soft wax stiffener; substitute for carnauba and beeswax	
Carnauba	bags and cases of varying weights	characteristic odor of new-mown hay on warming. Takes a fine hard gloss on rubbing with a cloth	manufacture of shoe and furniture polishes; lacquers, varnishes; phonographic records; carbon-paper coatings; hardening candles; leather finishes; electrical-insulating compositions; waterproofing textiles, wood, etc.; inks and ink compositions; mold lubricant; substitute for beeswax	Brazil wax
Ceresin	bags	plasticity	manufacture of candles; shoe, floor, and leather polishes; antifouling paints; wood polishes and fillers; incandescent gas mantles; paper size; waxed paper; lubricating compositions; waxes; figures and forms; toys, dolls; bottles for hydrofluoric acid; electrical insulations; general size; impregnating and preserving agent; substitute for beeswax in crayons, perfume pastes, pomades, cosmetics, salves, and other pharmaceutical products; matrix compositions in printing; printing forms; rubber mixtures; waterproofing textiles, textile dressing	cerin, cerosin, earth wax, mineral wax, purified ozokerite

Table 151 (Continued)

Wax	Containers and Weights in pounds	Special Features	Uses	Other Names
Insect Wax, or Chinese Wax	bags and barrels	fibrous crystalline structure	manufacture of candles; leather and furniture polishes; treating silk and cotton fabrics; sizing and glazing papers	insect wax, tree wax, vegetable spermaceti, wrongly called Japanese wax
Japan	cases 224, and other sizes	effloresces; tallowy odor	manufacture of furniture, shoe, and floor polishes; candles, wax matches; leather polishing and finishing compositions; substitute for beeswax	Japan tallow, vegetable wax of Japan; sumac wax
Montan	bags of various sizes	lustrous appearance; bituminous odor on friction or warming	manufacture of polishing preparations, paints, varnishes, tarry roofing compositions, waterproofing paints; shoe creams and polishes; phonograph records; insulating compositions; candles; soaps, adhesive pastes; finishing leather goods; increasing hardness of fat compositions; paper sizing; to raise the melting point of softer waxes; substitute for carnauba and beeswax	lignite wax
Ozokerite	bags of various sizes; stands 80–100	plasticity; mixture of hydrocarbons	manufacture of paints, varnishes, wood fillers, wood polishes; candles; lithographic and printing inks; dolls, toys; carbon paper, shoe polishes, incandescent gas mantles; pomades, cosmetics, pharmaceutical ointments; matrices in galvanoplastic work; crayons, waxed paper and cloth, linen and cotton sizing preparations; electrical insulating; leather finishing; lubricant, and sealing-wax compositions; process engraving and lithography; electrotyper's wax; rubber filler; substitute for beeswax, carnauba, ceresin	mineral wax, fossil wax, native paraffin ozacerite
Spermaceti	cases 50–60	odor and taste free when pure	manufacture of emulsions, cerates; candles, soaps, sweetmeats, candies, confectionery; cosmetics, ointments, pomades, toilet preparations; finisning textiles; in medicine	cetaceum

Table 152
SOLUBILITY OF WAXES

Wax	Acetone	Alcohol	Alkalies	Benzol	Carbon Bisulfide	Carbon Tetra-chloride
Beeswax	insoluble cold; soluble hot	soluble hot (76°C.)	saponi-fiable			soluble
Candelilla		soluble hot (63°C.)	saponi-fiable			soluble
Carnauba	insoluble cold; not very soluble hot	soluble hot (82°C.)	saponi-fiable			soluble
Ceresin		insoluble		soluble		
Chinese or Insect Wax		slightly soluble	saponi-fiable	soluble		
Japan	insoluble	soluble hot (76°C.)	saponi-fiable	soluble		soluble
Montan		soluble hot (70°C.)	saponi-fiable	soluble		soluble
Ozokerite		insoluble		soluble	soluble	
Spermaceti	insoluble cold; soluble hot	soluble hot (44°C.)	saponi-fiable		soluble	soluble

Wax	Chloroform	Dioxan	Ether	Hydro-genated Naphtha	Isopropyl Ether	Kerosene
Beeswax	insoluble cold; soluble hot	soluble hot	soluble		soluble	
Candelilla	soluble hot				soluble	
Carnauba	insoluble cold; soluble hot	soluble hot	insoluble cold; soluble hot		soluble	
Ceresin	not completely soluble			soluble	soluble	
Chinese or Insect Wax			slightly soluble		soluble	
Japan	soluble		soluble	soluble	soluble	
Montan	soluble	soluble hot			soluble	
Ozokerite			soluble		soluble	soluble
Spermaceti	soluble		soluble		soluble	

Wax	Naphtha	Toluol	Trichloro-ethylene	Tur-pentine	Xylol	Fusel Oil
Beeswax	insoluble		soluble	soluble		soluble
Candelilla			soluble	soluble		
Carnauba	insoluble cold; soluble hot		soluble			soluble
Ceresin	soluble	soluble	soluble		soluble	
Chinese or Insect Wax		soluble	soluble		soluble	
Japan	soluble	soluble	soluble	soluble	soluble	soluble
Montan	soluble	soluble	soluble		soluble	
Ozokerite		soluble	soluble	soluble	soluble	
Spermaceti	soluble		soluble	soluble		soluble

Table 153
PROPERTIES OF WAXES

Wax	Refractive Index	Specific Gravity	Melting Point °C.	Melting Point °F.	Direct Acid Number
Beeswax	1.4398–1.4451 at 75°C., 167°F.	0.959–0.975	62–70	145–158	16.8–20.6
Candelilla	1.4555 at 71.5°C., 160.7°F.	0.969–0.993	65–69	149–156	16
Carnauba	1.463 at 60°C., 140°F.	0.990–0.999	83–91	184–196	4–8
Ceresin	1.4415–1.4464	0.88–0.92	54–77	130–170	0
Insect Wax, or Chinese Wax		0.926–0.970	65–80	149–176	13
Japan	1.450 at 60°C., 140°F.	0.975–0.984	50–56	122–133	6, usually not over 20
Montan		crude 1.0	crude 76–92 distilled 72–77 refined 77–84	169–198 162–171 171–183	crude 25 distilled 73–85 refined 15–20
Ozokerite	1.4415–1.4464	0.85–0.95	58–100	136–212	0
Spermaceti	1.440 at 60°C., 104°F.	0.905–0.960	41–49	106–120	0.5–2.8, not usually over 0.5

Table 153 (Continued)

Wax	Fatty Acids, %	Saponification Number	Unsaponifiables, %	Ratio Number*	Iodine Number	Acetyl Number
Beeswax	46.77	88–96	52–56	3.6–3.8	8.3–11	15.2
Candelilla	29.0	46–65	74.0	4.7	14.0–36.8	
Carnauba	47.95	78–88	54–55	31	13.5	55
Ceresin	0	< 2	100	0	7–9	
Insect Wax, or Chinese Wax	51.5	80–93	49.5	29.3	1.4–2	
Japan	4–12	217–237	1.1–1.6	11–35	4–15	27–31.2
Montan	distilled 56–64 refined 11–15	crude 58 distilled 75–89 refined 70–80	crude 3.58 distilled 30–45	crude 3–3.5 distilled 0–0.3 refined 3–3.5	crude 16 distilled 10–15	
Ozokerite	0	0	100	0	7.8	
Spermaceti	52–53	121–135	51–54	124	2.6–3.8	2.6

$$* \frac{\text{Saponification No.} - \text{Acid No.}}{\text{Acid No.}}$$

Table 153 (Continued)

Wax	Dielectric Constant			Effective A–C Conductivity (Unit = 10^{-12} ohms cm)			Volume Resistivity (Unit = 10^{-12} ohms cm)		
	Initial ([a])	After Immersion([b])	Final ([c])	Initial ([a])	After Immersion ([b])	Final ([c])	Initial ([a])	After Immersion ([b])	Final ([c])
Beeswax	2.87–2.88	3.11–3.26	2.84–2.90	45–48	70–85	56–58	5–12	0.8–1.4	7–8
Candelilla	2.38–2.49	2.50–2.62	2.45–2.56	6.0–6.4	18	9.3–9.4	>290	>110	>630
Carnauba	2.66–2.83	3.84–4.19	2.82–2.83	4.9–5.3	240–300	74–75	60–120	0.3–4	270
Ceresin	2.16–2.24	2.29–2.32	2.28–2.29	<0.37	10–12	1.1	>290	400–>600	>900
Insect Wax, or Chinese Wax	2.94–2.98	3.62–3.89	2.95–2.97	125–137	189–210	120–123	1–2	0.03–0.16	1–2
Japan	2.96–2.99	3.15–3.20	2.91	9–12	33–34	13	6–7	0.40–0.75	21
Montan	2.62–2.67	2.88–3.05	2.68	29–31	45–55	25	>290	>270	>610
Ozokerite	2.37–2.43	2.50–2.55	2.40–2.47	10–12	19–24	15	30–40	7–9	30–40
Spermaceti	3.60–3.75	6–17	3.74–3.89	6.0–6.6	500–100,000	10–12	3–4	0.0006–0.0003	9–15

([a]) After drying over calcium chloride; ([b]) after 6-month immersion in 3.5% sodium chloride solution; ([c]) after redrying.

Table 154

Wax–Type Acids and Wax–Type Higher Alcohols

Name	Formula	Melting Point °C.	Specific Gravity at 15°C.	Soluble in	Occurrence
Cerotic Acid	$CH_3(CH_2)_{24}COOH$	77.8	0.836 at 79°C.	warm alcohol	free in beeswax, montan wax, carnauba; also as cerotate in insect wax and carnauba
Montanic Acid	$CH_3(CH_2)_{26}COOH$	83.0		methyl alcohol	free in montan wax
Melissic Acid	$CH_3(CH_2)_{27}COOH$	91.0			free in beeswax and montan wax
Palmitic Acid	$C_{16}H_{32}O_2$	62.2	0.846	alcohol; ether	as tripalmitin in Japan wax; as cetyl palmitate in spermaceti; as myricyl palmitate in beeswax
Lauric Acid	$C_{12}H_{24}O_2$	43.5			as laurin in Japan wax
Cetyl Alcohol	$C_{16}H_{33}OH$	50.0	0.810	alcohol; ether; benzol	as cetyl palmitate in spermaceti
Octadecyl Alcohol	$C_{18}H_{37}OH$	59.0		alcohol; ether	spermaceti
Ceryl Alcohol	$C_{26}H_{53}OH$	79.0			as ceryl cerate in Chinese insect wax
Myricyl Alcohol	$C_{31}H_{63}OH$	88.0		alcohol; ether	as myricyl palmitate in beeswax and carnauba
Anonymous Alcohol	$C_{24}H_{48}(OH)_2$	103.0			carnauba wax

Table 155

COMPOSITION AND UNSATURATION OF COMMERCIAL FATTY ACIDS

Average Percentage of Saturated Acids in	Capric, Caproic and Caprylic	Lauric	Myristic	Palmitic	Stearic	Higher Saturated
Coconut Fatty Acid, Distilled	ca. 13.5	ca. 45.5	ca. 20	ca. 15	ca. 5	—
Cottonseed Fatty Acid, Distilled	—	—	—	20–26	2–3	0.5–1.0
Cotton, High-Titre Fatty Acid, Distilled	—	—	ca. 1	40–45	4–5	—
Chinawood Fatty Acid, Split	Total about 2–3%					—
Helioline-D Fatty Acid, Distilled	—	—	—	4.5–9	0.5–1	—
Hempseed Fatty Acid, Distilled	—	—	—	10–12		?
Isoline Fatty Acid, Distilled	Total about 3–4%					—
Lauric Acid (Technical)	ca. 15	ca. 58	ca. 20	ca. 9	ca. 1	—
Linseed Fatty Acid, Distilled	—	—	—	2.7–3.8	5.4–7.7	—
Marine (Raw) Fatty Acid, Split	Total ca. 15–25%					—
Myristic Acid, Distilled	—	ca. 20	ca. 65	ca. 5		—
Palm Fatty Acid, Split	—	—	1–2	35–43	4–5	—
Perilla Fatty Acid, Distilled	Total about 6.5–8.0—mostly palmitic					—
Rapeseed Fatty Acid, Distilled	—	—	0–1.5	1–2		1–3
Ricinoleic Fatty Acid, Split	Total about 3–5%					—
Sardine Fatty Acid, Distilled	Total about 10–20%—mostly palmitic					—
Soybean Fatty Acid, Distilled	—	—	—	6.5–10	4.2–6.3	1–2
Sunflower Fatty Acid, Distilled	—	—	—	3–4	2.5–3.5	1–2

Table 155 (Continued)

Average Percentage of Unsaturated Acids in	Ricin-oleic	Oleic	Linolic	9/11 Octadeca-dienic	Lino-lenic	Elaeo-stearic	Clupan-donic
Coconut Fatty Acid, Distilled	—	ca. 6	3–4	—	—	—	—
Cottonseed Fatty Acid, Distilled	—	35–45	30–40	—	—	—	—
Cotton, High-Titre Fatty Acid, Distilled	—	22–25	25–30	—	—	—	—
Chinawood Fatty Acid, Split	—	5–15	—	—	—	80–85	—
Helioline-D Fatty Acid, Distilled	—	45–55	40–45	—	—	—	—
Hempseed Fatty Acid, Distilled	—	12–15	50–55	—	20–25	—	—
Isoline Fatty Acid, Distilled	—	6–9	40–45	40–45	—	—	—
Lauric Acid (Technical)	—	ca. 4	—	—	—	—	—
Linseed Fatty Acid, Distilled	—	6–12	35–47	—	37–42	—	—
Marine (Raw) Fatty Acid, Split	—	C16 30(?)	?	Containing jecoleic and therapic acid (?)		—	+
Myristic Acid, Distilled	—	ca. 10	—	—	—	—	—
Palm Fatty Acid, Split	—	40–50	8–12	—	—	—	—
Perilla Fatty Acid, Distilled	—	4.0–10.5	33–44	—	44–49	—	—
Rapeseed Fatty Acid, Distilled	—	30–40	10–20	Erucic 45–50	1–4	—	—
Ricinoleic Fatty Acid, Split	84–87	7–10	1–3	—	—	—	—
Sardine Fatty Acid, Distilled	—	?	?	Mostly jecoric (?) acid			+
Soybean Fatty Acid, Distilled	—	22–29	50–55	—	7–10	—	—
Sunflower Fatty Acid, Distilled	—	36–40	52–56	—	—	—	—

Table 155 (Continued)

Constants	Acetyl No.	Degree of Unsaturation				
		Thiocyanide No.	Iodine No. Hanus (1 hour)	Iodine No. Wijs (¾ hour)	Iodine No. Wijs (3 days) R-Rossman Method	Hexabromide No.
Coconut Fatty Acid, Distilled	—	5.5–12	8–13	—	—	—
Cottonseed Fatty Acid, Distilled	—	65–70	90–100	—	—	—
Cotton, High-Titre Fatty Acid, Distilled	—	40–50	60–75	—	—	—
Chinawood Fatty Acid, Split	—	85–90	—	<160	ca. 230–R 240–R	—
Helioline-D Fatty Acid, Distilled	—	80–88	120–140	—	—	—
Hempseed Fatty Acid, Distilled	—	85–100	155–165	—	—	4–17
Isoline Fatty Acid, Distilled	—	?	155–165	135–145	170–178–W	—
Lauric Acid (Technical)	—	2–4(?)	4–6	—	—	—
Linseed Fatty Acid, Distilled	—	Below 120	184–190	—	198–W	ca. 40
Marine (Raw) Fatty Acid, Split	?	105–110	160–170	—	—	?
Myristic Acid, Distilled	—	3–6(?)	5–10	—	—	—
Palm Fatty Acid, Split	—	40–50	45–60	—	—	—
Perilla Fatty Acid, Distilled	—	>130	185–200	—	—	>60
Rapeseed Fatty Acid, Distilled	—	80–83	100–109	—	—	4–7
Ricinoleic Fatty Acid, Split	153–156	86–93	87–90	—	—	—
Sardine Fatty Acid, Distilled	?	?	110–130	—	—	?
Soybean Fatty Acid, Distilled	—	80–85	128–138	—	—	5–11
Sunflower Fatty Acid, Distilled	—	78–88	125–135	—	—	—

Table 156

AVERAGE FATTY ACID CONTENT AND CONSTANTS OF FATS AND OILS

	CHEMICAL FORMULA	COCONUT	PALM KERNEL	TALLOW	LARD	PALM	OLIVE	COTTONSEED	CORN	PEANUT	SOYBEAN	SUNFLOWER	LINSEED	PERILLA	CASTOR	TUNG	WHALE	MENHADEN	SARDINE	HERRING
SATURATED ACIDS																				
CAPROIC	$C_6H_{12}O_2$	0.2	Trace																	
CAPRYLIC	$C_8H_{16}O_2$	8.0	3.0																	
CAPRIC	$C_{10}H_{20}O_2$	7.0	6.0																	
LAURIC	$C_{12}H_{24}O_2$	48.0	50.0																	
MYRISTIC	$C_{14}H_{28}O_2$	17.5	15.0	2.0	1.0	1.0	Trace	0.5									8.0	7.0	5.0	7.0
PALMITIC	$C_{16}H_{32}O_2$	8.8	7.5	30.0	28.0	42.5	9.0	21.0	7.5	7.0	6.5	3.5	5.0	7.5		4.0	11.0	16.0	14.0	8.0
STEARIC	$C_{18}H_{36}O_2$	2.0	1.5	21.0	11.5	4.0	2.3	2.0	3.5	5.0	4.5	3.0	3.5	Trace	2.0	1.5	2.5	1.0	3.0	Trace
ARACHIDIC	$C_{20}H_{40}O_2$						0.2	Trace	0.5	4.0	0.7	0.6	Trace							
BEHENIC	$C_{22}H_{44}O_2$																			
LIGNOCERIC	$C_{24}H_{48}O_2$					Trace			0.2	3.0	Trace	0.4								
UNSATURATED ACIDS																				
MYRISTOLEIC	$C_{14}H_{26}O_2$																1.5	Trace	Trace	Trace
PALMITOLEIC	$C_{16}H_{30}O_2$																17.0	17.0	12.0	18.0
OLEIC	$C_{18}H_{34}O_2$	6.0	16.0	45.0	58.0	43.0	82.5	33.0	46.3	60.0	33.5	34.0	5.0	8.0	8.6	15.0	34.0	27.0	10.0	9.0
LINOLEIC	$C_{18}H_{32}O_2$	2.5	1.0	2.0	3.5	9.5	6.0	43.5	42.0	21.0	52.5	58.5	61.5	38.0	3.5		9.0	Trace	15.0	13.0
LINOLENIC	$C_{18}H_{30}O_2$										2.3		25.0	46.5			Trace			
ELAEOSTEARIC	$C_{18}H_{30}O_2$															79.5				
RICINOLEIC	$C_{18}H_{34}O_3$														85.9					
C_{20} UNSATURATED	$C_{20}H_{2(20-n)}O_2$																5.0	20.0	22.0	20.0
C_{22} UNSATURATED	$C_{22}H_{2(22-n)}O_2$																12.0	12.0	19.0	25.0
CONSTANTS																				
SAPONIFICATION VALUE		261-264	240-250	196-200	195-200	196-206	185-200	192-200	188-193	185-192	189-194	189-194	189-196	188-197	175-183	189-195	185-195	189-193	189-193	179-194
IODINE NUMBER		8-10	16-23	35-44	50-69	48-58	74-94	100-115	116-130	83-95	124-148	120-136	179-204	185-206	82-86	160-170	110-136	148-185	170-190	130-144
TITER—°C		20-23	20-23	37-46	36-43	38-47	18-25	32-38	18-20	28-32	20-21	17-20	19-21	12-17	—	37-38	22-24	31-33	28-34	23-27

Every substance is unique in its chemical and physical properties. Therefore, no wax can replace another equally in all its characteristics. The following list which should be studied carefully gives most of the factors which must be considered in searching for a suitable substitute or alternative. Only those properties which are absolutely essential should be demanded, otherwise the search for a substitute will be greatly hampered, if not made futile.

Any substitute must be tried and tested before commercial use is attempted. Such tests should be made by a competent worker or consultant to avoid subsequent difficulties.

SUBSTITUTES

Availability
Homogeneity
Uniform Replacement
Grade or Purity
Form
Optical Properties
Odor
Taste
pH
Specific Gravity
Viscosity
Gelling or Thixotropic Tendencies
Slipperiness
Drying Qualities
Feel or Hand
Hygroscopicity or Efflorescence
Inflammability
Explosiveness
Toxicity
Effect on Skin, Hair or Fingernails
Edibility
Adsorption and Absorption
Heating Power
Thermal Changes
Effects of Pressure
Solvency
Plasticizing or Flexibilizing Properties
Melting Point and Titer
Boiling Point
Solubility
Hardness
Tenacity or Cohesion
Plasticity and Ductility
Flexibility
Elasticity
Length
Stickiness
Adhesion
Interaction With Other Materials
Stability
Bacterial Content
Emulsifiability
Surface Tension
Dispersing Properties
Electrical Properties
Thermal Conductivity

Lighting Power (Candle Power)
Handling and Working
Legal Restrictions

Wax	*Substitutes*
Bayberry Wax	Diethylene glycol distearate Japan wax
Beeswax	Ceresin with soap Coffee wax Flax wax "Flexo Wax" "Glaurin" "Isco 662, 663" "Norco Wax 36" "Prostearin" Sugar-cane wax
Cacao Butter	Borneo tallow Cetyl alcohol with mineral or vegetable oil Partially hydrogenated vegetable oils "Acrawax"
Carnauba Wax	Candelilla wax Green cotton wax Esparto wax Hydrogenated castor oil "Norcowax 350" Ouricury wax "Santowax M" Substituted stearamides "Stroba Wax" Sugar-cane wax
Cetyl Alcohol	Lanolin alcohols Monostearin Oleyl alcohol Stearyl alcohol

Esparto Wax	Carnauba wax
Flax Wax	Beeswax
Japan Wax	Bayberry wax "Hydrofol Wax No. 1" "Hydrofol Glycerides No. 50" "Isco Japan Wax Substitute" "Norco Wax 110"
Montan Wax	Lignite wax "Monten Wax" "Norco Wax 12A" Peat wax "Santowax"
Ozokerite	Carnauba with amorphous paraffin wax "Deetee Wax"
Paraffin Wax	Aluminum stearate Diglycol stearate Glyceryl tristearate Graphite Naphthalene Stearic acid Stearin
Spermaceti	Cetyl palmitate
Stearic Acid	Abietic acid Fatty acids Lauric acid Myristic acid Naphthenic acid Oxidized paraffin wax Palmitic acid Paraffin wax Rosin Tallow

Stearin	Glyceryl tristearate Paraffin wax Zinc stearate
Stearyl Alcohol	Cetyl alcohol
Sugar-Cane Wax	Beeswax
Yellow Wax	See Beeswax "Flexowax C" Soap with rosin

CHAPTER 7

wax formulary

In the formulas of various industrial products included in this chapter all types and quantities of waxes are used.

These formulas have been collected from volumes 1–9 of *The Chemical Formulary* edited by the author,* from the author's laboratory records, and from the *New Practical Formulary*.**

These formulas are given, without any warranty or liability, to show the almost infinite variety of wax applications and to serve as a primer for the formulation of useful commercial products.

ADHESIVES

Adhesive for Masks

Beeswax	52
Lanolin	24
Venice Turpentine	15
Castor Oil	9

Adhesive for Tissue Paper to Aluminum

For geological survey and other types of map work, base maps are very often printed on thin sheets of aluminum. Topographic maps of this type require the names of rivers, towns, etc., to be printed on thin tissue paper which is then fastened to the base map at the proper spot by an adhesive. It is very important in work of this kind that the tissue paper does not tear off, become wrinkled or discolored, so that photoreproduction is not made difficult or impossible.

* Published by the Chemical Publishing Co., Inc., New York, N. Y.

** Author: Mitchell Freman; publisher: Chemical Publishing Co., Inc., New York, N. Y.

Flexo Wax C Light*	40 g
Turpentine	57 cc

The wax is melted in a porcelain dish, the turpentine is added and the solution mixed well and bottled. It solidifies on cooling, forming a tacky wax which can be remelted for use.

Adhesive for Wigs

Dammar	20
Rosin	20
Beeswax	40
Venice Turpentine	20

Heat to 90°C. and stir until uniform; cast in sticks.

Adhesive Sealing Compound

(Universal)

Mix hot beeswax and Venice turpentine 1 to 1. Proportions may be varied according to needs and the mixture can be colored if desired. This is very good to attach glass to iron or wood, temporarily.

Adhesive to Join Metal to Glass or Wood

Formula No. 1

Black Pitch	3
Rosin	10
Yellow Wax	5
Powdered Brick	3

No. 2

Tar Resin	85
Plaster	10
Venice Turpentine	8

Adhesive Wax

Rosin	100
Paraffin Wax	10
Thin Mineral Oil	88

Adhesives for Tapes

Formula No. 1

Manila DBB	40
Solox (Anhydrous Alcohol)	65
Butanol	65
Paraffin Wax	25
Oleic Acid	6
Triethanolamine	2

Dissolve the manila DBB in solox and butanol. Heat this solution to 85°C. and hold it at that temperature throughout the process.

Heat the wax and oleic acid until they are molten and homogeneous. Add the triethanolamine with stirring. Add this mixture to the warm solution of manila DBB slowly and with continued agitation. A paste forms.

No. 2

Manila DBB	40
Solox (Anhydrous Alcohol)	65
Butanol	65
Ozokerite	25
Oleic Acid	6
Triethanolamine	2

Follow the same procedure as for formula No. 1.

* Glyco Products Co.

No. 3

Manila DBB	40
Solox	65
Butanol	65
Carnauba Wax	25
Oleic Acid	6
Triethanolamine	2

See procedure under formula No. 1.

No. 4

Manila DBB	40
Solox (Anhydrous Alcohol)	65
Butanol	65
Japan Wax	25
Oleic Acid	6
Triethanolamine	2

See procedure under formula No. 1.

No. 5

Manila DBB	40
Solox (Anhydrous Alcohol)	65
Butanol	65
Beeswax	25
Oleic Acid	6
Triethanolamine	2

See procedure under formula No. 1.

No. 6

Manila DBB	40
Solox (Anhydrous Alcohol)	65
Butanol	65
Crude Montan Wax	25
Oleic Acid	6
Triethanolamine	2

See procedure under formula No. 1.

No. 7

Manila DBB	20
Solox (Anhydrous Alcohol)	15
Butanol	15
Montan Wax	12½
Toluol	37½

Dissolve the manila DBB in the solox and butanol. Dissolve the montan wax in the toluol. Add the wax solution to the manila solution.

No. 8

Manila DBB	20
Solox (Anhydrous Alcohol)	15
Butanol	15
Japan Wax	12½
Toluol	37½

See procedure under formula No. 7.

Boat-Calking Wax

Formula No. 1

Yellow Beeswax	20
Beef Tallow	25
Lard	40

For technical purposes, B.Z. Wax A or other synthetic waxes may be substituted for beeswax.

No. 2

Beeswax	4
White Lead Paste (in Linseed Oil)	5

"Boltwood Wax"

(For Cementing Physical Instruments)

Shellac	40
Rosin	72
Venice Turpentine	8
Beeswax	60
Dry Talc	16
Dry Tin Oxide	16

Melt the rosin and the shellac. Heat to 200°C., then add the Venice turpentine and beeswax. Heat the mixture with stirring until it ignites spontaneously. Let it burn until the total mass has shrunk to about 40% of its original weight, then add the talc and tin oxide. This gives a tough, smooth, waxy cement easy to apply to delicate instruments.

Bottle-Sealing Wax

Formula No. 1

(First Quality)

Colophony (Rosin)	38
Thick Turpentine	20
Paraffin in Blocks (50–52°C.)	4
Whiting	10
Barite	10
Zinc Oxide	18

No. 2

(Best Quality; Insoluble in Alcohol)

Carnauba Wax	30
Beeswax	20
Paraffin in Blocks (50–52°C.)	20
Whiting	4
Barite	10
Chrome Orange	16

No. 3

(Second Quality)

Colophony	30
Thick Turpentine	22
Scale Paraffin Wax	6
Whiting	12
Barite	12
Chrome Yellow	18

No. 4

(Second Quality, Insoluble in Alcohol)

Montan Wax, Crude or Bleached	40
Japan Wax	10
Scale Paraffin Wax	15
Whiting	9
Barite	10
Zinc Oxide	16

No. 5

(Extraflexible; Insoluble in Alcohol)

Montan Wax	8
Japan Wax	20
Scale Paraffin Wax	52
Lithopone	20

No. 6

Scale Paraffin	70
Beeswax	10
Japan Wax	17
Oil-Soluble Dye	3

No. 7

Beeswax	5
Carnauba Wax	1
Paraffin	1
Minium	5
Whiting	2

No. 8

Rosin	5
Ceresin	10
Paraffin Wax	5
Japan Wax	10
Beeswax	30
Venice Turpentine	2

Brass-to-Glass Cement

Red Ocher	4
Rosin	30
Beeswax	6
Plaster of Paris	3

Brewers' Pitch

a. Rosin	160
Pale Rosin Oil	30
b. Rosin	168
Paraffin Wax	22
Linseed Oil	10

Cement for Glass or Metal

The following composition has a very low viscosity when melted and can be pressed into a thin film which will have excellent adherence when cold:

Piccolyte Resin*	5
Paraffin Wax	95

Cement for Stone, Porcelain, or Glass and Leather

Crude Rubber	9.1
Heavy Benzine	45.5
Japan Wax	13.6
Rosin	31.8

Glass-to-Metal Adhesive

Bleached Beeswax	2
Rosin	4
Venice Turpentine	1
English Red	4

Melt the wax and rosin and stir in the English red.

Heat- or Pressure-Sensitive Adhesive

Paraffin Wax	8
Vistanex U**	12
Balata Gum	14
Piccolyte Resin*	16
Acrawax C†	1

* Penn. Industrial Chem. Corp.
** Advance Solvents and Chem. Corp.
† Glyco Products Co.

Melt together and apply at about 325°F. This is not tacky at ordinary temperatures.

Heat-Sealing Adhesive

Low-Viscosity Ethyl Cellulose	20.0
Batavia Dammar A/E	40.0
Paraffin Wax	2.0
Toluol	116.0
Anhydrous Alcohol	14.5
Butyl Acetate	14.5

This is a label adhesive which dries on the labels yet is sufficiently thermoplastic to permit application to carton with a hot iron.

Laminating Adhesive for Glassine or Cellophane

Formula No. 1

Chlorinated Rubber	40–50
Resin	25–38
Paraffin Wax	2– 7
Plasticizer (e.g., Dibutyl Phthalate)	15–21

The resin particularly suitable for this purpose consists of 20–25% glycerol abietate and 5–13% β-coumarone. The paraffin wax preferably has a melting point of 143–145°F.

No. 2

Chlorinated Rubber (65–68% Chlorine)	46.0
Glycerol Abietate	19.0
Paraffin Wax	6.1

Dibutyl Phthalate	16.9
β-Coumarone	12.0

These ingredients are mixed and then dissolved in any suitable hydrocarbon, such as benzol, toluol, xylol, etc., to give a solution of the desired viscosity.

Linoleum and Tile Cement

1. Sika Seal*	82
2. Paraffin	9
3. Glycerin	9

Heat 1 and 3 to 80°C. and add 2, which has been melted, slowly, with vigorous stirring until emulsified.

Metal-to-China Putty

Gutta-Percha	2
Beeswax	1
Sealing Wax	3

Fuse the ingredients and warm the joints before applying the putty.

Odorless Barrel Sealer

Formula No. 1

High-Viscosity Petrolatum	70
Paraffin Wax (50–52°C.)	18
Beeswax	12

No. 2

Paraffin Wax (42–44°C.) or Petrolatum	98
Hard-Fat Tallow or Crude Palm Oil	2

* Sika Chem. Corp.

No. 3

Rosin	66
Paraffin Wax	30
Beeswax	3
Carnauba Wax	1

No. 4

Rosin	55
Paraffin Wax	43
Beeswax	2

No. 5

Rosin	80
Paraffin Wax	20

No. 6

Shellac	4
Dammar Resin	5
Alcohol	90
Castor Oil	1

This cannot be used for wine barrels because the denatured alcohol would spoil the taste and odor of the wine.

No. 7

Rosin (F-G)	46
Paraffin Wax (50–52°C.)	48
Beeswax	4
Castor Oil	2

Melt at not more than 100° C., adding the castor oil last, at about 70°C. Pour into forms floating on a cold water bath to avoid sticking to the forms. When a solid surface has formed, dip into the water to get quick cooling.

No. 8

Paraffin Wax (50–52°C.)	25.0
Linseed Oil	74.0

Oil-Soluble Black	0.8
Sodium Chlorobenzoate	0.2

Melt the paraffin, add the black and stir at about 85–90°C. until a clear solution has formed. Add the linseed oil in a thin jet and finally add the preservative.

Pour into cans near titer, covering the first by a second, quickly cooling layer.

Optical Glass-to-Frame Cement

Beeswax	120 g
Rosin	176 g
Castor Oil	60 cc

Warm together and mix until uniform. Melt before using.

Optical-Lens Cement

Canada Balsam	1
Beeswax	8
Red Mineral Jelly	4

Pliofilm Adhesive

Pliolite Resin*	100
Paraffin Wax (M.P. 150°F.)	10
High-Flash Naphtha	125
Carbon Tetrachloride	125

Mix the resin and the paraffin with about half the solvent in a dough mixer till smooth, then mix in the remainder of the solvent.

Alternatively, mix all the ingredients together in a closed-drum- or barrel-type agitator and keep in motion until the mix is smooth.

Preserve-Jar Sealing Wax

It washes off easily with hot water.

Paraffin Wax	35
Trihydroxyethylamine Stearate	3

Sealing Wax

Formula No. 1

Shellac	33
I. G. Wax (O. P.) **	5
Venice Turpentine	20
Talc	10
Pigment	25
Tricresyl Phosphate	5
Balsam	2

No. 2

Carnauba Wax	40
Paraffin Wax	20
Magnesia	15
Bronze Powder	25

No. 3

Carnauba Wax	30
Beeswax	20
Paraffin Wax	20
Whiting	4
Barium Carbonate	10
Pigment	16

No. 4

Rosin	30
Venice Turpentine	22
Paraffin Wax	6
Whiting	12
Barium Carbonate	12
Pigment	18

* Goodyear Tire and Rubber Co.
** General Dyestuffs Corp.

No. 5

Montan Wax	40
Japan Wax	10
Paraffin Wax	15
Whiting	9
Barium Carbonate	10
Pigment	16

No. 6

Paraffin Wax	70
Beeswax	10
Japan Wax	17
Aniline Dye	3

No. 7

Montan Wax	40
Japan Wax	10
Paraffin Wax	15
Whiting	9
Barium Carbonate	10
Pigment	16

No. 8

Rosin	34
Venice Turpentine	24
Paraffin Wax	6
Whiting	10
Barium Carbonate	10
Pigment	16

No. 9

Montan Wax	8
Japan Wax	20
Paraffin Wax	52
Pigment	20

No. 10

Rosin	30
Venice Turpentine	22
Paraffin Wax	6
Whiting	12
Barium Carbonate	12
Pigment	18

Shellac-Wax Cement for Laboratory Use

Rosin	35
Shellac	20
Beeswax	15
Fibrous Talc	0 to 30

Melt the rosin in a 6-inch hemispherical iron pan, add the shellac and beeswax with stirring, and heat the mixture over a hot gas flame, so that the temperature reaches 360°C. in about 6 minutes. At once, allow to cool spontaneously to about 275°C., add the talc, cool to about 90°C., and stir well as the mixture becomes viscous. Molds must be scrubbed thoroughly with household washing powder and water, rinsed, and then coated with very thick, smooth dextrine-water paste. When the cement mixture, well stirred, is almost too viscous too flow, pour into molds. The talc gives a broad apparent melting range; that is, when as much as 30 g talc is used, the cement, when heated, gradually softens to a highly viscous condition which is suitable where a sort of plastering is desired. If the time of heating of the mixture is too short, the cement is brittle; if too long, an excessive amount (over 20 g) is lost in the process.

Sticky Wax

Rosin	100
Talc	16
Lanolin	60
Paraffin	8
Saponified Wax	2

Melt together and, while stirring rapidly, add slowly a boiling caustic soda solution (10°Bé.). Stir until uniform.

Stoneware Cement

Paraffin Wax	30
Rosin	20
White Sand	25
Grainy Emery	15
Tripoli	15

Apply hot.

Vacuum-Holding Wax Adhesive

Beeswax	10
Rosin	10

Waterproof Adhesive

Vegetable Protein Gluc	100
Water	325
Wax Solution*	3

Wax Furniture Filler

Japan Wax	30
Paraffin Wax	10
Neutral Wool Fat	10
Rosin	40

Wax Putty

Beeswax	4
Oleostearin	2
Turpentine	1
Venice Turpentine	6

Wax Wood Filler

Paraffin Wax	50
Montan Wax	10
Rosin	40

Melt together.

Color the mixture either with mineral colors or coal-tar dyes.

Wine- or Liquor-Barrel Wax

Tallow	24
Paraffin	50
Japan Wax	5
Beeswax	5
Venice Turpentine	4
Rosin Oil	1
Talc	10

Wood Filler

Beeswax	10
Rosin	10
Sawdust	To suit

Melt the beeswax and rosin and stir in enough sawdust to obtain a hard mass on cooling.

AGRICULTURAL AND GARDEN SPECIALTIES

Bug Catcher

Rosin	60
Castor Oil	35
Ceresin	5

Caterpillar-Catching Bands

Formula No. 1

a. Smoked Sheets (Crude Rubber)	3

---

* Consists of:

Carbon Bisulfide	8
Carbon Tetrachloride	8
Paraffin Wax	1

Rapeseed Oil	17
Spindle Oil	40
b. Rosin	35
Crude Montan Wax	5

Soak and peptize the rubber in the oils *a*, heat the solution and dissolve *b* in it. Stir till heavy and apply.

No. 2

a. Rosin	350
Yellow Beeswax	10
Pitch	60
Linseed Oil	25
b. Heavy Turpentine	15

Melt *a* and thin with *b* in the absence of open flames.

No. 3

Rosin	30
Linseed Oil* Varnish	20
Yellow Beeswax	2

Place the melted and well-mixed glue on the bark of the tree; over it put a ring of cloth, fastened with wire, then place over that again a layer of glue, all around the stock.

Fly-Catcher Composition

Formula No. 1

a. Medium Dark or Light Rosin	6.60
b. Yellow Beeswax	0.30
c. Crude Rubber	5.00
Spindle Oil	4.75

Melt *a*, add *b*, and stir until it has melted up and mixed in. Avoid overheating.

Meanwhile, prepare *c* by heating the oil to 100°C., and adding the rubber in small portions, until all of it has completely dispersed. Add this rubber solution to the melted *a* and *b*.

No. 2

Yellow Beeswax	2
Rosin	60
Rapeseed Oil	38

No. 3

Refined Pine Resin	735
Ceresin	10
Nondrying Fatty Oil or Mineral Oil	255

Perfume: wax odor, honey odor or apple odor.

No. 4

Rosin	400
Pitch	200
Yellow Beeswax	20
Rapeseed Oil	250

Grafting Adhesives

The following formulas are for protecting the cuts made by grafting from the action of air, sun, and humidity. The adhesives used hot are more economical than those used cold, but somewhat more difficult to apply.

Formula No. 1

Black Pitch	125
Rosin	500
Yellow Wax	120
Tallow	110
Sifted Wood Ash	Sufficient

* Or rapeseed oil, or wool fat, when a longer catching period is desired.

No. 2

Black Pitch	50
Yellow Wax	10
Linseed Oil	20

This does not flow in the sun and can be applied in any weather.

No. 3

Neutral Wool Fat	22
Rosin	40
Ceresin (58–60°C.)	10
Beeswax	10
Rosin Oil	18

No. 4

Yellow Wax	12
Tallow	56
Rosin	12
Chalk	20
Castor Oil	Sufficient

Melt the wax, tallow, and rosin; stir in the chalk; and thin down with castor oil to the required consistency.

No. 5

Rosin	25
Japan Wax	10
Beeswax	25
Tallow	15
Venice Turpentine	25

No. 6

a. Rosin	350
Beeswax	10
Pitch	60
Linseed Oil	25
Venice Turpentine	15
b. Methanol	85

Melt up *a,* stir until cool, then add *b.*

No. 7

a. Linseed Oil	1
Turpentine Oil	4
Beeswax	3
Rosin	9
b. Methanol	1

Dissolve *a* cautiously; then thin with *b.*

No. 8

Castor Oil	¼
Rosin	5
Beeswax	1
Charcoal	¾
Glyceryl Monostearate	1

Melt and apply with brush. This excludes air and fungi; prevents drying out; and does not injure live tissues.

No. 9

Beeswax	1 lb
Linseed Oil	¼ pt
Rosin	5 lb
Powdered Charcoal	½ lb

No. 10

Beeswax	1 lb
Linseed Oil	½ pt
Rosin	5 lb
Fuller's Earth	½ lb

No. 11

Lanolin	22
Rosin	44
Ceresin	13
Beeswax	8
Japan Wax	2
Rosin Oil	9
Pine Oil	1

No. 12

Abietic Acid	5 lb
Crude Beeswax	1 lb
Linseed Oil	¼ pt
Powdered Charcoal	½ lb

No. 13

Rosin	5 lb
Crude Beeswax	1 lb
Linseed Oil	¼ pt
Powdered Charcoal	½ lb

Melt together and mix until uniform.

Insect Tree Bands

Rolls of corrugated paper are saturated with the following and wrapped around trees.

Mineral Oil	1½ lb
α-Naphthylamine	1 lb
Paraffin Wax	4 oz

Liquid Grafting Wax

Formula No. 1

Rosin	16 lb
Tallow	2 lb
Isopropyl Alcohol	3 qt
Turpentine	1 pt

Melt the rosin and tallow together, cool sufficiently, and dilute with the alcohol and turpentine. Keep tightly stoppered.

No. 2

Scale Wax	40
Paraffin Wax (135°F.)	47
Microcrystalline Wax (170°F.)	8
Anhydrous Lanolin	5

A portable lantern is available which will keep a small pot of this wax in a "just melted" condition. The wax must not be so hot as to injure plant tissues. The hardness of the wax can be controlled by adjusting the ratio of soft paraffin to hard paraffin.

Plant-Growth Regulator

Formula No. 1

Indole Butyric Acid	0.2 g
Wax Blend No. 1*	20.0 g
Pharmagel B**	2.0 g
Sorbitol	0.5 g
Sodium Bicarbonate	0.5 g
Water	80.0 ml

No. 2

Indole Butyric Acid	0.2 g
Wax Blend No. 1*	20.0 g
Polyvinyl Alcohol (RH–403)	2.0 g
Sorbitol	0.5 g
Sodium Bicarbonate	0.2 g
Water	80.0 g

Wax Coating for Fruits

Formula No. 1

Carnauba Wax	30.0 g
Oleic Acid	3.3 g
Caustic Soda	0.5 g
Triethanolamine	2.7 g
Borax Shellac Solution	60.0 cc
Water	260.0 g
Bentonite	0.3–2.6 g

No. 2

Carnauba Wax	30.0 g
Oleic Acid	3.3 g
Caustic Soda	0.5 g
Triethanolamine	2.7 g
Borax Shellac Solution	60.0 cc
Water	260.0 g
Gelatin	0.3–1.3 g

No. 3

Carnauba Wax	30.0 g
Oleic Acid	3.3 g
Caustic Soda	0.5 g
Triethanolamine	2.7 g
Borax Shellac Solution	60.0 cc
Water	260.0 g
Bleached Irish Moss	0.2–0.65 g

* Wax Blend No. 1.
** Swift and Co.

No. 4

Paraffin Wax	168.0
Beeswax	42.0
Oleic Acid	22.0
Sodium Bicarbonate	6.6
Sodium Chloride	2.2
Water	599.2

The soap must be formed simultaneously with the first stage of emulsification by the addition of the alkali, in a minimum amount of water, to the wax. These emulsions will readily coat fruits and vegetables with nonwaxy surfaces; but spreading agents are necessary when they are used for waxy fruits. Alkyl naphthalene-sulfonates are suitable spreading agents and also inhibit breaking of the emulsions.

BUILDING MATERIALS

Acidproofing Creamery Floors

Paraffin Wax (150°F.)	4
Turpentine	1
Toluol	16

Warm and stir until uniform. Pour into cans and allow to set. Spread on the floor and allow to penetrate for 24 hours. At the end of this time drive the residual layer into the concrete by heat. Do not use a free flame to avoid fire hazards; hot irons will be found safe and effective in forcing the paraffin into the pores and capillaries of the finish for some distance below the surface.

After treatment, give the floor a good waxing with any standard floor wax. As the wax film is worn away through use, replace it by a fresh coating using a polishing machine. This method of acidproofing creamery floors will not change the color of the finish appreciably.

Cement Preservative

Chinawood Fatty Acids	10 lb
Paraffin Wax	10 lb
Kerosene	40 gal

Soapstone-Table Preservative

Boiled Linseed Oil	1 gal
Turpentine	1 gal
Ceresin	4 oz

Apply hot.

Wax Release for Asbestos-Cement Pipe Molding

Stroba Wax*	7
Albacer*	7
Stoddard Solvent	186

Mix while heating until dissolved. Stir while cooling. Brush or spray on the mold before casting the pipe.

* Glyco Products Co.

CARBON PAPER, INKS AND CRAYONS

Benday Ink

This is really a form of transfer ink used with shading mediums (stippling films) in the process known as benday-ing. It consists of:

Tallow	1
Wax	3
Shellac	4
Soap	2
Asphaltum	4
Canada Balsam	1
Carbon Black	1

For use, reduce with castor oil to a working consistency.

Bottom and Shank Blacking

No. 1 Carnauba Wax (Cut with 46 oz Castile Soap in 11½ gal Water)	23
Garnet Shellac (Cut with 100 oz Borax in 13 gal Water)	26
Nigrosine (In 10½ gal Water)	21

Mix the three solutions and add water to make 50 gal.

Burnishing Ink

Formula No. 1

Crude Montan Wax	6½
Rosin	1
Stearic Acid (Titer 53/54°C.)	½
Potassium Carbonate	1
Water	100
Nigrosine Solution (12% in Water)	30

The wax, rosin, and stearic acid are melted together and added to the boiling potash carbonate solution with stirring. The finished cooled emulsion is dyed with the nigrosine solution.

No. 2

a.	Potash Carbonate	0.5
	Rosin	2.0
	Paraffin Wax	3.0
	Marseilles Soap	4.0
	Crude Montan Wax	15.0
	Water	65.0
b.	Borax	7.0
	Shellac	20.0
	Water	75.0
c.	Water-Soluble Nigrosine	6.0
	Water	25.0

Mix the three solutions after making them up separately.

Burnishing Wax

No. 3 North Country Carnauba Wax	9
Shellac Wax	4
Ozokerite Wax	1½
Montan Wax	1
Rosin	½
Paraffin Wax	½
Liquid Oil Black	3

Melt the waxes and add the liquid oil black and mold.

Carbon-Paper Inks

Formula No. 1

Black Carbon (C.P.A.)

Carnauba Wax	40
Mineral Oil	40
Carbon Black	10
Iron Blue	10

No. 2
Black Carbon

Carnauba Wax	30
Methyl Violet Base	2
Oleic Acid	8
Beeswax	3
Carbon Black	15
Lard Oil	42

No. 3
Black Universal Carbon

Extraheavy Mineral Oil	40
Oleic Acid	5
Carnauba Wax	37
Crystal Violet	3
Carbon Black	15

No. 4
Black Pencil Carbon

Mineral Oil	18
Oleic Acid	23
Rosin	7
Montan Wax	30
Lampblack	12
Milori Blue	10

No. 5
Blue Pencil Carbon

Carnauba Wax	30
Paraffin Wax	10
Milori Blue	25
Petrolatum	35

No. 6
Blue Pencil (Soluble Type)

Victoria Blue	10
Oleic Acid	30
Ceresin Wax	5
Carnauba Wax	25
Mineral Oil	30

No. 7
Hectograph Carbon

Crystal Violet	60
Carnauba Wax	9
Opalwax*	10
Mineral Oil	20
Lecithin	1

* Du Pont.

No. 8
One-Time Ink

Montan Wax	30
Paraffin Wax	5
Methyl Violet	1
Lampblack	15
Mineral Oil	49

No. 9

Carbon Black	20
Milori Blue	6
Violet Blue	5
Paraffin Oil	32
Crude Montan Wax	15
Double Bleached Montan Wax	3
Paraffin Wax (50/52°C.)	3

Violet

Methyl-Violet Base	5
Oleic Acid	10
Crude Montan Wax	60
Petrolatum	90

No. 10
Black

Use 75% of the following ingredients in proportions suitable for the grade desired.

Candelilla Wax	
Beeswax	
Crude Montan Wax	
Mineral Oil	
Oil-Soluble Toners	10%
Peerless Carbon Black	15%

Grind hot. It is a base formula which may be modified to suit conditions.

No. 11
Lead-Pencil Carbon

Carbon Black	10
Mineral Oil	40

Grind together until smooth. Add

Nigrosine Base	1⅔

Dissolved in

Oleic Acid	3⅓

Mix well and add a hot solution of

Paraffin Wax	40
Mineral Oil	30

Grind again and spread on paper at 40–50°C.

No. 12
Pencil

Glyceryl Monostearate	6
Oleic Acid	20
Oil-Soluble Dye	10
Petrolatum	14
Kerosene	7
Paraffin Wax	85
Carnauba Wax	8
Montan Wax	30

No. 13
Typewriter

Oleic Acid	11
Diglycol Oleate	2
Oil-Soluble Dye	3
Pigment	17
Petrolatum	38
Carnauba Wax	30

No. 14
Black

Petroleum Jelly	21 lb
Japan Wax	6 lb
Drop Black (Ground in Turpentine)	4 lb
Ceylon Graphite	4 lb
Soft Soap	1 lb
Oil Black Dye	2 oz

Melt the petroleum jelly, wax, and soap together, remove from the fire, and stir in the oil black when cooler, then grind to a fine paste with the others. Apply to paper either hot or cold with a fairly stiff brush; then wipe off the superfluous paste and hang up the papers with clips to dry.

No. 15

Raw Linseed Oil	2 gal
Tallow	8 lb
Lampblack	4 lb
Ceylon Graphite	2½ lb
Oil Black Dye	2½ oz

Warm with oil sufficiently to dissolve the tallow, stir in the oil black and mix well with the other ingredients.

Cattle Marker

Paraffin Wax	40
Tallow	20
Wool Fat	20
Ceresin (58/60°C.)	15
Oil-Soluble Color	1–2
Pigment (To Match Color)	As desired

Color for Blue Transfers

Formula No. 1

Rosin	50
Rosin Oil	20
Crude Montan Wax	10
Pigment	20

No. 2

Rosin	30
Rosin Oil	10

Crude Montan Wax	10
Toluol or Xylol	35
Pigment	15

Developing Ink

Good Letterpress Ink	1 lb
Beeswax	1 lb
Thin Litho Varnish	4 oz

Duplicating Ink

Formula No. 1

Hard Paraffin Wax	3
Bleached Montan Wax	3
Violet Blue	5
Milori Blue	6
Crude Montan Wax	15
Lampblack	20
Mineral Oil	32

No. 2

Methyl Violet Base	5
Oleic Acid	10
Bleached Montan Wax	10
Crude Montan Wax	60
Petrolatum	90

Grind in color mill.

Engraver's Black Stenciling Ink

Paraffin Wax	¾ lb
Pigment Black	10 oz
Kerosene	½ gal
Nitrobenzol	½ oz

Melt the wax. Stir in the pigment black until well dispersed and then add the kerosene and nitrobenzol to the cooling wax.

Finishing Ink

An ink sometimes used by etchers in cleaning up the plate (removal of shoulder) in the last stage of etching:

Black Litho Ink	2 lb
Beeswax	1 lb
Rosin	10 oz

Add sufficient turpentine to give the desired consistency. Apply the ink to the *warm* plate with a composition roller, manipulating in all directions until the surface of the etched relief design is thinly but thoroughly covered with ink.

Glass-Marking Crayon

Paraffin Wax	98 g
Stearin	2 g

Mix while it is melted; then add 10–15 g of an oil-soluble red, blue, black, or yellow dye. Let the liquid cool slightly and, before it sets to a hard mass, pour it into large paper soda straws, one end of which has been closed.

Gloss Ink

a.	Crude Montan Wax	15
	Rosin	2
	Soft Paraffin (40/42°C.)	3
	Potash Carbonate	½
	Marseilles Soap	4
	Water	65
b.	Shellac	20
	Borax	7
	Water	75
c.	Nigrosine NTL	5½
	Water	25

Make up separately *a, b,* and *c,* and then mix them together.

Hard Etching Ink

As its name implies, this is a hard, strong ink used where exceptional acid resistance is desired:

Black Litho Ink	1 lb
Beeswax	7 oz
Powdered Asphaltum	5 oz
Rosin	5 oz
Black Wax	3 oz

The black wax consists of:

Asphaltum	2 oz
Beeswax	5 oz
Stearic Acid	5 oz
Spermaceti	10 oz

Or, equal quantities of beeswax, rosin, and tar melted and thoroughly mixed may be used as a substitute for black wax.

Hard Wax Crayon (No. 1 Grade)

Base:

Double-Pressed Stearic Acid	65 lb
Paraffin Wax (138/140°F.)	35 lb

Add to 100 lb of base:

Color	Pigment	Quantity lb	oz
Black	Carbon Black	6	6
Magenta	Poster Cerise	25	
Blue	C.P. Prussian Blue	7½	
Ultramarine Blue	C.P. Ultramarine Blue	25	
Orange	C.P. Chrome Orange	25	
White	Lithopone	33–35	
Yellow	Primrose Yellow	12½	
	Chrome Yellow Light	12½	
Brown	Burnt Umber	18¾	
	Burnt Umber	6¼	
Raw Umber	Raw Umber	25	
Burnt Sienna	Burnt Sienna	17	10
Indian Red	Indian Red	18¾	
	Pure Red Oxide	6¼	
Venetian Red	Red Oxide	12½	
Rose Pink	Process Red	4	
	Lithopone	10	
English Vermilion	American Vermilion	28	3
Flesh Tint	Vermilion		8
	Lithopone	4	
Medium Yellow	Medium Chrome Yellow	25	
Golden Ochre	Mexican Yellow or Strong Yellow Oxide	8	11
Green	Green A or C.P. Milori Dark Green	12½	
Violet	Purple	12½	
Light Green	Extra Light Green	12½	
Cobalt Blue	Cobalt Blue	12½	
Dark Green	C.P. Peerless Dark Green	17	10
Olive Green	Chrome Yellow Medium	8	
	Black Lake	6	5
Red	Red No. 2097 (Lavenberg)	12½	

Hot-Metal Crayon

Formula No. 1
(White)

Quicklime	2
Zinc Oxide	4
Titanium Dioxide	70
Japan Wax	5
Stearic Acid	4
Paraffin Wax	15

No. 2

Chrome Green	75
Quicklime	2
Zinc Oxide	4
Stearic Acid	4
Paraffin Wax	15

Heat to 85°C. and mix thoroughly while hot. Force into forms and allow to cool.

Impregnating Wax for Hard Pressed Crayon

Double-Pressed Stearic Acid	10
Mutton Tallow	5
Japan Wax	10
Paraffin Wax (138/140°F.)	75

Soak the crayons in this mixture for 24 hours at 230/240°F.

Improving Printing Inks

Some waxes are used in certain printing inks because they cut down penetration, length, stickiness, and offset and increase lifting properties and hardness.

Carnauba wax is used to give luster, slip, and hardness; Japan wax, to prevent crystallization; beeswax, to prevent offset and crystallization and to slow down evaporation in solvent inks; in offset and lithographic inks, it lessens the tendency to emulsification and bleeding; spermaceti lessens offset.

Pigments, in printing inks, are wetted much more readily by the incorporation of 1% or less Acrawax.* This wetting causes a great increase in the viscosity of the ink. Thus thin inks can be thickened readily.

Increased Slip, Spread, and Antiblocking Effects

Ozowax** is being used very successfully to replace ozokerite in the same proportions in ink vehicles. It gives better spread, slip, and antiblocking effects.

Indelible Transfer Ink

Formula No. 1

Asphalt Emulsion	200
Para Cumarone Resin***	20
Ozokerite	20
Carbon Black	20

No. 2

Amberol†	150
Carnauba Wax	84
Ozokerite	32
No. 6 Litho Varnish	35
Blown Castor Oil	60

* Glyco Products Co.
** L. A. Salomon & Bros.
*** Varnish Grade.
† Rohm and Haas Co.

Cobalt Drier	1
Butyl Carbitol	10
Cadmium Red	215

Lithographic Crayon

Formula No. 1

Sodium Stearate	7
Beeswax	6
Carbon Black	1

No. 2

Beeswax	30
Tallow	25
Soap	20
Shellac	15
Lampblack	6

Heat in an enamelled pot to melt together. Then heat strongly until the vapors ignite. Allow to burn for a while and smother flame with the cover of pot. Take out a sample and test for elasticity. If not satisfactory, ignite again in the same way.

Magic Writing Pad

Cardboard is coated with the following composition and covered with a sheet of waxed or oiled paper or cellophane. When the top sheet is written on with a stylus or pencil the writing appears. When this sheet is lifted away from the coated cardboard the writing disappears.

Beeswax	4
Venice Turpentine	9
Lard	4
China Clay	3½
Carbon Black	1
Mineral Oil	2

The consistency may be varied by varying the proportions of liquids in the formula.

Marking Crayon

Ceresin	40
Carnauba Wax	35
Paraffin Wax	20
Beeswax	5
Talc	50
Chrome Green or Other Pigment	15

Marking Uncured Rubber Mixes

For general use, where the stock is hot, a mixture of 3 parts of montan wax, 1 part of beeswax, and 3 parts of chrome yellow is recommended. Where the stock receives no subsequent treatment and a mark which vanishes during vulcanization is desired, a mixture of 3 parts of montan wax, 1 part of beeswax, 3 parts of zinc oxide, and 1 part of dry white lead is recommended. For marking on wet stock, a mixture of 2 parts of paraffin wax (m.p. 140°F.), 1 part of tallow, and 2 parts of chrome yellow is recommended. Glycerol is suitable for lubricating the wooden or metal molds in which the crayons are formed.

Molded Drawing Crayon

Paraffin Wax	50
Mutton Tallow	40
Carnauba Wax	3
Pigment	7

Munsel-Type Wax Crayons

Base:

Double-Pressed Stearic Acid	65 lb
Paraffin Wax (138/140°F.)	35 lb

Add to 100 lb of base:

Color	Pigment	Quantity lb	Quantity oz
Middle Blue	Peacock Blue	17	
	C.P. Lincoln Toner		9½
Maximum Purple-Blue	C.P. Milori Blue	8	11
Middle Purple-Blue	Ultramarine Blue	1	
Maximum Green	Green No. 8791 (Kohnstam)	8	11
Middle Red	Van Dyke Brown	2	11
	Para Red	1	10
	Lithopone	2	2
Middle Green	Green No. 8791 (Kohnstam)	5	6
	Black Lake	2	2½
Middle Yellow	Mexican Yellow	10	2
	Black Lake	1	11
Maximum Yellow-Red	Chrome Orange	25	3
	Permatone Orange (United)	1	4
Maximum Purple	Magenta	4	12
Maximum Red	Red No. 1897 (Lavenberg)	11	
Maximum Red-Purple	Red Purple Lake No. 2128 (Lavenberg)	3	3½
Maximum Yellow	Primrose Yellow	12½	
	Chrome Yellow Light	12½	
Maximum Blue	Salisbury Blue (United)	11	
Black	Carbon Black	6	6
Gray	Lampblack	2	9
	Lithopone	25	13
Maximum Green-Yellow	Light Chrome Green	11	
Middle Green-Yellow	Light Chrome Green	5	6
	Black Lake	2	3
Middle Yellow-Red	Persian Orange	2	11
	Black Lake		8½
	Lithopone	5	6½
Middle Blue-Green	Peacock Blue	1	8½
	Green No. 10–7500 (United)		13
	Black Lake		9
Maximum Blue-Green	Peacock Blue	3	2
	Green No. 10–7500 (United)	1	9
Middle Purple	Purple Lake		4
	Red Lake		2
	Lithopone	7½	
Middle Red-Purple	Para Red	2	3
	Indian Red	1	2
	Black Lake	2	3
	Lithopone	5	6

Nonoffset Compound

No. 1 Lithographic Varnish	35
Soft Cup Grease	35
Paraffin Wax	10
Beeswax	20

Melt together, cool, and run in mill.

Photolitho Transfer Ink

For inking up images on photolithographic transfer paper, the following is recommended:

Palm Oil	1
Beeswax	1
Asphaltum	2
Burgundy Pitch	2
Black Litho Ink	8
Litho Varnish	2½

The ink is applied with a composition roller, and, if desired, may be thinned with oil of turpentine.

Photomechanical Ink (Autographic Transfer Ink)

This is used for writing on plain paper with a pen a text to be transferred to a litho stone or metal plate:

Marseilles Soap	10 oz
Acetic Acid	1 lb
Glycerin	20 lb
Water	100 lb
Shellac	12 oz
Tallow	10 oz
Yellow Beeswax	12 oz
Mastic	5 oz
Asphaltum	4 oz
Lampblack	3 oz
Water	125 oz

Pressed Drawing Crayon

Clay	70
Glue Solution	10
Pigment	20

Knead and force through hydraulic press into desired shapes.

Red Marking Crayon

Rhodamine B Stearate	1
Calcophen Red Y*	1
Stearic Acid	20
Beeswax	5
Paraffin Wax	16
Carnauba Wax	7

Dissolve colors in molten stearic acid. Then add the waxes, melt out, and pour into molds. Chill molds with cracked ice to remove the crayons.

Rolling-Up Ink

Letterpress Ink	1 lb
Asphaltum	4 oz
Rosin	3 oz
Beeswax	3 oz
Turpentine	10 oz

Heat the plate and apply a smooth film of ink to the plate by a composition roller.

Shoe-Edge Ink

Montan Wax	14
Caustic Potash	2
Nigrosine	3
Water	81

* American Cyanamid Co.

The dye is dissolved in the water and the caustic added and dissolved. Then the solution is stirred into the molten wax.

Shoe-Repair Crayons

Formula No. 1 (For Vici Kid)

Broken Victrola Records	2½
Ozokerite	1¾
Yellow Beeswax	1¾
Lump Oil Black	1½

No. 2

Japan Wax	8
No. 2 Carnauba Wax	1
Color	To suit

No. 3 (For Patent Leather)

Broken Victrola Records	7¼
Ozokerite	2¼
Lump Oil Black	3

No. 4 (For Gun Metal)

No. 2 Carnauba Wax	5
Ozokerite	18
Yellow Beeswax	18
Japan Wax	2
Hard Paraffin Wax	14
Oil Black	1

No. 5

Candelilla Wax	15
Stearic Acid	60
Ceresin Wax	7½
Ozokerite Wax	7½

Melt the ingredients together and stir until uniform. To 1½ oz of this molten base, add ¼ oz lampblack.

Soft Etching Ink

Preferred for delicate work and used in the French etching process (gillotage), this ink consists of:

Russian Tallow	6
Yellow Beeswax	5
Asphaltum	2
Black Litho Ink	1
Thin Litho Varnish	1

Solid Stencil Ink

Black

Black Rosin	20
Shellac	16
Bone Black	14
Rosin Spirit	2
Japan Wax	2
Tallow	1¾
Hard Yellow Soap	¾

Melt the rosin, shellac, and rosin spirit, add the tallow and wax, and stir well. Add the bone black and soap (sliced), stir until all dissolve, and incorporate; then turn out into the molds to set hard.

This ink is waterproof and will not rub off nor blur when the marking is exposed to moisture or rain.

Tailors' Marking Crayon

Carnauba Wax	11
Stearic Acid	2
Ceraflux*	76
Ozokerite	6
Terra Alba	5

* Glyco Products Co.

Textile-Marking Crayon

(Washable)

Formula No. 1

Barium Carbonate	25
Diglycol Stearate	25

Melt at the lowest possible temperature and stir until uniform; then pour into molds. Other colors are obtained by using colored pigments.

No. 2

Paraffin Wax	649
Stearic Acid	125
Ultramarine Blue HM	24
Emulsifier S–489*	91

Melt the paraffin and stearic acid at about 158°F.; stir in the ultramarine blue. Turn off the heat and add the S-489, while stirring thoroughly. Continue stirring until it begins to solidify. Place into molds and chill quickly.

Transfer Composition

Formula No. 1

Cumar Varnish	100
Ozokerite	12
Blown Rapeseed Oil	20
Lithographic Varnish (No. 6)	12
Butyl Carbitol	5
Cobalt Drier	2

Melt together and mix thoroughly; add color while in the molten state and pan.

In place of the butyl Carbitol any of the following may be used:

Pine Oil
Diethyl Oxalate
Monobutyl Tartrate
Nitrobenzol
Carbitol (Monoethyl Ether of Diethylene Glycol)
Butyl Oxalate

No. 2

Rosin	100
Beeswax	30
Gold Bronze or Pigment	30

No. 3

(Fugitive)

Rosin	100
Beeswax	10
Cobalt Drier	1
Gold Bronze	25

No. 4

(Water Fugitive)

Mutton Tallow	1
Cocoa Butter	1
Paraffin Wax	4
Rosin	6
Pigment	Sufficient

Transfer Ink

Formula No. 1

1. Ultramarine Blue	50
2. Gum Mastic	30
3. Beeswax	10
4. Petrolatum	10

Melt 3 and 4, work in 1, and mix with melted 2.

No. 2

(Emulsifiable)

Diglycol Stearate	20
Ethyl Cellulose	5

* Glyco Products Co.

Sodium Abietate	10
Pigment	10

No. 3

Nitrocellulose (½ sec.)	15
Phenol-Formaldehyde Resin	25
Beeswax	50
Acetone	50

Tusche

(Lithographic Drawing Material)

Yellow Beeswax	2
Mutton Tallow	2
Marseilles Soap	6
Shellac	3
Lampblack	1½

Wax Drawing Pastels

Black

Hard Soap	80
Crude Beeswax	60
Crude Spermaceti	28
Carbon Black	14
Burnt Umber	5
Prussian Blue	4

Melt the waxes and soap, mix in the pigments and grind until smooth; pour hot in molds; and plunge into cold water to set.

Red

Hard Soap	28
Saponified Japan Wax	28
Spermaceti	16
Carnauba Wax	2
Crude Beeswax	8
Orange Chrome Yellow	12

Method: as under Black.

Wax Offset Compound

1. Beeswax	22
2. Petrolatum Amber	20
3. Mutton Tallow	5
4. Paraffin Oil	22
5. Kerosene	10
6. High-Flash Naphtha	4

Melt 1, 2, 3 and 4 and stir until dissolved. Turn off heat and work in 5 and 6.

COSMETICS AND MEDICINAL PREPARATIONS

Absorption-Base Creams

Formula No. 1

Lanolin Absorption Base	25.0
Petrolatum	10.0
Mineral Oil	10.0
Beeswax	5.0
Monostearin	10.0
Moldex* (Preservatives)	0.1
Perfume	0.4
Water	39.5

All ingredients except water are melted together and cooled to 45°C. Water at 40°C. is slowly added with vigorous stirring which is continued until the mixture has reached 30°C. If the cream is to be remelted for filling, heating must be very slow and the temperature must not be allowed to rise to the point where the cream becomes completely liquid and is likely to separate. Use of a colloid mill is advisable in making absorption-base creams.

* Glyco Products Co

No. 2
Night Cream

Absorption Base	60.0
Water	40.0
Glycerin	10.0
Hard Paraffin Wax	5.0

Absorption Bases

Formula No. 1

Cetyl Alcohol	10
Anhydrous Lanolin	5
White Petrolatum	75
White Mineral Oil	15

No. 2

Cetyl Alcohol	15
Anhydrous Lanolin	10
White Petrolatum	45
White Mineral Oil	30

No. 3

Cetyl Alcohol	12
Anhydrous Lanolin	5
White Mineral Oil	63
Paraffin Wax	20

No. 4

Cetyl Alcohol	4
Anhydrous Lanolin	24
White Mineral Oil	17.5
Paraffin Wax	14.5

No. 5

White Petrolatum	95
Cholesterol	5

No. 6

Anhydrous Lanolin	7
White Petrolatum	84
Paraffin Wax	4
Cholesterol	5

No. 7

White Petrolatum	58
White Mineral Oil	12
Paraffin Wax	22
Cholesterol	8

Acid Creams

Formula No. 1

Pure Glycol Distearate	14.0
Mineral Oil	3.0
White Beeswax	2.0
Ceresin	1.0
Lanolin	1.0
Sodium Lauryl Sulfate	1.0
Citrus Pectin	0.5
Perfume	0.5
Citric Acid	2.0
Glycerin	6.0
Distilled Water	69.0

No. 2

Wool Wax	3.0
Lanolin	8.0
Stearyl Alcohol	2.0
Petrolatum	37.0
Glycerin	4.0
Lactic Acid	1.5
Distilled Water	44.5

Agar and Paraffin Confection

Alkanna Root	8	gr
White Soft Paraffin Wax	5	oz
Agar	2½	oz
Citric Acid	40	gr
Lemon Oil	16	min
Sucrose	to 8	oz

Digest the alkanna root in the melted soft paraffin on a water bath for fifteen minutes, strain on the agar, citric acid, and sucrose, previously powdered and mixed in a mortar. When cold add the oil of lemon.

Dose: 1 to 4 drams.

All-Purpose Cream

Formula No. 1

a. Paraffin Wax	13.6
White Petrolatum	3.5

	Mineral Oil	13.6
	Diglycol Stearate	13.6
b.	Water	53.6
	Triethanolamine	1.6
	Moldex*	0.1
c.	Perfume	0.4

Heat *a* to 70°C. and stir until complete solution is obtained. Heat *b* to the same temperature and add *a* to *b*, stirring continuously while cooling. Add *c* at about 55°C. Stir and pour at 45–46°C.

No. 2

Absorption Base	4.0
Soft Paraffin Wax	7.7
Lanolin	4.0
Liquid Paraffin	1.5
Water	20.0

No. 3

Absorption Base	15
Soft Paraffin Wax	40
Lanolin	7
Liquid Paraffin	3
Glycerin	8
Water	70

No. 4

Sorbitol Distearate	10
Mineral Oil	20
Lanolin	3
Glycerin	5
Beeswax	4
Ceresin Wax	10
Water	48

Almond Lotion

Paraffin Oil	8 oz
Beeswax	10 oz
Spermaceti	12 oz

Melt and keep the temperature at about 80–85°C., then add, while stirring, the following solution:

Boiling Water	¾ gal
Borax	¼ oz
Soap	14 oz

Add to this mixture, while stirring, 3 gal of water in which 3 oz of gum tragacanth were dissolved and which was heated to about 75°C. Remove from the fire and continue stirring until the mass gets cold. Let this mixture stand overnight and then add 5 gal of water in which 7 oz of gum tragacanth were dissolved. Also add to this mixture ½ gal of alcohol in which 9 dr of salicylic acid and 3 oz of benzaldehyde were dissolved.

Alum Cream

White Petrolatum	9
Paraffin Wax (52°C.)	6.5
Mineral Oil	14
Acimul*	12
Carbitol**	3
Saturated Alum Solution	5.5
Rose Water	50

Melt the oil and waxes at 70°C. Add them to the water, alum solution, and Carbitol of the same temperature with good stirring. Stir until cold enough to be just about pourable (50°C.). Perfume at 55°C.

* Glyco Products Co.
** Carbide and Carbon Chem. Corp.

Aluminum Chloride Cream

Polymerized Glycol Stearate	15
Spermaceti	5
Glycerin	3
Hydrated Aluminum Chloride Crystals	12
Water	To make 100

Analgesic Balm

Yellow Petrolatum	200 g
Paraffin Wax	50 g
Anhydrous Lanolin	20 g

Melt together and add:

Methyl Salicylate	10 g
Chloral Hydrate	1 g
Salicylic Acid	5 g
Menthol	1 g
Eucalyptol	1 g
Camphor	1 g
Turpentine	1 cc

Antichap Lipsticks

Antichap lipsticks are formulated from a suitable base with the addition of a substance or substances that are soothing and healing to the skin. Usually it is preferable to have a base containing lanolin and cacao butter.

Formula No. 1

Beeswax	40
Castor Oil	30
Cacao Butter	20
Anhydrous Lanolin	5
Mineral Oil (210/220)	5

No. 2

Beeswax	32
Castor Oil	33
Sesame Oil	22
Anhydrous Lanolin	13

No. 3

Beeswax	10
Anhydrous Lanolin	10
Mineral Oil (210/220)	43
Ceresin	37

No. 4

Mineral Oil (210/220)	55
Ceresin	45

The antichap lipstick base should contain some soothing substance. Eucalyptol, menthol, camphor or lavender used alone or in combination may help to heal a chapped lip. Benzocaine (U.S.P. XII) has an analgesic effect when applied locally and may be used in ointments in small amounts. It is believed that the addition of vitamin A and D to local applications may help heal traumatic tissue.

Antiperspirant Cream

Acid-Stable Glyceryl Monostearate	8½ lb
Spermaceti U.S.P.	4½ lb
Glycerin	2 lb
Distilled Water	10 lb
Methyl *p*-Hydroxybenzoate	6/7 oz
Aluminum Sulfate N.F.	7½ lb
Distilled Water	7½ lb
Titanium Dioxide	1 lb

Dissolve the aluminum sulfate in the 7½ lb of distilled water, then stir the titanium dioxide into the solution. Heat the glycerin, the 19 lb of distilled water, and the preservative in the mixing kettle at 90°C. Heat the monostearate and the spermaceti also at 90°C. and introduce into the glycerin-preservative solution with fairly vigorous agitation. Stir until the emulsion is well formed and almost cool; finally, add the aluminum sulfate mixture.

Aluminum chlorohydrate can be used in the preparation of antiperspirants.

Antiseptic Cream

Formula No. 1

Carbolic Acid	5 gr
Camphor	10 gr
Anhydrous Lanolin	1 oz
Soft Paraffin Wax	1½ oz
Cocoa Butter	½ oz

Melt the lanolin, paraffin, and cocoa butter together. Rub together the carbolic acid and camphor and when they are liquid add to the first mixture. Stir until cold.

No. 2

Emulgor A* (Polymerized Glycol Ester)	10.0
Stearic Acid	10.0
Paraffin Wax	15.0
Mineral Oil	15.0
Petrolatum	10.0
Oxyquinoline Sulfate	0.2
Water	To make 100

Antisunburn Cream

Stearyl Alcohol	1
Glycerol	5
Glyceryl Monostearate	10
Stearic Acid	5
Sunscreen	5
Triethanolamine	1
Vegetable Oil	2
Distilled Water	71

* Glyco Products Co.

Artificial Petroleum Jelly

Ceresin or Paraffin	15–20
White Mineral Oil	85–80

Aseptic Wax

Beeswax U.S.P.	875 g
Almond Oil	125 cc
Salicylic Acid	10 g

Melt the wax with the oil, stir, strain through muslin, and, with continued moderate heating for ½ hour, add the salicylic acid with stirring. Pour into jars while fluid.

Avocado Tissue Cream

Avocado Oil	20.0
Lanolin Absorption Base	10.0
Stearic Acid	5.0
Cetyl Alcohol	5.0
Beeswax	5.0
Almond Oil	25.0
Preservative	0.2
Perfume	0.8
Borax	1.0
Distilled Water	28.0

The water and borax are brought to 165°F. and are then added to the molten fats, oils, and waxes. Careful heating and constant stirring are essential, the perfume being added when the temperature has dropped to about 120°F.

Avocado-Oil Cleansing Cream

1. Cosmetic-Grade Hydrogenated Oil	11.0
White Beeswax U.S.P.	5.0
Triple-Pressed Stearic Acid	0.5
Sesame-Seed Oil	60.0
Avocado Oil	7.0
Antioxidant and Preservative	0.1
2. Distilled Water	15.3
Borax U.S.P.	0.5
3. Perfume	0.6

Melt 1 at 80°C., heat 2 to 75°C.

When 1 has cooled to 75°C., add 2 with constant stirring. Add 3 at 60°C. Pour at 55–50°C.

This cream is of the quick-liquefying type. It does not melt or run in the jar even in extremely hot weather. The quick-liquefying property is mechanical not thermal.

Black Mascara

Carbon Black	10.0
Beeswax	36.0
Carnauba Wax	9.0
Triethanolamine	13.5
Stearic Acid	31.5

Mix, heat gently, stir until dissolved, and run into molds.

Bleaching Cream

a. White Petrolatum	9.0
Paraffin Wax	6.5
Mineral Oil	14.0
Acimul*	12.0
b. Carbitol**	3.0
Lemon Juice†	5.5
Water	50.0

Melt *a*; mix *b*, heat to 70–75°C., and pour slowly into *a*, with stirring. Add perfume at 55°C. Pour at 50°C.

* Glyco Products Co.
** Carbide and Carbon Chem. Co.
† Lemon juice can be replaced by citric or lactic acid, etc., in proportionate amounts.

Brilliantine

Formula No. 1 (Jelly)

Spermaceti	14
Beeswax	6
Mineral Oil	100
Perfume	1
Color	To suit

Melt the waxes in the mineral oil. Strain and allow to cool to about 115°F. Add the perfume and stir until cold.

No. 2 (Solid for Hot Countries)

Ceresin Wax	2.00 lb
White Wax	0.10 lb
Light Mineral Oil	20.00 lb
Lemon Oil	0.32 oz
Geranium Oil	0.16 oz
Bergamot Oil	0.64 oz
Color	Sufficient

Brittle-Fingernail Cream

Almond Oil	25
Soft Paraffin	20
Water	35
Glycerin	5

Stearic Acid	5
Triethanolamine	4

This is applied at night and allowed to dry. In the morning the nails are rubbed (polished) with the following powder:

Tin Oxide	7
Talc	2
Zinc Oxide	1

Brushless Shaving Cream

Formula No. 1

Glyceryl Monostearate	12.0
Spermaceti	4.5
Mineral Oil	2.5
Glycerin	7.0
Water	73.5
Moldex*	0.1
Perfume	0.4

No. 2

Stearic Acid	12
White Mineral Oil	12
Paraffin Wax	5
Soap Flakes	3
Water	72

Burn Cream

Cetyl Trimethyl Ammonium Bromide	1.0
Sulfanilamide	3.0
Castor Oil	25.0
Beeswax	1.8
Wool Fat	1.8
Cetyl Alcohol	5.0
Glycerin	10.0
Water	52.4

Melt the castor oil, beeswax, wool fat, and cetyl alcohol at as low temperature as possible. Dissolve the bromide in the water with heat, mix with the oil, etc., at about 60°C., and stir till set. Then rub up the sulfanilamide with the glycerin and incorporate it, with thorough mixing into the cream.

Apply the cream with a knife blade or spoon previously sterilized by dipping for 2 minutes in boiling water or passing through a flame. Do not wash the burn or snip the blisters before applying the cream. Do not leave it on the burn for more than 2 days because prolonged application involves a slight risk of causing a sensitization dermatitis.

Burn-Treating Emulsion

Spermaceti	15 g
Soft Paraffin Wax	12 g
Liquid Petrolatum	12 cc
Sodium Lauryl Sulfate	1 g
Water	60 cc

Cacao-Butter Substitutes

Formula No. 1

Lanolin	6
Spermaceti	3
Olive Oil	1

No. 2

Paraffin Wax	1
Lanolin	1
Olive Oil	1

No. 3

Lanolin	9
Stearin	1
Triethanolamine Stearate	3

* Glyco Products Co.

No. 4

Sun-bleached Beeswax	2½
Hard Paraffin Wax	5
Lanolin	5
Spermaceti	35
Mineral Oil U.S.P.	100

Callus Softener and Remover

Castor Oil	65 cc
Paraffin Wax	60 g
White Soap	10 g
Sodium Thiosulfate	5 g

Melt the wax and soap together, then add the castor oil and the thiosulfate which has been pulverized to a fine powder. Pour out and let cool.

Apply a thin sliver or a small amount of the mixture to a piece of cloth and bandage overnight. Then soak feet in hot water and loosen the callus. Stubborn calluses may require a second or third treatment.

Camphor Cream

White Wax	2
White Petrolatum	38
Camphor	¾
Menthol	¾
Zinc Oxide	8
Boric Acid	12

Melt the wax in the heated petrolatum and stir into this the rest of the ingredients. Stir until the mixture congeals.

Camphor Ice

Formula No. 1

Camphor	175 g
Liquid Petrolatum	4 oz
White Wax	2½ oz
Cocoa Butter	1½ oz

No. 2

Powdered Camphor	20
White Beeswax	18
Soft White Petrolatum	15
Spermaceti	47

No. 3

Castor Oil	25 oz
White Beeswax	15 oz
Spermaceti	49 oz
Camphor Powder	10 oz
Ethyl Aminobenzoate	1 oz
Carbolic Acid	20 gr

Melt the castor oil, beeswax and spermaceti together and add the camphor and ethyl aminobenzoate. Stir until dissolved. Then add the carbolic acid and pour into molds.

Catarrh Cream

Menthol	2
Eucalyptol	3
Pumilio Pine Oil	3
White Beeswax	1
Hard Paraffin Wax	6
White Soft Paraffin Wax	85

The beeswax and white soft paraffin are melted together and stirred until nearly cold. The medicaments, previously mixed, are incorporated while the mass is still soft. Some form of closed mixer is desirable, as otherwise there is considerable loss of menthol.

Recommend on the tube and its carton the frequent use of the cream until the troublesome symptoms are relieved.

Chapped-Lip Stick

Camphor	12
Spermaceti	92
Peanut Oil	40
Diglycol Laurate	40
White Beeswax	16–20

Melt and mix at low temperature. Cool and pour into molds at the lowest possible temperature. The hardness can be varied by increasing or decreasing the amount of beeswax.

Chest Rub

Camphor	15
Menthol	5
Methyl Salicylate	5
Eucalyptus Oil	5
Anhydrous Lanolin	20
Soft Paraffin Wax	100

Melt and stir until uniform; pour into jars or tins at the lowest possible temperature.

Chest-Rub Salve

Brown Petrolatum	1 lb
Paraffin Wax	1 oz
Eucalyptus Oil	2 oz
Menthol Crystals	½ oz
Cassia Oil	⅛ oz
Turpentine	½ oz
Carbolic Acid	⅛ oz

Melt the petrolatum and paraffin wax together, then add the menthol crystals and stir till dissolved. Remove from the fire and, while cooling, add the oils, turpentine and acid. Pour into 1-oz tin boxes when it begins to thicken.

Chilblain Ointment

Phenol	1.0
Camphor	6.0
Peru Balsam	2.0
Soft Paraffin Wax	25.0
Hard Paraffin Wax	7.5
Anhydrous Lanolin	To make 100.0

Immerse the affected part in hot water at bedtime, dry it carefully, and apply the ointment; or rub in the ointment night and morning.

Cleansing Cream

Formula No. 1

Flaked White Beeswax U.S.P.	10
Ceresin Wax (M.P. 138°F.)	5
Mineral Oil (Viscosity 68/75)	65
Water	40
Borax	½
Perfume	Sufficient

The beeswax, ceresin, and a portion of the mineral oil are placed in a steam-jacketed kettle and the temperature in melting the waxes should be at no time in excess of 80°C. (The emulsifying power of beeswax is retarded or broken down when the temperature is above 80°C.) When the waxes have melted, the steam should be so regulated (this can be accomplished only by actual experiment) that when the balance of the mineral oil is added cold, the temperature will be approximately 57°C. It will take a little experimenting to gain the knack of properly controlling the kettle temperature in order to speed production.

When the melted waxes and mineral oil are at 57°C., the perfume oil should be added and the agitator started. The agitator should have only one propeller at the bottom, causing a complete vortex when agitating the melted waxes and oil and forming only a slight vortex when all the water and borax have been added.

In another kettle, the water and borax should be kept at exactly the same temperature, i.e., 57°C. It is important to continue the agitation while the water and borax are being added and while the curds form until the cream begins to thin down and becomes a homogeneous mass. After this, stirring should be continued for a few minutes only. When the cream must stand before passing into the filling machine, it might be necessary to agitate it to destroy any crust caused by surface cooling.

No. 2

Sorbitol Monooleate	5
Sorbitol Distearate	10
Beeswax	7
Mineral Oil	32
Lanolin	3
Water	43

This combination gives a moderately firm cream with a fine-grained texture.

No. 3

Beeswax	6.0
Spermaceti	2.0
Petrolatum	5.0
Raisin-Seed Oil	50.6
Mineral Oil	10.0
Rancidex* (Preservative)	0.1
Water	25.0
Borax	0.6
Triethanolamine	0.2
Perfume	0.5

Melt the fats and oils at 160°F. Bring the water and borax in a separate kettle to the same temperature. Emulsify by adding the water to the oils at 160°F., with constant stirring. Add perfume oils at 115°F.

No. 4

1. Mineral Oil	80
2. Spermaceti	30
3. Glycosterin*	24
4. Water	90
5. Glycerin	10
6. Perfume	To suit

Heat 1, 2 and 3 to 140°F. and stir into it slowly 4 and 5 heated to the same temperature. Add perfume at 105°F.; stir slowly until cold; after allowing to stand for 5 minutes, stir until smooth and pack.

* Glyco Products Co.

No. 5

White Mineral Oil	280
Spermaceti	80
Glycosterin*	80
Neutral Diglycol Laurate	20
Water	360
Carbitol**	40
Perfume	5

No. 6

1. Mineral Oil	78
White Wax	5
Spermaceti	28
Special Trihydroxyethylamine Stearate	20
2. Perfume	1
3. Glycerin	4
Water	92

Heat 1 and 3 separately to 200°F.; then add 1 and 2 slowly, stirring thoroughly. When the cream begins to set, add and stir in the perfume. Allow to stand overnight. Stir thoroughly the next morning and package.

This cream will not sweat oil during hot weather.

No. 7

Paraffin Oil	350
Spermaceti	130
Diglycol Stearate	100
Propylene Glycol	40
Distilled Water	380

No. 8

Paraffin Oil	310
Beeswax	20
Spermaceti	110
Triethanolamine Stearate	80
Glycerin	16
Perfume	4
Distilled Water	360

* Glyco Products Co.
** Carbide and Carbon Chem. Corp.

Cod-Liver-Oil Ointments

Formula No. 1

Beeswax	10
Spermaceti	10
Cod-Liver Oil	80

No. 2

Yellow Beeswax	20
Petrolatum	50
Cod-Liver Oil	40

No. 3

Cod-Liver Oil	100
Beeswax	20
Glycerin	20
Petrolatum	30
Lanolin	30
Water	50

No. 4

Yellow Beeswax	10
Triethanolamine Stearate	10
Lanolin	25
Cod-Liver Oil	75

Cold Cream

Formula No. 1

Diglycol Stearate	65
Paraffin Wax	25
Spermaceti	40
Liquid Paraffin	135
Glycerin	25
Distilled Water	210

No. 2

Diglycol Stearate	47.5
White Petrolatum	25.0
Paraffin Wax	32.5
Liquid Paraffin	82.5
Borax	4.0
Distilled Water	308.5

No. 3

Diglycol Stearate	30.00
White Petrolatum	25.00
Paraffin Wax	55.00

Liquid Paraffin	11.25
Borax	7.50
Stearic Acid	3.75
Distilled Water	367.50

No. 4

Diglycol Stearate	61.25
White Petrolatum	26.25
Paraffin Wax	8.75
Liquid Paraffin	141.25
Distilled Water	262.50

No. 5

Diglycol Stearate	50.0
White Petrolatum	37.5
Paraffin Wax	27.5
Liquid Paraffin	75.0
Potassium Carbonate	5.0
Borax	7.5
Distilled Water	297.5

No. 6

Diglycol Stearate	35
Paraffin Wax	5
Spermaceti	5
Liquid Paraffin	150
Borax	5
Beeswax	15
Distilled Water	285

No. 7

Stearic Acid	2	oz.
Anhydrous Lanolin	1½	oz.
White Beeswax	1	oz. 2 dr.
Mineral Oil	2	oz. 4 dr.
Triethanolamine	5	dr.
Glycerin	1	oz. 2 dr.
Water	7	oz.

Melt together the lanolin, stearic acid, and beeswax in the mineral oil and heat to 70°C. Add the triethanolamine to the water and heat to 70°C. Stir the triethanolamine solution with a high-speed mechanical stirrer and add gradually the melted ingredients. Stir till a smooth emulsion is obtained. Then add the glycerin and a suitable perfume. Pour into jars while warm.

For a better-quality cream, all or part of the mineral oil can be replaced by suitable animal or vegetable oils, e.g., olive or grapeseed oil.

No. 8

1. Borax	2
Water	54
2. Glyco-Wax A*	20
White Beeswax	26
White Mineral Oil	120
3. Perfume	1

Dissolve the borax in the water, heating to 150°F. Melt in another pot the beeswax, Glyco-Wax A, and white mineral oil and keep at about 150°F.; add with stirring 3/7 parts lily of the valley (or other perfume). Add 2 to 1 slowly with thorough stirring; continue stirring until cool enough to pour.

Softer creams can be prepared by increasing the amount of water in the formula.

If creams are packed when too warm, the finished products will not look so well as if they are poured when cooler. The best time for packing is just before the cream begins to set.

* Glyco Products Co.

No. 9

Stearic Acid	30
Anhydrous Lanolin	20
White Beeswax	16
White Mineral Oil	33
Triethanolamine	3.8
Carbitol*	16
Water	95

Melt the stearic acid, lanolin, and beeswax in the mineral oil and heat to about 70°C. Prepare in a separate kettle a boiling solution of the triethanolamine and water and add to this the hot solution of waxes. Stir vigorously until a creamy emulsion is obtained and add the Carbitol to which the perfume has been added. Continue stirring until homogeneous and the product has reached the proper consistency. Pour into jars while still warm.

No. 10

1. Diglycol Stearate	14	lb
2. Paraffin Wax	2	lb
3. Mineral Oil	3¾	gal
4. White Petrolatum	6	lb
5. Water	6	gal
6. Perfume Oil	5½	fl oz

Melt 1, 2, 3 and 4 at 170°F. Heat 5 to 180°F. and add while mixing. Allow the mixer to run until the batch is completely emulsified. Then allow to cool to 125°F., add 6, and mix at low speed. Allow to cool without stirring to 105°F. and pour into jars.

* Carbide and Carbon Chem. Co.

No. 11

a. Diglycol Stearate	7
Beeswax	3
Spermaceti	1
Paraffin Wax	1
Mineral Oil	30
b. Borax	1
Water	58

Melt *a* to 80°C., add *b*, which has previously been heated to 80°C., to *a* slowly with constant stirring. When the temperature has dropped to 50°C., add perfume with stirring and pour at 45°C.

If a cream of greater consistency is desired, part of the mineral oil can be replaced with petrolatum.

No. 12

White Beeswax	12
White Petroleum Jelly	12
Peach Kernel Oil	50
Rose Water	25
Borax	1
Perfume	To suit

No. 13

Cetyl Alcohol	10
Liquid Paraffin	10
White Petrolatum	80
Water	60

Transparent, soft, white cream.

No. 14

Cetyl Alcohol	100
White Beeswax	150
Vegetable Oil	200
Mineral Oil	250
Cholesterin	5
Borax	10
Water	260
Perfume	25

No. 15

a.	White Ceresin	27
	Spermaceti	25
	Stearic Acid	21
	Undecylic Acid	1
	White Mineral Oil	87
b.	Borax	5
	Distilled Water	36

Melt *a* and add with agitation to the hot solution *b*. Stir until cold.

Concentrated Day Cream

Stearic Acid	180
Spermaceti	55
Cetyl Alcohol	45
Potassium Hydroxide (50° Bé)	20
Triethanolamine	5
Glycerin	150
Water	525

Preservative and perfume oil to be added in the usual proportions.

The stearin and spermaceti are melted at approximately 85°C. and then added to the remainder (except for the cetyl alcohol), which has also been heated to 85°C. Heating is continued for approximately 15 minutes while stirring, then stirring is continued until cool. When the temperature is down to approximately 50°C., the cetyl alcohol and finally the preservative and perfume oil are added.

Corn Salve

a.	Refined Pine Rosin	8
	Glyceryl Monoricinoleate	2
	Larch Turpentine	10
	Yellow Beeswax	16
b.	Salicylic Acid	8
c.	Peru Balsam	8

Melt *a;* dissolve *b* in the melted base, and add *c*. Stir until homogeneous, cool, and pour.

Corn-Plaster Base

a.	Rosin Oil	30
	Copaïba Balsam	40
	Larch Turpentine	20
	Pale Rosin	40
	Yellow Beeswax	12
b.	Ethyl Ether	100
c.	Sheet Rubber	100
	Ethyl Ether	500

Melt *a* and stir until homogeneous. Strain through a fine-mesh cloth into a wide-mouth metal bottle, and add *b*.

Separately, digest the rubber solution *c* in a wide-mouth glass bottle, and keep tightly closed. Shake daily and use as soon as a homogeneous and pourable solution has formed. This takes about 3 weeks. Add *c* to the wax-resin solution *a* and *b*, and add enough ether to make 800 g.

Corpse Finishing Cream

Glycol Stearate	12
Glyceryl Tristearate	5
Rose Oil	2
Glycerin	3
Water	78
Titanium Dioxide	1

Corpse Wound Filler

a.	Yellow Beeswax	5
	Paraffin	5
	White Petrolatum	15

b. Soap Flakes	2
Water	5

Cosmetic Antiperspirants and Deodorants

Formula No. 1

Monostearin	170
Spermaceti	90
Glycerin	39
Distilled Water	380
Preservative (Moldex*)	1
Aluminum Sulfate	150
Distilled Water	150
Titanium Dioxide	20
Duponol ME**	5

No. 2

Glyceryl Monostearate	15
Spermaceti	5
Methenamine†	3
Glycerin	8
Water	69

Cosmetics for the Eyes (Mascara)

Formula No. 1

Triethanolamine Stearate	30
High-Melting Paraffin Wax	40
Beeswax	12
Lanolin	8
Lampblack	10

Melt, mix, and mill the ingredients and then cast or extrude them into stick or tablet form.

This gives the usual black mascara. Where a dark-brown mascara is desired, the following formula will serve:

No. 2

White Beeswax	300
Montan Wax	100
Stearic Acid	300
Triethanolamine	130
Lampblack	20
Burnt Umber	150

Melt the waxes and grind in the color in a warm mill. Stir in the ethanolamine and pour into molds.

Cake cosmetics are very easy to apply, so long as they are made with the proper oil-in-water emulsifying agents of appropriate wetting properties. They are, therefore, still the most popular representatives of their class. Even so, formulation is not without problems. The product has to adhere to the eyelashes when the brush puts it there, it must dry so as not to stain the lower eyelid and, when dry, must not peel off and fall as dust. Soaps possess all these qualities, but to avoid irritation, the milder triethanolamine stearate is used.

Consistency, gloss, and reduced solubility are imparted to the final product by means of substances of higher melting point, such as wax, ceresin, etc. Resins and gums may also be added without danger of incompatibility.

* Glyco Products Co.
** Du Pont.
† Hexamethylene tetramine.

Cream for Brittle Nails

Triethanolamine	2.0
White Petrolatum	1.5
Beeswax	0.5
Anhydrous Lanolin	0.5
Water	15.0

This cream is to be applied at night and during the day if possible.

Cream Rouge

Formula No. 1

Cetyl Alcohol	3
Cocoa Butter	3
White Beeswax	3
Spermaceti	5
Lake Color	3
Eosin	2
Petroleum Jelly	50
Viscous Absorption Base	20
Tetrahydrofurfuryl Acetate	10
Perfume and Preservative	To suit

The manufacturing routine is the same as for lipsticks. This preparation must be passed through a colloid mill.

No. 2

1.	Spermaceti U.S.P.	10
	Cocoa Butter	6
	White Beeswax, U.S.P.	6
	Absorption Base	40
	Cetyl Alcohol	6
	White Petrolatum U.S.P.	100
2.	Lake	6
3.	Eosin (Tetrabromofluorescein Soluble)	4
	Tetrahydrofurfuryl Acetate	20
4.	Methyl *p*-Hydroxybenzoate	3/16
	Perfume Oil	1

Heat 3 to 145°F. and add to 1 and 4 previously heated to 160°F., when a clear solution results. Add 2 with continuous agitation and cooling. Pass the resultant mass through a colloid mill. Finally remelt the mass at 145°F. and add the perfume oil.

Croquignole Wax for Kinky Hair

White Petrolatum	18¾ lb
Paraffin Wax	1½ lb
Liquid Drug and Cosmetic Red No. 18	4 oz
Perfume Oil	⅜ oz

Melt the waxes, remove from the fire and add the color, finally add the perfume oil.

The red No. 18 is prepared by dissolving 1 dr of dry color in 1 pt of white mineral oil.

The perfume oil is prepared by mixing the following:

Benzyl Propionate	12 oz
Rose Oil	2½ oz

Cuticle Creams

Formula No. 1

White Petroleum Oil	87.750
Paraffin Wax (M.P. 125°F.)	9.000
Menthol	3.000
Thymol	0.025
Oil-Soluble Red Color	To suit

No. 2

Anhydrous Lanolin	12.0
Distilled Water	12.0
Lecithin	0.5
Cream or Lily-White Petrolatum	55.5
Light Mineral Oil	20.0
Perfume	To suit

Since these cuticle creams are essentially hydrocarbon oils which are difficult to rinse off thoroughly, their use should be directed after—not before—the application of liquid polish. Thus any impairment of luster and adhesion of the subsequently applied polish will be avoided.

No. 3

Pale-Yellow Petroleum Jelly	1½ oz
Deodorized Coconut Oil	1 oz
Hard Paraffin	1 dr
Stearic Acid	2 dr
Hydrous Lanolin	1 dr
Water	2 dr
Borax	5 gr

Day Cream with Vitamin F

Beeswax	15
Spermaceti	2
Cocoa Butter	3
Lecithin	2
Emulsifier (Borax-Wax Emulsion)	2
Ethyl *p*-Hydroxybenzoate (Preservative)	3
Vitamin F Concentrate	2
Water	80
Perfume	To suit

Melt the first five ingredients by heating to 55°C. Heat the water also to 55°C. and add it slowly to the melted mixture, stirring constantly. Just before solidification, stir in the vitamin F, the preservative, and the perfume oil.

* Glyco Products Co.

Deodorant Cream

Formula No. 1

Acimul*	15
Glycerin	3
Spermaceti	5
Water	60
Titanium Dioxide	2
Aluminum Sulfate	15

Make up the cream omitting the aluminum salts. When the cream is cold, add the salts, stirring until smooth. Mill if convenient.

No. 2

Paraffin Wax	10
Mineral Oil	10
Deomul	15
Petrolatum	5
Aluminum Sulfate	10
Water	50

Heat the waxes and oils to 70°C., add 30 parts water. Cool to 45°C., and stir in the salt dissolved in 20 parts water.

No. 3

1. Glyceryl Monostearate	15
2. Spermaceti	3
3. Glycerin	5
4. Titanium Dioxide	2
5. Methenamine	5
6. Mineral Oil	3
7. Water	67
8. Perfume	To suit

No. 4

1. Acimul*	15
2. Aluminum Chloride	10
3. Spermaceti	5
4. Glycerin	3
5. Water	67
6. Perfume	To suit

Dissolve 2 in about 20 or 25 parts of 5, and warm to about 50°C. Place 1, 3, 4 and remainder of 5 in one kettle, and bring to about 90°C. Remove from the fire, place on the mixer, make emulsion by stirring rapidly at first, and more slowly until the product temperature drops to about 50°C., at which temperature increase agitation speed and add aluminum chloride solution a little at a time until all is added. Continue stirring, add perfume at about 45°C. It may be necessary to homogenize or mill this cream if it is grainy or lumpy. If a white cream is wanted, 2 to 5% titanium dioxide may be added.

No. 5

Beeswax	8 oz
Liquid Petrolatum	24 oz
Sodium Borate	100 gr
Benzoic Acid	20 gr
Salicylic Acid	400 gr
Hot Water	16 oz

Melt the wax and oil and heat to about 160°F. Dissolve the other ingredients in the water, heat to the same temperature as the wax solution, and pour it into the latter, beating briskly until the cream is formed. A comparatively high temperature of the solutions and a small amount of stirring result in a glossy cream.

Deodorant Pencil

Formula No. 1

Zinc Phenolsufonate	5.00
Zinc Oleate	10.00
Aluminum Palmitate	7.50
Parachol*	20.00
Glyco Wax*	40.00

No. 2

Zinc Phenolsulfonate	10.00
Zinc Oleate	10.00
Aluminum Palmitate	7.50
Parachol*	30.00
Glyco Wax*	30.00
Titanium Dioxide	15.00

Rub the first three ingredients to fine powder and add to the liquefied wax the Parachol mixture. Stir until just before solidification and pour into molds.

Dermatological-Cream Base

Cetyl Alcohol	15
Beeswax	1
Sodium Lauryl Sulfate	2
Propylene Glycol	10
Water	72

Melt the cetyl alcohol and wax, add the propylene glycol, and heat to about 65°C. Dissolve the sodium lauryl sulfate in the water heated to about 65°C. Add the oil phase to the water phase slowly and with constant but gentle stirring. Continue to stir until

* Glyco Products Co.

an ointmentlike consistency results.

This base was studied with about fifty commonly used dermatological ingredients and found to be compatible with most. Although some ingredients, like menthol, soften the base and others harden it slightly, the only cases of frank separation are 70% cod-liver oil and 1% acriflavine. Lead acetate T.S. and Burow's solution do not invert the emulsion. The base has an apparent pH of approximately 7 and is stable at 0–50°C. Autoclaving at 15 lb pressure causes separation of the waxes.

Eczema Preparation

Lanolin	200
Petrolatum	200
Beeswax	50
Phenol	5
Camphor	10
Eucalyptus Oil	50
Salicylic Acid	10
Perfume	To suit

Electric Shaving Aid

The preparation is intended to impart rigidity to the hairs prior to shaving with a blade or electric shaver:

Paraffin Wax	25.5
Petrolatum	10.0
Stearic Acid	30.0
Triethanolamine	5.0
Lanolin	3.0
Ichthyol	1.5
Castor Oil	20.0

Shape into solid sticks. It will deposit a thin film on the skin.

Emollient Cream

Stearyl Alcohol	4.0
Beeswax	1.5
Diethylene Glycol	10.0
Gum Karaya	1.5

Emulsified Ointment Base

Sodium Lauryl Sulfate	0.5
Cetyl Alcohol	8.0
Cocoa Butter	6.5
White Petrolatum	20.0
Water	65.0

Eye Shadow

The ideal eye shadow is fairly hard at room and body temperature, but has enough lubricant to yield a thin film. Stickiness and drag must be avoided. The following are representative formulas for eye shadow:

Formula No. 1

Hydrogenated Oil	45
Mineral Oil	10
Beeswax	20
Paraffin Wax	5
Ozokerite	5
Color	15

No. 2

Short-Fiber Petroleum Jelly (43–48°C.)	23
Ceresin (64°C.)	21
Spermaceti	3
Mineral Oil (65–75)	53

No. 3

White Beeswax	4.5
Spermaceti	9.5
Lanolin Absorption Base	13.0
White Soft Paraffin Wax	73.0

From 78 to 85% of such a base is used with 12–15% of zinc oxide (to bring out the color) and 2–5% of a suitable mineral or earth pigment or cosmetic lake, e.g., 3% middle green lake, plus 2% gold-bronze. The metallic powders (gold bronze and aluminum) impart a pleasantly scintillating effect, but the most important ingredient is the background brightening agent, zinc oxide or titanium dioxide.

No. 4

Stearyl Alcohol	15
Ceresin Wax	15
Liquid Paraffin	40
Pigment	30

No. 5

Mineral Oil	10
Hydrogenated Oil	45
Beeswax	20
Paraffin Wax	5
Ozokerite	5
Color	15

Mix, heat gently, and stir until dissolved.

No. 6

Mineral Oil	5 lb
Lanolin	2 lb
Petrolatum	1 lb
Beeswax	1 lb
Paraffin	2 lb
Perfume Oil	4 oz

Color with any of following combinations:

Blue

Ultramarine Blue	2 lb
Zinc Oxide	2 lb

Light Green

Zinc Oxide	3 lb
Green Lake	1 lb

Gray

Ultramarine Blue	1 lb
Carbon Black	1 lb
Zinc Oxide	2 lb

Brown

Burnt Umber	3 lb
Zinc Oxide	1 lb

Dark Green

Green Lake	3 lb
Zinc Oxide	1 lb

Violet

Violet Lake	1 lb
Zinc Oxide	3 lb

Heat the colors and wax mixture and grind in an ointment mill; pack by pouring hot.

Eyebrow Pencils

Apart from those methods which serve to preserve the eye region in good physical condition, actual beauty treatment is now practiced on a very considerable scale. Coloring of the eyebrows, painting of the eyelashes and shading of the eyelids are now important phases of face cosmetics, the greatest attention being devoted to the first operation. Coloring of the eyebrows or their simulation after complete shaving is effected with colored wax pencils. Ordinary pure charcoal pencils tend to cause falling out and drying of the hair.

Ingredients used in preparing the wax pencils are white wax, benzoated tallow, cocoa butter, petroleum oil and olive oil. The pigments are lampblack, umber, and ochre. Large manufacturers find it economical to use pigment-grinding machines and other equipment of the most modern design. The base is made up from 110 g fine petroleum oil, 60 g white ceresin, 15 g white wax, 240 g benzoated tallow, and 1 g coumarin. The fatty base is thoroughly ground with the pigments, the molten base being gradually stirred into the very finely powdered pigment contained in a mortar. After thorough trituration, the mixture is again warmed, digested for about ½ hour on a water bath, and again allowed to cool. As soon as the mass begins to thicken, it is again vigorously stirred and forced through a fine-mesh sieve by applying powerful pressure with the pestle. Lumps and impurities are retained on the sieve. The preparation which passes through the mesh is then again thoroughly mixed, with gentle heating before casting. The mass should be neither too hot nor too fluid when being cast, since settlement of the insoluble pigment will result in lack of uniform coloration. Oil-soluble dyestuffs will be used in exceptional cases only. According to another process, the melt is prepared from 2 parts cocoa butter, 2 parts ceresin, and 1 part olive oil. Into this is stirred 0.6 part dyestuffs (i.e., about 10% of the total gross weight), which has previously been ground with a little olive oil.

As soon as the mass has reached the state when it can just be cast, it is emptied into metal molds. As a rule, these impart the required taper to the pencils, but if this is not the case, they are tapered after removing from the molds and wrapped in thick metal foil while leaving the points exposed.

Formula No. 1

Mineral Oil	100
White Ozokerite (60/62°C.)	55
Beeswax	15
Benzoated Tallow	240
Coumarin	1

The color for this is, for 1,000 parts of the base:

Medium Brown:

Umber	225
Mahogany	150

Dark Brown:

Umber	150
Mahogany	150
Kassel Brown	50

Blue-Black:

Zinc Oxide	150
Dark Ultramarine Blue	145
Lampblack	5

No. 2

(Brown)

Burnt Sienna	80
Burnt Umber	100
Hard Paraffin	420
Yellow Soft Paraffin	400

No. 3

White Wax	6.0
Ceresin	1.5
Lanolin	2.0
Petrolatum	2.0
Olive Oil	3.0
Pigment	To suit
Talc	To suit

Melt the wax and ceresin together, add the lanolin and petrolatum, and, when all are melted, incorporate the oil. Triturate the pigment with talc and mix it with the fatty mixture. Pour the finished product into suitable molds.

The pigment to be used depends on the color desired. Animal charcoal, carmine, lampblack, sienna, umber, and many of the anilins, such as rhodamin, eosin, etc., are used for coloring.

No. 4

Beeswax	15
Mineral Oil	25
Carnauba Wax	35
Pigment	25

Mix, heat gently, stir until dissolved, and run into molds.

No. 5

Cholesterin	30
Cocoa Butter	30
Ozokerite (Ceresin)	54
White Mineral Oil (65/75 Viscosity)	15
Lampblack	39

Melt the first four ingredients together and, while cooling, grind in the lampblack thoroughly and run into the usual molds.

Fingernail Polish

Formula No. 1

Light Burgundy Rosin	160
Yellow Beeswax	60
White Ceresin	500
Kieselguhr	630
Zinc Oxide	370
Liquid Paraffin	80

Melt together the rosin and the waxes, add the liquid paraffin, then the powders, stirring vigorously while continuing to heat until a homogeneous mixture is obtained. Allow to cool, pass through a grinding mill; finally pulverize and sieve.

No. 2

Beeswax	15.0
White Ceresin	2.5
Bleached Carnauba Wax	2.5
Liquid Paraffin	60.0
Cetyl Alcohol	2.0
Water	23.0
Borax	2.0
Aluminum Stearate	20.0

Melt the fats and waxes, add the borax solution, and finally the stearate. Cool while stirring. When the paste thickens, add a suitable perfume and incorporate 0.5 part fish-scale essence suspended in 5 parts amyl alcohol.

No. 3

Carnauba Wax	12
Beeswax	8
Tripoli	25
Powdered Gum Sandarac	4
Tin Oxide	30
Liquid Paraffin	60

Foundation Creams

(Powder Creams; Liquid Make-Up)

Glyceryl monostearate greaseless creams are especially well suited for use as tinted foundation and make-up bases, since they do not have the high degree of oiliness associated with cold creams or petrolatum and are free from the tendency to roll on the skin as stearate creams sometimes do. Since relatively large amounts of pigment must be included for covering and tinting, the cream itself should be relatively soft. Red, yellow, and brown oxides are most satisfactory for tinting, as they do not bleed. Titanium dioxide gives good hiding power; zinc oxide, because of its ability to react with soaps and destroy the emulsion, should not be used. The amount of titanium dioxide used will depend on the hiding power desired. The proportion of colors necessary to attain a desired shade is dependent on and is increasing with the amount of titanium dioxide used.

Foundation creams can be manufactured by first preparing the cream base, adding the white and colored pigments while the cream is still soft, and then passing the mixture through a colloid mill. Better color dispersion (freedom from color streaks) is generally obtained, however, by passing the solidified cream, containing pigment, through a roller mill.

Formula No. 1

Glyceryl Monostearate	20.0
Spermaceti	5.0
Glycerin	5.0
Moldex*	0.1
Water	66.9
Pigments	3.0

Melt the glyceryl monostearate and spermaceti at 160°F. Add the glycerin, Moldex, and water previously heated to 160°F. and stir until cool. Add the pigment and pass through a roller mill.

No. 2

a.	Lanolin Alcohols	3.5
	Cocoa Butter	5.2

* Glyco Products Co.

Lanolin U.S.P.	5.5
Cetyl Alcohol	12.0
Beeswax	2.0
Spermaceti	10.0
Glyceryl Monostearate	32.0
b. Water	225.0
Moldex* (Preservative)	0.75
Perfume	To suit

Melt *a* at 70°C.

Heat *b* to 70°C.

Add *a* to *b* and stir until the mixture is uniform.

No. 3

Stearic Acid	18.0
Cetyl Alcohol	0.5
Cocoa Butter	0.7
Glycerin	8.0
Potassium Hydroxide	1.0
Titanium Dioxide	2.0
Perfume	0.5
Water	69.0

Four-Purpose Cream

Beeswax	8.0
Triethanolamine	0.4
Paraffin Wax	7.0
Cetyl Alcohol Flakes	11.0
Duponol C** or Diglycol Laurate	1.0
Mineral Oil	26.0
Water	46.6

Frost-Bite Prevention Creams

Formula No. 1

Lanolin	10
Cocoa Butter	15
Olive Oil	25
White Beeswax	22
Water	25
Borax	1½
Benzoic Acid	½
Perfume	1

Mix the fat, oil and wax at about 49°C., add the solution of borax and water which should be about 10° warmer. Stir the whole mixture until the temperature has dropped to 45°C. The last ingredients are the perfume and the benzoic acid which should be dissolved in a little alcohol to make a more "elegant" preparation.

No. 2

Almond Oil	425
Lanolin	185
Beeswax	62
Borax	4½
Spermaceti	62
Distilled Water	300

Germicidal Cream

a. White Petrolatum	12.00
Rose Oil (Mineral Oil)	9.00
White Beeswax	2.25
b. Water	6.00
Borax	1.50
c. *p*-Chloro-*m*-Cresol	1.50

Heat *a* and *b* to 80°C. and then mix them together. When the emulsion is formed, add *c*, also at 80°C. Stir constantly until cold.

Glycerin Cold Cream

a. White Wax	80
Spermaceti	80
Peanut Oil	300
Petrolatum	300

Melt.

* Glyco Products Co.
** Du Pont.

b. Glycerin	120
Water	120
Borax	10

Warm up to 90°C.; pour into melted *a*.

Add when cool:

Fresh-Odor Perfume	20

Greasy Base for Theatrical Make-Up

Mineral Oil (Medium Viscosity)	110
White Ceresin	60
White Wax	15
Benzoinated Tallow	235

Greasy Night Cream

a. White Paraffin Oil	2500
Scale Wax	500
Bleached Beeswax	500
Anhydrous Lanolin	500
b. Distilled Water	3000
Triethanolamine	75
Borax	35

Melt *a* together at 75°C.; add *b* to *a* at the same temperature. Stir until cold.

Haemorrhoid Ointment (Pile Ointment)

Yellow Petrolatum	53
Anhydrous Lanolin	30
Yellow Beeswax	5
Ethyl· Aminobenzoate	5
Bismuth Subgallate	5
Thymol Iodide	2

Melt the yellow petrolatum, lanolin, and beeswax together and allow to cool. Mix the three powders and triturate with a portion of the ointment base until smooth. Then add gradually the remainder of the base and mix until the ointment is homogeneous. This ointment must not come in contact with iron as discoloration will result. Only porcelain or wooden utensils should be used.

Hair Cream

Formula No. 1

Glyceryl Monostearate	10
Neutral Mineral Oil	20
Beeswax	3
Castor Oil	5
Water	50
Preservative and Perfume	Sufficient

No. 2

Triethanolamine	1.25
Stearic Acid	6.00
Spermaceti	3.00
Mineral Oil	25.00
Distilled Water	65.00
Perfume	Sufficient

Dissolve the triethanolamine, stearic acid, and spermaceti in the mineral oil on a water bath. While still warm, run in the water previously warmed; stir vigorously until quite homogeneous and cool. Finally work in the perfume, taking care not to occlude air.

No. 3

High-Viscosity Mineral Oil	48.0
Spermaceti	1.0
Glycerin	1.2
Quince Mucilage	0.5
Stearic Acid	0.5
Perfume Compound	0.6

Sodium *p*-Oxy-benzoate	0.2
Fatty Acid Sulfonate	2.0
Distilled Water	46.0

The quince mucilage is made from 3 parts quince seed dissolved in 97 parts distilled water. All fats and waxes, together with the sulfonated fatty alcohol, are melted at about 60°C. and then added to the aqueous balance of the mixture, which has also previously been heated to 60°C. The addition is made under constant stirring and when cold, the perfume is added.

Another formula is based on a solution of sodium alginate in water, 2.5 parts sodium alginate being dissolved in 100 parts distilled water, to which 1 part perfume compound dissolved in 5 parts alcohol is added. A small quantity of calcium citrate is then triturated in 90 parts distilled water and added to the sodium alginate solution. After 2 hours, the mixture will thicken, the proportion between sodium alginate and calcium citrate being the determining factor of viscosity.

No. 4

Stearic Acid	12
Ceresin	17
White Mineral Oil	72
Triethanolamine	5½
Distilled Water	71
Deodorized Kerosene	17
Diethyleneglycol Ethyl Ether	5

Dissolve the triethanolamine in the water and heat to 70°C. Melt the fats at 70°C. and pour slowly, with constant stirring, into the aqueous solution. Continue stirring until cool, then stir in the kerosene and the diethyleneglycol ethyl ether.

No. 5

White Wax	10
Liquid Paraffin	125
Borax	1
Distilled Water	14

Allow the wax to dissolve in 60 parts of liquid paraffin, then stir in the remainder of the liquid paraffin. Dissolve the borax in the distilled water, add to the previous solution, and stir the cream so formed very thoroughly.

Hair Fixers or Straighteners (Waxy)

Formula No. 1

Ceraflux*	40
Glyco Wax A*	10
White or Yellow Petrolatum	100
Rosin	40

Melt together until clear and stir until uniform. Pour into jars while melted.

No. 2

Beeswax	10
Yellow or White Petrolatum	100

* Glyco Products Co.

Paraflux	40
Flexoresin B1*	40

Method as given for formula 1. Formula 1 will give a very light-colored product if white petrolatum and FF rosin are used. Harder or softer product may be obtained by slight variation of the formulation.

No. 3

Beef Suet	16 oz
Yellow Beeswax	2 oz
Castor Oil	2 oz
Benzoic Acid	10 gr
Perfume	Sufficient

Melt the suet and wax, add the castor oil and acid, allow to cool, add perfume.

Hair Pomade

Formula No. 1

Cholesterol	1
Cetyl Alcohol	3
Anhydrous Lanolin	5
Paraffin Wax	20
Liquid Petrolatum	30
White Petrolatum	50
Water	70

No. 2

Cholesterol	1
Cetyl Alcohol	3
Anhydrous Lanolin	5
Paraffin Wax	20
Liquid Petrolatum	27
White Petrolatum	50
Water	106

No. 3

Tallow	390
White Ceresin	150
White Beeswax	50
Pale Rosin	200
White Mineral Oil	200
Perfume	10

Hair Pomade and Straightener

Yellow Petrolatum	6
Yellow Beeswax	1
Refined Peanut Oil	2
Perfume Oil	Sufficient

Melt the petrolatum in a double boiler and add the other ingredients. Do not add the perfume until the mixture begins to solidify around the edges. This is an amber product. If a water-white product is desired, use white petrolatum and white beeswax and substitute equal parts of deodorized castor oil and mineral oil for the peanut oil. By adding more wax and less of the oils, a product can be obtained that will straighten the most unruly hair.

Hand Lotion

Formula No. 1

Spermaceti	2.00
White Beeswax	0.25
Glycerin	6.00
Soap	3.00
Borax	0.50
Sodium Benzoate	0.20
Quince Seed	3.40
Alcohol	4.00
Water	80.00
Perfume	0.50
Preservative	0.15

* Glyco Products Co.

Add the quince seed to two thirds of the water in which the preservative has first been dissolved. Allow to soak for 24 hours and strain. Add the glycerin, borax, and sodium benzoate to the remainder of the water and stir until dissolved. Melt and add the spermaceti and white wax, stir rapidly for 1 or 2 hours; then add the soap. Add the mucilage next and finally add the perfume dissolved in the alcohol, in a very thin stream.

No. 2

Water	1 gal
Triethanolamine	2 oz
Glycerin	3 oz
Stearic Acid	3½ oz
Beeswax	1½ oz
Absorption Base	4 oz
Mineral Oil	2 oz
Karaya Gum Solution	2 oz
Perfume	¼ oz

Heat the oils and waxes in one container and the water, glycerin, and triethanolamine in another container to approximately the same temperature, about 80°C. Mix the two solutions and stir thoroughly. Add the karaya gum solution, which consists of 2 oz of powdered karaya gum to 1 gal of water, and continue stirring until cold. This lotion can also be used as a powder base.

Hand-Protective Cream

Formula No. 1

Beeswax	5.0
Glyceryl Monostearate	12.5
Hydrous Wool Fat	5.0
Sodium Silicate	5.0
Ammonium Hydroxide (10% Solution)	0.5
Petrolatum	72.5

If desired, this formula can be modified by the addition 5% by weight of latex to produce a rubbery film on the skin.

No. 2

(Recommended for Prolonged Contact with Water)

White Wax	10
Hydrous Wool Fat	5
Sulfonated Olive Oil	10
Petrolatum	75

Healing Plaster

Beeswax	2
Pine Tar	4
Rosin	4

Melt together and, when almost cool, mold into sticks.

Apply by heating the wax, coating a thin layer on a piece of muslin, and placing over the injury.

Hormone-Cream Base

Anhydrous Lanolin	22.0
White Petrolatum	40.0
Yellow Petrolatum	6.0
Cholesterol	1.0
Heavy Mineral Oil	20.0
Lecithin	0.5
Distilled Water	10.0
Perfume	0.5

Industrial Hand-Protection Creams

Formula No. 1

The following preparation has a pH of 5.4 and is, therefore, recommended where there is prolonged contact of the hands with soapy water:

Beeswax	10.0
Lanolin	5.0
Glyceryl Monostearate	12.5
Stearic Acid	2.0
Petrolatum	75.5

No. 2

The following preparation to waterproof the skin is recommended where there is prolonged contact with water:

Beeswax	10
Lanolin	5
Sulfonated Olive Oil (75%)	10
Petrolatum	75

No. 3

Glyceryl Monostearate	12.0
Beeswax	12.0
Anhydrous Lanolin	6.0
Cholesterol	1.0
Sodium Silicate	5.0
Ammonium Hydroxide (10% Solution)	0.5
Water	To suit

Melt the wax, glyceryl monostearate, lanolin and cholesterol in one pot. Add the sodium silicate and ammonium hydroxide solutions to the water previously heated in another pot. Stir the aqueous solution into the wax mixture.

This is a clean, nongreasy preparation that dries on the skin and does not rub off. Its use is indicated, therefore, in dry work, as a protection against dust-borne irritants, or when soiling materials by the protectives must be avoided.

Formula 1 is smeared on the skin, while 2 and 3 are liquids and are applied with a brush or swab.

No. 4

Beeswax	5.0
Glyceryl Monostearate	12.5
Lanolin	5.0
Sodium Silicate	5.0
Ammonium Hydroxide (10%)	0.5
Petrolatum	72.5

No. 5

Glyceryl Monostearate	8
Magnesium Stearate	14
Beeswax	3
Petrolatum	10
White Mineral Oil	5
Water	60

No. 6

Hard Paraffin Wax	50
Yellow Petrolatum	400
Peanut Oil	200
Yellow Beeswax	100
Anhydrous Lanolin	60
Pale Rosin	40
Water	140
Borax	10

No. 7

Lanolin	2
Glycosterin*	19
Dextrin	4
Glycol Bori-Borate	3
Water	72

* Glyco Products Co.

Dissolve the last four ingredients by warming to 60°C., with stirring. Melt the first two ingredients together and add the solution, stirring slowly but well. A little perfume may now be stirred in.

This gives an excellent cream to protect hands while working with greases, lacquers, etc.

Insect-Repellent Cream

Beeswax	3–4
Citronella Oil	15
Spirits of Camphor	8
Cedar-Wood Oil	8
White Petrolatum	60

Melt the petrolatum and beeswax, then add the balance, and stir until smooth.

Liniment or Rheumatic Cream

Formula No. 1

Raisin-Seed Oil	9
Methyl Salicylate	15
Diglycol Stearate	15
Water	61

No. 2

Avocado Oil	10
Methyl Salicylate	15
Diglycol Stearate	22
Water	60

Heat the diglycol stearate with the oil at 70°C. until dissolved. While mixing, add warm methyl salicylate and then pour the mixture slowly, with rapid mixing, into the water heated to 70°C.

Lip Pomade

Mineral Oil	1 gal
White Petrolatum	2 lb
White Ozokerite	5 lb
White Beeswax	2 lb
Perfume	1 oz
Color	To suit

Lip Rouge

Liquid Paraffin	7
White Wax	5
Spermaceti	5
Carmine No. 40	2
Perfume	To suit

Rub down the carmine in a cold mortar to a smooth paste with some of the liquid paraffin, then add the remainder of the liquid paraffin, and mix well. Melt the wax and spermaceti and add the carmine mixture. Mix well and run through an ointment mill.

Lipstick Base

Formula No. 1

Beeswax	150
White Ceresin	300
Anhydrous Lanolin	50
Petroleum Jelly	250
Cetyl Alcohol	50
Mineral Oil	100

No. 2

Beeswax	300
White Ceresin	70
Anhydrous Lanolin	80
Petroleum Jelly	300
Cetyl Alcohol	50
Mineral Oil	50

No. 3

Beeswax	250
White Ceresin	50
Anyhdrous Lanolin	70

Petroleum Jelly	400
Cetyl Alcohol	50
Spermaceti	90
Mineral Oil	50

No. 4

Beeswax	300
Anhydrous Lanolin	80
Petroleum Jelly	300
Cetyl Alcohol	50
Paraffin Wax	80
Mineral Oil	50

No. 5

Beeswax	300
Anhydrous Lanolin	60
Petroleum Jelly	300
Cetyl Alcohol	60
Spermaceti	50
Mineral Oil	50

No. 6

Beeswax	150
Castor Oil	40
White Ceresin	300
Anhydrous Lanolin	60
Petroleum Jelly	300

The proportion of lakes and pigments depends naturally on the shade, with a 15% minimum and 25% maximum—seldom more.

Lipstick Removers

Formula No. 1

Beeswax	40
Stearic Acid	20
Mineral Oil (Medium Cosmetic Grade)	40

No. 2

Cetyl Alcohol	5
Beeswax	15
White Ceresin	40
Mineral Oil	40

Lipsticks

Formula No. 1

Cetyl Alcohol	240
White Petrolatum	370
Ceresin	240
Colored Lakes	120
Perfume	30

No. 2
(The Color Develops on Application)

a. Stearic Acid	9
Eosin	1
b. Stearic Acid	30
White Beeswax	40
Paraffin Wax (50/52°C.)	20
White Ceresin Wax	20
Mineral Oil	60
Anhydrous Lanolin	5
c. Anhydrous Lanolin	15
White Ceresin Wax	50
Spermaceti	5
Mineral Oil	60

Add to the wax composition *b* or *c*, 2 parts of the eosin stearate, mix well, and add a bit of alcohol to disperse the dye better. The alcohol evaporates quickly.

No. 3

Petrolatum	15 oz
Beeswax	10 oz
Spermaceti	400 gr
Carmine	6 dr
Perfume	To suit

Melt and stir. Allow to cool before adding perfume. Pour into molds.

No. 4
(Indelible)

Stearoricinol	28
Mineral Oil	4
Anhydrous Lanolin	2
Petrolatum	2
Paraffin Wax	8
Beeswax	8

Bromo "Acid"	1
Lake Colors	5
Perfume Oil	1

By varying the colors correspondingly different shades may be obtained.

No. 5

(For Theatrical Use)

Petrolatum	4 lb
Paraffin Wax	2 lb
Mineral Oil	1 lb
Carnauba Wax	6 oz
Lanolin	8 oz
Lake Colors	1 lb
Perfume	As desired

The same lake-color mixtures as are used in the greasy cream rouges are suggested to secure the various shades. The procedure is the same.

No. 6

Paraffin	2
White Vaseline Oil	3
White Beeswax	1
Ozokerite Ceresin	3
Titanium Dioxide	1

Use a suitable red dye for coloring.

No. 7

Carnauba Wax	1½
Beeswax	7½
Ozokerite	2½
Spermaceti	2
Pure Cetyl Alcohol	½
Petrolatum	5
Lanolin	5½
Bromosolv (a 4% Solution of Tetrabromo Fluorescein)	25
Perfume	½

Melt the waxes, lanolin, petrolatum, and cetyl alcohol. Stir occasionally to assist solution. Add the Bromosolv and perfume and continue stirring until a clear solution results. Pour into molds and chill.

Liquefying Cream

Formula No. 1

Mineral Oil	7
Ceraflux*	3
Petrolatum	2

Melt together at 220°F. and stir at room temperature until cold; add perfume; pour into jars while liquid, but at the lowest possible temperature.

This cream will not sweat oil in hot weather.

No. 2

(Medium Translucent)

Light or Medium Mineral Oil	50
Paraffin	18
White Petrolatum	23
Spermaceti	9

No. 3

(Hard Type for Hot Climates)

Spermaceti	5 lb
White Petrolatum	8 lb
Ozokerite	5 lb
Liquid Petrolatum	1½ gal

Liquefying Massage Skin Food

Lanolin	4
Cocoa Butter	4

* Glyco Products Co.

Almond Oil	4
Coconut Oil	8
Yellow Soft Paraffin Wax	12
Hard Paraffin Wax	4
Perfume	As desired

Liquid Almond Cream

Oil of Sweet Almonds	1 lb
Spermaceti	2 lb
Beeswax	2 lb
Powdered Castile Soap	3 lb
Borax	2 lb
Quince Jelly	1 lb
Alcohol	1 pt
Water	4 pt

Melt the spermaceti and wax together. Dissolve the soap and borax in hot water. Mix these together and add the balance of ingredients. Stir and filter through cloth.

Liquid Lip Make-Up

a. Triethanolamine Oleate	15
Glycerin	10
Water	10
b. Beeswax	10
White Mineral Oil	40
Oil-Soluble Scarlet	2½
c. Perfume (Eau de Cologne Oil, Rose, Fougère)	As desired

Emulsify by adding *b* to *a* at 50–60°C. Stir till cold and add perfume.

Liquid Lip Stain

Triethanolamine Oleate	15.0
Glycerin	10.0
Water	10.0
Beeswax	10.0
White Mineral Oil	40.0
Oil-Soluble Scarlet	2.5
Perfume	To suit

Make-Up Remover

Ceraflux*	2½
Mineral Oil	8
White Petrolatum	10
	No perfume

Mascara

Formula No. 1

Glyceryl Monostearate	10
Triethanolamine	5
Stearic Acid	15
White Petrolatum	20
Beeswax	25
Gelatin	2
Water	13
Carbon Black	10

Make a solution of the gelatin in the water, and add the triethanolamine. Melt the wax, glyceryl monostearate, petrolatum, and stearic acid; add the pigment; and then the hot gelatin solution. Mill to get a uniform product.

No. 2
(Soapless—Poured)

Triethanolamine	14
Stearic Acid	20
Oleic Acid	5
Ricinoleic Acid	5
Carnauba Wax	30
Ozokerite	15

Massage Cream

Spermaceti	10.0
Solid Paraffin	15.0

* Glyco Products Co.

Mineral Oil	45.0
Lecithin	1.5
Cholesterin	0.5
Borax	1.0
Water	30.0
Perfume	As required

Lecithin and cholesterin can be best dissolved in the liquid or melted fats and waxy matter. The melted mass is permitted to cool at 40°C. and the hot solution of borax in water is poured first in small portions and then in larger portions into the fused mass while stirring thoroughly. Then it is stirred cold.

In the case of a vanishing cream, it is somewhat more difficult to work in the lecithin. The easiest method is to dissolve the lecithin in the melted stearic acid (overheating should be prevented) and to mix the potash solution into it by stirring in the usual way. Saponification and emulsification might be affected by the lecithin. If any oil is permitted in the vanishing cream, lecithin is ground fine with warm mineral oil (1 part of lecithin to ½–1 part of mineral oil), so that a mass is produced that can be distributed. As soon as the cream has been mixed and while it is still warm, the warm lecithin oil is stirred thoroughly into it. The whole cream is stirred cold.

Mentholated Adhesive Tape

Yellow Beeswax	100
Rosin	750

Melt together; mix until uniform; cool until it thickens slightly and then mix in:

Menthol	50

Apply to cloth or rubberized tape.

Metol Dermatitis Ointment

Many photographers show an idiosyncrasy to metol, which causes dermatitis of their fingers. Other organic chemicals used in photography may have a similar effect on the skin. The following ointment is recommended:

Lanolin	10
Soft Paraffin Wax	15
Ichthyol	5
Boric Acid	15

Rubber gloves should be worn if at all practicable.

Migraine Pencil

Stearic Acid	70
Menthol	30

Melt together on a water bath and cast in molds.

Milk of Almond

Formula No. 1 (Synthetic)

Paraffin Oil (Viscosity 30–32)	350

Triethanolamine Stearate	80
Beeswax	20
Perfume	10
Distilled Water	500

No. 2

Sweet Almond Oil	50
Pulverized Almond Oil Soap	15
Spermaceti	10
Beeswax	10
Diethylene Glycol	25
90% Alcohol	50
Distilled Water	840

Milkers' Hand Lubricant

Formula No. 1

Hard Paraffin	320
Ceresin	280
White Mineral Oil	400

No. 2

White Mineral Oil	600
White Ceresin	200
Anhydrous Lanolin	200

No. 3

High-Viscosity White Mineral Oil	60
White Ozokerite (60/62°C.)	20
Neutral Wool Fat	20

Melt the ingredients on a water bath, mix thoroughly, and pour into containers near the filter point of the "fat."

Modern Cold Creams

The term "cold cream" is properly limited to traditional beeswax-borax, oil and water creams, but it can also be applied to any white cream that contains sufficient water to exert a cooling effect on the skin bec.use of water evaporation. The regular type of cold cream generally contains 25–35% water. Small amounts of water tend to favor the formation of water-in-oil emulsions. In modern cold creams the water content may be 40– over 60%, depending on the other ingredients present. Both glyceryl monostearate and diglycol stearate, which are similar chemically to naturally occurring fats, are used as emulsifying agents and require no additional emulsifying aids. Because of their fatlike qualities they may be subject to attack by certain molds. Preservatives should always be included in formulations using fatty acid esters, particularly where shelf life is an important factor.

Formula No. 1

Glyceryl Monostearate	12.0
Beeswax	3.0
Spermaceti	3.0
Mineral Oil	30.0
Glycerin	8.0
Water	43.5
Moldex (Preservative)	0.1
Perfume	0.4

All glyceryl monostearate creams may be made by either of two alternative procedures. According to the simpler but not necessarily better method, all of the ingredients, with the exception of perfume, are heated together to about 85°C.

until the waxes, fats and oils have been completely melted and float on the water as a liquid mixture. Stirring is now started and continued until the mixture becomes a smooth cream. Unless particular care is taken in this procedure, some wax lumps may remain undispersed. A more usual method involves melting the waxes and oils in one pot, heating the water with the water-soluble components in another pot, and mixing the two together at the same temperature, i.e., 70°C. Stirring is continued until a smooth cream results and the perfume is added at 45°C., with stirring.

For creams with different consistency, texture, oiliness, etc., proportions may be varied within relatively wide limits. The oil generally should not exceed 30% of the total; glyceryl monostearate may, in special cases, go as high as 25%. Where the oil content is relatively high, passage of the cream, while it is still warm, through either a colloid mill or a homogenizer almost always improves the product.

No. 2

Glyceryl Monostearate	14.0
Petrolatum	6.0
Ozokerite	2.0
Mineral Oil	25.0
Glycerin	5.0
Water	47.5
Moldex*	0.1
Perfume	0.4

The type of petrolatum has a definite effect on the quality of the cream: long-fiber petrolatum gives an oilier cream, while the preparation containing the short-fibered grade will show greater drag.

No. 3

Glyceryl Monostearate	12.0
Petrolatum	9.0
Paraffin Wax	6.0
Mineral Oil	14.0
Glycerin	3.0
Water	55.5
Moldex*	0.1
Perfume	0.4

Paraffin wax generally used in cosmetic formulation is the 125°–127°F. grade; higher melting varieties give a somewhat harder cream. Paraffin wax hardens the cream yet at the same time permits it to melt quickly on the skin.

Modern Tissue Cream

In kettle 1 place:

Parachol* (Absorption Base)	12 lb
Mineral Oil	3 lb
Lanolin	2 lb
Sun-Bleached Beeswax	1 lb
Cetyl Alcohol	1 lb

Heat gently to 150°F.

In kettle 2 place:

* Glyco Products Co.

Water	16 qt

Heat gently to 150°F. Add the water to kettle 1 and stir gently. Let cool, stirring occasionally till creamy. Add 10 to 12 drams perfume. Mix well and fill jars with a spatula when cold.

Moorhaf's Bone Wax

Iodoform	20
Spermaceti	40
Sesame Oil	40

Heat the ingredients slowly to 100°C. and allow the mixture to cool while stirring. For use as a bone plug it is heated to 50°C.

Mosquito Cream

Glyceryl Monostearate	11.0
Cedar-Leaf Oil	4.0
Pennyroyal Oil	4.0
Linalyl Acetate	3.0
Gasoline	5.0
Menthol	0.5
Phenol	2.0
Glycerin	5.0
Water	65.5

Put the glyceryl monostearate into the water, add the glycerin, and bring the mixture to a boil with constant stirring. Keep stirring and, when the temperature drops to about 45°C., add the rest of the ingredients which have been mixed together.

Mosquito-Repellent Cream

Formula No. 1

a. White Wax	50 g
Spermaceti	50 g
b. Borax	4 g
Ammonia (0.96)	40 g
c. Water	510 cc
Wheat Starch	1 g
Gelatin	4 g
Sodium Benzoate	½ g

Make up the cream as usual, pouring *b* into *a*, then add the solution *c* which is to be made up before (soak cold, then warm to clear solution; if necessary, pour through a fine sieve), stir thoroughly, stop heating, stir until cooled, and add:

Eucalyptus Oil	50 cc

No. 2

White Mineral Oil	16 oz
Beeswax U.S.P.	4 oz
Spermaceti	1 oz
Distilled Water	8 oz
Borax	30 gr
Butyl Salicylate	1½ oz

Moth-Protection Tablets

Naphthalene	225
Camphor	75
Ceresin	50

Melt together and then add

Hexachloroethane	50
Pine-Needle Oil	5

Dip cardboards into the mixture while fluid.

Mustache Pomade

Glycosterin	4
Gum Arabic	8
Water	31

Castile Soap	12
Beeswax	30
Diethylene Glycol	12
Perfume	To suit

Warm until the wax has melted and mix vigorously until uniform.

Mustache Wax

Propylene Glycol	100
Hard Soap	100
White Beeswax	250
Trigamine*	10
Gum Arabic	50
Water	480
Perfume and Color	To suit

Mustard Ointment

Mustard Oil	30 min
Menthol	40 gr
Camphor	80 gr
Yellow Wax	110 gr
Lard	4 oz

Melt the wax and lard together and, while cooling, stir in the remaining ingredients.

This is a very satisfactory rubefacient preparation to be used in sprains, rheumatism, etc.

Nail and Cuticle Softener and Conditioner

Castor Oil	2 lb
Anhydrous Lanolin	4 oz
White Petrolatum	2 lb
Beeswax	4 oz
Perfume	¼ oz

Red Alkanet Solution to Produce the Desired Shade.

* Glyco Products Co.

Melt the beeswax and fats at a very low heat. Add a few drops of the oil-color solution until the proper tint is obtained. Strain the mixture through a fine cheesecloth to remove any foreign particles. Add the perfume to the mixture when sufficiently cool. The consistency of the product can be altered by varying the amount of the beeswax.

The red alkanet solution is prepared as follows:

Oil-Soluble Red Alkanet	½
Hot Mineral Oil	12

Mix together.

Nail Cream

a. White Beeswax	15.0
White Ozokerite	2.5
Bleached Montan Wax	2.5
Mineral Oil	60.0
Cetyl Alcohol	2.0
b. Water	23.5
Borax	1.5
c. Aluminum Stearate	20.0
d. Fish Silver Paste (10% Dispersed in Amyl Acetate)	5.0

Melt *a* and add *b* at about 70°C. Stir thoroughly and add *c*. Stir until the emulsion is almost cold and add *d*.

Nail-Polish Paste

Formula No. 1

A good formula for a nail polish in paste form contains

100 parts of light-colored rosin, 60 parts of stearin, 60 parts of yellow beeswax and 200 parts of ceresin wax. These ingredients are melted together on a water bath and then 300 parts of white petrolatum are mixed in. Then a well-mixed mixture of 200 parts of washed kieselguhr, 140 parts of zinc oxide, and 100 parts of tin oxide is stirred into the waxy base. Before the mixture is removed from the water bath, coloring matter is added; for example, alkanna pink, and 15—20 parts of perfume. These ingredients must be added shortly before the mass becomes solid and is poured into containers.

No. 2

Ceresin	6
Olein	44
Precipitated Chalk	50
Color and Perfume	To suit

Neutral Cleansing Cream

1. Mineral Oil	80
2. Spermaceti	30
3. Glyceryl Monostearate	24
4. Water	90
5. Glycerin	10
6. Perfume	To suit

Heat 1, 2 and 3 to 140°F. and stir into it slowly 4 and 5 heated to the same temperature. Add perfume, at 105°F., stir slowly until cold; after allowing to stand for 5 minutes, stir until smooth and pack.

A four-purpose cream that cleans, nourishes, stimulates, and acts as a powder base is made as follows:

Mineral Oil	3 pt
White Petrolatum	½ lb
(Heat to 140°F.)	
Water	4½ pt
Glycerin	5 oz
Preservative	½ oz

Heat to 140°F. and add slowly with stirring to the oil mixture. As the temperature falls, a gelatinous mass forms at 120°F. one ounce perfume oil is added while stirring and the gelatinous mass changes to a white cream. Slow stirring is continued until cold. This cream may be packed either in tubes or in jars.

It can be modified by various coloring agents and perfume to obtain specialty creams. Since it is neutral, viosterol, or gland or hormone extracts can be incorporated.

Nongreasy Cold Cream

Formula No. 1

Glyceryl Monostearate	22
Petrolatum	16
Paraffin Wax	12
Mineral Oil	30
Water	98

Heat the first four ingredients to 170°F. and stir together. Then slowly, with stirring, pour in the water which has been heated to the same temperature. Stir thoroughly and then allow to stand (hot) until the air bubbles are gone. Add perfume and stir; pour at 110–130°F. Cover the jars as soon as possible.

No. 2

Stearic Acid	16
Glycerin	48
Mineral Oil	12
Paraffin Wax	2
Stronger Ammonia Water	4
Water	64
Perfume	3/4

Nongreasy Liquid Cleansing Cream

1. Beeswax	1.5
2. Spermaceti	6.5
3. Cherry-Kernel Oil	6.0
4. Glycosterin*	4.0
5. Water	122.0
6. Alcohol	3.0
7. Galagum	1.0
8. Borax	3.0
9. Perfume	3.0
10. Glycerin	4.0

Melt together 1, 2 and 3. Heat while stirring, 4, 5, 7 and 8 together until uniform. Mix these two solutions, stirring until uniform. Stir in 6, 9 and 10 and mix until uniform.

Nonstaining (Nonleaking) Mineral Oil

(Laxative)

White Soft Paraffin Wax	2
Mineral Oil U.S.P.	6

Warm together and stir until uniform.

* Glyco Products Co.

Nourishing Cream

Formula No. 1

1. Beeswax	15
Mineral Oil	45
Anhydrous Lanolin	12
Glyco-Wax A*	15
2. Water	25
Borax	1 1/4
Benzoate of Soda	1/2
3. Perfume	1/2

Heat 1 and 2 separately to 200°F., then add 1 to 2 slowly with stirring in an emulsifier or beater. When the cream begins to set, add the perfume. Allow to stand overnight; stir the next morning and package.

This cream has exceptional penetrating power and is absorbed very readily by the skin.

No. 2

Cetyl Alcohol	50
Lanolin	50
Cocoa Butter	50
Cholesterin	10
White Beeswax	50
Arachis Oil	100
Mineral Oil	150
White Petrolatum	240
Water	280
Perfume	20

No. 3

White Beeswax	9
Spermaceti	3
White Petroleum Jelly	35
Benzoated Lard	18
Lanolin	4
Liquid Paraffin	9
Distilled Water	21
Borax	1

No. 4

Beeswax	6.0
Paraffin Wax	12.5

Petrolatum	8.0
Mineral Oil	58.0
Water	15.0
Potassium Carbonate	0.3
Borax	0.2

The fats and oils are melted at 65–70°C. The liquid solution of potassium carbonate and borax, previously brought to the same temperature, is then added under constant stirring. If lecithin, vitamins, etc., are also used, they and the perfume oil are dissolved in a small amount of mineral oil and added to the cream after the temperature has dropped to about 35°C.

No. 5

Paraffin Wax	2½ lb
Beeswax	5 lb
Pure Cetyl Alcohol	½ lb
Anhydrous Lanolin	5 lb
White Mineral Oil	8⅝ lb
Peanut Oil	10 lb
Borax	¼ lb
Water	17½ lb
Preservative (e.g., Methyl *p*-Hydroxybenzoate)	1½ oz
Perfume	4–6 oz

Melt together the waxes, lanolin, alcohol, and oils and keep at 60°C. Dissolve the borax in the water and heat to the same temperature. Add the borax solution to the melted wax-oil mixture and stir vigorously. When emulsification is complete, stir more slowly until cold. Add the perfume at 35°C.

Ointment Base

Glyceryl Monostearate	15
Cetyl Alcohol	15
Glycerin	35
Diethylene Glycol	35

Oriental Balm

Formula No. 1

Paraffin Wax	20
White Petrolatum	30
Eucalyptus Oil	12
Peppermint Oil	3
Camphor Oil	3
Menthol	3
Nutmeg Oil	3
Cinnamon Oil	1

No. 2

Paraffin Wax	20 g
White Petrolatum	30 g
Eucalyptus Oil	10 g
Peppermint Oil	3 g
Camphor Oil	3 g
Menthol	3 g
Clove Oil	6 drops
Peru Balsam	11 drops
Nutmeg Oil	5 g

Melt the first two ingredients, stir in the rest, and pour into small glass jars to cool. This balm may be used to alleviate headaches, rheumatism, colds, etc.

Peanut-Oil Massage Cream

This is excellent for undernourished parts, cold feet, stiff joints, and as an aid to any flabby parts of the body.

White Beeswax	20
Spermaceti	20
Peanut Oil	45

Lanolin	12
Water	45
Borax	3
Camphor	1

Melt the wax over a water bath, add the lanolin and oil, and heat to 80°C. Dissolve the camphor in the oil with gentle heat; dissolve the borax in the water and heat to 80°C. At this temperature, add the water to the wax-oil mixture at the same temperature with vigorous stirring; a motor stirrer is the best.

Peroxide Cream

a. Paraffin Wax	4
Mineral Oil	10
Lanolin	6
Acimul*	15
b. Hydrogen Peroxide (100 Volume)	5
c. Water	60

Heat *a* to 70°C. and stir till completely clear. Heat *c* to 70°C. and add *a* to *c* slowly, stirring in one direction only. At 55°C., add *b* and perfume. Stir and pour at 50°C.

If 5 parts of the water are replaced with 5 parts Carbitol,** the cream can be poured at 45°C.

Petroleum Jelly

Formula No. 1

White Petrolatum	20
Ceresin	3
Paraffin Wax	4
White Light Mineral Oil	73

No. 2

White Petrolatum	10
Paraffin Wax	6
Ceresin	3
Ozokerite	2
Cetyl Alcohol	6
White Heavy Mineral Oil	73

Mix all solids. Heat until melted and dissolved. Heat the oil in a separate container 5–10° above the temperature needed to melt the solids. Mix all ingredients and stir well. Pour in containers and allow to solidify without stirring.

Poison-Gas Ointment

Powdered Boric Acid	44 gr
Eucalyptus Oil	15 min
Soft Paraffin Wax	to 1 oz

This should be spread on the smooth side of strips of white lint 10 × 4 in., stored folded across the middle to keep the active side of the dressing clean. The strips are opened out and laid on the skin with a slight overlap and kept in position with the lightest bandaging; the dressing should be changed, if possible, night and morning.

Polishing of Pharmaceutical Tablets, Pills, Etc.

Pharmaceutical tablets are generally polished by tumbling

* Glyco Products Co.
** Carbide and Carbon Chem. Corp.

with a carnauba wax—beeswax mixture in a suitable solvent, such as acetone. This mixture of natural waxes can be replaced by glyceryl tristearate which will not spot or discolor colored or black coatings. It also effects a considerable economy because of lower cost.

In this connection, anhydrous ethyl acetate or deodorized naphtha is suggested as a solvent.

Pore or Blackhead Cream

For the removal of blackheads and cleansing of the pores, a pore cream or blackhead cream is used. Applied over areas of large pores after the face has been moistened with warm water, it is left on for a few minutes and then washed off with warm water. The face is then rinsed with ice-cold water and an astringent lotion applied with cotton.

Salicylic Acid	2.5
Starch	25.0
Zinc Oxide	35.0
Castor Oil	7.5
Soft White Paraffin Wax	30.0

Grind the three powders thoroughly; then add slightly warmed paraffin to the oil and powder after triturating in a warmed mortar.

Powder Cream

Absorption Base	73.5
Soft Paraffin Wax	49.0
Ozokerite	10.0
Liquid Paraffin	7.5
Water	30.0
Powder	80.0

The powder is ground into the finished emulsion.

Pressing Oil

White Petrolatum	30 lb
Paraffin Wax	1¼ lb
Imitation Bergamot Oil	⅜ oz
Liquid Food Drug and Cosmetic Yellow No. 3	1¾ oz

Melt the petrolatum and wax. Remove from the fire, add the color, and then the bergamot oil.

The yellow No. 3 is prepared by dissolving 1 dr of dry color in 1 pt of white mineral oil.

Prickly-Heat Preparation

Menthol	2 gr
Camphor	10 gr
Eucalyptus Oil	3 min
Paraffin Wax	To make 1 oz

Warm together and mix until uniform.

Rouge

Hydrogenated Castor Oil	56
Beeswax	16
Mineral Oil	8
Anhydrous Lanolin	4
White Petrolatum	4
Lake Colors	10
Eosin	2

Salt-Containing Cosmetic Cream

Polymerized Glycol Stearate	10
Stearic Acid	10
Paraffin Wax	15
Mineral Oil	15

Petrolatum	10
Sodium Chloride	6
Water	To make 100

Scabies Preparation

Sodium Benzoate	0.1
Spermaceti	10.0
Sodium Lauryl Sulfate	1.0
Stearyl Alcohol	10.0
Cetyl Alcohol	3.0

Skin Protective

Beeswax	5.0
Glyceryl Monostearate	12.5
Hydrous Wool Fat	5.0
Sodium Silicate	5.0
Ammonium Hydroxide (10% Solution)	0.5
Petrolatum	72.5

The ammonium hydroxide is incorporated to prevent the precipitation of the sodium silicate. The same formula with 5% latex gives a product which deposits a rubberlike film on the skin. The base is melted, the latex added with stirring, and the mixture stirred constantly until it congeals.

Soft Tissue Cream with Cholesterin Base

Absorption Base	30
Lanolin	5
Water	55
White Beeswax	10

Melt the wax and lanolin, add the base, and stir in the water (warm).

The consistency can be adjusted by changing the wax content to suit.

Solid Brilliantine

Yellow Paraffin Jelly	12 lb
White Beeswax	1 lb
Oil-Soluble Red Color	4 gr
Perfume	To suit

Solid Perfume

Paraffin Wax	500
Melt on Water Bath and Add	
Perfume Oil	50
Menthol	½

Mix well and pour into containers at 160°F.

Sports Cream

Formula No. 1

Absorption Base	24
Cetyl Alcohol	1
Liquid Paraffin	5
Lanolin	3
Glycerin	3
Water	64

No. 2

Absorption Base	24.0
Cetyl Alcohol	1.2
Liquid Paraffin	10.5
Glycerin	3.0
Water	56.3
Hard Paraffin Wax	5.0

No. 3

Glycosterin*	12
Mineral Oil	10
Cetyl Alcohol	3
Water	100

Heat to 60°C. and mix vigorously till emulsified.

* Glyco Products Co.

Stick Salve

Coconut Oil	10
White Beeswax	30
Lard	22
Lanolin	35

The base should be carefully melted and agitated and the pulverized medicament thoroughly mixed into the melted mass before it congeals.

Sulfathiazole Creams

Formula No. 1

Glyceryl Monostearate S	12.0
Yellow Wax U.S.P.	2.0
Glycerin	6.0
Corn Oil	4.0
Water	76.0
Sulfathiazole	5.0

No. 2

Sulfathiazole	5
Sodium Lauryl Sulfate	1
Stearyl Alcohol	10
Cetyl Alcohol	3
Spermaceti	10
Glycerin	10
Water	61

Sulfonamide Chewing Wafer

(For Mild Tonsillitis)

Sulfanilamide	2 g
Sodium Bicarbonate	2 g
Sulfathiazole	4 g
Paraffin Wax	16 g
Peppermint Oil	10 min

The solids are ground into a fine powder and, under constant stirring, are added to the melted wax and peppermint oil, poured into a mold, and divided into twenty tablets.

The patient chews the tablets as he would chew gum. One tablet is administered every 2 hours. Subjectively there is an almost immediate soothing effect on the inflamed mucous membranes. Resolution of the inflammatory process usually occurs in 2 or 3 days.

Sulfonamide Ointments

Formula No. 1

(Washable Oil-in-Water Base)

Mineral Oil	64.00
Water	23.45
Beeswax	5.00
Sulfonamide	5.00
Triethanolamine	2.00
Methocel*	0.30
Sodium Bisulfite	0.20
Aerosol OT**	0.05

No. 2

(Washable Oil-in-Water Base)

Water	71
Cetosten	15
Glycerin	5
Sulfonamide	5
Duponol C†	2
White Wax	2

Sunburn Ointment

Peanut Oil
Spermaceti
Beeswax

Heat and mix, then add:

Calcium Acetate
Water

* Dow Chem. Co.
** Am. Cyanamid Co.
† Du Pont.

Sunburn Preventive Ointment

An excellent ointment for sunburn may be prepared by gently heating and mixing:

Peanut Oil	60
Spermaceti	8
White Beeswax	7

Then adding:

Calcium Acetate	5
Distilled Water	20

Sun-Tan Cream

Ceresin Wax	15.0
Beeswax	8.0
Lanolin	10.0
Petrolatum	10.0
Mineral Oil	56.5
Perfume	0.5
Color	Sufficient

Melt the waxes; add the lanolin and petrolatum. Mix, strain, and perfume. Then add a sufficient brown color.

Suppository Base

Formula No. 1

Beeswax	1.5
Cocoa Butter	4.5
Cod-Liver Oil	4.0

No. 2

Triethanolamine Stearate	4
Beeswax	3
Cod-Liver Oil	20

No. 3

White Wax	2.50
Hard Paraffin Wax	5.00
Wool Fat	5.00
Spermaceti	35.00
Liquid Paraffin To make	100.00

Tablet-Polishing Solution

Formula No. 1

White Beeswax	2 oz
Carnauba Wax	1½ oz
High-Test Lead-Free Gasoline	1 gal

Melt the waxes together and while hot stir into the gasoline which has been previously heated to the boiling point on a water bath.

If gasoline is objectionable because of its inflammability, carbon tetrachloride may be used as the solvent.

Be certain that there are no open flames around, as this mixture is highly inflammable. Use the polishing solution hot. Add 1 oz of the polishing solution and rotate the tablets in the canvas polishing pan until the solvent has completely evaporated and the tablets begin to take on a luster. Then add about ½ oz additional hot polishing solution and rotate the tablets as before. Add third and last coat of ¼ oz hot polishing solution and rotate the tablets until completely polished. Do not rotate the tablets too long while polishing as there is a tendency for the coating to chip off. Do not blow air on the tablets during the polishing process.

Some firms apply small amounts of powdered talc from an ordinary commercial shaker box to the rotating tablets after the last coat of polishing solution has completely dried.

No. 2

Carnauba Wax	1–2
Bareco 170 Wax*	1

Melt together and mix.

Toothache Gum

Yellow Beeswax	60
Venice Turpentine	10
Gum-Mastic Powder	10
Ethyl Aminobenzoate	5
Dragon-Blood Powder	10
Clove Oil	5

Melt the beeswax and turpentine together and add the gum mastic. Stir until dissolved. Then add the ethyl aminobenzoate and, when dissolved, the dragon blood. Stir until cooled to about 50°C., then add the clove oil, and mold into sticks.

Translucent Jelly Cream

Stearic Acid	6
Spermaceti	15
White Petrolatum	30
Mineral Oil	49
Perfume Oil	To suit

Melt the stearic acid and the spermaceti, add the petrolatum and when melted stir in the mineral oil which has first been heated. When almost set, stir in perfume.

Tropical Ulcer and Wound Ointment

Clean the ulcer with a weak phenylmercuric nitrate solution and apply a water-in-oil emulsion consisting of:

Sulfathiazole	12
Calcium Oleate	2
Beeswax	3
Cod-Liver Oil	60
Water	40

Sterilize the ingredients separately and work into a paste under aseptic conditions. Then apply sulfanilamide-impregnated dressings on top.

In some cases, 10% sulfanilamide paste in vitaminized oil has been used with satisfactory results.

Ulcer Salve

Ethyl Aminobenzoate	3
Paraffin	10
Petrolatum	20

Spread on gauze and apply to ulcer.

Vanishing Ointment Base

Diglycol Stearate	12
Glycerin	3
Spermaceti	3
Distilled Water	42

No. 2

Stearic Acid	13
Diglycol Stearate	20
Water	67

Vegetable Vanishing Cream

Hydrogenated Cottonseed Oil	10
Sesame Oil	50
Water	30
Beeswax	8
Lanolin	2

* Bareco Oil Co.

To this formula 1–3% sorbitol (83%) with or without 0.2% borax may be added to give a firm, fine-grained stable cream. The oil constituents are melted together at 170°F. and the water-soluble substances are then mixed with the oil at the same temperature and mixing continued while cooling.

DENTAL PREPARATIONS

Carving Wax for Tooth Forms

Carnauba Wax	80
Ceresin	128
White Wax	64
Zinc Oxide	48
Saturated Solution of Acriflavine in Alcohol	1

Dental Modeling Waxes

Formula No. 1

Paraffin Wax	2
Carnauba Wax No. 1	2
Glyceryl Tristearate	5
Beeswax	1

No. 2

Yellow Beeswax	16
Wheat Starch	8
Thick Turpentine	4
Olive Oil	1
Venetian Red	1

This plastic mass softens at 55°C.

No. 3

Manila Copal	15
Dammar Resin	15
Stearin (53/54°C. Titer)	1
Ceresin (58/60°C.)	2
Peru Balsam	1
Precipitated Barium Sulfate	20
Carmine	To suit

Melt the first five ingredients together, work in the barium sulfate, dye with some carmine, and stir down to very little above the solidification point, where plates can be poured without danger of separation of the barium sulfate.

Dental Molding Wax

Formula No. 1

Paraffin Wax	75.00
Japan Wax	10.00
Beeswax	15.00
Thymol	0.05
Menthol	0.05

No. 2

Carnauba Wax	5
Beeswax	40
Kauri Copal	15
Ozokerite (58/60°C.)	10
Soft Paraffin (40/42°C.)	30

Dental Waxes

(Sticky Waxes)

These are suitable for temporary cementing small pieces of plaster or similar materials together in order to make molds from broken plaster impressions; also to hold small pieces of metal together for investment before soldering.

Formula No. 1

Rosin	16
Yellow Beeswax	8
Vermilion	1

Melt on a water bath and stir together, pour on a glass slab and roll with wet fingers into pencils or pour in molds.

No. 2

Yellow Beeswax	4
Rosin	1
Gum Dammar	1

No. 3

Gum Dammar	1
White Beeswax	4
Light-Yellow Rosin	7

No. 4

Stearic Acid	1
Paraffin Scale Wax	2
Glyceryl Tristearate	1
Carnauba Wax	2
Ethylene Glycol Glyceryl Stearate	2

Dental-Impression Wax

Formula No. 1

Stearic Acid	19
Ethyl Cellulose	1
Paraffin Wax	40
Beeswax	40

No. 2

White Manila Copal	30
Pale Rosin	30
Refined Carnauba Wax	10
Stearic Acid	5
Talc	75
Peru Balsam	3

No. 3

Paraffin Wax	90
Ceresin	39
Beeswax	40
Venice Turpentine	30
Japan Wax	20

Dentist's Molding Mass

Soft Manila Copal	30
Light, French Rosin	30
Refined Carnauba Wax	10
Stearin	5
Peruvian Balsam	2½

Gold-Inlay Impression Waxes

Formula No. 1

Yellow Beeswax	10
Gum Dammar	10
Yellow Ceresin	20
Hard Paraffin (120°F.)	30
Carnauba Wax	30
Dye	To suit

Melt the beeswax, ceresin, paraffin, and carnauba wax in a porcelain dish on a water bath, add the gum dammar in small portions, and stir constantly until a uniform mass is obtained. Remove from the fire and add the dyestuff.

Gold-inlay waxes should be colored deeply. Lampblack or an oil-soluble aniline dye is best suited for this purpose. (Red and blue Cerasine or Sudan Red aniline dyes are recommended.)

No. 2

(Price's)

Stearic Acid	110
Paraffin Wax	10
Beeswax	15
Tamarack (Larch Turpentine)	10
Gum Dammar	110

No. 3

(National Bureau of Standards)

Carnauba Wax	25
Paraffin Wax (55°C.)	60
Ceresin	10
Beeswax	5

Oil-soluble aniline dye may be added.

No. 4

Light Carnauba Wax	10
Paraffin Wax (70°C.)	35
Pure Beeswax	55

Melt the carnauba wax on a low flame, avoiding foaming; add the paraffin and, after complete melting, add the beeswax.

Pink Base Plate Wax

Formula No. 1

White Beeswax	50
Paraffin Wax	25
Whole Alkanet Root	1

Melt the waxes and add the alkanet root. Leave on the fire until the desired shade of pink is obtained, strain through cheesecloth into tin molds, about 1/32 or 1/16 in. thick.

Coat the molds with a film of glycerin. To polish the sheet wax, pass between the rubber rollers of a wash wringer.

No. 2

White Beeswax	40
Gum Turpentine	10
Cotton-Seed Oil	3
Vermilion	4

No. 3
(To Be Used in Hot Weather.)

White Beeswax	20
Crude Turpentine	4
Cottonseed Oil	1
Vermilion	2

No. 4
(To Be Used in Cold Weather.)

White Beeswax	20
Crude Turpentine	6
Cottonseed Oil	2
Vermilion	2

No. 5
(Hard-Base Plate Wax)

Yellow Beeswax	50
Gum Mastic	6
Prepared Chalk	3
Vermilion	4

No. 6

Rosin	1
Ceresin	3
Paraffin Wax	6

Toothache Gums

Paraffin Wax	188 gr
Burgundy Pitch	3⅔ oz
Oil of Cloves	1 dr
Creosote	1 dr

Melt the paraffin and pitch together; when nearly cool, add the rest, and stir well. Make the finished mass into small pills or turn it out in forms of small cones or cylinders.

Wax for Partial Impressions

Rosin	1
Yellow Beeswax	16

EMULSIONS

Acid Wax Emulsions

Formula No. 1

Polymerized Glycol Distearate	5
Candelilla Wax	20
Phosphoric Acid (85%)	5
Boiling Water	70

No. 2

Polymerized Glycol Distearate	15
Glycerin Monostearate	15
Paraffin Wax	6
Hydrochloric Acid	9
Water	55

Acrawax Emulsion

Acrawax*	28
Stearic Acid	6

Melt together and add slowly, with good stirring a mixture of:

Boiling Water	62
Monoethanolamine	4

Antitack Emulsion
(For Hycar)

Ceresin Wax	25
Morpholine Oleate	4
Water	71

Warm until fluid and mix vigorously until emulsified. Spray this emulsion on surfaces coated with Hycar latex to prevent sticking.

Beeswax Emulsion

Beeswax	90
Ammonium Linoleate	12
Water	500

Candelilla-Wax Emulsions

Formula No. 1

Candelilla Wax	60
Myristic Acid	15
Emulsifier S489*	19
Water	To make 375

This gives a very translucent, almost clear emulsion which dries to a transparent film.

No. 2

Candelilla Wax	60
Myristic Acid	13
Morpholine	10
Water	To make 375

This produces a very thin emulsion which is translucent and dries to a transparent film.

No. 3

Refined Candelilla Wax	60
Myristic Acid	13
Emulsifier S489*	10
Water	To make 375

This gives a dark but very translucent, thin emulsion with high wax content, which dries to a transparent film.

No. 4

Candelilla Wax	60
Oleic Acid	16
Morpholine	8
Water	To make 375

Very thin, translucent and dries to a transparent film.

No. 5

Candelilla Wax	40 g
Abietic Acid	8 g
Isopropanol	10 cc
Ammonia	15 g
Borax	1 g
Water	To make 400 cc

This gives a thin, amber translucent emulsion, which dries to a transparent film.

No. 6

Candelilla Wax	30

* Glyco Products Co.

Flexo Wax C*	10
Oleic Acid	8
Ammonia	15
Butanol	10
Water	To make 400

This gives a translucent, thin emulsion which dries to a transparent film.

No. 7

Candelilla Wax	20
Flexo Wax C*	10
Oleic Acid	11
Morpholine	4
Water	To make 225

No. 8

Candelilla Wax	20 g
Oleic Acid	3 g

Heat to 95–100°C. and then run in the following heated to 95–100°C., mixing with a high-speed stirrer:

Triethanolamine	4 g
Borax	2 g
Water	185 cc

No. 9

Candelilla Wax	20 g
Oleic Acid	3 g

Use the method under formula 8, adding:

Triethanolamine	3 g
Borax	1½ g
Water	185 cc

No. 10

a. Candelilla Wax	30
Triethanolamine	7
Red Oil	5
b. Borax	2½
Water	20
c. Water	150

* Glyco Products Co.

Method as in esparto wax emulsion (see p. 553).

No. 11

Candelilla Wax	23
Ammonium Linoleate	3
Water	232
Ammonia	2

Melt the wax at 100°C. Thin the ammonium linoleate by working in 30 cc water a little at a time. Heat the balance of water to 95–100°C., and add the ammonium linoleate solution. Boil the whole and add to the melted wax, a little at a time, while stirring vigorously. Finally add the ammonia and stir until cooled to 50°C.

Carnauba-Wax Emulsions

Formula No. 1

Carnauba Wax	60
Myristic Acid	16
Oleic Acid	4
Emulsifier S489*	19
Water	To make 375

This gives a very clear, light colored almost transparent emulsion, which dries to a transparent film.

No. 2

Carnauba Wax	60
Oleic Acid	16
Morpholine	11½
Water	To make 375

This makes a very light-colored, translucent emulsion of

low viscosity which dries to a transparent film with good gloss.

No. 3

Carnauba Wax	60
Linseed-Oil Fatty Acids	13
Emulsifier S489*	17
Water	To make 375

This gives a very thin, almost transparent emulsion which dries to a transparent film with good brightness.

No. 4

Carnauba Wax	260.00
Trihydroxyethylamine Oleate	48.50
Triethanolamine	28.00
Borax	11.11
Caustic Soda (50%)	2.22
Water	650.22

No. 5

Carnauba Wax	90
Ammonium Linoleate S	12
Boiling Water	620

Ceresin-Wax Emulsion

Formula No. 1

1. Ceresin Wax	16
2. Aminostearin	4
3. Water	80

Heat 1 and 2 to 85°C. and add slowly, with vigorous mixing, to the water at 90–100°C.

No. 2

Ceresin Wax	90
Stearic Acid	10
Triethanolamine	5
Water	200

* Glyco Products Co.

Colored, Liquid Wax Emulsions

Formula No. 1

Black

Crude Montan Wax	7
Dark Rosin	1
Potash Carbonate	1
Water	120
Water-Soluble Nigrosine	4 }
Water (80–90°C.)	12 }

No. 2

Carnauba Wax	10
Bleached Montan Wax	5
Marseilles Soap	3
Potassium Carbonate	3
Water	160
Water-Soluble Nigrosin	0.1– 0.4
English Red	15
Chrome Orange	6

Emulsify the melted waxes with the alkali (soap) solution; with formula 1, the dye solution is added last. With formula 2, the dye is dissolved in the emulsification water. The pigments are then worked in on a pigment mill.

Esparto-Wax Emulsion

a. Esparto Wax	15
Triethanolamine	3
Red Oil	1⅞
b. Borax	1¼
Water	10
c. Water	32½

Heat *a* to 95–100°C. and mix well; run into *b* heated to 95–100°C., beating vigorously; when a smooth gel is formed add *c* heated to 95–100°C. and beat vigorously until cool.

Coloring Wax Emulsions

There are oil-in-water and water-in-oil emulsions. Oil-in-water emulsions are best colored with water-soluble dyes. The water is colored before emulsification. If the emulsion is to be colored after completion, the color is dissolved in as little water as possible and the concentrated dye solution is added to the emulsion with vigorous stirring. The following colors are recommended. The proportions are from 1 lb to 400 gals, down to 116 to 1,200 gal, depending on the depth of shade desired.

Pink	—Rhodamine B Extra
Green	—Cyanine Green
Golden Yellow	—Metanil Yellow
Canary Yellow	—Tartrazine
Blue	—Alizarine Blue
Red	—Cloth Red
Heliotrope	—Violamine
Opal	—Fluorescein
Black	—Nigrosine

Water-in-oil emulsions are best colored with oil-soluble-colors, which are dissolved in the oils before emulsification. Recommended are:

Yellow	—Azo Oil Yellow
Red	—Azo Oil Red
Black	—Oil Black
Orange	—Azo Oil Orange
Blue	—Alizarine Oil Blue
Violet	—Alizarine Oil Violet
Green	—Oil Green

With the exception of the black, 1 lb colors 200 gal., while 1 lb black will color 50 gal.

Halowax Emulsions

Formula No. 1

Water	3 lb
Halowax*	1 lb
Stearic Acid	53 g
Triethanolamine	27 g

No. 2

Water	3 lb
Halowax*	3 lb
Stearic Acid	108 g
Triethanolamine	54 g

No. 3

Water	3 lb
Halowax*	13 oz
Halowax Oil No. 1000*	3 oz
Stearic Acid	108 g
Triethanolamine	54 g

Melt the wax and stearic acid together and stir. Heat the water and triethanolamine until they start to boil. Add the wax to the water and stir with an electric stirrer and then run through a colloid mill. A good emulsion is obtained if it is cooled quickly after coming from the colloid mill.

Japan-Wax Emulsion

a. Japan Wax	36 g
Trihydroxyethylamine Stearate	4 g
b. Boiling Water	360 cc

* Halowax Corp.

Warm *a* to 65°C. and add *b* to it slowly, with vigorous stirring.

Laundry-Calendering Wax Emulsion

Mix 33 parts of paraffin wax with 3 parts of oleine and pour this mixture into a solution of 0.6 part of strong ammonia in 63.4 parts of water, heated to 160°F.

Montan-Wax Emulsion

Montan Wax	20
Oleic Acid	5

Heat to 100°C. and add the following heated to 95–100°C. slowly with good mixing:

Triethanolamine	8
Borax	2
Water	185

Opal-Wax Emulsion

a. Opalwax*	15
Aminostearin	5
b. Water	75
Sulfatate**	5

Heat *a* to 100–105°C. and add to it slowly, while stirring by a high-speed mixer, *b* heated to 95°C. A smooth emulsion results that may be used as a furniture polish.

* Du Pont.
** Glyco Products Co.

Ouricuri-Wax Emulsion

Formula No. 1

Ouricury Wax	13.2
Diglycol Oleate	2.5
Triethanolamine	2.2
Borax	1.0
Water	80.0

Belt the wax and diglycol oleate at 80°C., and make a solution of the other ingredients at the same temperature. Add the solution to the melt slowly while mixing vigorously with an electric stirrer.

No. 2

Ouricury Wax	90
Stearic Acid	10
Triethanolamine	5
Water	200

Ozokerite Emulsion

Ozokerite	90
Ammonium Linoleate	14
Water	400

Paraffin-Wax Emulsion

Formula No. 1

Paraffin Wax	120 g
Stearic Acid	12 g

Melt together and, while stirring vigorously, add the following heated to 55°C.

Ammonia (26°Bé.)	6 cc
Water	182 cc

Stir until uniform.

No. 2

Glyceryl Monostearate	5 g
Water	150 cc

Heat and stir vigorously until uniform. Pour into this slowly, while stirring strongly:

Melted Paraffin Wax	40 g

No. 3

a. Diglycol Stearate	10
Paraffin Wax	40
b. Water	250
Granulated Sugar	½
Ammonia (26°Bé.)	1

Melt *a* and add *b* (warmed to 40°C.) with vigorous stirring.

No. 4

a. Paraffin Wax	12
b. Stearic Acid	5
c. Trigamine*	3
d. Water	80

Melt *a* and *b* at 65–70°C. and add *c* and *d* at the same temperature slowly, while mixing with a high-speed agitator.

No. 5

(Nonalkaline)

Paraffin Wax	25
Glycol Stearate	5

Melt together and while stirring vigorously add

Boiling Water	175

No. 6

Paraffin Wax	90
Stearic Acid	10
Triethanolamine	5
Water	120

No. 7

Paraffin Wax	90
Stearic Acid	10
Water	125
Morpholine	3

* Glyco Products Co.

Resin-Wax Emulsions

Formula No. 1

Batavia Dammar	12.5
Beeswax	12.5
Morpholine	1.0
Stearic Acid	1.5
Water	100.0

Melt the beeswax and add the dammar, stirring until the mixture is homogeneous. Stir into this mixture the morpholine and the stearic acid. Add this mixture slowly with stirring to the water which has been heated to 85–95°C.

This emulsion remains stable and can be diluted with water without breaking.

No. 2

Batavia Dammar	25.0
Paraffin Wax	25.0
Oleic Acid	5.0
Morpholine	2.5
Water	250.0

Melt the wax and sift in the dammar. When all is melted, add the oleic acid and stir until the mixture is homogeneous. Do the same with the morpholine. Keeping the mixture hot and continuing the stirring, add the water slowly. The water should previously be heated to 85–95°C.

No. 3

Batavia Dammar	37.5
Besewax	37.5
Oleic Acid	8.0
Morpholine	4.0
Water	375.0

Follow the method under formula No. 2.

Santowax Emulsions

Formula No. 1

Santowax M*	50.00
Stearic Acid	3.10
Turpentine	50.00
Triethanolamine	0.94
Water	400.00

Dissolve the Santowax and stearic acid in the turpentine, heating to about 85°C. Dissolve the triethanolamine in 100 g of soft water and heat to 85°C. Add the solution of triethanolamine to the hot turpentine solution with rapid mixing. Stir for about 2 minutes, then add 300 g of soft water at room temperature. Continue to stir for 10 minutes or more. This gives a smooth, fairly fluid, creamy-white suspension. Follow the same procedure with formula No. 2.

No. 2

Santowax M* or Santowax 9523*	50.00
Turpentine	50.00
Stearic Acid	2.00
Triethanolamine	1.23
Water	400.00

Sulfathiazole Emulsion

Sulfathiazole	5
Triethanolamine	2
Distilled Water	24
Beeswax	5
Liquid Paraffin	64

* Monsanto Chem. Co.
** Glyco Products Co.

Synthetic-Wax Emulsion

Acrawax**	10
Ethanolamine Oleate	1
Water	84

Heat to 95–100°C. with vigorous mixing. Continue stirring until temperature drops to 75°C. Then add

Methyl Ethyl Ketone	5

Triethanolamine-Wax Emulsions

The water method of emulsification has been developed for use particularly with waxes and other semisolids, such as greases and asphalts, and for the preparation of wax polishes. It gives very good results, however, also in other emulsion problems and is a convenient method whenever stearic acid is preferable to liquid fatty acids.

The following tabulation presents suggestions for the formulae to be adopted for waxes. The proportions of the ingredients are given on a weight basis.

Water Method

Type of Wax	Wax	Stearic Acid	Tri-ethanol-amine	Water
Beeswax	88	9	3	300
Carnauba	87	9	4	400
Paraffin	88	9	3	300
Lanolin	80	15	5	200
Japan	85	12	3	400

In this method, the wax or oil is emulsified by means of a water solution of the soap which is made from the triethanolamine and stearic acid. The water is measured out into a container or kettle which can be heated. The triethanolamine is then stirred into this and then the stearic acid is added. On heating, the acid gradually melts and can be stirred into the water to give a smooth soap solution, and the temperature is raised to just below the boiling point. The wax is now melted in a separate container and its temperature brought to 85—95°C. It is then added to the water solution and the whole at once stirred vigorously to obtain a good emulsion. Stirring is continued gently until the product has cooled.

Wax-Emulsion Spray

Montan Wax	160
Ammonium Linoleate	40
Water	600

Dissolve the ammonium linoleate in the water and heat to 95—100°C.; pour the wax into the solution slowly, stirring vigorously. While still hot, run through a colloid mill. A stable emulsion that sprays easily results.

FOOD PRODUCTS

Artificial Honeycombs

a.	Yellow Beeswax	45 kg
	Paraffin Wax (52–54°C.)	40 kg
	Ceresin	10 kg
	Japan Wax	5 kg
b.	Beeswax-Perfume	About 25 g
	Honey-Perfume	Small amount

Melt *a* with steam, add *b*, and mix thoroughly. Pour out into shallow dishes. Rewarm on the water bath and roll to give it the honeycomb shape.

Cheese-Coating Wax

The addition of 1% polyethylene glycol distearate to a combination of waxes used for coating cheeses causes the wax blend to wet and adhere to the damp cheese surfaces. In addition to the cheese industry, this addition will be helpful also in other cases where waxes are in contact with damp surfaces.

Chewing-Gum Bases

Formula No. 1

Ester Gum	30

Coumarone Resin	45
Dry Latex	15
Paraffin Wax (M.P. 180°F.)	10

No. 2

Dry Jelutong	80
Gutta Siak	15
Paraffin Wax (M.P. 176°F.)	5

No. 3

Pontianac Gum	425
Gutta Katian	400
Gutta Soh	75
Candelilla Wax	60

No. 4
(Bubble-Gum Base)

Washed Pontianac Gum	425
Washed Gutta Katian	400
Washed Gutta Soh	75
Candelilla Wax	10

The mixed gums and wax are heated until the total batch contains only 8–9% moisture.

Coating for Sausages

Paraffin Wax	35.0
Rosin	62.8
Whiting	2.2

Increasing the Melting Point of Butter

Butter, for use in hot climates, is mixed with melted glyceryl tristerate (an edible product), in various amounts to increase its melting point.

Wax Coatings for Cheese
Formula No. 1

Aged cheese is wiped to remove a portion of the oil, mold, and slime on the surface and is then coated with a mixture of 85% amorphous wax melting at 135–145°F. and 15% mono- and diglycerides. The coating adheres firmly to the surface and appears to absorb the moisture.

No. 2

Wash the cheese, which should not be less than 10 days old, with warm water containing 1% ordinary blue lime; dry thoroughly; and dip in paraffin wax (M.P. 130–135°F.) heated to 300°F. and containing:

Amber Petrolatum	5.0–10.0
Petrolatum Ceresin	1.0– 2.5
Oil-Soluble Carmine Red Paint	2.0– 3.0

LEATHER AND FURS

Belt Adhesion Compound
Formula No. 1

Rosin	25
Anhydrous Wool Fat	20
Blown Rapeseed Oil	20
Tallow	25
Scale Wax (50–52°C.)	10

Melt all ingredients together and mix until uniform.

No. 2

Cumarone Resin	35–40
Bleached Montan Wax	5–10
Train Oil	55

Belt Dressing
Formula No. 1

Wool Fat	50
Mineral Oil (0.885)	20
Paraffin Wax (56–58°C.)	10

Yellow Ceresin (58–60°C.)	5
Second-Pressing Castor Oil	10
Degras	5

No. 2

Resin	40
Train Oil	10
Blown Cottonseed Oil or Sperm Oil	15
Paraffin Scale Wax (48–52°C.)	15
Mineral Oil (0.905)	20

No. 3

a. Neutral Wool Fat	30
Tallow	20
b. Amorphous Graphite	10
Castor Oil	10

Melt *a;* stir it into the fused *b.* Press. The product is soft and salvelike.

No. 4

Rosin	65.0
Tallow	6.0
Stearic Acid	1.0
Scale Wax	20.0
Castor Oil	2.0
Rosin Oil	0.5
Lanolin	4.2

Belt-Preserving Grease

Formula No. 1

Beef Tallow	50
Paraffin Wax (50–52°C.)	10
Turpentine	10
Very Fluid Saturated Steam Cylinder Oil	50
Ground Rosin	10

Melt the first three ingredients on a steam bath, add the oil, stir until cooled, and stir in the rosin.

No. 2
(Liquid)

Beeswax	4
Pitch	5
Resin	3
Neatsfoot Oil	13

Melt the beeswax first and add the other ingredients in the order named.

Black Harness Oil

Ceresin	2
Ozokerite	2
Cod Oil	32
Herring Oil	26
Whale Oil	14
Pale Oil No. 100	22
Nigrosine Base	2

Mix all components, except the dye, until melted and dissolved. Add the dye; mix thoroughly and let it cool.

Black Leather Fat

Formula No. 1

Paraffin Wax	8
Raw Wood Fat	2
Crude Montan Wax	4
Gray Carnauba Wax	2
Fat-Soluble Nigrosine	1
Train Oil	8
Mineral Oil	60

No. 2

Paraffin Wax	4
Raw Wood Fat	1
Crude Montan Wax	3
Fat-Soluble Nigrosine	0.3
Train Oil	4
Mineral Oil	28

No. 3

Paraffin Wax	5.20
Crude Montan Wax	3.90
Fat-Soluble Nigrosine	0.39
Train Oil	5.40
Mineral Oil	32.00

Black Padding Wax

North-Country Carnauba Wax	40

Green Ozokerite	2
Paraffin Wax	58
Rosin	2
Oil-Soluble Black Dye	7

Calf-Leather Cleaner and Polish

Water	20 gal
Potassium Oleate	7½ lb
Trisodium Phosphate	½ lb
Yellow Beeswax	6 lb
Yellow Carnauba Wax	6 lb
White Spirits, Turpentine, or Benzol	4½ gal
Pine Oil	½ gal
Terpineol	1 qt

Dissolve the first three ingredients and bring to a boil; melt together the balance and run into the first solution, while stirring vigorously.

Conservation of Shoe Soles

Linseed Oil	50–60
Paraffin	40–50

Melt by heating to 80°C. Treat the soles with this mixture after thorough cleaning, two or three times in 4—6 weeks.

Fungicide for Leather

Formula No. 1
(Dubbing)

Tallow	50.0
Neatsfoot Oil	40.0
Paraffin Wax	9.0
Aluminum Stearate	1.0
p-Nitrophenol	0.8
p-Chloro-*m*-Xylenol	0.8
Tetrachlorophenol	0.8

Warm together and mix until uniform.

No. 2
(Dubbing)

Paraffin Wax	20
Stoddard Solvent	68
Cyclohexanol	10
DDT	2

Warm together until dissolved.

No. 3

Salicylanilide	2.2
Isopropyl Alcohol	25.0
Paraffin Wax	33.0
Stoddard Solvent	39.8

Warm gently and mix until dissolved. Impregnate the leather, squeeze, drain, and dry.

Government Harness Dressing

Neatsfoot Oil	1 gal
Bayberry Tallow	2 lb
Beeswax	2 lb
Beef Tallow	2 lb
Castor Oil	1 qt
Lampblack	1 oz

Hardener for Shoe Soles

Formula No. 1

Beeswax	12.50
Turpentine	12.50
Castor Oil	12.50
Wood Tar	3.25
Linseed Oil	125.00

Heat up the mixture to get a homogeneous oil.

Apply hot on dry soles.

No. 2

a. Asphalt	60
Ceresin	40
Train Oil	200
Linseed Oil	700
b. 10% Rubber in Benzene	30

Dissolve *a* hot. Cool. Add *b*. Mix.

No. 3

Tallow	500
Beeswax	500
Rosin	30
Linseed Oil	15

Apply hot.

No. 4

Beeswax	25
Ozokerite or High-Melting Petrolatum	40
Rosin	35

Leather Dubbing

Ceresin	9
Aluminum Stearate	1
Neatsfoot Oil	40
Tallow	50

Leather Finishes

Formula No. 1

Castile Soap	15 lb
No. 1 Yellow Carnauba Wax	5 lb
Candelilla Wax	5 lb

Dissolve the soap in 8 gal water by boiling for 1 hour, adding water as it is lost by evaporation; then add the melted wax and continue boiling for 2 more hours; finally, add water enough to make 10 gal. Strain the mass through four layers of cheesecloth.

No. 2

Carnauba Wax	300 lb
Dry Castile Soap	40 lb
Water	50 gal

Boil the soap in the water for 30 minutes; add the wax and boil again for 30 minutes. Add water to make 150 gal. Mix this with casein-shellac solutions for leather finishes.

Leather Grease

Scale Wax (50–52°C.)	8
Bleached Montan Wax	7
Wool Fat	10
Pale Train Oil	50
Distilled Spindle Oil	25
Oil-Soluble Dye	1–2

Leather Oiling and Finishing

Formula No. 1

Neutral Wool Grease	8
Dark Petrolatum	4
Paraffin Wax	4

No. 2

Petrolatum	16
Beeswax	2

Leather Stuffing

Ozokerite	6
Paraffin Wax	8
Rosin Oil	40
Mineral Oil	48

Leather-Stuffing Grease

Formula No. 1

Crude Montan Wax	12
Ceresin	3
Neutral Wool Fat	5
Car Waste Oil	50
Birch Tar	3
Train Oil	27

No. 2
(Yellow)

Yellow Ozokerite-Ceresin (58–60°C.)	17
Distilled Spindle Oil	83

No. 3
(Yellow)

Yellow Scale Wax (50–25°C.)	10
Bleached Montan Wax	10
Pale Rosin	5
Distilled Spindle Oil	75

No. 4
(Brown)

Crude Montan Wax	10
Dark Rosin	10
Scale Wax	5
Distilled Spindle Oil	75

No. 5
(Black)

Crude Montan Wax	12
Dark Rosin	10
Scale Wax	3
Distilled Spindle Oil	73
Oil-Soluble Nigrosine	2

No. 6
(Black)

Crude Montan Wax	15
Yellow Ozokerite-Ceresin	3
Crude Wool Fat	5
Oil-Soluble Black Leather Dye	2
Distilled Spindle Oil	75

No. 7

Ozokerite	3
Paraffin Wax	5
Neutral Wool Fat	10
Mineral Oil	22
Train Oil	60

Liquid Burnishing Wax for Shoe Soles

Carnauba Wax	20.0
Turpentine	20.0
Oil-Soluble Black Dye	3.0
Duponol W.E.*	5.0
Ferric Acetate	6.0
Glacial Acetic Acid	0.2
Water	45.8

Reduce the ferric acetate to a powder and dissolve it in the acetic acid and water mixture. Dissolve the Duponol in this solution and heat to about 170°F. Melt the carnauba wax and pour into the turpentine which has been previously heated to above 180°F.; dissolve the black dye in this mixture; and then add this solution to the first solution, while agitating vigorously. Allow to cool with continued stirring.

Neutral Shoe Cream

Carnauba Wax	6
Ceresin Wax	3
Candelilla Wax	3
Turpentine	5
Lemon-Grass Oil	½
Light Mineral Oil	1
Diglycol Stearate	5

Heat together below 90°C. and stir till clear. Pour slowly, while stirring with high-speed stirrer into

Boiling Water	100

Stir until the temperature falls to 70°C. The finished product is snow white, applies easily, and rubs to a high gloss.

Old Leather Freshener
(Book Covers)

Beeswax	1
Cedarwood Oil	1
Anhydrous Lanolin	8
Dry Cleaners' Naphtha	12

* Du Pont.

Melt the wax and lanolin; add the cleaning solvent (fire hazard!) and stir until cool. Add the oil while the mixture is cooling and pour into bottles for storage.

Shake before using. Treat the leather thoroughly, but do not allow the mixture to reach the leaves of the books. Do not use on calfskin, which is better treated with milk followed by petrolatum.

Paste Shoe Polish

Carnauba Wax	25 g
Ceresin	10 g
Triethanolamine	1 cc

Melt together and stir; add the following slowly:

Turpentine	30 cc
Deodorized Kerosene	5 cc
Gasoline	30 cc

For a black polish, add 1% oil black dye.

For a brown polish, add brown iron oxide sufficient to color.

Pour into cans at 35°C. and allow to set.

Polishing Wax for Lacquered Heels

Paraffin Wax (50–52°C.)	27.5
Pale Spindle Oil	13.5
Finest Powder Chalk	59.0

Shoe and Leather Dressing

No. 1

White Laundry Soap	½ lb
Water	1½ gal
Carnauba Wax	1¾ lb
Turpentine	½ gal

Dissolve the soap and wax in the water by heating, with frequent stirring. Remove from the fire and add slowly the turpentine, with constant stirring. Keep stirring until cold and a soft paste is formed. If a firmer paste is desired, reduce the water in the formula to 9 pt. Pack in tubes or boxes.

Apply lightly to the leather and, after a short time, rub up with a brush and a polishing cloth to a high luster.

No. 2

Carnauba Wax	⅞ lb
Turpentine	¼ gal
Soft Soap	⅜ lb
Water	⅝ gal

Dissolve the soap in boiling water, add the wax, and stir well. When all is in solution, remove from the fire and add the turpentine slowly, with constant stirring. Continue stirring until cold.

Apply as No. 1.

Shoe Finishers' Black Stick Wax

Candelilla Wax	9
Rosin	1
North-Country Carnauba Wax	32
Oil-Soluble Black Dye	6
Carbon Black	¼
Paraffin Wax	1

Shoe Wax Finish

Orange Shellac	14
Alcohol	60

Carnauba Wax	2
Paraffin (50–52°C.)	1
Turpentine	23

The shellac is dissolved in the alcohol either in the cold or with the aid of heat. The waxes are melted and thinned with the turpentine. The shellac solution is warmed to about 122°F. and the warm (140°F.) wax solution is slowly stirred into the alcohol solution. The mixture is allowed to cool, with constant stirring, and may be colored with suitable dyes.

The wax stains are either solutions of wax in turpentine or emulsions of wax in water. The emulsions are stabilized with alkalies, such as ammonia or potash. They are colored with water-soluble dyes. The proper selection of waxes insures the desired gloss, harder waxes giving a higher gloss than so-called soft waxes. The stain may have the following composition:

Shoe Wax Stain

Carnauba Wax	4
Refined Ozokerite	2
Paraffin (50–52°F.)	2
Sudan Yellow Dye	2
Thinner	90

When preparing a turpentine stain, the waxes are melted and thinned. The dye is best dissolved in the other waxes before adding the ozokerite. The thinner should be warmed before adding it to the molten waxes. The mixture is allowed to cool, while stirring constantly to avoid a separation of the components.

A water-emulsion stain may have the following composition:

Shoe Emulsion Stain

Double-Bleached Montan Wax	7.0
Potassium Oleate	3.0
Caustic Potash	0.8
Water-Soluble Dye	2.0
Water	87.2

The potash, the soap, and the dye are dissolved in the boiling water. The boiling solution is then slowly poured into the wax melt, adding a small portion at a time. Care must be taken that a drop in temperature does not cause separation. It is essential to stir constantly while cooling.

Shoe-Bottom Dressing

Bleached Montan Wax	10
White Paraffin or Scales (50–52°C.)	10
Oil-Soluble Anilin Dye	2
Turpentine Oil or Substitute	54

Shoemaker's Brushing Wax

Yellow to Tan Carnauba Wax	63
Lacquer Benzine (0.78–0.80)	37

Shoemaker's Hard Wax

Rosin	8
Ester Gum	2
Crude Montan Wax	30

Paraffin Wax	45
Stearin Pitch	10
Beeswax	5
Oil-Soluble Color	To suit

Shoemaker's Polishing Wax

Yellow Carnauba Wax	26–28
Low-Titer Olein	10– 7
Abrasive (Tripoli, Bole, Clay, etc.)	64–65

Shoemaker's Soft Wax

Rosin	5
Paraffin Wax	65
Japan Wax	5
Stearin Pitch	20
Beeswax	5
Oil-Soluble Color	To suit

Shoemaker's Thread Wax

Formula No. 1

Rosin	76.5
Tallow	8.5
Yellow Beeswax	15.0

Melt and stir until uniform. Pour into containers at the lowest possible temperature.

No. 2

Rosin	54 lb
Beeswax	6 lb

Melt and stir until uniform. Cool to 150°F. and add

Benzene	9½ gal

No. 3
(Sticky Wax)

X Grade Rosin	74
Glaurin* (Plasticizer)	1
Petrolatum	20
Water	5

No. 4

Candelilla Wax	2
Rosin	55
Burgundy Pitch	20
Rosin Oil	4
Lard	3
Heavy Mineral Oil	1

* Glyco Products Co.

No. 5

Refined Ozokerite	70
Beeswax	30
Linseed Oil	5–10

Shoe-Sole Paint

Formula No. 1
(Brown)

Double Bleached Montan Wax	25
Beeswax	25
Paraffin	25
Linseed Oil	2
Red or Brown Iron Oxide	22
Oil-Soluble Brown Dye	1

No. 2

Venetian Red	1 lb	4 oz
Fast Brown	2 lb	8 oz
Gum Arabic	2 lb	12 oz
Water	43 lb	12 oz

Make a mucilage with the gum, using water; make a paste with the pigment and then thin down, using more water. Make a solution of the dyestuff, mix the pigment-water mixture with the mucilage; then add the balance of water.

No. 3

Red Iron Oxide	2.0
China Clay	13.5
Yellow Ochre	11.0
Middle Chrome Yellow	3.0
Gum Arabic	9.5
Water	61.0

No. 4

Red Iron Oxide	4.0
China Clay	13.5

Yellow Ochre	12.0
Gum Arabic	9.5
Water	61.0

No. 5

Red Iron Oxide	16.0
China Clay	13.5
Gum Arabic	5.0
Water	61.0

Shoe-Sole Stain

Shellac	13
Casein	4
Trigamine*	5
Water	130

Warm and stir until dissolved. Then mix in:

Water-Soluble Nigrosine	5

Continue heating and add:

Crude Montan Wax	6
Carnauba Wax No. 3	2
Paraffin Wax	5
Stearic Acid	2

Soft Leather Polish

a. White Mineral Oil	20
Turpentine	15
Carbon Tetrachloride	5
Beeswax	3
Oleic Acid	4

Stir until dissolved and add:

b Triethanolamine	1.5
Water	31.5

Mix *a* and stir until the ingredients are dissolved, then add *b*. Stir until a smooth emulsion is formed.

Softening and Flexibilizing Dyed Furs

Diglycol Stearate	10 lb
Water	12 gal

Boil together and mix until uniform. Cool to room temperature and add to it 107 gal of dye-bath solution. Dye in the usual way. If salts are present in the dye-bath, the addition of a wetting agent, such as Wetanol* is necessary.

Stuffing Fat Liquors for Leather

Formula No. 1

Bleached Montan Wax	10
Fish Train Oil	3
Paraffin Oil	35
Wool Fat	25

No. 2

Crude Montan Wax	10
Ceresin (64–70°C.)	16
Train Oil	30
Spindle Oil	44

Waterproof Boot Dressing

Spermaceti	3 oz
Raw India Rubber	6 dr
Tallow	8 oz
Lard	2 oz
Amber Varnish	5 oz

Waterproofing Leather

Formula No. 1

Fish Oil	36.5–54.5
Alizarin Oil or Pine Oil	0– 4.0
Paraffin Wax	36.0–50.0
Diglycol Stearate	7.5–10.0
Ammonia	0.5– 2.0

Impregnate the leather with this emulsion at 45°C.

No. 2

Gutta-Percha	2
Boiled Rapeseed Oil	8

* Glyco Products Co.

Yellow Wax	6
Pig Fat	25
Venice Turpentine	60
Spermacetic	1

No. 3

Linseed Oil	100
Gutta-Percha	10
Copal Varnish	A little

Wax Finish for Furs

Dissolve 3 to 6 oz of paraffin wax in 1 gal petroleum cleaning solvent.

Approved cleaning solvent is preferable because of its safety during ordinary handling.

Precaution: Paraffin separates from the petroleum solvent below 70°F. At 15°F., it is completely chilled out of the solvent.

This finish is used for the saturation of dry-cleaned furs to replace any oils removed and to make them water repellent. It is also sponged or sprayed on materials that are lifeless or lusterless after cleaning and drying to produce high gloss.

Wax for Leather Strips

Lanolin	15
Paraffin Wax	25
Montan Wax	30
Water-White Rosin	10
Gypsum	20
Caustic Soda (30°Bé.)	2

Melt the lanolin, paraffin, montan wax and rosin together; add the soda. Mix well and add the gypsum, with constant stirring.

Wax Stick to Cover Small Defects of Shoe Leather

Formula No. 1

(Black)

Crude Montan Wax	28
Gray Carnauba Wax	7½
Shellac Wax	7½
Beeswax	20
Paraffin	20
Linseed Oil	2
Nigrosine Stearate	15

No. 2

(Brown)

Double-Bleached Montan Wax	25
Beeswax	25
Paraffin	25
Linseed Oil	2
Red or Brown Iron Oxide	22
Oil-Soluble Brown Dye	1

Weatherproofing Leather Harness

Formula No. 1

Beeswax	20
Paraffin Wax	5
Varnish	75

No. 2

Benzol	20
Paraffin Wax	5
Stearin	3
Color	To suit

METAL TREATMENT

Air-Seal Grease for Glass Joint

Melt together beeswax and castor oil in proportions to suit this use. It should be tacky and not too stiff.

Aluminum Lubrication

Aluminum antiseize compounds of particular interest in the lubrication of delicate airplane parts and mechanisms that must remain free moving under varying conditions of temperature can be made by using diglycol stearate as a base wax. Typical formulae are listed:

		Temperature Limits °C.
Diglycol Stearate S-125	80	
Glyceryl Monoricinoleate or		-35 to 70
Dibutyl Phthalate, or		-55 to 70
Tricresyl Phosphate	20	-20 to 70

The wax and the plasticizer are heated together until complete solution is obtained. A 35% alcohol solution is made of the resultant blend which can then be brushed, sprayed, or wiped on the metal parts. This gives a thin flexible film having high lubricity that will not flake at lower temperatures, nor flow off to volatize at the higher temperature limits. By varying the amount and type of plasticizer, a wide range of lubricants, having other interesting properties, is possible.

Aluminum Protection

As a protective coating for aluminum castings against corrosion, particularly where such parts are to be used at low temperatures causing condensation of water vapor or at temperatures below 32°F., the following mixture is recommended:

Diglycol Stearate	10
Flexo Wax C*	20
Paraffin Wax	40

The waxes are melted together until a uniform blend is obtained. Seven parts of this blend are then dispersed in forty parts of water heated to 200°F., and stirred until cool. The castings can then be dipped into the dispersion. The use of this aqueous dispersion has eliminated fire hazard and cost of maintaining high temperatures.

This wax blend has replaced, with equal satisfaction, solvent-oil-grease slushes as a protective coating on ferrous and nonferrous metals. Hazardous solvents are replaced with water,

* Glyco Products Co.

effecting worth-while economies. In addition to giving a firm, flexible film as the protective coating that will not smear off or collect dirt and grime, this wax-blend dispersion tends to inhibit electrochemical action resulting in corrosion whereas greases do not. It is of interest to note that a wax will seek out rust, whereas oils seem to flow around corroded areas and thus fail at points where they are most needed.

Ammunition Lubricant

Carbon Tetrachloride	90
Ceresin Wax (176°F.)	9
Aluminum Stearate	1

The aluminum stearate is first dissolved in the wax at approximately 260°F., with thorough stirring. The wax is then cooled and dissolved in the solvent at approximately 130°F.

In applying the lubricant, it is maintained at about 120°F.

Antirust Compound

Paraffin Wax	6
Asphaltic Still Residue (About 1000 visc. at 210°F.)	94

Ball-Bearing Grease

Paraffin Wax (40–42°C.)	16
Refined Ozokerite	4
Spindle Oil, Containing Paraffin (Visc. 2° E. at 50°C.)	80

Carriage Greases

Formula No. 1

(Yellow)

a. Bleached Montan Wax	20
Rosin	5
b. Pale Mineral Oil	30
c. Caustic Soda (24°Bé.)	5½
d. Pale Mineral Oil	42

Melt *a,* dilute with *b,* saponify with *c,* dilute with *d.*

No. 2

(Natural Color)

a. Crude Montan Wax	18
Rosin	3
b. Dark Paraffin Oil	30
c. Caustic Soda (36°Bé.)	5
d. Mineral Oil	46

Use the same method as under No. 1.

Chain Lubricant

Formula No. 1

Stearin	85
Beeswax	5
Carnauba Wax	5
Japan Wax	5

No. 2

Stearin	85
Beeswax	2½
Carnauba Wax	5
Japan Wax	5

Pour at the lowest possible temperature and allow to cool slowly and undisturbed.

Cock-Valve Grease

(For Benzine, Gas, Organic Solvents, below 180°C.)

Tallow	700
Paraffin Wax	80
Solid Sodium Hydroxide	110

Solid Potassium Hydroxide	20
Aluminum Sulfate	20
Calcium Oxide	30
Magnesium Chloride	30

Cogwheel Lubricant

Formula No. 1

Ceresin	5 lb each
Talc	5 lb each
Boiled Linseed Oil	5 lb each
Blown Rape-Seed Oil	5 lb each
Crude Wool Fat	40 lb
Graphite	40 lb

No. 2

Ozokerite	7
Blown Colza Oil	10
Tallow	20
Neutral Wool Fat	15
Carbon Tetrachloride or Other Solvent	8
Graphite Flakes	40

This liquid, leaves a tacky, solid mass on the wheel after cooling.

Collector Lubricant (For Electric Motors)

Formula No. 1

Paraffin Wax (56–58°C.)	19
Neutral Tallow	14
Neutral Wool Fat	40
Finest Graphite Flocks	31

Put only a thin layer of the mixture on the collectors, after removing the brushes and using a dry woolen rag.

No. 2

a. Tallow	24
Mineral Oil	66
Castor Oil	6
Ceresin	18
b. Finest Graphite	6
Finest Cooper Powder	½

Melt *a* and disperse *b*.

Dynamo-Brush Lubricant

Ceresin	20
Acid-Free Tallow	10
Neutral Wool Fat	10
Castor Oil	10
Petrolatum Oil	50

Melt together and add enough organic solvent (heavy benzoline, naphtha or tetralin).

Foundry Mold (Casting) Powder

a. Finest Powder Limestone	25.0 kg
b. Crude, Powdered Montan Wax	750.0 g
Oil-Soluble Sudan Yellow	12.5 g
Limestone Powder	200.0 g

Heat *a* and *b* to 140°C. in an iron kettle until homogeneous.

Cool and grind thoroughly. Sift as fine as possible, e.g., through miller gauze.

Graphite Grease

Ceresin	70
Tallow	70

Heat together at 80°C. and work in

Graphite	30

Graphite Lubricants

Formula No. 1 (Stick Form)

Ozokerite	30
Beeswax	13
Rosin	10

Castor Oil	5
Graphite	33
Carbon Black	9

The rosin and castor-oil mixture is stirred over direct heat until a clear liquid results, which is poured into the waxes that have been melted by hot water or steam heat. The pigments are stirred in, heating is continued, and when a smooth cream is obtained, the heating is stopped and cooling allowed to proceed until it becomes viscous. Then it can be poured into cold molds of the hinge type.

No. 2
(Paste)

Paraffin Wax	10
Candelilla Wax	10
Graphite	20
Naphtha	59
Lavender Oil	1

The waxes are melted by hot water or steam and poured into the naphtha, which has been heated to the same temperature as the wax and in a similar manner.

The graphite and perfume oil are stirred into this and kept hot until a smooth slurry results, when heating can be stopped. When cooling has proceeded to the point where stirring is difficult, the mixture can be poured into tin cans to cool and set.

No. 3
(Liquid)

Asphalt (By-Product)	11
Asphalt (Gilsonite)	6
Paraffin Wax	10
Naphtha	25
Graphite	20
Carbon Tetrachloride	25
Oil-Soluble Nigrosin	3

The natural asphalt is first crushed and heated until it can be stirred; then the by-product asphalt is poured in and heating continued until a homogeneous liquid results. The paraffin is then added and after it has melted, the container is removed from the flame and the naphtha stirred in. When it is homogeneous, the black liquid is passed through a screen to remove coarse minerals always present in gilsonite. In the meantime, the graphite and nigrosine are soaking in the carbon tetrachloride, and when both containers are at room temperature, the contents are mixed and are ready for packaging. In view of the fact that graphite forms aggregates or lumps in mixtures, where water is absent, it might settle to the bottom on long standing and not be redispersed with vigorous shaking. However, the paraffins present in this mixture have a thickening effect and offset settling.

No. 4

Rosin	7
Crude Montan Wax	2
Naphtha	2
Caustic Soda	1
Water	70
Nigrosine	3
Graphite	12
Lampblack	3

Melt together, mix well, and pour into cardboard molds.

Grease for Pressure Lubricants

Refined Montan Wax	22
Oleic Acid	5
Engine Oil (6–7°E. at 50°C.)	67
Lime Hydrate	3

Greaseless Lubricating Pencil

Formula No. 1

Beeswax	100
Graphite Powder	50–100

Melt together and stir until a little above room temperature. Pour into molds and allow to cool.

It is useful for lubricating hinges, automobile doors, etc., as it will not run off and stain.

No. 2

Beeswax	80
Diglycol Stearate	20
Graphite Powder	100–200

Melt together and stir until just cold enough to pour. Pour into molds and allow to set.

No. 3

Petrolatum	10
Paraffin Wax	10

Gum for Parting the Punch from the Die

Beeswax	1
Rosin	½
Venice Turpentine	¼

Hot-Bearing Grease (For Heavily Loaded, Hot Bearings)

Tallow Fatty Acid	14.0
Refined Montan Wax	3.0
Engine Oil (6–7°E. at 50°C.)	79.5
Lime Hydrate	1.5
Caustic Soda (40°Bé.)	3.0

Laboratory Stopcock Lubricants

Formula No. 1

Standard Viscous Oil (No. 24 or 32)	80
Beeswax	20

Heat the oil to 100–120°C., add the beeswax, and mix thoroughly. No. 24 gives a lubricant of medium body; No. 32 gives a very firm, viscous product. Standard Viscous Oil is manufactured by the Standard Oil Co. of California.

No. 2

Petrolatum	100
Beeswax	10
High-Grade Red Rubber Tubing (Small Pieces)	20

Heat together for several hours at 150°C. until the mixture is homogeneous. Pour into a wide-mouth jar.

No. 3

White Petrolatum	77.5
Paraffin Wax	15.0
Pale Rubber Latex	7.5

Heat on an oil bath for 6 hours at 130°C.

Lubricant

Technical Caustic Soda (38°Bé.)	3
Crude Montan Wax	15
Dark Mineral Oil	82

Lubricating Compound for Drawing Aluminum

Formula No. 1

Diglycol Stearate	6
Glycowax A*	2
Water	56

The entire mass is heated to about 180°F. and then stirred at moderate speed until uniformly dispersed. Small parts, such as collapsible tube blanks, are dipped into the hot dispersion while in a wire basket and then centrifuged to remove excess lubricant. If a microscopic protective film is not required on the finished work, the Glycowax A may be replaced with additional diglycol stearate.

No. 2

1. Diglycol Stearate	10
2. Paraffin Wax (145°F.)	60
3. Boiling Water	400

Melt 1 and 2 and then add 3 slowly, with vigorous mixing.

* Glyco Products Co.

Lubricating Compound for Tin Stamping

Diglycol Stearate	1
Water	100

The diglycol stearate s is dispersed by heating while stirring in the water. This is applied prior to stamping and is considerably cheaper than the old wax lubrication method. It permits stamping of lacquered metal without fracturing the coating.

Lubricating Grease

Formula No. 1 (Light-Colored)

Powdered Lime Hydrate	5
Oleic Acid	10
Doubly Bleached Montan Wax	25
Pale Spindle Oil	80

No. 2 (Dark)

Rapeseed Oil or Deodorized Train Oil	5
Technical Caustic Soda (10%)	10
Crude Montan Wax	25
Mineral Oil (0.910)	260

Metal-Etching Stop-Off

Formula No. 1

Wax	4
Rosin	2
Pitch	1

This coating is used as a general stop-off and is made by

melting the ingredients in a kettle and then adding 4 oz of finely powdered asphalt.

The following stop-off is also recommended particularly as being very hard and offering good resistance to strongly acid etching solutions:

No. 2

Black Pitch	1
Japan Wax	2
Rosin	1
Gum Dammar	1

Add to these, melted together in a kettle, 1 oz of turpentine.

A series of etchants is available but usually dilute solutions of nitric acid are used on metals other than gold or platinum. Strong acids are not suitable because of the lifting of the stop-off by evolution of gas resulting in nonuniform and dull etching.

Metal-Stamping Lubricant

Diglycol Stearate	1
Water	100

The wax is dispersed by heat in the water and is stirred until cool. The dispersion can be sprayed on lacquered metal sheets or, in semiautomatic operations, brushing or spraying is satisfactory. This operation, carried out prior to stamping, is a considerable improvement over the old hot paraffin-wax method by eliminating spotting and a greasy surface of the finished lacquered stampings, in addition to considerable reduction in operating and material costs.

Plastic-Mold Lubricant

Bright-Drying Carnauba-Wax Emulsion	2 oz
Pine Oil	5 drops

Spray this into the mold before use for easy ejection of molded articles.

Porcelain-Mold Oil

Stearic Acid	24
Ozokerite	1
Paraffin Wax	3
Heavy Mineral Oil	82

Rod Lubricant

Yellow Ceresin	25
Sperm Oil	25
Tallow	50

Melt together.

Yellow Ceresin	1
Refined Spindle Oil	3–8

Melt at low temperature.

Soldering Paste

Formula No. 1

Cerasin	10
Stearic Acid	30
Rosin	22
Anhydrous Zinc Chloride	38

The first three substances are

melted in a steam-heated kettle and when liquid, the zinc chloride is stirred in. It will result in a granular structure.

No. 2

Ammonium Chloride	29
Sulfur Powder	33
Rosin	2
Paraffin Wax	9
Tallow	21
Salt	5
Balsam	1

The paraffin and tallow are melted in a steam-heated kettle and while molten the other solids are stirred in; the balsam is added after the mixture has been removed from the heat.

Stopcock Grease

Petrolatum	16
Pure Gum Rubber	8
Paraffin	1

Stir while warming gently until all are dissolved.

Tripoli Buffing Stick

Double-Pressed Stearic Acid	30
Edible Tallow	25
Paraffin Wax	25
Tripoli Flour	20

A buffing or polishing paste may be made, using this formula with the addition of a small amount of turpentine and of water, to bring to the consistency desired.

MISCELLANEOUS

Artificial Beeswax

Formula No. 1

Carnauba Wax	½
Paraffin Wax	9½
Ceres Yellow Color	1/16

Melt the carnauba wax and paraffin in separate vessels and pour the paraffin into the carnauba wax. Stir until thoroughly mixed. To the melt, add the ceres yellow and continue stirring until the mixture is perfectly homogeneous.

Cast in cakes of convenient size in metal molds. Lycopodium may be used to prevent the wax from sticking to the mold.

No. 2

Glyceryl Stearate	20
Beeswax	8
Japan Wax	10

No. 3

Japan Wax	20
Indene or Paracumarone	5
Glyceryl Tristearate	1

Melt together and stir until quite thick.

Batik Wax

Formula No. 1

Rosin	70
Beeswax	20
Japan Wax	10

No. 2

Rosin	60
Beeswax	15
Ceresin	15

No. 3

Rosin	60
Japan Wax	20
Tallow	20

No. 4

Paraffin Wax	60
Ceresin	10
Japan Wax	30

Bed Wax

(To tighten the cover cloths of feather beds)

Formula No. 1

Beeswax	10
Rosin	1
Caustic Soda (30°Bé.)	1½
Hot Water	47½

Melt first two ingredients, saponify, and stir with the water to obtain a white, smooth cream.

No. 2

Rosin	2
Ceresin	3
Paraffin	20
Japan Wax	5
Beeswax	20

Beeswax Substitute

Formula No. 1

Octadecane Amide	30
Mineral Oil (100 Vis. at 100°F.)	10
WW Rosin*	60

No. 2

Rosin	60
Mineral Oil	10
Stearamide	30

* WW = water-white.

Cable-Impregnating Compound

Formula No. 1

Yellow or Brown Beeswax	50
Paraffin Wax	50

No. 2

Rosin	10
Ozokerite	5
Ceresin	30
Montan Wax	10
Stearin Pitch	15
Paraffin Wax	20

Cable Wax

Formula No. 1

Bleached Ozokerite	25
Paraffin Wax	70
Rosin	5

No. 2

Paraffin Wax	70
Japan Wax	2
Carnauba Wax	2
Rosin	18

Candle Coatings

Dissolve 0.05 g Sudan color in a melt of 25 g stearic acid and add a melt of 75 g paraffin wax. Where extreme fastness of light, sublimation, and aging is required, 2 g of fast candle color is used instead of 0.05 g Sudan color.

Candle String

(Summer)

Pale Paraffin Pitch (44/46°C.)	70
Wool Fat	20
Yellow Ozokerite	5
Pale Rosin	5

Candle String

(Winter)

Paraffin Pitch (38/40°C.)	30
Paraffin Pitch (44/46°C.)	40
Second-Quality Ozokerite	5
Crude Wool Fat	25

The wicks are made from strings, unbleached yarn, cotton strings or jute.

The production is easiest in winter (quick, cheap cooling). Colors: Yellow and black.

Carnauba-Wax Substitute

Shellac Wax	85
Beeswax	10
Sal Dammar	5

Cartridge Wax

Formula No. 1

Beeswax	1
Copaiba Balsam	1

No. 2

Carnauba Wax	70
Beeswax	20
Ceresin	10

Carving Wax

(For statuettes and models)

Formula No. 1

Stearic Acid	6
Ceraflux*	24
Carnauba Wax	1
Terra Alba	75

No. 2

Bleached Montan Wax	40
Barium Sulfate	60
Red Iron Oxide	0.5

Condenser Impregnating Wax

Ethyl Cellulose	68
Carnauba Wax	2
Piccolastic Resin**	10
Acrawax C*	20

Coopers' Wax

Paraffin Wax	50
Tallow	30
Beeswax	5
Venice Turpentine	5
Talc	10

Defoaming Agent

Paraffin Wax	1 oz
Liquid Petrolatum	1 oz
Octyl Alcohol	2 fl oz

Warm together and stir until uniform.

Electrotypers' Waxes

Formula No. 1

Beeswax	5½
Paraffin Wax	3
Burgundy Pitch	¾
WW Rosin***	½
Zinc Oxide	1½

Melt together the waxes and resins and add the zinc oxide slowly with good mixing.

No. 2

Ozokerite	63½
Beeswax	31¾
Graphite Powder	4¾

* Glyco Products Co.
** Penn. Industrial Chem. Corp.
*** WW = water-white.

No. 3

Beeswax	85
Burgundy Pitch	5
Turpentine	10

No. 4

Ozokerite	95
Graphite Powder	5

No. 5

Green Ozokerite	33
Paraffin Wax	50
WW Rosin*	16
Petrolatum	1/8

No. 6

Brown Ozokerite	90
Graphite Powder	2
Pine Pitch	8
Rosin Oil	1/4

No. 7

Ozokerite (Green, Austrian)	100
Pitch	20
Beeswax	10

No. 8

Ozokerite (Green, Austrian)	50
Candelilla Wax	6½
Pitch	10
Beeswax	5

Engravers' Transfer Wax

Beeswax	3
Tallow	3
Fir Balsam	1
Olive Oil	1

Melt together and stir while cooling.

Ethyl Cellulose Phonograph-Record Stock

Ethyl Cellulose (N 100)	485.0
Vinsol**	100.0
Dutrex 25***	67.0
Octyl Phenol	4.8
Carnauba Wax	3.0
Whiting	888.0
Slate Dust	444.0
Carbon Black	13.0

* WW = water-white.
** Hercules Powder Co.
*** Shell Chem. Co.
† Glyco Products Co.

Fire Kindler

Crude Paraffin Wax	30
Rosin Pitch	10
Wood Flour	60

Compress strongly into bricks.

Flexible Wax

Methyl Abietate	10
Modified Hydrocarbon Wax	90

Heat together and stir until homogeneous. The finished product has a softening point of 58°C. and a melting point of 67°C.

Gasoline Line and Airplane Parts Lubricant

(Insoluble in water, gasoline, oils and solvents)

Acrawax†	3
Glyceryl Monoricinoleate (S125)	10

Golf-Club Grip Wax

Yellow Beeswax	1 oz
Rosin	1 dr

Melt together, mix, and run into molds.

Hand-Dipped Candles
(White)

Paraffin Wax (135/137°F.)	90 lb
Double-Pressed Stearic Acid	10 lb

To make colored candles, add any one of the following amounts per 100 lb of the above white composition.

(Red)

Oil Red GRO National	1½ oz

(Rose)

Oil Red GRO National	3½ g

(Med. Green)

Quinolene Yellow Base	21 g
Quinizarine Green Base	2½ g

(Yellow)

Quinolene Yellow Base	1½ oz

Hardened Stearic-Acid Wax

Stearic Acid	75.0
Magnesium Oxide	5.3

Heat with stirring for ½ hour at 130—150°C. Pour at lowest temperature.

Horsehoof Grease

Formula No. 1

Paraffin Wax	10
Wool Fat	30
Spindle Oil	20
Cresylic Acid Emulsion	1

No. 2

Hog Fat	40
Cod-Liver Oil	20
Beef Tallow	15
Wool Fat	5
Bone or Charcoal	5
Potash Soap	5
Water	10

No. 3

Train Oil	10
Crude Wool Fat	15
Crude Montan Wax	20
Dark Mineral Oil	50–55
Preservative	0.2–0.5
Oil-Soluble Nigrosine	1

Horsehoof Healing Grease

Ceresin (58/60°C.)	15
Cresylic Acid Emulsion	10
Refined Spindle Oil	75

Horseshoe Grease

Formula No. 1

Neutral Wool Fat	5.2
Spindle Oil	2.0
Soft Tar	0.5
Paraffin Wax	0.3
Water	2.4

No. 2

Crude Montan Wax	20
Crude Wool Fat	15
Train Oil	5
Black Color	1
Dark Paraffin Oil	59

Illumination Candles

50–52°C. Paraffin	79.0
Stearin	19.5
Bleached Carnauba Wax	1.5

Imitation Chocolate for Display

Barium Sulfate	60
Crude Montan Wax	40
Sienna	0.5–0.7
Vanillin	0.1–0.2

To obtain a better-smelling wax base, it is recommended to use the following:

100 g of melted crude montan wax are poured over 5 g of finely powdered Sumatra gum benzoin. The mass is stirred for 2—3 hours on a boiling wa-

ter bath, strained hot through a cloth, and the filtrate is to be used in the formula.

Ironing Wax

Formula No. 1

Paraffin Wax	8
Stearic Acid	1
Beeswax	2

Melt together and form into cakes of any desired shape and size.

Rub the cake over the hot iron, then rub the iron on a piece of canvas or unbleached muslin until smooth.

No. 2

Ceresin	10
Stearic Acid	30
Japan Wax	5
Beeswax	5

Japan-Wax Substitute

Formula No. 1

Hydrofol Glycerides 45*	85
Palm Oil	10
Palm-Oil Fatty Acids	5

No. 2

Hydrofol Glycerides 45*	87
Crude Beeswax	8
Palm-Oil Fatty Acids	5

Jewelers' Molding Wax

Formula No. 1

Cumar Resin	5
Beeswax	20
Carnauba Wax	5
Diglycol Stearate	5
Montan Wax	35
Ceresin Wax	30

This wax gives sharp moldings and is tough and sufficiently elastic. It burns out easily without leaving a residue.

No. 2

Diglycol Stearate	12
Acrawax B**	14
Beeswax	16
Ceresin Wax	30

These are melted together and stirred until homogeneous. This wax gives sharp definition, is tough and elastic. It burns out easily and leaves no residue.

Lanette Wax SX Replacement

Cetyl Alcohol	90
Sodium Lauryl Sulfate	10

Laundry Ironing Wax (Germany)

Japan Wax	44
Tallow	30
Hydrogenated Fish Oil	10
Monopol Soap†	8
Sodium Perborate	8

Melt together and mix well all ingredients but the perborate. Cool to 47°C. and add the perborate, with stirring. Pour into forms.

---
* Werner G. Smith Co.
** Glyco Products Co.
† Jacques Wolf and Co.

Lost Wax

Carnauba Wax	8
Albacer*	2
Burgundy Pitch	3
Venice Turpentine	4

Match-Head Composition

Powdered Sulfur	1 lb
Manganese Dioxide	2 lb
Ferric Oxide	½ lb
Powdered Glass	1¼ lb
Powdered Glue	1 lb
Potassium Bichromate	0.8 lb
Potassium Chlorate	6¾ lb
Powdered Paraffin Wax	1 oz

This should be mixed with great caution as it is explosive.

Modeling Compounds

Formula No. 1

Stearin	25
Gum Dammar	50
Powdered Soapstone	85
Carmine	Enough to color

Melt the stearin on a water bath, add the gum dammar and, when melted, stir in the powdered soapstone, tinted with the carmine.

No. 2

Stearic Acid	20
Oleic Acid	4
Gum Copal	19
Seed-Lac	17
Powdered Soapstone	40

No. 3

Manila Copal	30
Light-Colored Rosin	30
Carnauba Wax	10
Stearic Acid	5
Powdered Soapstone	75
Carmine	Enough to color

* Glyco Products Co.

No. 4

Glyceryl Oleate	10
Red Oil	50
Crude Beeswax	20
Castor Oil	15
Powdered Pipe Clay	To suit

No. 5

Tallow	19
Gum Mastic	30
Crude Beeswax	3
Ozokerite	2
Paraffin Wax	4
Gypsum	12
Pipe Clay	60

No. 6

Plastic Clay	46
Cup Grease	24
Paraffin Wax	11
Rosin Oil	1

Modeling Waxes

Formula No. 1

Beeswax	10
Mastic	10
Ceresin	7½
Hard Paraffin Wax	15
Tallow	65
Sulfur Flowers	90
Calcium Sulfide	42½
Kaolin	120

Grind the powders into the melted waxes. Add a mineral pigment, e.g., Armenian bolus (4%). Stir thoroughly and cool.

No. 2

Bleached Montan Wax	10
Paraffin Wax (40–42°C.)	20
Yellow Petrolatum	2
Wool Fat	15
Finely Powdered Clay	50
Mineral Color	20

No. 3

Venice Turpentine	90
Rosin	16

Beeswax	60
Tallow	14
Thin Mineral Oil	4
Color	To suit

No. 4

Beeswax	4
Venice Turpentine	9
Lard	4
China Clay	3½

No. 5

Gum Mastic	3
Beeswax	3
Ozokerite	2
Paraffin Wax	4
Tallow	19

Melt together and, keeping hot, work in

Sulfur Flowers	22
Gypsum	12
Pipe Clay	33
Mineral Pigment	4

Modeling Wax for Brass Foundries

Burgundy Pitch	5
Beeswax	50
Lard	5
Venice Turpentine	5

Modeling Waxes for Engravers

Formula No. 1
(Winter Wax)

Syrian Asphalt	15
Gum Mastic	30
Beeswax	40

No. 2
(Summer Wax)

Syrian Asphalt	60
Gum Mastic	30
Amber	30
Beeswax	120

No. 3

Yellow Beeswax	1
Tallow	1
Burgundy Pitch	2

Moisture Preventing on Glass

Beeswax	75
Japan Wax or Paraffin	24
Glycerol	1–2

Melt and stir until cold.

Mold Lubricant for Portland Cement-Asbestos Composition

Stroba Wax	3.5
Albacer*	3.5

Dissolve in hot Stoddard solvent and stir the mixture until cool, giving a colloidal gel. Apply to the mold with a brush.

Molding Composition

Zinc Oleate	147
{Made from Distilled Oleic Acid 133; Zinc Oxide 17}	
Japan Wax	33
Mineral Oil	40
Sulfur	107
White Kaolin	68
Pigment	6–7

Natural Resin-Wax Compounds

Petrolatum Wax	45
Pale East India or Batu or Black East India Gum	45
Paraffin Wax	10

* Glyco Products Co.

Melt the waxes together and add the resin, stirring until the mixture is uniform. Allow to cool.

Such mixtures are of higher melting point and are more water-resistant than similar masses made with ester gum instead of the natural resins.

Neutral-Colored Rubber Mat

Reclaimed Rubber	38.00
Smoked Sheets	4.00
Sulfur	0.75
Diphenylguanidine	0.25
Zinc Oxide	1.00
Clay	27.00
Whiting	25.25
Medium Process Oil	3.00
Paraffin Wax	0.50
Stearic Acid	0.25

Perbunan Laboratory Tubing Compound

Perbunan	100.0
Diphenyl Phthalate	20.0
Zinc Oxide	5.0
Stearic Acid	2.0
Paraffin Wax	1.5
Sulfur	1.8
Soft Brown Factice	50.0
Semireenforcing Gas Black (Gastex)	25.0
Semireenforcing Carbon Black (for Example, P-33)	25.0
Benzothiazyl Disulfate (Altax)	1.5
Butyl Amine	0.25

Cure: 45 minutes at 287°F.

Pure Beeswax Candles

They are made from the pure wax and range down to combinations as low as 40% beeswax, 50% paraffin and 10% stearic acid.

Purifying Lignite, Ozokerite or Montan Wax

Crude Wax	100
Benzol	220

Warm and stir until dissolved. Add

Alcohol	400–800

Filter and wash residue with

Alcohol	1200
Benzol	300

Bleach by heating at 105–115°C. for 7 hours, using

Potassium Dichromate	1
Sulfuric Acid (48%)	10
Nitric Acid (4%)	10

Recording (Phonograph) Wax

Formula No. 1

Stearic Acid	84

Melt and add slowly with stirring:

Litharge	8½

Boil off water at 220–230°F. Stir to prevent caking at bottom of kettle. When solution is complete, add slowly (by sifting in):

Soda Ash	7

When a drop cools to a clear mass, reaction is complete.

Drive off all gas, froth and water by heating up to 270°F.

If a brown wax is desired add

Stearin Pitch	2

If a black wax is wanted add some oil-soluble nigrosine to the brown formula.

No. 2

Distilled Montan Wax	60
Litharge	4½
Soda Ash	4
Paraffin Wax	30

Follow method under Formula No. 1.

No. 3

Polymerized Vinyl Chloride (120–130°C.)	21
Novolac (92–96°C.)	21
Powdered Slate	49
Gas Black	4½
Pitch	5½
Calcium Stearate	2
Montan Wax	2

Red Butyl Inner-Tube Stock

Butyl Rubber	100.0
Red Iron Oxide	5.0
Zinc Oxide	5.0
Lime	3.0
Thionex*	1.0
Polyac*	0.5
Akroflex C*	1.0
Sulfur	2.0
Paraffin Wax	1.0
Acrawax C**	1.0
Calcium Silicate	25.0
Burgess Iceberg Pigment†	60.0

Rubber Flooring Composition

Crepe Rubber	100
China Clay	30
Lithopone	30
Paraffin Wax	2½
Sulfur	2½
Nonox N.S.	1
Woodflour‡	160
Whiting	50
Zinc Oxide	10
Stearic Acid	2
Vulcatex Blue (B.S.)	4
Vulcafor F.	1

Cure: 20–30 minutes at 141°C.

Rubber Stair Tread

Reclaimed Rubber	32.000
Whiting	24.625
Mineral Rubber	12.000
Clay	24.000
Iron Oxide	4.000
Zinc Oxide	1.000
Paraffin Wax	1.000
Stearic Acid	0.250
Sulfur	0.875
Mercaptobenzothiazole	0.250

Rubber-Paraffin Embedding Composition

To 200 g of paraffin add 2 g of rubber and 0.5 g of beeswax. Heat at 105° for 16 hours with occasional stirring. The most suitable rubber is "Heveatex" which is 37% rubber latex in aqueous solution with a small amount of ammonium hydroxide as a preservative. It is air-dried in thin layers and cut into small pieces. The method is superior to paraffin for any tissue which varies in density

* E. I. du Pont de Nemours Co.
** Glyco Products Co.
† Burgess Pigment Co.
‡ White softwood fir, 200 mesh.

For trichrome staining, it is excellent, since it permits cutting the entire specimen without fraying. Specimens can be re-embedded in celloidin.

Rubber-Wax Mixture

Rubber can be introduced into waxes or high-boiling oils by heating the molten wax or oil at 120 to 130°C., stirring and introducing rubber latex in a fine stream at a rate which allows the water in the latex to boil off. Heating and stirring are continued until all the water is out of the mix. Up to 4% by weight of rubber can thus be introduced into molten paraffin wax, yielding a very viscous mass. The rubber is finely disseminated throughout the oil or wax. In waxes, the rubber serves to give the product additional strength and cuts down brittleness. The rubber can be vulcanized by the addition of vulcanizers.

Saddler's Wax

Formula No. 1

Beeswax	10
Venice Turpentine	1
Rosin	2

No. 2

Bleached Beeswax	8
Rosin	6
Olive Oil	1

No. 3

Bleached Beeswax	6
Rosin	3
Tallow	1

Sculptors' Modeling Wax

Formula No. 1

Burgundy Pitch	1
Beeswax	10
Lard	1
Venice Turpentine	1

No. 2

Burgundy Pitch	2
Beeswax	16
Lard	1

The lard can be replaced by tallow.

Sealing Wax for Candle Decorations

Rosin	50
Ruby Shellac	3
Gypsum	1

Ski Wax

For Ironing In:

Paraffin Wax (125°F.)	60
Ceresin	16
Palm Oil	14
Talc	10

For Rubbing In:

Paraffin Wax (125°F.)	60
Refined Wool Grease	6
Carnauba Wax	4
Montan Wax	18
Rosin	12

Slow-Flowing Wax (High Melting)

Prepare by mixing equal parts of methyl abietate and

aluminum palmitate, stirring to a paste. Add ethylene dichloride sufficient to thoroughly dissolve with heat. To this add paraffin with constant stirring. This turns into a gel which solidifies on cooling. On reheating, the gel reappears at temperatures up to 350°F. By varying the portions of wax, the characteristics may be changed.

Soft Elastic Wax

This wax has proved valuable in the chemical laboratory, especially for making joints which must be airtight and somewhat flexible. It does not harden on exposure to air.

Yellow Beeswax	100
Finely Powdered Red Mercuric Oxide	10
Venice Turpentine	10
Oil-Soluble Red Dye	3

The ingredients are melted together, stirred thoroughly to distribute the mercuric oxide and allowed to cool. The hardness of the wax may be altered by varying the amount of the Venice turpentine used. The dye is used only to give a pleasing color and may be omitted.

"Soluble" Waxes

The following formulae will give waxes which are readily dispersed in hot water to form stable, white emulsions of different viscosities. These are useful in the textile industry for sizing, finishing, etc., in the polishing industries for furniture, floor and automobile polishes, for leather finishing, paper sizing and finishing, etc.

Carnauba Wax

a. Carnauba Wax	90
Oleic Acid	10
Trigamine*	10
b. Caustic Soda (50%)	6
Water	24

a. is melted. *b* is mixed together and heated. It is added to *a* slowly with stirring and the whole stirred until it begins to solidify. One part disperses easily in five parts of boiling water.

Paraffin Wax

a. Paraffin Wax	50
Stearic Acid	14
b. Trigamine*	7

Melt *a* at about 70°C. Turn off the heat and add *b,* stirring well. Continue stirring till it begins to solidify. One part of the mixture disperses easily in 2 parts of hot water.

Japan Wax

a. Japan Wax	50
Stearic Acid	14
b. Trigamine*	7

Procedure as for paraffin wax. Emulsion obtained by dispersing 1 part in 8 parts of hot water.

* Glyco Products Co.

Standard Candle Formula

Paraffin Wax	60
Double-Pressed Stearic Acid	35
Beeswax	5

These are melted together and agitated to insure complete blending. When melted, an oil-soluble dye of the desired hue is added. Then the mass is poured into molds and cooled. The dye should be carefully selected to eliminate bleeding or fading. Many good dyes are available. It may be desirable to make up known strengths of dyes in blocks of paraffin by merely adding the dye to the melted wax and then pouring in molds, forming blocks of uniform size. This permits easy storing and facilitates the complete blending of the color when introduced into the melting kettle.

A better grade of candles is made by increasing the amount of stearic acid and decreasing the amount of paraffin, or vice versa.

Sticky Nonoxidizing Wax

Formula No. 1

Hydrogenated Rosin	90
Flexowax C*	5
Partially Polymerized Petroleum	5

No. 2

Hydrogenated Rosin	80
Flexowax C*	10
Partially Polymerized Petroleum	10

Stuffing-Box Packing

Kaolin	69
Beeswax	8
Rubber	23

Work together in a rubber mill. As a filler, you may add the following:

Barium Carbonate	4
Calcium Carbonate	1

Synthetic Rubber Thread (Neoprene)

Neoprene GN†	100
Sodium Acetate	1
Stearic Acid	2
Ceresin Wax	½
Age Rite White‡	1
Philblack (Carbon Black)**	18
Calcined Magnesia	2
Zinc Oxide	4

Tailor's Wax

Ceresin	15
Paraffin Wax	15
Japan Wax	25
Beeswax	45

Thread Wax

Beeswax	40
Japan Wax	10
Paraffin Wax	150

* Glyco Products Co.
† Imperial Chem. Ind.
‡ R. T. Vanderbilt Co.
** Philips Petroleum Co.

Tissue Embedding and Sectioning Compositions

Formula No. 1

Paraffin Wax	90
Rosin	10

No. 2

Paraffin Wax	90
Rubber	2
Cumar	10

No. 3

Paraffin Wax	90
Clear Nevillite Resin*	10

Upholsterers' Wax

For rubbing the inside of the ticking cover before the down is put in so the feathers do not pierce the fabric.

Yellow Wax	50
Venice Turpentine	8
Burgundy Pitch	1

Vinyl Phonograph-Record Stock

Formula No. 1

Vinylite VYHH**	80
Vinylite VYNW**	20
Dutrex 20***	20
Lead Silicate	4
Carnauba Wax	0.5–1
Armac T† (Static Eliminator)	0.5

Mix all ingredients together. Fuse on two-roll mill or Banbury mixer. Sheet out and cut into biscuits. Mold in a conventional manner.

No. 2

Vinylite VYHH**	20
Dutrex 20***	11
Lead Silicate	1
Atomite‡	61
Slate Dust	1
Carbon Black	1
Carnauba Wax	1

Wax Composition to Protect Parts of Surface of Painted Tiles Against the Adhesion of Glaze

Colophony	10
Bleached Montan Wax	20
Soft Ozokerite	40
Soft Paraffin (48–50°C.)	30

Wax Lighting Tapers

Paraffin Wax (40–42°C. or 42–44°C.)	65–85
Ceresin (58–60°C.)	30–10
Beeswax	2–3
Thickened Turpentine	3–2

Wick thirty loose cotton threads together for a size of 2–4 mm wound on wire.

Wax Molds

Paraffin Wax	10
Beeswax	2¼
Carnauba Wax	2

Melt together and mix until uniform. Pour into a box or other form (the inside of which has been painted with a soap solution). When cool, tap out.

* Neville Co.
** Carbide and Carbon Chem. Corp.
*** Shell Chem. Co.
† Armour and Co.
‡ Thompson-Weinman and Co.

Cut out and engrave the desired design. With this, plaster casts can be made.

Wax Solvent

Cellosolve*	10
Ethyl Acetate	45
Xylol	45

Wax Strings for Foundry Work

Bleached Montan Wax	10
Beeswax	10
Paraffin Wax (40–42°C.)	20
Wool Fat	5
Oil-Soluble Nigrosine	½

Wax-Fruit and Novelty Composition

Paraffin Wax	60
Microcrystalline Wax	4
Albacer**	16
Hydrogenated Castor Oil	6
Glyceryl Tristearate	4
Stearic Acid	9
Beeswax	1

White Carnauba Wax

A white carnauba wax having a melting point of 80.8°C. can be obtained from fat-gray carnauba wax, melting at 82.1°C., by fractional crystallization. A by-product consisting of a wax-paraffin residue, melting at 60.2°C., is also obtained. Carnauba wax, partially purified with the aid of paraffin, is dissolved in ten times its weight of benzine at 55°C. The benzine fraction used is that boiling between 80 and 130°C. Turbidity appears at 50°C. The mixture is allowed to cool with gentle stirring to 16°, when crystallized wax is filtered off. It is freed from solvent by steam distillation. A yellow wax-paraffin mixture is obtained from the filtrate as a by-product. Cooling cannot go much below 16°C., as paraffin separates from benzine solution at about 12°C.

Wood and Fiber Lubricant

Wood and nonmetallic surfaces of the type of synthetic fibers are difficult to lubricate since the lubricant usually causes swelling or remains entirely on the surface. This is true even when the friction is against metal. A suitable lubricant for this purpose can be prepared from 1 oz camphor, sufficient castor oil to make a thin paste, and 2.5 lb of finely shaved paraffin wax. The camphor is first dissolved in the castor oil, with slight heating if necessary; the paraffin wax is then added and the whole well mixed until solution is complete (eventually with further heating). The mixture is cooled and stored in airtight containers.

* Carbide and Carbon Chem. Corp.
** Glyco Products Co.

PAPER

Chlorinated-Rubber Paper Coating

(For Glassine)

Parlon* (125 cp)	21.7
Teglac Z-152**	5.4
Diamyl Phthalate	8.7
White Ceresin Wax	1.3
Toluene	62.9

Duplicating Machine Stencils

The wax composition used should not be brittle and hard, but must be tough and elastic.

The paper used is a very fine, thin, fibrous, but tough tissue-paper.

Impregnation: Very thin.

Suggested Mixture:

Beeswax	20
Soft Ozokerite (60/62°C.)	15
Paraffin Wax (50/52°C.)	30

Melt together; mix; apply hot.

Mothproof Paper

Eucalyptus Oil	1
Naphthalene	42
Paraffin Wax	25
Naphtha	11
Cedar-Wood Oil	1

Spray or paint the paper with a warm solution.

Overprint Paper Coating

Ethyl Cellulose (20 Seconds)	10.0
Nevindene R-1 Resin	4.0
Dibutyl Phthalate	2.0
Tollac or Toluol	67.0
Isopropanol	17.0
Paraffin Wax (130°F.)	0.5

All the ingredients are combined and stirred in a covered vessel or churned until dissolved.

This glossy coating is scratch resistant.

Paper Finish

Clay	300 lb
Water	20 gal
Talc	18 lb
Wax Emulsion†	12 gal
Casein Glue‡	25 gal

The ingredients are boiled together, until the wax is emulsified, and sufficient water

* Hercules Powder Co.

** Am. Cyanamid Co.

† The wax emulsion is made up as follows:

Carnauba Wax	50 lb
Water	50 gal
Castile Soap	12 lb

‡ The casein glue may be made up as follows:

Casein	100 lb
Water	50 gal
Borax	17 lb
Ammonia (26° Bé.)	1 qt

is added to bring the volume to 50 gal.

Another formula, for making a wax emulsion for flint papers, is as follows:

Laundry Soap	7 lb
Carnauba Wax	50 lb
Water	12½ gal

Boil with live steam till thoroughly emulsified (generally 4–5 hours). Cool to 35°C. and add 2 lb 26° ammonia. Make up to 50 gal with cold water. Allow to stand before using, as the emulsion seems to improve with age.

A beeswax emulsion suitable for friction-calendered papers is made as follows:

Yellow Beeswax	150 lb
Castile Soap	28 lb
Water	25 gal

Dissolve the soap in the water and add the wax. Melt and stir till emulsified and smooth. Add water to make 150 gal.

In the manufacture of coated boxboard, a wax emulsion is sometimes added to the casein-coating mixture to make the coating more flexible and to improve the bending and folding properties of the coated board. Such an emulsion is usually made with Japan Wax.

Japan Wax	75 lb
Stearic Acid	17 lb
Water	25 gal

Heat till the waxes are melted. Cool slightly and add 10 lb of borax dissolved in 10 gal of water. Boil till a smooth emulsion is formed, cool, add 1 qt of 26° ammonia, and make up to 75 gal.

Waterproof Coating for Paper Labels

Ethyl Cellulose	60
Paraffin Wax	12
Toluene	150
Methyl Ethyl Ketone	25

Waterproofing Paper

Formula No. 1

Trihydroxyethylamine Stearate	4½
Stearic Acid	½
Water	100

Boil and mix until smooth; pour into this slowly while stirring vigorously:

Paraffin Wax (Heated to 90–100°C.)	30

Stir until cool.

Use 1 part of this emulsion to 5–10 parts of warm water.

No. 2

Carnauba Wax	20
Ceresin Wax	5
Petrolatum	10
Rosin	5

Melt together and apply to paper with a special brush coater.

No. 3

Paraffin Wax	20
Ceresin Wax	10
Congo Gum	20
Rosin	10
Rubber Latex	2
Titanium Dioxide	38

Melt the gums and waxes, add the rubber, filter, and blend while molten.

No. 4

(Millboard)

Cellulose Waste Fibers	100

are mixed with following emulsion:

Paraffin Wax (Match)	1.50
Turpentine	0.75
Aluminum Oxide	12.00
Naphthenic Acid	15.00
Water	70.00

Press between rollers heated to 50°C.

No. 5

To meet the frequent need to waterproof and strengthen paper without the use of paraffin or wax, the following simple method is worth keeping in mind:

Using 1 part of glycerin by weight to 1 part by weight of gelatin and 4 parts by weight of water, cover both sides of the paper with the warmed solution. Then after a few minutes, before it is fully dry, place the paper in a solution of 7½ parts by weight of formaldehyde with 50 parts by weight of water.

Waterproofing Paper and Fiberboard

Paraffin (M.P. about 130°F.)	22.5
Trihydroxyethylamine Stearate	3.0
Water	74.5

The paraffin is melted and the stearate added to it. The water is heated to near boiling and then vigorously agitated with a suitable mechanical stirring device, while the mixture of melted wax and emulsifier is slowly added. The resulting mixture is cooled with stirring.

The paper or fiberboard is coated on the side which will be in contact with water. This is quickly heated to the melting point of the wax, which then forms a continuous film that does not soak into the paper which is preferentially wetted by the water. This method works most effectively on paper-pulp molded containers and is much cheaper than dipping in melted paraffin as only about one tenth as much paraffin is needed. In addition, the outside of the container is not greasy, and can be printed on after treatment which is not the case when dipped into melted wax.

Wax Emulsions for Paper Coating

Formula No. 1

Yellow Beeswax	360 lb
Caustic Potash	2 lb
Ammonium Hydroxide	8¼ lb

Boil with stirring for 1 hour. Add 150 gal water and shut off heat.

No. 2

Ceresin	200 lb
Stearic Acid	200 lb
Caustic Potash	9 lb
Ammonium Hydroxide	13 lb
Water	200 gal

Boil with stirring for 1 hour; bring up to 800 gal with water while stirring.

No. 3

Carnauba Wax	50 lb
Water	50 gal
Soap	12 lb

The soap is dissolved in the water and brought to boiling. The wax is added and boiling continued until all is emulsified. The emulsion is preferably stirred continuously until cold. The soap may be any good grade of washing soap free from rosin.

No. 4

Yellow Laundry Soap	7 lb
Carnauba Wax	50 lb
Water	12½ gal

Boil with live steam till thoroughly emulsified (3–4 hours).

No. 5

Crude Montan Wax	9.0
Paraffin Wax (50–52°C.)	1.0
Castile Soap	0.8
Borax	0.2
Distilled Water	50.0

No. 6

Montan Wax	6.0
Castile Soap	1.9
Distilled Water	40.0

No. 7

Marseilles Soap	2.0
Crude or Bleached Montan Wax	6.5
Water (Free of Calcium)	40.0

No. 8

Carnauba Wax	50 lb
Stearic Acid	7 lb
Borax Soap	10 lb
Water	To make 50 gal

This is added to a clay-casein mixture to produce a coating on paper which can be readily friction calendered to a high finish.

Waxed-Paper Impregnant

Paraffin Wax (150°F.)	186
Yellow Carnauba Wax No. 1	14

PHOTOGRAPHY

Litho Washout Solution

Asphaltum	2 lb
Beeswax	3 oz
Stearin	2 oz
Transfer Ink	1 oz
Lavender Oil	2 oz
Japan Liquid Drier	4 oz

Mix thoroughly by heating, then add 1 gal of turpentine.

Use for strengthening litho images on metal plates for offset printing.

Lithographers' Protective Hand Cream

Anhydrous Lanolin	20
Paraffin Wax	6
Paraffin Oil	7
Glycol Stearate	12
Water	55

Melt the lanolin glycol stearate, and wax; add the oil; then add the water while agitating rapidly until cold.

This gives protection against water solutions of moderately concentrated acids.

Lithographic-Plate Wax

Turpentine	16
Powdered Asphalt	3½
Beeswax	2½

Dissolve the wax in the warm turpentine. Then add the asphaltum and stir until completely dissolved.

Photo Print Paste or Wax

White Beeswax	1 oz
Dammar Varnish*	200 min
Turpentine	1 oz

Dissolve with gentle heat in a double boiler and bottle for use.

POLISHES AND ABRASIVES

Aluminum Polish

Powdered Stearic Acid or Paraffin Wax	10
Magnesium Oxide	40
Precipitated Calcium Carbonate	30
Red Iron Oxide	20

Auto Polishes and Cleaners

Formula No. 1

Carnauba Wax	27 lb
Spermaceti	50½ lb
Ozokerite	44 lb
Kerosene or V.M.P. Naphtha	32 gal
Finest Diatomaceous Silica	65 lb
Cream Tripoli	131 lb
Bentonite	16 lb
Pigment or Dye	To suit
Water	13¼ gal
Borax	3¼ lb
Trisodium Phosphate	3¼ lb
Finest Grade Soap Chips	16¼ lb
Water	13¼ gal

* Dammar Varnish

Gum Dammar	1 oz
Turpentine	2½ oz

No. 2

Yellow Wax	10
Air-Floated Tripoli	18
White Spirit	19
Soft Soap	½
Water	2½

Melt the wax in a double pan and add the powder slowly; keep stirring while slowly adding the white spirit. Dissolve the soft soap in the water and add to the mix with constant stirring. On cooling, this forms a soft paste.

No. 3

Kieselguhr	30
Tripoli	5
Paraffin Wax	4½
Carnauba Wax	½
Varnolene*	30

Tint with iron oxide.

Automobile, Floor, and Furniture Polish Paste

Formula No. 1

Albacer**	61
Paraffin Wax	80
Candelilla Wax	5
V. M. and P. Naphtha	102
Tetralin	4
Hexalin	12

Melt the waxes together and stir until complete solution is obtained. Turn off the flame and add the mixed solvent slowly, stirring thoroughly. Pour at 25–30°C.

No. 2

Albacer**	20
Paraffin Wax	30
Japan Wax	5
Diglycol Stearate	4
Beeswax	5
V. M. and P. Naphtha	65
Turpentine	20

The procedure is the same as in formula No. 1

These pastes polish easily and give a high luster. For shoe polishes, etc., suitable oil-soluble colors can be dissolved in the waxes.

No. 3

Carnauba Wax	8
Montan Wax	5
Ceraflux**	13
Naphtha	40
Turpentine	40

Warm on a water or steam bath until the waxes have melted; stir until thick and pour into tight-closing cans.

No. 4

Yellow Beeswax	6
Ceraflux**	16
Carnauba Wax	27
Montan Wax	8
Naphtha	89
Turpentine	10
Pine Oil	3

Melt together and pour into cans. Do not disturb until solidified. This makes an excellent auto polish of great durability and luster. Variations

* Standard Oil Co. of N. J.
** Glyco Products Co.

can be made to suit individual requirements.

Automobile Polish

Formula No. 1

Carnauba Wax	9
Beeswax	4
Ceresin Wax	4
Naphtha	75
Stearic Acid	7
Triethanolamine	2½
Water	75
Abrasive	25 to 60

No. 2

Sulfonated Olive Oil (30%)	10
Mineral Oil	30
Pumice Powder	5
Amyl Alcohol	10
Powdered Carnauba Wax	5
Water	40

Mix the oils, wax, and alcohol; add the water slowly with good stirring; mix in the other ingredients.

No. 3

1. Carnauba Wax, N.C.	500
Paraffin Wax (50–52°C.)	200
Beeswax	100
Rosin	150
2. Turpentine Oil	4000
Tetralin	800
Methyl Hexalin	800
3. Potash Carbonate	100
Soft Soap	50
Water	5000
4. Neuburger Chalk	600
Methanol	100

Melt 1, with stirring; add 2 which is warmed up to 55°C., cool for a while, add 3, and finally 4.

No. 4

Carnauba Wax	12
Bleached Montan Wax	4
Beeswax	3
Ozokerite–Ceresin (56–58°C.)	6

From this melted mixture of waxes take according to the season:

Waxes	6–8
Heavy Benzoline or Turpentine	80
Petrolatum	12–14

The cooled finished product should be a stiff-liquid jelly.

No. 5

a. Gray Carnauba Wax	15
Beeswax or Japan Wax	5
Glycowax*	5
b. Koalin or Floated Chalk	30–35
c. Solvent (Turpentine, White Spirit, or a Mixture of Both and 10 Parts of Kerosene)	30
d. Black Soap	1
Water	7–10

Melt *a,* stir in *b* very thoroughly; add slowly *c,* with continuous agitation.

Add hot *d* and stir until a paste forms on cooling.

No. 6

a. Bleached Montan Wax	3.0
Carnauba Wax Residues	4.0
Yellow Beeswax	1.5
Japan Wax	1.5
Glycowax*	1.0
b. Potash Carbonate	0.2
Borax	0.3

* Glyco Products Co.

Black Soap	0.5
Water	35.0
c. Kerosene	10.0
Turpentine	10.0

Melt *a* at about 100°C. and add *b* of equal temperature, with very good agitation. Watch closely to prevent foaming over (carbon dioxide formed!) and keep boiling until the saponification is finished. Then add *c*. Stir until cold.

No. 7

a. Carnauba Wax	500
Paraffin Wax (50–52°C.)	200
Beeswax	200
Rosin	150
b. Turpentine Substitute	4000
Tetralin or Methylhexalin	600–800

No. 8

Ceresin	6
Hard Paraffin Wax	10
Doubly Bleached Montan Wax A	12
Turpentine or Substitute	72

No. 9

Paraffin Wax	8
Carnauba Wax	24
Turpentine	68

No. 10

White Mineral Oil	35
Beeswax	5
Carbon Tetrachloride	10
Kieselguhr	3
Oleic Acid	3
Potash Carbonate	½
Ammonia	½
Water	43

No. 11

a. Bleached Montan Wax	8 g
Paraffin (40–42°C.)	8 g
Refined Ozokerite	2 g
b. Infusorial Earth	35 g
Refined Spindle Oil	13 cc
White Spirit	13 cc
c. Turpentine Oil or Substitute	21 cc

No. 12

a. Bleached Montan Wax	8 g
Double-Bleached Montan Wax	5 g
Olein	2 cc
b. Potassium Carbonate	2 g
Glycerin (28°Bé.)	3 cc
Boiling Water	40 cc
c. Yellow Clay or Bentonite	To suit
d. Turpentine Oil or White Spirit	22 cc

Melt *a*, add hot (boiling) *b*, then *c*; cool and add *d*.

No. 13
(Wax-Solvent Type)

Carnauba Wax	4.5
Ceresin Wax	5.5
Candelilla Wax	7.5
Paraffin Wax	12.0
Naphtha	60.5
Turpentine	10.0

No. 14

Carnauba Wax	4.0
Yellow Beeswax	2.0
Paraffin Wax	2.0
Kieselguhr	15.0
Naphtha	30.0
White Spirit	5.0
Triethanolamine Stearate	5.0
Water	37.0

No. 15

Carnauba Wax	3.0
Yellow Beeswax	2.0
Paraffin Wax	1.5
Tripoli	12.0
Bentonite	3.0
Naphtha	34.0
Kerosene	3.0
Olein	4.0
Triethanolamine Stearate	5.0
Water	36.0

No. 16

a.	Gray Carnauba Wax	10
	Bleached Montan Wax	3
	Japan Wax	5
	Scale Wax (50–52°C.)	2
b.	Black Soap	1½
	Water	10
c.	Solvent (Turpentine, White Spirit, etc.)	60

Melt *a* at 95—100°C., prepare the hot soap solution *b*.

Add *c* with good agitation, and stir until it begins to thicken (at about 48°C.).

No. 17

Carnauba Wax	5
Beeswax	5
Ceresin Wax	5
Stearic Acid	2
Soap	2
Varnolene*	45
Water	10

No. 18

Paraffin Wax	55
Candelilla Wax	20
Rezo Wax B	25
Microcrystalline Wax	20
Turpentine	45
Mineral Spirits	205

Melt together at 90°C.

No. 19

(Powdered)

Mineral Oil	5
Kerosene	10
Diglycol Laurate	1
Silica Dust	½
Kieselguhr	4
Tripoli	1

No. 20

Indusoil (Tall Oil)	40
Carnauba Wax	44
Beeswax	20
Ceresin	20
Naphtha	370
Triethanolamine	13
Water	370
Bentonite	123

* Standard Oil Co. of N. J.

No. 21

Mineral Oil	35.0
Naphtha	5.0
Stearic Acid	3.6
Ethylethanolamines	1.4
Water	55.0

Mix together the mineral oil, naphtha, and stearic acid and heat to 122°F. Mix the water and amine until a clear solution is formed. Add the oil-naphtha-stearic acid solution to the water and amine. Stir during the addition and continue stirring until cool.

Shake the polish before use.

Axe- or Hammer-Handle Wax

Formula No. 1

White Beeswax	5½
White Rosin	½
White Lead	4
Dammar Varnish	⅛

Melt the beeswax; crush, melt, and stir in the rosin; add white lead while stirring and finally pour in the dammar varnish. While still in a liquid state, pour into small paper bags which serve as molds.

No. 2

White Rosin	10 lb
Paraffin Wax	2 lb
White Lead	2 oz
Linseed Oil	½ lb

The finished product looks like beeswax, but is lighter in color. The rosin and paraffin are melted and mixed and allowed to cool somewhat before stirring in the white lead and linseed oil to prevent foaming.

Bakelite Polish

Formula No. 1

Montan Wax	3
Tallow or Bone Fat	2
Diglycol Oleate	1
Fine Asbestos Fibers	12

No. 2

Montan Wax	1½
Tallow or Bone Fat	1½
Diglycol Stearate	2
Mineral Oil	12
Vienna Chalk	To suit

Barrel-Tumbling Polish

Formula No. 1

Beeswax	½ lb
Carnauba Wax	1½ lb
Turpentine	1 gal

Shave the wax into chips, place in a suitable container, add the turpentine, and dissolve.

No. 2

Stearic Acid	4oz
Carnauba Wax	8 oz
Turpentine	1 gal

Dissolve same as in No. 1.

No. 3

Stearic Acid	4 oz
Spermaceti	4 oz
Paraffin Wax	4 oz
Carnauba Wax	4 oz
Flushing Oil	4 oz
Beeswax	2 oz
Turpentine	1 gal

No. 4

Precipitated Chalk	2
Turkelene*	8
Hydrogenated Castor-Oil Wax	4
Sawdust	56

Battleship Linoleum Polish

Ceresin Wax	27.2
Carnauba Wax	25.0
Rezo Wax B	2.2
Naphtha	72.0

Belt Wax

Rosin	25
Paraffin Wax	15
Japan Wax	10
Tallow	50
Lanosyn**	20
Venice Turpentine	10

Brass Polish

Triethanolamine Oleate	1 lb
Potassium Oleate	4 oz
Naphtha	1 gal
Ammonia (28°Bé.)	7 fl oz
Silica (300-Mesh)	3½ lb

Mix thoroughly the silica and naphtha, then stir in the two

* Glyco Products Co.
** J. Landau Co.

oleates, and finally the ammonia. This makes a superior metal polish which gives a very fine finish on brass. If carefully prepared, there will be no separation of the ingredients on long standing.

Bright-Drying Floor-Wax Emulsion

Formula No. 1

Diethylaminoethanol	2.7
Oleic Acid	6.7
Carnauba Wax	30.0
Water	300.0

No. 2

Ethylethanolamines	2.5
Oleic Acid	6.8
Carnauba Wax	30.0
Water	300.0

The wax is melted with the fatty acid, usually red oil (oleic acid), in a steam-jacketed kettle which is equipped with a stirrer. The mixture is held at about 194°F. and stirred until homogeneous. The amine is then added slowly and agitation is continued until the mass is uniform. The water is brought nearly to the boiling point in a separate kettle and added in small portions with continuous stirring. Each portion must be fully incorporated before the next portion is added. The mass gradually thickens and then thins on the addition of more water. After the thinning point is reached, the remainder of the water may be added rapidly. Agitation is essential until the finely dispersed globules of wax solidify.

In order to improve the anti-skid and leveling properties of these waxes, a shellac solution may be added to the thinned emulsion. This solution is prepared in a jacketed kettle by dissolving 5 parts shellac in 25 parts water and 0.8 part 28°Bé. ammonia (33% by weight) at not over 120°F. The solution should be filtered before use.

Borax may be substituted for 25% of the amine in the soap without influencing gloss or tackiness. Streakiness is decreased by the use of borax. An excess of amine over the amount required causes streakiness and loss of stability. An excess of oleic acid decreases streakiness, but also decreases gloss and stability and causes tackiness. Other natural and synthetic waxes and resins may be substituted for carnauba wax.

Buffing Nickel Polish

Double-Pressed Saponified Stearic Acid	86
Paraffin Wax	16
Edible Tallow	10
Japan Wax	3
Silex	376

Carnauba-Shellac Polish

Carnauba Wax	10
Petromix No. 5*	4
Butyl Cellosolve**	1
Water	65
B. Solution†	20

Carnauba Wax-Resin Emulsion Polishes

Formula No. 1

Carnauba Wax No. 1	45
Gum Dammar A	15
Manila Gum DBB	5
Caustic Soda	1
Water	420
Triethanolamine	4
Oleic Acid	9

The film is dull. Slight rubbing gives good gloss.

No. 2

Carnauba Wax No. 1	45
Gum Dammar A	15
Manila Gum DBB	5
Caustic Soda	1.5
Water	420
Triethanolamine	4
Oleic Acid	9

The film is the same as No. 1.

No. 3

Carnauba Wax No. 1	30
Gum Dammar A	30
Caustic Soda	0.9
Water	422
Trigamine‡	9.9
Oleic Acid	7

The film is the same as No. 1.

No. 4

Carnauba Wax No. 1	30
Gum Dammar A	30
Caustic Soda	2
Water	422
Trigamine*	9.5
Oleic Acid	7

The film is the same as No. 1.

No. 5

Carnauba Wax No. 1	30
Gum Dammar A	30
Caustic Soda	2
Water	422
Trigamine*	19
Oleic Acid	14

The film is the same as No. 1.

No. 6

Carnauba Wax No. 1	33.5
Gum Dammar A	26.5
Caustic Soda	1.5
Water	420.0
Triethanolamine	2.0
Oleic Acid	4.0

The film is glossy.

No. 7

Carnauba Wax No. 1	47.0
Gum Dammar A	13.0
Caustic Soda	0.6
Water	420.0
Triethanolamine	4.0
Stearic Acid	9.0

The film is glossy.

No. 8

Carnauba Wax No. 1	47.0
Gum Dammar A	13.0
Caustic Soda	0.6
Water	420.0
Triethanolamine	4.0
Oleic Acid	9.0

The film is glossy.

No. 9

Carnauba Wax No. 1	30
Caustic Soda	1
Water	420
Triethanolamine	4
Oleic Acid	9

---

* L. Sonneborn

** Carbide and Carbon Chem. Co.

† B Solution

Dry Powdered Shellac	20
Water	79
Ammonium Hydroxite	1

Heat to boiling. Makes a clear solution.

‡ Glyco Products Co.

Run Boea (Hard Bold Amber)	30

The film is glossy.

No. 10

Carnauba Wax No. 1	30.0
Caustic Soda	0.5
Water	420.0
Triethanolamine	4.0
Oleic Acid	9.0

The film is glossy.

Cream Buffing Wax

Carnauba Wax	4
Yellow Beeswax	1¼
Yellow Ozokerite	¾
Paraffin Wax	2
Mineral Spirits	18¾
Diatomaceous Silica	2¾
Water	164½
Trihydroxyethylamine Stearate	6

Put the water in a steam-jacketed kettle, add the stearate, and heat to 165°F. Melt the waxes and mineral spirits at 165°F. on a water bath and stir in the silica. Then add the hot wax solution to the water solution slowly with vigorous agitation. Mix occasionally until cold.

This compound is used with flexible-shaft mechanical buffers to produce a high polish on automobile finishes. It is non-scratching and leaves a protective wax film. It is a heavy cream emulsion which is very stable.

Dance-Floor Polishes

Formula No. 1

Batavia Dammar	1
Carnauba Wax	1

No. 2

Batavia Dammar	1
Paraffin Wax	1

No. 3

Pale East India Resin	1
Carnauba Wax	1

No. 4

Pale East India Resin	1
Paraffin Wax	1

No. 5

Batu Bold Scraped	1
Carnauba Wax	1

No. 6

Batu Bold Scraped	1
Paraffin Wax	1

No. 7

Melt

a. Paraffin Scale Wax	12
Yellow or Red Oil-Soluble Dye	25–30
b. Talc	80
Yellow Ochre	8

Mix *a* and *b* thoroughly, cool, and pulverize.

No. 8

Ceresin	10
Stearic Acid	10
Paraffin Wax	80
Pine-Needle Perfume	To suit

No. 9

(Discs)

Paraffin Wax (50–52°C.)	100
Ozokerite (68–70°C.)	5

Color with oil-soluble color yellow and cast in discs.

No. 10

Stearic Acid	20
Paraffin Wax	50
Zinc Stearate	½
Talcum	30
Pine-Needle Oil	To suit

No. 11

Ceresin	44
Stearic Acid	12
Scale Wax	140
Carnauba Wax	4
Oil-Soluble Color	To suit

No. 12

Stearic Acid	20
Ceraflux*	50
Talc	30

No. 13

Stearic Acid	5
Ceresin	5
Soft Paraffin Wax	40
Talc	50

Emulsion Polish

1. Carnauba Wax	10
2. Petromix No. 5**	6
3. Deodorized Kerosene	34
4. Water	50

Melt 1 and 2 together. Add 3 and heat to 180°F. Heat the water to 200°F. and add to the melt, with stirring.

Ferrotype Polish

(For Metal Photographic Plates)

Beeswax	40 gr
Carbon Tetrachloride	4 oz

Floor Polish

Formula No. 1

Carnauba Wax	2
Ceresin	2
Turpentine	3
Gasoline	3

Melt the waxes by heating in a vessel placed in hot water, add the turpentine and gasoline, and cool the mixture as rapidly as possible, while vigorously stirring to produce a smooth creamy wax.

No. 2

Turpentine	1 pt
Beeswax	4 oz
Aqua Ammonia (10%)	3 fl oz
Water	1 pt

Mix the beeswax and turpentine and heat them by placing the vessel in hot water until the beeswax dissolves. Remove the mixture from the source of heat, add the ammonia and water, and stir vigorously until the mass becomes creamy.

No. 3

a. Carnauba Wax	15
Montan Wax	5
Pale Rosin	5

Melt on a water bath, put out the fire and add:

b. Turpentine Oil or Substitute	20

At the same time, prepare:

c. Potassium Carbonate	5
Hard Soap	5
Hot Water	45

and pour in a thin jet into *a* plus *b,* with stirring. Keep at 55–60°C. Stir continuously,

* Glyco Products Co.
** L. Sonneborn.

add a yellow dye, and then pour into cans.

	No. 4	No. 5	No. 6
Carnauba Wax	15	15	15
Cumarone Resin	6	10	4
Oleic Acid	4	5	5

Melt together and while hot add the following mixture brought to a boil.

Triethanolamine	4	5	5
Borax	6	6	3
Water	25	25	185

These polishes dry dull, but on rubbing will give a hard glossy finish.

No. 7

Esparto Wax	2¾ lb
Bleached Montan Wax	1½ lb
Bleached Beeswax	1¼ lb
White Ceresin Wax	¾ lb
Paraffin Wax (135–140°F.)	2 lb
Dyestuff	Trace
Turpentine	2 gal
Cassia Oil	4 oz

No. 8

a

Double-Bleached Montan Wax	8
Bleached Montan Wax	3
Paraffin (50–52°C.)	19
Thinner	70

b

(Yellow or Orange)

Fat-Gray* Carnauba Wax	4
Yellow Ozokerite	2
Yellow Paraffin (48–50°C.)	24
Thinner	70

* Dye with 0.02% Sudan Yellow G.

No. 9

Paraffin Wax (50–52°C.)	50
Ceresin (58–60°C.)	10
Carnauba Wax	40

Melt these and dissolve:
in summer, 7–9 parts in 93–91 parts of turpentine;
in winter, 6–7 parts in 94–93 parts of turpentine.

No. 10

Paraffin Wax (50–52°C.)	18 g
Carnauba Wax	5 g
Ceresin (58–60°C.)	2 g
Pale Rosin	4 g
Stearin	1 g
Potassium Carbonate	2 g
Caustic Soda (38°Bé.)	0.5 cc
Water	66 cc

Boil and stir until smooth.

No. 11

Carnauba Wax	15
Paraffin Wax	26
Ceresin	32
Benzine	170–180

Color to suit with any oil-soluble color.

No. 12

Carnauba Wax	60
Paraffin Wax	104
Ceresin	128
Turpentine	600
Naphtha	100

No. 13

Carnauba Wax	30
Rosin	6

Heat to 140°C., cool to 100°C. and add the following, with vigorous stirring, after heating to 95–100°C.

Soap Flakes	10
Turpentine	1
Water	270

No. 14

Yellow Beeswax	5 lb
Paraffin Wax	4 lb
Soap Chips	3 lb
Stearic Acid	3 lb
Turpentine	3 gal
Salts of Tartar	1½ lb
Water	3½ gal

Dissolve the salts of tartar and soap in boiling water. Melt the waxes in another container and heat to 200°F.; then add the boiling water soap solution slowly, with vigorous stirring, until homogeneous. Turn off the heat and run in the turpentine slowly, with good stirring. Pack in cans when cold.

No. 15

Carnauba Wax	5 lb
Ozokerite	5 lb
Turpentine	1 gal
Gasoline	5 gal

Heat gently until the waxes dissolve completely. Cool quickly.

No. 16

Paraffin Wax	16000 g
Carnauba Wax	3000 g
Yellow Beeswax	1000 g
Turpentine	46000 cc
Yellow 1435 (Dye)	20 g

No. 17

Paraffin Wax	16000 g
Carnauba Wax	3000 g
Yellow Beeswax	2000 g
Turpentine	25000 cc
Yellow 1435 (Dye)	20 g

No. 18

Paraffin Wax	16000 g
Carnauba Wax	2500 g
Yellow Beeswax	1500 g
Turpentine	30000 cc
Yellow 1435 (Dye)	20 g
Amyl Acetate	100 cc

No. 19

Powdered Scale Wax	20–30
Talcum	70–80
Ochre	To suit

Mix thoroughly and sift.

No. 20

Powdered Stearic Acid	80
Powdered Ceresin	15
Powdered Hard Soap	5

No. 21

A stable emulsified floor wax can be obtained with 40% of a mixture of 1 part of beeswax, 1 part of paraffin, 1% of soap, 0.33% of soda ash, and 2.6–2.7% of methyl cellulose. The soap and soda ash are dissolved in boiling water, and an aqueous 5% solution of methyl cellulose added. This is warmed and the melted wax added. The product is then stirred while cooling to give a good emulsion.

No. 22

A wax which spreads well and is very resistant to wear is prepared according to the following formula: 63 parts of a 5% aqueous solution of methyl cellulose are mixed into 1 part of potassium carbonate and 3 parts of soap dissolved in 37 parts of boiling water. The mixture is cooled to 70°C. and 100 parts of a melted mixture containing 1 part of beeswax, 1 part of paraffin, and 8 parts of ceresin added. Heating is continued for ½ hour, the mixture is stirred well, the amount of water which has evaporated is replaced, and 96 parts of white spirits or turpentine substitute are gradually added. The whole is stirred until cool.

No. 23

Heavy Panoline Oil	165.0
Mineral Seal Oil	75.0
Paraffin Wax	13.5
Carnauba Wax	1.5
Varnolene*	24.0

No. 24

Carnauba Wax	30
Beeswax	15
Ceresin Wax	15
Paraffin Wax	120
Varnolene*	340

No. 25

Ozokerite	2.5
Paraffin Wax	25.0
Rezo Wax B	8.5
Turpentine	35.0
Mineral Spirits	35.0

Melt the waxes on a water bath and then add the solvents slowly while stirring. Fill into containers and allow to stand undisturbed until gelled.

No. 26

Beeswax	25
Diglycol Stearate	10
Bleached Montan Wax	25
Rosin	3
Ozokerite (60–62°C.)	17
Ammonia (0.91)	70
Water	850
Water-Soluble Dye**	0.3–0.5%

No. 27

Turpentine	1 pt
Beeswax	1 lb

Melt the beeswax and work in the turpentine.

No. 28

Beeswax	500
Water	400
Calcium Carbonate	100
Water	4000

Boil the first three ingredients. Thin with the water. To 3 kg of the resulting mixture, add:

Kassel Brown (Dye)	125
Potash Carbonate	250
Water	500

Boil, with stirring. Apply when still warm, putting it on the furniture in a thin layer and polishing with a brush.

No. 29

Double Bleached Montan Wax	5
Paraffin Wax (50–52°C.)	5
Diglycol Stearate	1
Water	67
Ammonia (0.910)	2
Alcohol	20

* Standard Oil Co. of N. J.
** *Yellow:* Metanil Yellow, Extra X. *Orange:* Orange II Concentrated.

Cut the soap and waxes into small pieces and take up to a boil in the water and the ammonia, stirring thoroughly. When the mass is homogeneous, cool with stirring and add the alcohol when cooled sufficiently. Keep agitating until it becomes viscous.

•

No. 30

Carnauba Wax	2
Bleached Montan Wax	6
Beeswax	5
Paraffin Wax (52–54°C.)	14

Melt.
Add:

Linseed Oil (or Varnish)	3

When the temperature is 43–45°C., add:

Turpentine	70

Furniture and Leather Polish

Melt together 20 oz of yellow beeswax and 20 oz of Japan wax. Dissolve 5 oz of hard soap and 1 oz of pearlash (technical potassium carbonate) in 3½ pt of boiling water. Mix the hot solution with the molten waxes, stirring well until a uniform mixture is obtained. Finally stir in 3½ pt of boiling water. Color it with alkanet extract dissolved in the soap solution.

Furniture, Lacquer Enamel and Polished Nickel or Chromium Wax Polish

Carnauba Wax	12
Beeswax	10
Benzine	40
Turpentine	13
Xylol	20
Paraffin Oil	5

Furniture or Automobile Polish (Fluid)

Formula No. 1

1. Albacer*	15
Trigamine* Stearate	5
2. Water	75
Sulfatate*	5

Melt 1 and heat 2 to 85°C. Add 1 to 2 slowly, with rapid agitation. Continue stirring till cool. This gives a heavy-bodied liquid emulsion.

If 3 parts of boiling water are added to 1 part of this emulsion and stirred thoroughly, a fluid emulsion is obtained.

No. 2

Carnauba Wax	120
Kerosene	50
Stearic Acid	15
Oleic Acid	3
Benzaldehyde	6
Triethanolamine	8

Melt the carnauba wax, stearic acid and oleic acid. Remove from the fire and add the kerosene and benzaldehyde. Stir thoroughly. Now add a hot solution (80–85°C.) consisting of 240 parts of water

* Glyco Products Co.

and the given amount of triethanolamine. Stir well until a smooth emulsion is formed; then add sufficient hot water to make 825 parts. Continue stirring until nearly cool; then bottle. This polish will form a permanent emulsion which is stable at all temperatures.

Furniture Polish

Formula No. 1

Yellow Ceresin	3 lb
Japan Wax	1 lb
Beeswax	2 lb
Raw Linseed Oil	4 gal
Turpentine	1 gal
Paraffin Oil	1 gal
Water	7 gal
Carbonate of Potash	3 oz
Animal-Fat Soap Chips	1 lb

Mix thoroughly.

No. 2

Carnauba Wax	30
Beeswax	15
Ceresin Wax	15
Turpentine	26
Naphtha	24
Stearic Acid	8
Triethanolamine	4
Water	65

No. 3

a. Carnauba Wax	60
Turpentine	60
Stearic Acid	2
b. Trihydroxyethylamine Stearate	12
Water	62

Heat *a* and *b* in separate vessels to 200°F. and run *b* into *a* slowly, with vigorous stirring. Stop when homogeneous.

No. 4

Carnauba Wax	6
Paraffin Wax	9
Ceresin	2
Naphtha	43
Turpentine	4
Stearic Acid	1
Trihydroxyethylamine Stearate	4½
Water	130

Use the same method as for No. 3

No. 5

Carnauba Wax	10
Beeswax	4
Ceresin Wax	4
Naphtha	80
Stearic Acid	8
Triethanolamine	4½
Water	200

Melt the waxes and stearic acid and add the triethanolamine at about 90°C. Add the naphtha slowly so that a clear solution is maintained. Using a water- or steam-jacketed kettle prevents overheating and also caking of the waxes on the sides of the container. Add the boiling water to the naphtha solution, stir vigorously until a good emulsion is obtained, and then slowly until the emulsion is cold.

No. 6

a. Yellow Beeswax	13 g
Yellow Ozokerite	2 g
b. Thinner (White Spirit)	75 cc
c. Alkali Solution (Water: Ammonia (0.91 = 85:15)	10 cc

Melt *a,* add the warmed *b* to the clear solution, then add *c* in a thin jet, stirring thoroughly.

No. 7

Bleached Carnauba Wax	6
Japan Wax	3½
Paraffin Wax	1½
Turpentine	12
White Curd Soap	3
Pale Rosin	2
Water	30
Clovel	Trace

No. 8

Light Mineral Oil	1 gal
Carnauba Wax	2½ oz

Heat until the wax is dissolved.

No. 9

Indusoil (Tall Oil)	43
Carnauba Wax	161
Beeswax	80
Casein	80
Turpentine	139
Naphtha	128
Triethanolamine	20
Water	349

No. 10

a. Yellow Carnauba Wax	3.20
Beeswax	1.30
Ceresin	1.30
Stearic Acid	2.60
b. Triethanolamine	1.50
c. Naphtha	10.00
Veegum*	
Water	79.35

Melt *a* on a water bath. Heat *b* and add slowly to *a,* with stirring. Add *c* to this mixture slowly when the temperature is about 90°C.

Add the Veegum to the water slowly, continuously agitating until smooth. Heat this liquid gel, add it to the naphtha-wax solution, and agitate rapidly until cool.

Gold Polish

Yellow Carnauba Wax	9.5
Triple-Pressed Stearic Acid	9.5
Superfine Red Rouge	0.6
Alundum (38x600x) **	80.4

Melt the mix and cast into bars.

Inlaid Floor Polish

a. Carnauba Wax	6
Bleached Montan Wax	4
Low-Grade Ozokerite	2
Paraffin Wax (50–52°C.)	4
b. Heavy Benzine	160
Turpentine	24–20

Melt *a,* thin with *b.* By using more or less solvent, the product can be adapted to various requirements.

Knife-Polishing Stones

Formula No. 1

a. Paraffin Wax (50–52°C.)	10
Ozokerite	5
b. Diglycol Laurate	10
c. Powdered Pumice	30
Emery	25
Lime Green	20

* R. T. Vanderbilt Co.
** Norton Co.

Melt *a*, add *b*, and stir in slowly *c*. Cool till the paste sets and pour into molds.

No. 2

a.	Stearic Acid	10
	Japan Wax	10
b.	Oleic Acid	10
c.	Finely Powdered Pumice	60
	Chrome Oxide	10

Treat as in No. 1.

No. 3

a.	Paraffin Wax (50–52°C.)	10
	Stearic Acid	10
	Japan Wax	8
b.	Powdered Pumice	25
	Kieselguhr	25
	Yellow Ochre	22

Stir *b* into the melted *a*. Continue stirring till the paste sets and pour into forms.

No. 4

a.	Paraffin Wax (50–52°C.)	25
	Montan Wax	10
b.	Whiting	40
	Emery	20
	Carbon Black	5

Procedure as in No. 3.

Leather Dressing

Tallow	70
Petroleum Jelly	3½
Diglycol Stearate	13
Beeswax	9
Rosin	2
Water	2

Leather Polish

Carnauba Wax	11
Turpentine	16
Stearic Acid	3
Oil-Soluble Nigrosine	2
Triethanolamine	1
Water	66
Water-Soluble Nigrosine	1

Dissolve the water-soluble nigrosine in the water, add the triethanolamine and stearic acid, and heat to boiling. Stir until a smooth soap solution is obtained. In a separate container, melt the carnauba wax in the turpentine and add the oil-soluble nigrosine. When this solution has reached a temperature of 85—90°C., add it to the soap solution. Stir vigorously to obtain a good dispersion of the wax and then stir slowly until the emulsion is cold.

Linoleum Polish

Formula No. 1

Carnauba Wax	1 lb
Paraffin Wax	1 oz
Yellow Wax	7 oz
Turpentine	1 gal

No. 2

Carnauba Wax	1 lb 6 oz
Ceresin Wax	1 lb 6 oz
Petroleum Spirits	8 lb

Melt the two waxes together and stir in the petroleum spirits. Then grind the wax.

No. 3

(Red)

Carnauba Wax	1½
Ceresin Wax	1½
Dry Venetian Red	½
Petroleum Spirits	6½

Liquid Floor Wax

Formula No. 1

Heat 10 lb of Glyco Wax B* and 2 lb beeswax with 30 lb naphtha or kerosene until dissolved. Cool and stir thoroughly when thickening begins. Color yellow or orange with an oil-soluble dye. This may be made thicker or thinner by varying the amount of wax.

No. 2

Carnauba Wax	1 lb
Paraffin Wax	¼ lb
Raw Linseed Oil	½ pt
Turpentine	½ pt

Melt together and heat to 85°C. Add a gasoline-kerosene mixture (1:2) or naphtha with thorough stirring to make a total volume of 2 gal. Add 1 oz of ammonium linoleate and stir thoroughly until cool. If the mix fails to emulsify, heat up again and stir well until cool.

No. 3
(Nonslippery)

Candelilla or Carnauba Wax	1 lb
Japan Wax	½ lb

When melted, add gradually, with stirring, 1 gal of light-gravity paraffin oil. Mix with 4 gal more of the paraffin oil and 1 gal of turpentine substitute (petroleum spirit). Kerosene or turpentine will do, but the odor will not be pleasant, unless some essential oil is added to hide the odor. The compound will not be slippery. This mixture can also be used as a reviver or polish for any varnished surface.

No. 4

Santowax OSA**	7.00
Beeswax	1.00
Paraffin Wax	2.00
Stearic Acid	1.00

Melt together.

Add to the melted waxes, with stirring:

Morpholine	0.25
Water	88.75

No. 5

Morpholine	0.5
Water	78.0
Stearic Acid	1.5
Santowax OSA**	7.0
Paraffin Wax	2.0
Beeswax	1.0
Turpentine	10.0

No. 6

Morpholine	0.5
Water	68.0
Stearic Acid	1.5
Santowax OSA**	7.0
Paraffin	2.0
Beeswax	1.0
Turpentine	20.0

Liquid Leather Polish

Crude Montan Wax	10
Carnauba Wax	3
Candelilla Wax	3

* Glyco Products Co.
** Monsanto Chem. Co.

Ozokerite	1
Paraffin Wax	5
Diglycol Oleate	25
Water	70
Turpentine	90
Dye	5

Liquid Shoe Polish

Carnauba Wax	30	g
Paraffin Wax	45	g
Ozokerite	10	g
V. M. and P. Naphtha	215	cc
Turpentine	20	cc
Stearic Acid	20	g
Triethanolamine	7½	g
Water	650	cc
Oil-Soluble Black	15	g
Water-Soluble Black	5	g

Liquid Ski Waxes

Formula No. 1

Carnauba Wax	4
Montan Wax	12
Linseed-Oil Varnish	84

No. 2

Refined Montan Wax	15
Ceresin	3
Turpentine-Oil Substitute	82

No. 3

Rosin	30
Ceresin	25
Tallow	55

No. 4

Talc	10
Palm Oil	14
Ceresin	16
Paraffin Wax	60

No. 5

Tallow	125
Rosin	275
Montan Wax	400
Turpentine Oil	200

No. 6

Rice Starch	40
Tallow	125
Larch Turpentine	260
Yellow Wax	500

No. 7

(Sohm's Ski Wax)

Ozokerite	55
Tallow	15
Rosin	30

All these waxes may be thinned with turpentine oil to the desired fluidity.

No. 8

(Gliding Wax)

Paraffin Wax	60
Ceresin	16
Tallow	14

Melt together and add turpentine to suit.

No. 9

(Gliding Wax)

Black Ozokerite	55
Rosin	30
Tallow	15

Melt together and add turpentine to suit.

No. 10

Paraffin Wax	30 g
Bleached Montan Wax	80 g
Rosin	20 g
Japan Wax	20 g
Wood-Tar Oil	10 cc
Turpentine Oil	10 cc
Yellow Dyestuff	Enough to color

No. 11

(White Wax Polish)

Paraffin Wax	16 g
Light Carnauba Wax	3 g
White Beeswax	1 g
Turpentine	46 cc

No. 12

(Liquid Wax Polish)

Paraffin Wax	50 g
Ozokerite	5 g
Carnauba Wax	100 g
Turpentine Oil	750 cc
Benzoline	94 cc
Camphor Oil	2 g
Amyl Acetate	3 cc

No. 13

(For Gliding)

Paraffin (50–52°C.)	60
Ceresin (60°C.)	16
Tallow or Palm Oil	14
Talcum	10

No. 14

(For Climbing)

Paraffin (40–42°C.)	50
Rosin	20
Wool Fat	15
Wood Tar	15

Liquid Sporting-Shoe Polishes

Formula No. 1

Carnauba Wax, N.C.	3 g
Crude Montan Wax	2 g
Paraffin (50–52°C.)	3 g
Oil-Soluble Black Dye*	3 g
Thinners	75 cc
Refined Spindle Oil	14 cc

No. 2

Carnauba Wax, N.C.	3 g
Crude Montan Wax	2½ g
Paraffin (50–52°C.)	2½ g
Oil-Soluble Black Dye*	3 g
Thinners	72 cc
Petrolatum Oil	11 cc
Sardine Train Oil	6 cc

No. 3

Carnauba Wax, N.C.	4 g
Crude Montan Wax	2 g
Paraffin (50–52°C.)	2 g
Oil-Soluble Black Dye*	3 g
Thinners	70 cc
Refined Spindle Oil	19 cc

Liquid Stove Polishes

Formula No. 1

Crude Montan Wax	5.0
Rosin	1.5
Paraffin Wax (40–42°C.)	2.5
Diglycol Stearate	2.5
Potassium Carbonate	3.5
Graphite	50.0
Bone Black	20.0
Oil-Soluble Nigrosine	3.0
Turpentine or Substitute	97.0
Water	115.0

Melt the waxes together with the oil-soluble dyestuff, then thin out with turpentine and mill in the graphite and lamp-

---

* Black Dyes
Nigrosine Base 51017
Nigrosine Base 4322
Nigrosine Base LJF
Nigrosine Base SRN
Nigrosine Base SR
Nigrosine Base C

Dissolve as follows:

a. Olein	1
Raw Montan Wax	1
Nigrosine Base	1

Warm together and stir.

or

b. Stearin	2
Nigrosine Base	1

black. The stearate, rosin and potassium carbonate are heated together until homogeneous, and added slowly with high-speed agitation.

No. 2

a.	Crude Montan Wax	2.0
	Beeswax	1.0
b.	Turpentine	64.0
	Graphite	20.0
	Lampblack	3.0
c.	Ammonium Linoleate	2.5
	Water	8.5

Melt *a*. Make suspension *b*; thin *a* with *b*.

Heat *c* to almost boiling and add to *a* and *b*.

No. 3

a.	Crude Montan Wax	2
	Carnauba Wax	2
	Rosin	1
	Potash Carbonate	2
	Water	72
b.	Water-Soluble Nigrosine	1
	Graphite	20

Boil *a* until homogeneous. Stir in *b* gradually until smooth.

No. 4

Crude Montan Wax	2
Rosin	1
Carnauba Wax	2

Heat to 90°C., with stirring, and add slowly:

Caustic Potash	2
Boiling Water	86
Nigrosine	3

Keep on the heat and agitate vigorously until uniform. Cool and work in

Graphite Flake	5
Lampblack	3

Mix thoroughly until uniform.

No. 5

a.	Water-Soluble Nigrosine	1
	Hard Soap	3
	Water	32
b.	Paraffin Wax	3
	Ozokerite	1
	Oil-Soluble Nigrosine	1
c.	Turpentine	20
d.	Graphite	39

Melt *b*, thin with *c*; make a paste of this wax solution with *d*.

Boil *a* separately, and add *b*, *c*, to do this, with good stirring.

No. 6

a.	Crude Montan Wax	2
	Rosin	1
	Carnauba Wax	2
b.	Potassium Carbonate	2
	Water	83
c.	Lampblack	3
	Graphite Flakes	5
d.	Water-Soluble Nigrosine	3–4

No. 7

Rosin Oil	2
Black Montan Wax	2
Carnauba Wax	2
Water	82
Caustic Potash	2
Nigrosine	3
Graphite	5
Lampblack	3

The oil and waxes are carefully melted. The potash and nigrosine are dissolved in the water and gradually stirred into the wax mixture at 85 to 95°C. The resultant product is agitated thoroughly during the cooling, when the graphite and lampblack are uniformly incorporated. The caustic potash forms a soap with the rosin and aids emulsification and retards settling of the black.

No. 8

Flaky Graphite	1000
Lampblack	50
Crude Beeswax	10
Crude Montan Wax	100
Paraffin Scales (50–52°C.)	30

Melt together.

Fat-Soluble Nigrosine	5
Naphtha	To suit

No. 9

Colloidal Graphite	20
Paraffin Wax	13
Lacquer Benzoline (White Spirits)	67

Liquid Wax Polish

Formula No. 1

Heavy Panoline	165
Mineral Seal Oil	75
Paraffin Wax	13½
Carnauba Wax	1½
Varnolene*	24

No. 2

Bleached Montan Wax	5
Ozokerite	2
Paraffin Wax	6
Carnauba Wax	2
Turpentine	80
Diglycol Oleate	3

Melt the waxes at 100°C. and then cool to 85°C. Add the turpentine and diglycol oleate. Stir till cold.

* Standard Oil Co. of N. J.
** Glyco Products Co.

Metal Polish Cake

Crude Montan Wax	25
Double-Bleached Montan Wax	5
Chromium Oxide	70

Melt together and stir until cool.

Metal Polish for Chromium or Silver

O-Dichlorobenzol	400 cc
Water	70 cc
Strong Ammonia	1 cc
Saturated Solution of Castile Soap	2 cc
Levigated Alumina	35 g

If carefully made, a stable emulsion results.

For brass or copper polish, substitute precipitated chalk for levigated alumina. If a more abrasive polish is desired substitute kieselgurh for the alumina.

Metal Polish Paste

Formula No. 1

a.	Ceraflux**	8
	Ozokerite	2
	Oleic Acid	40
b.	Finest Whiting	50

Melt *a* and work in the well-sifted *b*.

No. 2

a.	Paraffin Wax (50–52°C.)	5
	Stearic Acid	16
	Oleic Acid	29
b.	Finest Whiting	20
	Magnesium Carbonate	30

Combine as in No. 1

No. 3

a. Paraffin or Scale Wax (50–52°C.)	17
Refined Montan Wax	3
Refined Spindle Oil	30
b. Finest, Sifted Whiting	46
Red Iron Oxide	4

Combine as in No. 1.

No. 4
(For Chrome and Nickel)

a. Paraffin Wax (50–52°C.)	10
Refined Montan Wax	2
Refined Spindle Oil	38
b. Kieselguhr	49
Red Iron Oxide	1

Combine as in No. 1.

No. 5

Scale Wax (50–52°C.)	5
Ceresin (58–60°C.)	3
Stearin	15
Distilled Olein	17
Spindle Oil	10
Floated Calcium Carbonate	30
Neuburg Chalk	15
Chrome Green	5

Melt the first five ingredients and stir in the abrasive powder.

Military Leather Dressing

Yellow Beeswax	65
Acrawax C*	2
Turpentine	33

Military Leather Paste Polish

Carnauba Wax	18
Candelilla Wax	2
Japan Wax	10
Paraffin Wax	2
Turpentine	20

* Glyco Products Co.

Natural Resin–Wax Polishes

Formula No. 1

Pale East India Gum	18
Paraffin Wax	8
Carnauba Wax	6
Turpentine	68

Melt the paraffin and carnauba, and sift in the pale East India. Keep at 150°C. until the mixture is homogeneous. Add the turpentine. Allow to cool with occasional stirring.

This has a buttery consistency. Apply a small amount on a clean rag and rub until dry. A quick-drying, hard, glossy film results.

No. 2

Pale East India Gum	18
Carnauba Wax	15
Paraffin Wax	17
V. M. and P. Naphtha	70

Melt the paraffin and carnauba and sift in the pale East India. Keep at 150°C. until the mixture is homogeneous. Add the V. M. and P. naphtha and allow to cool, stirring occasionally.

This polish is more viscous than No. 1. By applying with a clean rag and rubbing until dry, a quick-drying, hard, glossy film results.

No. 3

Pale East India Gum	23
Beeswax	6
Carnauba Wax	20
Montan Wax	8
V. M. and P. Naphtha	89
Turpentine	10
Pine Oil	3

Melt the waxes and sift in the resin. Keep at 150°C. until the mixture is homogeneous. Add the solvents and allow to cool, stirring occasionally.

This is a liquid at room temperature with pasty particles precipitated from it, but when applied with a clean rag, it gives a quick-drying, hard, glossy film.

No. 4

Pale East India Gum	18
Carnauba Wax	15
Paraffin Wax	17
Turpentine	35
V. M. and P. Naphtha	35

Melt the wax and sift in the resin. Keep the temperature at 150°C. until the mixture is homogeneous. Add the solvents and allow to cool, stirring occasionally.

This has about the same consistency as Formula No. 2. When applied with a clean rag, rubbing until dry, a quick-drying, hard, glossy film results.

Nickel, Chrome, Copper, and Brass Polish

Calcined Kieselguhr	40
Whiting	35
Red Iron Oxide	10
Paraffin (50–52°C.)	15

Melt and grind in a mill.

Nonrubbing Bright-Drying Floor Polishes

These polishes are mixtures of aqueous Manila-resin solutions with carnauba-wax emulsions. The following resin solution is recommended:

Formula No. 1

Ground Manila DBB or Loba C	240
Monoethanolamine	45–50
Water	2090

Dissolve the monoethanolamine in 700 parts of water, stir, and add the resin. Heating the mixture to 50—55°C. helps the resin to dissolve more rapidly. Finally, add the remainder of the water and allow to cool. Strain.

This resin solution is mixed with a carnauba-wax emulsion to obtain the final product. Established wax-emulsion formulae are satisfactory, but the following may be used if desired:

Carnauba Wax	240
Triethanolamine	28
Linoleic Acid	45
Borax	18
Water	2412

Melt the wax at 90—95°C.

In a separate container, dissolve the triethanolamine in 200 g of water, stir at 90°C., and add the linoleic acid. Stir the mixture 5 minutes at 85–90°C. and then add the resulting cloudy viscous solution to the wax. Stir the wax-soap mixture at 90–95°C. for 5 minutes.

Dissolve the borax in 50 g of boiling water, add, and stir until a hot drop off a spatula is clear (14–15 minutes). Then add the remainder of the water cautiously at 95–100°C. When the mixture starts to thin out, add the water more rapidly. Allow to cool to room temperature and add water to compensate for the evaporation loss.

Mix the resin solution, and the wax emulsion in the desired proportions. Stir the emulsion and add the resin solution. Continue stirring to obtain a homogeneous polish.

No. 2

Refined Carnauba Wax	15.0
Durez 420 Resin*	15.0
Oleic Acid	4.5

Heat to 300–325°F. until fluid. Cool to 210°F., add the following at the boiling point:

2-Amino, 2-Methyl-1-Propanol	1.0
Sodium Hydroxide	0.4
Water	5.0

Stir until the mixture is clear and jellylike. Add the following slowly, with good agitation:

Hot Water	210.0

Cool rapidly and add:

Casein Solution (10%)**	25.0

No. 3

Carnauba Wax	18
Pine Oil	1½
Soap	1½
Hot Water	3

Cook till transparent.

Sodium Silicate	2¼
Hot Water	3

Cook 5 minutes. Make up to 4 qt.

No. 4

Carnauba Wax	8½ lb
Candelilla Wax	16½ lb
Borax	2 lb
Duponol† W.A. Paste	5½ lb

* General Plastics Corp.

** Casein Solution

Casein	10.0
Borax	1.5
Pine Oil	0.2
Preservative	0.2
Water at 160°F.	15.0

Mix thoroughly and add:

Water at 160°F.	15.0

Mix thoroughly and add:

Ammonia in Water, 7% of 28%	1.0
Water at 160°F.	70.0

Mix until smooth.

† Du Pont.

Dissolved in

Water	2 gal

Cook till smooth. Make up to 50 gal with boiling water.

In a separate kettle, cook

Gum Loba C	30 lb
Water	30 gal
Ammonia (26°)	150 oz

until the gum is completely dissolved and add to the wax emulsion. Make up the volume of the finished mixture to 100 gal.

No. 5

Candelilla Wax	64
Oleic Acid	24
2-Amino-2-Methyl-1-Propanol	11
Water	To make 500

Very translucent, thin emulsion which dries to a transparent, bright film.

No. 6

Boiled Water	2½ gal
Dry Bleached Shellac	6¼ lb
Borax	1¾ lb

Melt together:

Esparto Wax	5 lb
Emulsifier	2 lb

Stir 2 gal boiling water into the molten wax, cool, and add the shellac solution while stirring.

No. 7

1. Prepare a carnauba-wax emulsion as follows:

Yellow Carnauba Wax No. 1	30
Water	210
Trigamine*	5
Oleic Acid	3.5
Caustic Soda	0.25

Melt the wax and oleic acid together in a boiling water bath until a homogeneous, clear liquid is obtained. Stir in the Trigamine.

Dissolve the caustic soda in 25 parts of water. Heat it to boiling. Add this solution slowly and with stirring to the wax mixture still in the boiling water bath. A gel forms; then add the remainder of the water, also preheated to boiling, slowly and with stirring.

2. Prepare a Manila gum solution as follows:

Manila DBB Gum	30
Water	158
Ammonium Hydroxide (28–29% NH_3)	12

Powder the resin and mix it with the water and ammonia. Agitate the mixture until the resin has dissolved. Strain the solution.

3. Prepare the finished polish as follows:

Manila-Resin Solution (2)	39
Carnauba-Wax Emulsion (1)	34

Mix the two well, using high-speed stirring. The polish dries

* Glyco Products.

to a hard, glossy, waterproof film.

No. 8

Carnauba Wax	13.20
Oleic Acid	1.50
Triethanolamine	2.10
Borax	1.00
Water	108.00
Shellac	2.20
Ammonia (28%)	0.32

Melt the wax and add the oleic acid. The temperature should not be above 90°C. Using a hot-water or steam-jacketed kettle maintains a good temperature and prevents wax caking along the sides of the container. Add the triethanolamine slowly, stirring constantly. The solution should be *clear* at this point. Dissolve the borax in about 1 pt of boiling water and add to the wax solution to obtain a clear jellylike mass. Stir for about 5 minutes. Add 92 lb water, previously heated to boiling, slowly with constant stirring. An opaque solution should be obtained. Cool. Add 16 lb water to the shellac and then the ammonia and heat until the shellac is in solution. Cool. Add this to the wax solution and stir well to obtain an even mixture.

No. 9

Carnauba Wax	11.9
Paraffin Wax	1.3
Oleic Acid	3.0
Triethanolamine	1.9
Hot Water	50.0
Cold Water	50.0

Melt the wax; add the oleic acid and triethanolamine. With the temperature at 95°C., add small (5-part) portions of boiling water, stirring after each addition and continuing until a heavy translucent paste is formed. Add the remaining boiling water and stir until homogeneous. Add the cold water; cool the complete dispersion to room temperature. If desired, all of the water may be added at boiling temperature.

Due to the possible variation in the composition of some of the suggested ingredients and to the varying methods and conditions which may attend the compounding, modifications to meet individual requirements may be necessary.

No. 10

Carnauba Wax No. 2	120.0
Oleic Acid	13.7
Trigamine*	19.0
Caustic Soda (47%)	3.5
Water	843.5

Melt the wax in the oleic acid, add the Trigamine and heat to 90°C. Then add the caustic soda and 25 cc of water at about 95°C. Stir until gell

* Glyco Products.

is formed. Then add the rest of the water which must be boiling. Stir thoroughly till cool.

No. 11

Hydromalin*	138
Yellow Carnauba Wax No. 1 or No. 2	250

Heat at 120—140°C. for ½ hour.

Cool to 100—105°C.

Add to this slowly with stirring:

Boiling Water	280

Stir to a smooth paste then add slowly with a good stirring:

Boiling Water	1500

Keep as close to 100°C. as possible for 15 minutes with rapid stirring. Then stir till cool.

No. 12

Aquamel*	15
Carnauba Wax	25
Water	225

Heat the wax and Aquamel together till a clear jelly is obtained. Then add the water which must be as near boiling as possible. Stir thoroughly till cool.

No. 13

1. Make a carnauba-wax emulsion as follows:

Yellow Carnauba Wax No. 1	30
Water	210
Triethanolamine	3
Oleic Acid	5

Melt the wax and the oleic acid in a boiling water bath. When the mixture is homogeneous, add the triethanolamine and stir until it is all uniformly incorporated.

Heat the water to boiling and add it slowly, portion by portion, permitting each addition to be well agitated before the next addition. As the first additions of water are made, the mixture thickens considerably and becomes gellike. As more water is added, the mixture becomes creamy. When still more water is added, it breaks into an emulsion and then becomes a watery product. During the gel and cream stage, it is very important to add the water carefully and agitate efficiently to obtain a homogeneous mass. It is just as detrimental to add the water too slowly as too fast. Equally as important is the maintenance of high temperatures during the process.

2. Make a Manila-resin solution as follows:

Manila DBB	30
Water	158
Ammonium Hydroxide (28–29% NH_3)	12

Powder the resin and agitate it with the water and ammonia

* Glyco Products Co.

until all the resin is dissolved. Strain the solution.

3. Prepare the finished polish as follows:

Carnauba-Wax Emulsion	34
Manila DBB Solution	39

Mix the two well, with stirring. A glossy, waterproof polish is obtained.

No. 14

Yellow Carnauba Wax No. 1	30.0
Thermally Processed Congo	30.0
Sodium Hydroxide	0.5
Water	420.0
Triethanolamine	4.0
Oleic Acid	9.0

Melt the wax and add the Congo at 125–150°C. Hold until the mixture is homogeneous; then put the mixture in a boiling water bath. Add the oleic acid slowly and with vigorous stirring. Do the same with the triethanolamine. Dissolve the sodium hydroxide in 25 parts of water. Heat the solution to its boiling point and add it slowly to the mixture still in the boiling water bath, with vigorous stirring. When the mixture is thoroughly homogeneous, add the remainder of the water (also at its boiling point and with vigorous stirring).

This polish dries quickly to a hard, glossy, waterproof film. It is especially valuable for polishing floors.

Nonslippery Linoleum Wax

Carnauba Wax	15 }
Montan Wax	5 }
Light Rosin	5 }
Turpentine	20
Potash Carbonate	5 }
Diglycol Stearate	5 }
Water	45 }
Alkaline Yellow Dye	To suit

Melt the first three ingredients on the water bath, stir in the turpentine, and add the hot solution of the balance in a thin jet with stirring. Keep at 55–60°C. on the bath until completely "bound." Pour at the lowest possible temperature.

Norwegian Klister (Ski) Waxes

Formula No. 1

Rice Starch	40
Tallow	125
Larch Turpentine	260
Yellow Wax	500

No. 2

Paraffin (40–42°C.)	60
Rosin	12
Wool Fat	6
Carnauba Wax	4
Montan Wax	80

No. 3

Ozokerite	55
Rosin	35
Refined Spindle Oil	10

No. 4

Paraffin	70
Rosin	15
Wool Fat	10
Carnauba Wax	5
Montan Wax	15

No. 5

Ozokerite	5
Rosin	4
Train or Spindle Oil	1½

No. 6

Crude Montan Wax	120
Paraffin	30
Wool Fat	20
Seal Train Oil	15
Hard Tallow	10
Rosin	5
Wood Tar	3

No. 7

Paraffin	1
Tallow	1½
Rosin	2½
Ozokerite	15

No. 8

Wool Fat	10
Ceresin	90

Paste Leather Polish

Bleached Montan Wax	10
Crude Montan Wax	5
Candelilla Wax	4
Carnauba Wax	2
Paraffin Wax	5
Diglycol Stearate	3
Water	50
Turpentine	70
Oil-Soluble Color	5

Paste Wax Polish

(Auto Wax)

Formula No. 1

Carnauba Wax	10.0
Candelilla Wax	5.0
Yellow Ozokerite	9.5
Yellow Beeswax	8.5
Stearic Acid	0.5
Pine Oil	0.5
Wood Turpentine	2.0
Kerosene	44.0
Mineral Spirits	20.0

No. 2

(Floor and Furniture Wax)

Carnauba Wax	4.0
Candelilla Wax	2.0
Yellow Ozokerite	3.0
Yellow Beeswax	3.0
Paraffin Wax	10.0
Pine Oil	1.0
Wood Turpentine	10.7
Mineral Spirits	55.1
V. M. and P. Naphtha	11.2

Melt the waxes in a steam bath or steam-jacketed kettle. Add the solvents and heat until clear, if necessary. Cut off the heat and cool to about 130°F. Pour into containers and allow to solidify uncovered.

Polish for Cellulose Lacquer

Any hard wax polish can be used as a base and to this abrasive compounds, such as silica, air-floated tripoli powder, pumice flour, kieselguhr, etc., are added. A suitable wax polish base is:

Beeswax	5
Carnauba Wax	6
Xylol	10
Benzine	20
Paraffin Oil	2½
Turpentine	6½

The proportion of abrasive will depend on the type used and the degree of cutting action required. The abrasive should be mixed with a proportion of the paraffin and turpentine to wet it well and then mixed thoroughly with the wax base.

Polish for Cord

Paraffin Wax (50–52°C.)	10
Japan Wax	5
Caustic Potash	2
Water	To desired consistency

Boil together until saponified.

Polish for Leather Furniture

Paraffin Wax (50–52°C.)	20
Ozokerite/Ceresin (58–60°C.)	5
Neutral Wool Fat	5
Beeswax	10
Carnauba Wax	10
Turpentine Oil	150

Color similar to that of furniture. Pour at 40—45°C. into jars.

Polish for Patent Leather

Yellow Wax or Ceresin	3 oz
Spermaceti	1 oz
Turpentine	11 oz
Asphaltum Varnish	1 oz
Borax	80 gr
Frankfort Black	1 oz
Prussian Blue	150 gr

Melt the wax, add the borax, and stir until smooth. In another pan, melt the spermaceti; add the varnish, previously mixed with the turpentine; stir well; and add to the wax; add the colors as the last.

Polish for Piano Keys

Montan Wax	30
Tallow	20
Diglycol Laurate	100
Micro-Asbestos	1200

Polishing Compound for Lucite and Plexiglas

Carnauba Wax	10
V. M. and P. Naphtha	10
Santomerse No. 3* (Emulsifying Agent)	5
Silica Smoke (1,000 mesh)	8
Rough (Red Iron Oxide)	2
Boiling Water	65

The carnauba wax is melted and poured into the heated naphtha (89—95°C.) with vigorous agitation. The emulsifying agent is dissolved in the water. The wax solution is then poured, while both solutions are still quite hot, into the water solution with vigorous agitation. The mixture is continuously stirred until cooled to room temperature. While still hot, the powdered silica and the rouge are added.

Polishing Dye
(Polishing Ink)

a.	Water	80.0
	Marseilles Soap	3.0
	Carnauba Wax	4.0
	Japan Wax	2.0
b.	Water	20.0
	Methyl Cellulose	1.0
c.	Turpentine Oil	1.0
	Citronella Oil	0.3

* Monsanto Chem. Co.

Boil *a;* when cooled, add *b* which has been boiled separately. When cooled, add *c.*

Polishing Plastic and Hard-Rubber Articles

Formula No. 1

Carnauba wax may be satisfactorily replaced with Albacer* where a nonaqueous solvent is employed. 1 lb of Albacer dissolved by heat in 2 qt of Turkelene* gives a nongreasy, high polish on Catalin and similar plastic articles. Approximately 3–4 oz in the standard 1½ bushel tumbling barrel is usually sufficient. The work is tumbled for about 16 hours either in sawdust or in wooden pegs which are slightly wet with the polishing solution. Using Turkelene* in place of oleic acid gives a much less greasy film and is not subject to rancidity. If a volatile solvent is desired, the Albacer may be dissolved by heat in approximately 1 gal of naphtha or a high-boiling solvent, such as Varnolene.** If 1 pt of butanol is added to the solvent, more of the wax will remain dissolved at room temperature. Wax which separates in fine particles should be redispersed by stirring before adding the proper quantity to the contents of the tumbler barrel.

When a solvent or oily vehicle is undesirable, a very satisfactory aqueous liquid preparation is a bright-drying carnauba wax emulsion. About 4 oz to each 1½ bushel barrel containing either sawdust or, preferably, wooden pegs, is ordinarily sufficient.

No. 2

Stearic Acid	8
Montan Wax	2
Lauryl Sulfonate	2
Iron Oxide	38

Polishing Wax

Formula No. 1

Montan Wax	15
Carnauba Wax	5
Candelilla Wax	2
Paraffin Wax	3
Japan Wax	1
Turpentine	75

No. 2

Carnauba Wax	10
Candelilla Wax	15
Paraffin	75

No. 3

Bleached Beeswax	2
Carnauba Wax	6
Halowax No. 1013†	4
Turpentine	20
Stearic or Oleic Acid	3½
Triethanolamine	1½
Water	63
Perfume	To suit

* Glyco Products Co.
** Standard Oil Co. of N. J.
† Halowax Corp.

Melt the three waxes and add the turpentine with agitation. Prepare the soap solution by heating the triethanolamine dissolved in the water almost to the boil and adding the molten stearic or liquid oleic acid. The soap may also be prepared by heating the triethanolamine and water with stearic acid in powdered form. Run the first solution into the second with agitation. Continue to agitate until cool. If properly made, the emulsion is stable.

No. 4

Paraffin Wax	28
Ozokerite	6
Carnauba Wax	3
Yellow Beeswax	4
Turpentine	60

No. 5

Paraffin Wax	15
Candelilla Wax	3
Ester Gum	2
Acrawaax B*	10
Varnolene†	70

Melt the solids together on a water bath until dissolved and then work in the Varnolene.

No. 6

Glyceryl Abietate	10
Paraffin Wax	20
Candelilla Wax	3
Acrawaax B*	10
Varnolene†	80

No. 7

a. Paraffin Wax (50–52°C.)	5
Diglycol Stearate	16
b. Oleic Acid	25
c. Whiting	20
Magnesium Carbonate	30

Melt *a,* add *b,* and then *c* in small portions.

No. 8

Santowax M††	15
Barnsdall Special Wax (M.P. 160–165°F. No. 2 Color)	8
Carnauba Wax	8
Candelilla Wax	6
Paraffin Wax	8
Solvasol No. 5	48
Turpentine	5
Pine Oil	2

No. 9

Santowax M††	15
Barnsdall Special Wax	8
Carnauba Wax	8
Candelilla Wax	6
Paraffin Wax	8
Turpentine	55

No. 10

Santowax OSA††	6
Candelilla Wax	14
Beeswax	2
Paraffin Wax	4
Turpentine	74

Pour at 43°C.

No. 11

Santowax PSA††	8
Opalwax‡	13
Paraffin Wax	6
Turpentine	73

Pour at 92°C.

* Glyco Products Co.
† Standard Oil Co. of N. J.
†† Monsanto Chem. Co.
‡ Du Pont.

No. 12

Santowax P*	8
Candelilla Wax	8
Paraffin Wax	6
Beeswax	2
Turpentine	76

Pour at 120°C.

The waxes are first melted together. Refined grades must be used as crude waxes always contain varying amounts of dirt, part of which would settle out when the waxes are melted, but another part would remain in fine suspension in the wax mixture to give an unsatisfactory product. The temperature should be kept as low as possible, not only to conserve steam, but to speed up the process so that the solvents can be added as soon as the waxes are melted and mixed. The kettle used is preferably equipped with an automatic mixer, as thorough mixing is absolutely necessary. Melting and mixing of the waxes can be accomplished at 190°F. As soon as the molten waxes are well mixed, the heat is turned off and the solvents added with rapid stirring. Odorants and dyes are then added as soon as the solvent-wax mix appears to be homogeneous, and the rapid stirring continued so that the mix will cool as quickly as possible. Refrigeration is often employed to hasten cooling and insure proper setting.

Rapid cooling makes the product smooth and prevents granules from forming. Shortly before the mixture is ready to congeal, it is drawn off into containers and allowed to solidify uncovered. The temperature at which it should be drawn off will probably be about 125°F., but this can be determined for the individual mix by experiment, by measuring the approximate solidifying point of samples withdrawn from the batch at a temperature safely above this. As soon as the paste has solidified so that it will not be disturbed by handling, the covers could be fitted on tightly to prevent further evaporation of solvent.

The room in which the process is carried out must be well ventilated because of the escape of fumes during the mixing and congealing of the polish. It must also be remembered that although the petroleum solvents referred to here are not classed as inflammable, their hot vapors are combustible and due caution must be exercised to prevent a flame or spark coming near them.

* Monsanto Chem. Co.

Polishing Wax for Rotating Discs

Montan Wax	3
Tallow	3
Stearic Acid	4
Yellow Mineral Oil	24
Viennese Chalk To desired consistency	

Pure Carnauba Polish

1. Carnauba Wax	10
2. Petromix No. 5* (Emulsifier)	4
3. Butyl Cellosolve**	1
4. Water	85

Melt 1 and 2; then add 3. Add 4 at 200°F., with constant stirring.

Razor Hone Paste

Washed Emory or Carborundum Powder	6
Freshly Rendered (400–600 Mesh) Beef Suet	2
Yellow Beeswax	1

Melt the suet and the wax then incorporate the emery or carborundum. Cool and cut into small bars.

Razor-Strop Preparations

Formula No. 1

Acidless Tallow	250
Caput Mortuum	750

Melt the fat and work in the iron pigment.

No. 2

Finest Powder Emery	30
Tin Oxide	15
Paris Red	15
Natural Petrolatum	30
Beeswax	30

No. 3

Natural Petrolatum	82
English Red (Iron Oxide)	18

Run through a mill.

Rubbing-Type Paste Wax

Formula No. 1

Carnauba Wax	150
Paraffin Wax	166
Beeswax	76
Durez 420 Resin†	34

Heat to 300–325°F. to melt. Add the following:

Mineral Spirits	1280
Turpentine	32

Stir and cool to 160°F. Dump into containers.

No. 2

Carnauba Wax	14.0
Beeswax	14.0
Paraffin Wax	14.0
Stearic Acid	5.0
Amino Methylpropanol	2.5
Water	80.0

The first four ingredients are melted together on a boiling water bath. The amino methylpropanol is stirred into the wax melt, the temperature of which should not exceed 100–105°C. at this point, and the boiling

* L. Sonneborn.
** Carbide and Carbon Chem. Corp.
† General Plastics Corp.

water is added with vigorous agitation.

No. 3

Glyco Wax A*	10
Naphtha or Kerosene	30

Heat the wax with the naphtha or kerosene until dissolved. Cool and stir thoroughly when thickening begins. Color yellow or orange with an oil-soluble dye. This may be made thicker or thinner by varying the amount of wax.

No. 4

Carnauba Wax	60 oz
Dry Castile Soap	15¼ oz
Oleic Acid	2¾ oz
Caustic Soda	1½ oz
Water	To make 4 gal

Heat the wax, soap, and acid to 100°C. and, while stirring rapidly, add the caustic dissolved in the water, heated to 80°C., slowly. A stable creamy emulsion results.

Rubless Bright-Drying Floor-Polish Emulsion

Formula No. 1

a. Candelilla Wax	12.5

Heat to 200°F.

b. Triethanolamine	1.8
Linolenic Acid	1.7
Borax	1.0
Water	83.0

Heat to boiling.

Pour *a* into *b*, stirring well.

No. 2

Fine-Melt Congo	120
Caustic Soda	3
Morpholine	12
Water	500
Water	100
Carnauba Wax	20
Stearic Acid	6
Triethanolamine	3
Water	100
Water	400

To the 500 parts of water, add the sodium hydroxide and morpholine. Heat this solution to 85–90°C. With continued stirring, add the fine-melt Congo slowly, taking about 15–20 minutes to add the entire amount. It is very important that the resin is ground to powder. Hold at 85–90°C. for ½ hour, during which time add 100 parts of water. The resin should now be completely dissolved.

In a separate container, melt together the carnauba wax and the stearic acid, and add slowly to the Congo mixture, continuously stirring. Then add the triethanolamine and about 100 parts of hot water slowly, keeping the temperature around 80°C. Add the remaining 400 parts of water more quickly and allow the polish to cool to room temperature. Add water to compensate for evaporation loss and then filter the polish.

* Glyco Products Co.

This polish exhibits excellent gloss, good leveling properties, and very good water resistance.

No. 3

Fine-Melt Congo	120
Caustic Soda	5
Morpholine	6
Water	500
Water	100
Carnauba Wax	20
Stearic Acid	6
Triethanolamine	3
Water	100
Water	400

Same method as in No. 2.

This is slightly less glossy and water resistant than No. 2.

The use of a protective colloid like casein is found to increase compatibility and the drying time of Congo polishes.

Rubless Floor Wax

a. Carnauba Wax	60 g
Paraffin Wax	9 g
Oleic Acid	15 g
b. Borax	3 g
Ammonium Hydroxide	9 g
Water	18 g
c. Dewaxed Shellac	15 g
Ammonium Hydroxide	9 g
Water	372 g

Melt *a* over direct heat, oil bath, or water bath in a glass, stainless-steel, or enameled vessel at about 95°C.

Dissolve the borax in the water, heat to 80–85°C., stir in the ammonium hydroxide and add *b* to *a,* with thorough stirring, to give a dark, nearly transparent, viscous, water-in-oil emulsion. Heat *c,* with stirring, until the shellac dissolves completely, and the temperature is 95–100°C. Have this ready, and begin to add it, a little at a time, to the mixture of *a* and *b*. Stir rapidly and thoroughly as each portion of shellac is added, adding 15–20 cc at a time. When about 50–100 cc of shellac solution has been added, sharp decrease in viscosity will be noted. Then add the balance of the shellac solution rapidly, with very thorough mixing. This results in phase inversion to wax-in-water emulsion and should yield a semitransparent product. Continue to stir until cool. Cooling may be hastened by a cold-water bath. The wax emulsion now contains 20% solids and may be diluted with 340 g cold water to give 850 g of wax dispersion containing 12% solids, the usual concentration for use.

Shoe Cream

Formula No. 1

1. Trigamine Stearate	25
Beeswax	10
Candelilla Wax	30
Carnauba Wax	40
Turpentine	20
2. Water	500

Heat 1 to 200°F. and in a separate pot, heat 2 to 200°F.

Run 1 into 2 slowly while stirring vigorously. This gives a light neutral cream. If a colored cream is desired, dissolve some oil-soluble dye in the wax mixture while it is melting. A thicker cream suitable for tubes is made by decreasing the water content.

No. 2

Crude Montan Wax	15
Soft Paraffin Wax	20
Rosin	4–6
Potassium Carbonate	3
Water	95
Oil-Soluble Color	
{Mandarin Yellow, Metanil Yellow}	1 each

No. 3

Double-Bleached Montan Wax	15
Soft Paraffin Wax	20
Rosin	4–6
Potassium Carbonate	3
Water	95
Oil-Soluble Color	
{Mandarin Yellow, Metanil Yellow}	0.1 each

No. 4

Crude Montan Wax	10
Double-Bleached Montan Wax	4
Rosin	3
Potassium Carbonate	2½
Hard Soap	½
Borax	½
Water	100
Water-Soluble Nigrosine	2

No. 5

High-Temperature (75 to 80°C.) Pouring

Wax Mass:

Crude Montan Wax	20
Paraffin Wax (50–52°C.)	10
Rosin	5

Saponifying Solution:

Potassium Carbonate	6
Water	60

Dye Solution:

Nigrosine T.	5.5
Water	50
Final Additional Water	50

No. 6

Low-Temperature (40 to 50°C) Pouring High-Grade Cream

Wax Mass:

Crude Montan Wax	10.00
Beeswax	4.00
Carnauba Wax	1.00
Japan Wax	5.00
Shellac Wax	1.25

Saponifying Solution:

Potassium Carbonate	1.5
Boiling Water	75.0
Borax	0.5
Diglycol Stearate	1.0
Palm-Oil Soap	1.5

Dye Solution:

Water	50
Nigrosine	5
Final Additional Water	150

No. 7

Crude Montan Wax	10.00
Double-Bleached Montan Wax	3.00
Rosin	2.00
Paraffin Wax (50/52°C.)	2.00
Potassium Carbonate	2.20
Castile Soap	0.30
Sodium Tetraborate	0.30
Water-Soluble Nigrosine	2.00
Metanil Yellow Dye	0.08
Water	95.

Pour at 65°C.

No. 8

Crude Montan Wax	15
Paraffin Wax (40/42°C.)	20
Rosin	5

Potassium Carbonate	3
Water	96
Metanil Yellow	0.9
Water-Soluble Yellow Mandarin Dye	0.8

Pour at 40—45°C.

No. 9

Double-Bleached Montan Wax	15.00
Paraffin Wax (40–42°C.)	20.00
Rosin	5.00
Potassium Carbonate	3.00
Water	95.00
Metanil Yellow	0.05
Water-Soluble Yellow Mandarin Dye	0.10

Pour at 40—42°C.

Shoe Stain:

a. Crude Montan Wax	15.0
Rosin	2.0
Paraffin Wax (40–42°C.)	3.0
Potassium Carbonate	0.5
Castile Soap	4.0
Water	65.0
b. Shellac	20.0
Sodium Tetraborate	7.0
Water	75.0
c. Water-Soluble Nigrosine	5.5
Water	25.0

a, b, and *c* are prepared separately and then mixed together.

No. 10
(Tropical)
Black

Crude Montan Wax	4.50
Nigrosine Base	1.20
Gray Carnauba Wax	5.50
Paraffin Wax (54–56°C.)	8.00
Yellow Ozokerite (66°C.)	1.30
Candelilla Wax	1.30
Shellac Wax	1.20
American Turpentine	77.00

Mahogany

Gray Carnauba Wax	8.50
Yellow Ozokerite	3.50
Paraffin Wax (54–56°C.)	4.50
Shellac Wax, Refined	0.50
Sudan Brown 5B	0.10
Sudan Red 5B	0.05
Turpentine	83.00

No. 11

a. Crude Montan Wax	18
Japan Wax	2
Carnauba Wax	4
Rosin	2
b. Water	260
98% Potash	6
Water-Soluble Nigrosine	12

Heat *a* and *b* separately to 95—100°C. and add *b* to *a* while stirring vigorously with an electric mixer.

No. 12

Acrawax B*	10
Paraffin Wax	15
Candelilla Wax	3
Ester Gum	2
Varnolene**	54

Gives a very fair paste which has a good consistency, spreads well, and polishes well.

No. 13

Candelilla Wax	50
Diglycol Stearate	30
Stroba Wax*	25
Varsol**	250
Water	225–300

Melt the first three ingredient together and, after remov-

* Glyco Products Co.
** Standard Oil Co. of N. J.

ing the flame, stir in the Varsol. Add the water (at 90°C.) slowly, with vigorous stirring, to the wax-solvent mixture.

No. 14

Carnauba Wax	20
Paraffin Wax	12

Heat to 200°F. and add to this slowly, with good stirring, while heating on a steam table.

Turpentine	65
Carbon Black No. 1	2½
Oil-Soluble Black Dye	½

Stir until uniform.

No. 15

Fat-Gray Carnauba Wax	6 g
Bleached Montan Wax	7 g
Paraffin (50–52°C.)	11 g
Ozokerite	2 g
Dyestuff	2 g
Thinner (Turpentine Oil, or Substitute or a Mixture of Both)	72 cc

No. 16

Candelilla Wax	30 g
Paraffin Wax	45 g
Ozokerite	10 g
Turpentine	20 cc
V. M. and P. Naphtha	80 cc
Oil-Soluble Black	15 g

No. 17

Crude Montan Wax	4.0
Japan Wax	0.3
Paraffin Wax (50–52°C.)	1.0
Carnauba-Wax Residue	2.0
Carnauba Wax	0.5
Rosin	0.7
Beeswax	0.3
Potassium Carbonate	1.15
Water	30.0
Turpentine	10.0

No. 18

Gray Carnauba Wax	7 g
Crude Montan Wax	7 g
Paraffin (50–52°C.)	12 g
Oil-Soluble Black Dye	3 g
Thinner (Turpentine Oil, or Substitute, or a Mixture of Both)	51 cc
Petrolatum Oil	20 cc

No. 19

Mixture 1

Carnauba Wax	8
Montan Wax	8
Paraffin Wax	4

These are saponified in a hot solution of:

Potash	3
Water	50

Warm water is added to make up for evaporation.

Mixture 2

No. 1 Polish Black	4
Water	25

Mill together in a color mill until thoroughly dispersed.

While Mixture No. 1 is hot, add Mixture No. 2 slowly and with constant stirring. As it cools, the mass will slowly set to a paste. Before it is too stiff for flowing, pour into suitable containers and set aside until cold.

No. 20

Carnauba Wax	2
Paraffin Wax	4
Ceresin Wax	2
Beeswax	2
Glyceride	4

Silica Black	3
Water	17
Turpentine	6
Glucarine B*	5

Melt the first five ingredients together. Then stir in the silica black and add the water while heating and stirring until homogeneous. Remove from the flame, let cool a little, and stir in the turpentine and Glucarine B. Continue stirring till the mass thickens to the proper consistency.

Shoe Polish

Formula No. 1

Albacer*	2
Glyceride Composition	5
Paraffin Wax	11
Candelilla Wax	4
Oil-Soluble Black Dye	9/10
Turpentine	9
V. M. and P. Naphtha	21
Sovasol No. 5**	20

The waxes are melted and the color dispersed in the melted mass. The solvents are added and the solution poured at 200°F. If the solution is not kept at this temperature, the surface of the cooled paste will be rough looking and the paste will not be homogeneous. This will give an excellent polish with a very high, smooth surface luster.

No. 2

(Base)

Paraffin Wax (50–52°C.)	2.50
Pale Carnauba Wax	0.65
White Beeswax	0.80
Turpentine	10.00

Dyes for this Formula

Brown:

Brown 1435	2
Brown 3148	10
Orange in Pieces	2½

Pale Brown:

Brown 1435	1
Brown 3148	6
Orange in Pieces	1½

Dark Brown:

Brown 1435	1½
Brown 3148	13
Orange in Pieces	3

Reddish Brown:

Brown 1435	6
Brown 20162	6
Brown 17484	6
R. L. Ponceau	1

No. 3

Beeswax	1 lb
Ceresin Wax	1 lb
Carnauba Wax	6 oz
Turpentine	3 pt
Yellow Soap	6 oz
Oil-Soluble Black Aniline	Enough to color
Water	Sufficient

Shave the soap and dissolve in the smallest possible quantity of water by means of heat; melt the waxes together; add the turpentine and stir well;

* Glyco Products Co.
** Socony-Vacuum Co.

then add the aniline dye and stir in the soap solution, continuing to stir until cold.

No. 4
(Black)

Carnauba Wax	4989
Solvent Naphtha	7000
Stearic Acid	1360
Oil-Soluble Nigrosine	907
Diglycol Laurate	200
Triethanolamine	450
Water	29937
Water-Soluble Nigrosine	453

Dissolve the water-soluble nigrosine in the water, add the triethanolamine, and, while hot, the stearic acid and diglycol laurate. Take up to a boil.

Add to this the solution of carnauba wax and the oil-solution nigrosine in the naphtha at 85–90°C. Stir until cold.

No. 5
(Nonalkaline)

Carnauba Wax	6
Ceresin Wax	3
Candelilla Wax	3
Turpentine	5
Lemon-Grass Oil	½
Light Mineral Oil	1
Diglycol Stearate	5

Heat together below 90°C. and stir until clear. Pour slowly while stirring with a high-speed stirrer into

Boiling Water	100

Stir until the temperature falls to 70°C. The finished product is snow white, applies easily, and rubs to a high gloss.

If a paste polish is desired, reduce the water to about 50–60 parts. If a colored paste is desired, dissolve an aniline dye in the water used.

No. 6

a. Carnauba Wax	5
Ceresin Wax	15
Rosin	30
Paraffin Wax	60
Naphtha	165
b. Glue	35
Water	75
Soft Soap	3
Starch	1/10

Warm *a* and *b* separately and stir until uniform; then mix together while stirring vigorously.

No. 7

a. Beeswax	100
Paraffin Wax	20
Gray Carnauba Wax	8
Naphtha	80
Turpentine	70
b. Water	200
Calcined Potassium Carbonate	6
Glue	35

Method as under formula No. 6.

No. 8

a. Yellow Carnauba Wax	500 g
Carnauba-Wax Residue	500 g
Bleached Montan Wax	500 g
Paraffin Wax (50–52°C.)	200 g
Rosin	150 g
b. Water	8500 cc
Caustic Potash	300 g
Olive-Oil Soap	100 g
c. Turpentine Oil or Substitute	1500 cc

Melt *a*, saponify with *b*, stir until cool, and add *c* shortly before solidified.

No. 9

Crude Montan Wax	6 g
Carnauba Wax	3 g
Ozokerite (58–60°C.)	2 g
Candelilla or Shellac Wax	3 g
Paraffin Scales (50–52°C.)	14 g
Nigrosine Base	3 g
Turpentine	20–30 cc

No. 10

Esparto Wax	2½ lb
Crude Montan Wax	¾ lb
Beeswax	1 lb
Paraffin Wax (M.P. 118°F.)	½ lb
Paraffin Wax (M.P. 135°F.)	2½ lb
Lampblack	¾ lb
Turpentine	2 gal
Cassia Oil	1 oz
Lemon-Grass Oil	6 oz

Shoe Polish and Preservative

Carnauba Wax	2
Beeswax	2
Neatsfoot Oil	1

Heat on a hot-water bath (not over fire) till melted, and then add turpentine until a soft paste is obtained when the mixture is cold. This should be applied to the clean, dry leather with a rag or a piece of waste and rubbed hard until no more polish is absorbed. Polish with a clean cloth. A higher polish will be obtained by reduction of the proportion of oil, but the leather will not be so well preserved.

Shoemaker's Waxes

Soft Black:

Rosin	5
Paraffin Wax	70
Beeswax	5
Asphaltum	20
Black Color	To suit

Hard Black:

Rosin	10
Paraffin Wax	45
Beeswax	5
Asphaltum	10
Black Color	To suit

Soft Yellow:

Rosin	10
Paraffin Wax	85
Beeswax	5
Oil-Soluble Yellow Color	To suit

Hard Yellow:

Rosin	10
Paraffin Wax	55
Beeswax	5
Oil-Soluble Yellow Color	To suit

Silver Polish

a. Paraffin Wax (50–52°C.)	8
White Mineral Oil	45
b. Magnesium Carbonate	35
Calcium Carbonate	12

Ski Wax

Formula No. 1

Wood Tar	70
Diglycol Stearate	20
Carnauba Wax	10

No. 2
(For Cold Weather)

Wood Tar	50
Paraffin Wax	25
Turpentine	5

No. 3
(For Melting Weather)

Montan Wax	17
Wool Fat	18
Paraffin Wax	10
Rosin	28
Ozokerite	25
Spindle Oil	5
Wood Tar	2

No. 4
(For Collapsible Tubes)

Montan Wax	10
Rosin	5
Beeswax	5
Rosin Oil	30
Linseed Oil	10
Spindle Oil	20
Wood Tar	20

No. 5
(Climbing and Gliding)

Paraffin Wax (50–53°C.)	40
Crude Montan Wax	15
Neutral Wool Fat	15
Rosin	10
Mineral Oil	15
Wood Tar	5

No. 6
(Climbing Klister)

Crude Montan Wax	17
Neutral Wool Fat	18
Paraffin Wax	10
Rosin	28
Ozokerite	25
Mineral Oil	5
Wood Tar	2

No. 7
("Skate")

Crude Montan Wax	10
Rosin	12
Paraffin Wax	30
Ozokerite	20
Petrolatum	8
Wood Tar	5

No. 8
(Climbing)

Rosin	10
Ceresin	5
Montan Wax	5
Paraffin	5
Japan Wax	10
Beeswax	5
Tallow	5
Lanosyn*	5

No. 9
(Gliding)

Rosin	5
Ceresin	20
Montan Wax	7½
Ceraflux**	5
Japan Wax	7½
Wool Fat	5

No. 10
(Climbing and Gliding)

Crude Montan Wax	30
Rosin	10
Wool Fat	25
Wood Tar	8
Paraffin (52–54°C.)	20
Ozokerite-Ceresin	7

Apply on dry ski. Rub in until shiny, using a cork or the palm of the hand. For climbing, rub in some wax transverse to the axis of ski. Polish this down before gliding.

No. 11

Crude Montan Wax	18
Paraffin Wax	60
Ozokerite	4
Wool Fat	6
Rosin	12

Melt together and add turpentine oil to the desired consistency.

* J. Landau Co.
** Glyco Products Co.

No. 12
(Ascension Wax)

Ceresin	10
Paraffin Wax	20
Wool Fat	28
Rosin	15
Montan Wax	27

Melt together and add turpentine to the desired consistency.

Solvent Polish for Furniture and Floors

Formula No. 1

Santowax PSA*	8
Opalwax**	8
Paraffin Wax	8
Mineral Spirits	76

No. 2

Santowax OSA*	2
Beeswax	2
Paraffin Wax	2
Mineral Spirits	94

Stove Polishes

Formula No. 1

Paraffin Wax (40–42°C.)	3.5
Crude Ozokerite	2.0
Turpentine	40.0
Rosin	0.7
Lampblack	4.0
Graphite	24.0
Ammonium Linoleate	4.0
Water	20.0
Oil-Soluble Nigrosine	0.8
Potassium Carbonate	1.0

No. 2

Carnauba Wax	1.5
Self-Emulsifying Wax (Japsol)	2.0
Diglycol Stearate	2.0
Colloidal China Clay	4.0
Rosin	1.5
Potassium Carbonate	1.5
Graphite	18.0
Lampblack	5.0
Water-Soluble Nigrosine	1.0
Water	63.5

Several self-emulsifying waxes, suitable for inclusion in such a formula, are now commercially available. The diglycol stearate also helps to improve the suspension and homogeneity of the preparation, while the colloidal clay tends to prevent caking of the graphite and lampblack.

No. 3

Cumarone Resin	2
Soft Paraffin Wax	2
Crude Montan Wax	5–6
Potash Carbonate	4
Water	200
Nigrosine Base BB	10
Lampblack	24
Graphite	56

No. 4

Ceresin (56–58°C.)	8 kg
Oleic Acid	10 kg
Spindle Oil	12 kg
Neuburg Chalk	20 g
Quartz Flour	10 kg
Powdered Emery	25 kg
Carborundum	15 kg

No. 5

Ceresin (56–58°C.)	3
Oleic Acid	17
Natural Petroleum Jelly	4
Spindle Oil	6
Powdered Quartz or Pumice	45
Chrome Oxide	25

* Monsanto Chem. Co.
** Du Pont.

No. 6

a.	Crude Montan Wax	5.0
	Ceraflux*	1.5
	Beeswax	0.5
b.	Oil-Soluble Nigrosine	0.5
c.	Graphite	40.0
	Lampblack	2.5
	Turpentine	50.0

Melt *a,* and dissolve *b* in the melted *a* with stirring.

Prepare suspension *c* and stir it slowly into the melted wax-color mixture. Let it solidify in cans.

No. 7

a.	Crude Montan Wax	2.5
	Rosin	0.8
	Japan Wax	0.7
b.	Oil-Soluble Nigrosine	0.6
c.	Potassium Carbonate	1.5
	Water	53.9
d.	Graphite	28.0
	Soot	12.0

Melt *a.* Dissolve *b* in the melted *a.* Boil *c* and stir in *d* to a smooth suspension. Mix $c+d$ into $a+b$, with good stirring. Put into cans while hot.

No. 8
(Solid)

Emery	25
Powdered Quartz	25
Scale Wax (52–54°C.)	10
Distilled Oleic Acid	10
Spindle Oil	10

No. 9

Flake Graphite	33.0
Lampblack	1.7
Beeswax	4.0
Ceresin (58–60°C.)	1.0

* Krebs Pigment and Color Corp.

Tripoli Buffing Stick

Double-Pressed Saponified Stearic Acid	30
Edible Tallow	25
Paraffin Wax	25
Tripoli Flour	20 (or as much as will be absorbed)

A buffing or polishing paste may be made with addition of a small amount of turpentine and water to bring to the desired consistency.

Tripoli Composition

Stearic Acid	55
Edible Tallow	2
Oleostearine	5
Rosin	9
Petrolatum	40
Japan Wax	1
Flint	315
Double-Ground Tripoli Flour	93
Ponolith*	2

Valve-Grinding Compounds

Formula No. 1

Ceresin	5
Glass Powder	40
Pale Cylinder Oil	55

No. 2

Double-Bleached Montan Wax	5
Crude Montan Wax	25
Abrasive	70

No. 3

Ceresin	3
Oleic Acid	17
Petrolatum	4
Spindle Oil	6
Powdered Quartz or Pumice	45
Powdered Chrome Oxide	25

No. 4

Ceresin	8
Oleic Acid	10
Spindle Oil	12
Neuburg Chalk	20
Powdered Quartz	10
Powdered Emery	25
Silicon Carbide	15

No. 5

a. Ceresin	5
Tallow	2
Distilled Oleic Acid	30
Mineral Oil	10
b. Powdered Abrasive*	53

Melt *a* at 80–90°C. Mix in *b* in small portions. Stir until cooled to sufficient viscosity.

Water-Resistant Emulsion Floor Polishes

Formula No. 1

a. Carnauba Wax	11.2
Oleic Acid	2.4
Morpholine	2.2
Water	67.0
b. Shellac	1.5
Morpholine	0.2
Water	15.5

Melt the carnauba wax carefully with the oleic acid and maintain the temperature closely at 90°C. Stir until well mixed, add the morpholine, and stir constantly until the whole mass is quite clear. In the meantime, bring the water to a simmer in a separate kettle. Add it slowly to the hot wax mixture with steady stirring, making certain that each small portion is well incorporated before further addition. The mixture becomes increasingly viscous and should be of the appearance of petrolatum when two thirds of the water have been added. After this stage has been reached, the mixture begins to thin out and the remainder of the water may be added rapidly. The total time for adding the water is 30–40 minutes. A steam-jacketed kettle and a hand-operated paddle, or slow-speed, large-blade propeller are recommended for successful production.

Allow the mixture to cool. While slowly stirring, add the shellac solution. This has been made by warming together the morpholine, water and shellac indicated in *b* and filtering if necessary.

Another morpholine dry-bright polish is formulated as follows, using paraffin as a substitute for part of the carnauba wax:

No. 2

Carnauba Wax	64.8
Paraffin Wax	7.2
Morpholine	14.4
Oleic Acid	15.3

* Emery, Quartz, Tripoli, Carborundum, Slate Powder, Iron Oxide, Diatomaceous Earth, Chrome Oxide, Graphite, Viennese Chalk.

Boiling Water	430.0
Cold Water	100.0

By great care in the gradual addition of boiling water up to the point of emulsion inversion, a polish is prepared which is exceptionally translucent.

No. 3

Rosin	454.0
Oiticica Oil	760.0
Soybean Oil	18.0
Ouricury Wax	4.5
Kerosene (62°)	152.0
Naphtha (54°)	182.4
Liquid Drier	1.5

Heat the rosin and oiticica oil to 300°C. Then remove from the heat. Add the soybean oil with good mixing. When cooled to 290°C., add the wax and stir well. Add the kerosene in portions while stirring until the temperature has fallen to 165°C. Then add the naphtha slowly, with stirring, and finally the drier.

Wax Polish for Varnished Metals

Two types of waxing composition are in use: one suitable for application immediately after varnishing and containing 16% ceresin wax, 10% calcium stearate, 30% petroleum and 44% mineral cleaning oil; the other a wax cleaning composition, containing a mild abrasive powder, 3% calcium stearate, 7% ceresin wax, 2% iron oxide, 0.5% carbolic acid, 53% petroleum and 1.5% concentrated ammonia.

Wax Polishes for Wood

Formula No. 1

Bleached Montan Wax	70
Soft Ozokerite	10
Paraffin Wax (50–52°C.)	20

No. 2

Paraffin Wax (50–52°C.)	80
Ozokerite	20

No. 3

Crude Montan Wax	50
Soft Ozokerite	20
Paraffin Wax (50–52°C.)	30

No. 4

a. Double-Bleached Montan Wax	4
Beeswax	5
b. Soft Soap	1
Potassium Carbonate	3
Water	82
c. Ammonia (0.910)	5

Melt *a,* boil up with *b,* and stir until a smooth emulsion has formed. Cool. Add *c.*

No. 5

Crude Montan Wax	5
Beeswax	4
Soft Soap	1
Potassium Carbonate	3
Water	87

Boil all ingredients together until a smooth emulsion has formed. Stir cold.

White-Kid Cleaner

a. Gray Carnauba Wax	50
Bleached Montan Wax	50

	Paraffin Wax (50–52°C.)	20
	Japan Wax	20
	Rosin	15
	Diglycol Stearate	10
	Potassium Carbonate	32
	Water	600
b.	Turpentine	150
	Benzene	100
	Trichloroethylene	150
	Acetone	50

Boil *a,* with stirring, until a homogeneous emulsion has been formed. Cool to 40°C. and add *b.* Stir until cooled.

White-Shoe Polishing Stick

Flora Carnauba Wax	4
Stearic Acid	4
Paraffin Wax	17
Bleached Montan Wax	16
China Clay	9
Titanium Dioxide	1

Wood-Button Polish

Turpentine	120 cc
White Wax	120 g

Melt and with stirring add:

Alcohol	50 cc

PROTECTIVE AND DECORATIVE COATINGS

Acidproof Tank Lining

Pitch	75
Plaster of Paris	9
Yellow Ochre	9
Beeswax	15
Litharge	3

Melt the pitch and beeswax over an electric burner or a very low flame and add all the other ingredients. Apply this mixture while hot with a stiff brush; then let cool.

Alberene Stone-Top Dressing

Paraffin Wax	1,000 g
Kerosene	700 cc
Gasoline	600 cc

Melt the paraffin over a water bath and cool to about 65°C. Pour in the kerosene slowly with vigorous stirring. Allow to cool to about 50°C. and add the gasoline with thorough mixing. Keep in a stoppered bottle, away from flame.

This mixture should be heated over a water bath until it becomes liquid, and applied to the stone by means of a cloth. The surface of the stone must not be moist and too cool. After the thin layer of the dressing has set, it should be polished with a dry soft cloth.

Black Tire Finish

Gilsonite	20
Acrawax B*	20
Paraffin Wax	40
Candelilla Wax	20
Carbon Black	6
Twitchell Base**	5
Varnolene†	190

* Glyco Products Co.
** Emery Industries.
† Standard Oil Co. of N. J.

Coloring Meerschaum Pipe Bowls

Beeswax	50
Olive Oil	50
Triethanolamine	15

The meerschaum pipes are immersed in the liquid which is slowly heated to boiling and maintained at this temperature for 15–30 minutes. Pipes so treated will color very rapidly.

Electrical Insulation

Formula No. 1

Carnauba Wax	1 lb 14 oz
Yellow Beeswax	4 oz
Venice Turpentine	6 oz
Gum Obsidian	6 oz
Sulfur	2 lb 8 oz

Cook until thoroughly uniform.

This wax should have a melting point of 285°F. and a flash point of 499°F.

No. 2

Reclaimed Rubber	40.0
Smoked Sheets	10.0
Whiting	34.3
Mineral Rubber	10.0
Zinc Oxide	2.5
Paraffin Wax	1.5
Stearic Acid	0.5
Sulfur	1.0
Captax	0.2

Etch Resist

Asphaltum	12½
Beeswax	4½
Ceresin Wax	58
Stearic Acid	25

In etching glass, it is used to block off portions which one desires to keep unetched.

Etching Filler

White Beeswax	10
French Chalk	5

Melt together.

Use as a filler for etched lines in metal to make them more distinctive.

Fruit-Protective Coatings

Formula No. 1

Carnauba Wax	13.6
Oleic Acid	2.9
Morpholine	2.5
Water	81.0

No. 2

Paraffin Wax	25.0
Oleic Acid	13.1
Triethanolamine	6.9
Water	315.0

Ignition-Cable Covering

Formula No. 1

Smoked Sheet	100.0
Vistanex B 100*	67.0
Polystyrene Resin	33.0
Zinc Oxide	10.0
Zinc Salt of Lauric Acid	1.7
SRF Black**	6.7
Dipolymer Oil	1.7
Antioxidant	3.3
Paraffin Wax	6.7
Clay	97.5
Mercaptobenzothiazole	1.0
Tetramethylthiuramdisulfide	0.5
Litharge	1.0
Sulfur	2.5

* Advance Solvent and Chem. Corp.
** Am. Aniline Products.

No. 2

Smoked Sheet	100.0
Vistanex B 100*	67.0
Polystyrene Resin	33.0
Zinc Oxide	10.0
Zinc Salt of Lauric Acid	1.7
SRF Black**	6.7
Dipolymer Oil	1.7
Antioxidant	3.3
Paraffin Wax	6.7
Whiting	97.5
Mercaptobenzothiazole	1.0
Tetramethylthiuramdisulfide	0.5
Litharge	1.0
Sulfur	2.5

Inside Coating for Wood Barrels

a. Yellow Wax	40
Rosin	200
b. Iron Oxide	40
c. Molding Gypsum	10

Melt *a,* then stir in *b,* finally *c.* Apply the liquid, hot mixture with a brush.

Moistureproof Coating

Aroclor 5460†	96
Paraffin Wax	4

Overprint Varnishes

Formula No. 1

Batavia Dammar A/E	5850 lb
China-Wood Oil	520 gal
Linseed Oil	115.5 gal
Kerosene	397 gal
Cobalt-Drier Solution (6%)	675 lb
High-Melting Paraffin Wax	71 lb
Oil Length, Gallons	8.9
Drying Time, Hours	4
Gloss	High

No. 2

Batavia Dammar A/E	6300 lb
China-Wood Oil	930 gal
Kerosene	497 gal
Cobalt-Drier Solution (6%)	472 lb
High-Melting Paraffin Wax	157 lb
Oil Length, Gallons	14.8
Drying Time, Hours	4
Gloss	Good

Heat the resin and oil to 450°F. Add the kerosene, then the drier, and finally the paraffin wax.

These varnishes dry to water-white, flexible films with maximum marproofness.

Paint and Varnish Remover

Formula No. 1

Amylene Dichloride	80
Alcohol	40
Naphtha	20
Diglycol Stearate	2

No. 2

Benzol	50
Methanol	25
Acetone	15
Gasoline	10
Paraffin Wax	2½

Paint Remover

Formula No. 1

Cellulose Acetate (PH-1)	4.0
Sulfonated Castor Oil	7.0
Paraffin Wax	1.5
Lactic Acid	3.0
Methyl Ethyl Ketone	87.0
Water	10.0

* Enjay Co.
** Am. Aniline Products.
† Monsanto Chem. Co.

No. 2
(Airplane Camouflage, U. S. Army)

	%
Methylene Chloride	76.5
Methyl Alcohol	6.5
Cellosolve*	4.0
Methyl Cellulose	2.0
Sodium Petroleum Sulfonate (e.g., Sulfatate B**)	5.0
Paraffin Wax	3.0
Water	3.0

No. 3

Paraffin Wax	3
Wood Spirit	30
Acetone	25
Benzol	20
Carbon Tetrachloride	15
Xylol	10

The paraffin wax is dissolved in the benzol-carbon tetrachloride-xylol mixture and the wood spirit-acetone mixture added. This precipitates the wax and forms a paste when it is set.

No. 4

a. High-Melting Paraffin Wax	5
Trichloroethylene	95
b. Aluminum Naphthenate	5
Trichloroethylene	95

Make solutions *a* and *b*, let stand for 24 hours separately, and then mix.

Porcelain-Enamel Patching

White Rosin	40
Carnauba Wax	20
Titanium Dioxide	40

Melt the first two ingredients and sift into it the titanium dioxide. Mix well.

Protective Coating for Aluminum Castings and Zinc- and Tin-Plated Parts

Diglycol Stearate	1
Ozokerite	2
Paraffin Wax	4
Water	40

Melt the three waxes together and keep at about 65–70°C. Heat the water up to about the same temperature, or a little above, and add slowly, with high-speed agitation, to the melted waxes to obtain an emulsion.

Protective Coating for Beer Cans

Asphalt	100
Montan Wax	10
Cumarone Resin	½–2

Thin to suit with a mixture of equal parts of toluol and carbon tetrachloride.

Protective Coating for Hydrofluoric Acid Containers

Beeswax	1
Paraffin Wax	4

* Carbide and Carbon Chem. Co.
** Glyco Products Co.

Protective Coatings for Stone-Top Tables

Formula No. 1

High-Melting Paraffin Wax	1 lb
Kerosene	1 qt
Raw Linseed Oil	2 qt
Lampblack	¼ oz

No. 2

High-Melting Paraffin Wax	½ lb
Carbon Tetrachloride	1 qt
Naphtha	1 qt
Lampblack	¼ oz

Apply and allow the solvent to evaporate. Rub down with a soft rag. Apply several coats for best results.

Raw-Rosin Carnauba Varnish

Raw M. Rosin	100 lb
China-Wood Oil	20 gal
Soybean Oil	4 gal
Carnauba Wax	1 lb
Kerosene (62°)	4 gal
Naphtha (54°)	48 gal
Concentrated Varnish Drier No. 1007	1 gal

Melt the rosin in the china-wood oil; run to 570°F. and pull from the fire.

Allow to stand until the temperature registers 580°F., then chill back with the 4 gal of cold soybean oil and stir well.

Add the soybean oil all at once and continue stirring until the temperature begins to fall.

When the temperature has dropped 20°, add the carnauba wax; stir until dissolved. Then allow to stand, with frequent stirring, until cooled to 500°F., at which point, thin with the kerosene.

After the kerosene is in, stir frequently while cooling to 330°F. Then thin with the naphtha and finally add the concentrated varnish drier.

This varnish is fairly pale; has a good luster; dries in about 8 hours; and is extremely waterproof.

It is not neutral; consequently it must not be used with lead and zinc pigments.

It is designed especially for refrigerator cars; bathroom wall and kitchen woodwork, but it is not long enough in oil for a floor varnish.

Remover for Lacquers of Oil–Chlorinated Rubber and Oil Paints

Formula No. 1

Scale Paraffin Wax (40–42°C.)	15
Benzol	45
Toluol	40

No. 2

Scale Paraffin Wax (40–42°C.)	20
Benzol	60
Hexalin	20

Rubbing Compound

(For Paint, Lacquers, Etc.)

1. Carnauba Wax	42 lb
2. Beeswax	18 lb
3. Ceresin	18 lb
4. Varnolene*	3 gal
5. Water	3 gal
6. Triethanolamine	8 oz
7. Stearic Acid	2 lb
8. Tripoli	24 lb
9. Pumice	15 lb

Melt 1, 2, 3, and 7 with 4. Heat 5 and 6 to 90°C., add to the wax mixture and stir till emulsified. Then add 8 and 9 and stir till cool.

Submarine Grinding Oil

No. 887

Menhaden Fish Oil	32 gal
Manganese Dioxide	6 lb
China-Wood Oil	16 gal
Litharge	24 lb
Raw H. Rosin	50 lb
Japan Wax	2 lb
Kerosene (46°)	48 gal

Put the manganese dioxide in the cold menhaden oil, stir well, then heat slowly to 560°F.

At about 320°F., the manganese dioxide will begin to give off oxygen, and at 420–480°F., the mixture will froth strongly. Therefore, proceed cautiously.

When a temperature of 560°F. has been reached, usually in 2–3 hours, pull from the fire, add the cold wood oil, and stir until thoroughly mixed. By then the temperature will have dropped to 460–440°F.

Now begin adding the litharge, about 2 lb at a time, sprinkling it on the surface of the oil, allowing it to remain for ½ minute and then stirring for 1 minute.

Whip down the foam, if necessary, before adding more litharge, then continue the additions until all of the litharge is in.

Put the kettle back on the fire, heat to 480°F., pull from the fire, and add the rosin, a little at a time, stirring constantly until the rosin is in solution. The addition of rosin will cause some foaming as all of the litharge will not have been taken up.

When frothing ceases, put the kettle back on the fire, heat to 480°F., and cook at 480–520°F. until the litharge is taken up. Then pull from the fire, add the Japan wax, stir for 2 or 3 minutes, and thin with the kerosene.

This dark-colored, waterproof oil is used in roof paints, freight-car and smokestack paints.

* Standard Oil Co. of N. J.

Tarnish and Corrosion Prevention

Ozokerite	20
Paraffin Wax	40
Diglycol Stearate	10

Dispersing 7 parts of this wax mixture in 40 parts of hot water will yield an emulsion which is applied preferably by spraying, leaving a thin, transparent, adhesive and flexible film. Protection is obtained from tarnish or corrosion of ferrous and nonferrous metal surfaces. For instance, the flat ground steel surfaces and the nickeled intaglio portion of candy molds are both protected effectively. The coating tolerates extremes of high room temperature and very low freezing temperature. A recent report indicates that waxy coatings on steel machinery and parts are much better than grease slushing because waxy coatings apparently seek out the spots particularly susceptible to corrosion or which have already begun to corrode. Greases do not always inhibit the electrochemical action resulting in corrosion, whereas waxes do. Where the coating must be removed, boiling water, hot water and soap, or solvents may be used.

Waterproof Awning Paint

Crude Beeswax	1 lb
Rosin	1 lb
Nondrying Vegetable Oil	0.5 gal
Outdoor House Paint	1.25 gal
Volatile Mineral Spirits (Painters' Naphtha)	3.75 gal

Melt the beeswax and rosin in the heated vegetable oil. Add to the paint with stirring; when cooled sufficiently, dilute with naphtha. Allow to dry a week between coats. Do not fold for storage until thoroughly dry.

Wax Furniture Filler

Japan Wax	30
Paraffin Wax	10
Neutral Wool Fat	10
Rosin	40

The waxes are simply melted together.

Wax "Lacquer"

a.	Orange Shellac	14
	Alcohol	60
b.	Carnauba Wax	2
	Paraffin (50–52°C.)	1
	Turpentine	23

Heat *a* to 50°C. on a steam bath and *b* to 60°C. separately.

Add *b* to *a* with thorough stirring; agitate until cold.

Wine-Tank Coating

Gilsonite	25
Paraffin Wax	75

Melt together until uniform and apply very hot.

SOAPS AND CLEANERS

Abrasive Cleaner

a.	Scale Wax (50–52°C.)	5
	Ceresin	5
	Oleic Acid	8
	Spindle Oil	5
	Petroleum	3
b.	Neuburg Chalk	25
	Artificial Bitter Almond Oil	1/10

Melt *a* and mix thoroughly with *b*. Fill into cans while hot.

Carbon-Paper and Ink Stain Cream Remover

Lanolin	2.7
Carbowax 4000*	19.1
Carbowax 1500*	19.1
Triethanolamine	3.0
Oleic Acid	5.3
Water	19.1
Terpineol	0.2
Multi-Cel 1000**	31.4
Sodium Bisulfite	0.1

Cigarette (Nicotine) Stain Remover

Beeswax	10.0
Paraffin Wax	5.0
Mineral Oil	46.0
Powdered Pumice	8.0
Borax	0.5
Water	30.0
Perfume	0.5

Dustless Sweeping Mixture

Sand	50
Sawdust	25
Mineral Oil	20
Beeswax	5

* Carbide & Carbon Chem. Corp.
** Tamms Silica Co.

Melt the wax in the oil. Then stir in the sand and sawdust. Add a green oil-soluble dye to give the mixture a pleasing appearance.

Floor-Sweeping Compound

Formula No. 1

Sawdust	20 lb
Fine Sand	6 lb
Paraffin Wax	1½ oz
Light Mineral Oil	10 fl oz

Melt the paraffin and add the oil. Work this into the sawdust; then add the sand and mix well.

No. 2

Fine Sand	35
Pine Sawdust	40
Paraffin Oil	15
Water and Dye (If coloring is desired)	10

Some commercial compounds are colored with iron oxide or other pigments and some contain naphthalene flakes, paraffin wax, etc. Essential oils, such as oil of eucalyptus, oil of sassafras, etc., are sometimes added to impart a pleasant odor to the compound or to mask any unpleasant odor of the ingredients used. Pine oil, a small amount of creosote oil, or some other compound can be used as a disinfectant.

The water-wax emulsion type of sweeping compound is an outgrowth of the development of the water-emulsion floor waxes. In this type, the mineral oil is replaced with waxes, resins, water, and emulsifying agents. Instead of a thin film of oil on the floor, a film of wax is deposited. The wax type of sweeping compound is intended for use on floorings, such as linoleum, rubber, asphalt tile, mastic and polished wood, that may be affected by oils.

Glass-Windshield Cleaner

Triethanolamine Oleate	2
Kerosene	15
Beeswax	4
Carnauba Wax	1
Celite* (Light Abrasive)	16

Ink-Removing Cream

Carbowax 4,000**	50.0
Water	50.0
Propylene Glycol	50.0
Triethanolamine	8.0
Oleic Acid	14.0
Anhydrous Lanolin	7.0
Terpineol	0.5
Mild Abrasive (Multi-Cel 1000†)	82.0
Sodium Bisulfite	0.3

* Johns-Manville.
** Carbide and Carbon Chem. Co.
† Tamms Silica Co.

Parquet-Floor Cleaner

a. Refined Montan Wax	15
Ceresin	15
Soft Paraffin Wax	15
Stearic Acid	15
b. Turpentine	864
Tetralin	52
Benzol	24

Melt *a* and thin with *b*. Stir till cold. Perfume with pine-needle oil.

Saddle Soap

Formula No. 1

Soap Powder	15
Water	72
Neatsfoot Oil	5
Beeswax	8

Dissolve the soap in hot water. Heat the neatsfoot oil and wax until melted and pour the hot mixture into the soap solution. Stir until it begins to thicken; then pour into cans.

No. 2

Palm-Oil Soap Chips	8
Water	24
Beeswax	1½
Neatsfoot Oil	1¼
DuPont Orange No. 110 (2% solution)	10

Dissolve the soap chips in hot water and add the color. Melt the wax with the neatsfoot oil and add to the soap solution. Stir until the mix begins to thicken and then pour into cans.

Soap Superfatting and Neutralizing Compound

Beeswax	700
Lanolin	3000
Stearic Acid	400
Borax	62
Distilled Water	3000

The beeswax and half of the lanolin are heated together to about 80°C. The borax is dissolved in the water, which is boiling hot and the solution stirred into the melted fats and boiled for some time. The mixture is then allowed to cool to about 60°C., with stirring. The stearic acid is melted in another suitable vessel and the balance of the lanolin added to it. This is added to the first mixture at 50°C. and stirred. About 2% of the compound may be used as a superfatting agent with soap chips.

Typewriter Cleaner

Yellow Petrolatum	2–5
Anhydrous Lanolin	12–15
Bleached Montan Wax	10
Soft Paraffin Wax (40°C.)	20
Finely Ground Clay	50
Earth Color	3–13

Waterless Hand Cleaner

Formula No. 1

a. Glyceryl Monostearate	1.75 oz
Sodium Hydroxide	2.50 oz
Water	3.00 qt

Heat to boiling.

b. Stearic Acid	2.50 oz
Mineral Oil	2.50 oz
Paraffin Wax	11.50 oz
Spermaceti	7.00 oz
Lemenone (Perfume)	0.20 oz

Heat in a double boiler until all is melted.

Pour solution *b* into *a*, with constant stirring, until cold. This will make a gallon.

TEXTILE COMPOUNDS

Composition for Stiffening Shoe Tips

Glue	250
Water	250
Portland Cement	300
Ceresin	100
Coal-Tar Pitch	100

Concentrated Finishing Compound

(For Cotton Piece Goods)

Sulfonated Tallow (75%)	22–26
Japan Wax	12–15
25% Trisodium Phosphate Solution	20–24
Water	50–60

Emulsify the Japan wax in a separate vessel.

Mix the tallow, one third of the emulsified Japan wax, and the required amount of trisodium phosphate solution until thoroughly blended.

Add the remainder of the Japan-wax emulsion, agitate, and heat; it is best not to boil.

Stir until a creamy mix is obtained.

Concentrated Rosin Size

Rosin	70
Soda Ash	7
Beeswax	2
Water	21

Boil together until a sample solidifies on cooling. This may be shipped in the solid state and dispersed in hot water when needed.

Cotton Corset Filling Composition

Formula No. 1

Corn Starch	56 lb
Paraffin-Wax Emulsion	1½ gal
China Clay	5 lb
Glycerin	8 lb
Water	100 gal
Bluing	To suit

No. 2

Corn Starch	35 lb
White Dextrin	12 lb
Paraffin-Wax Emulsion	8 lb
Sulfonated Castor Oil	4 lb
Water	To make 65 gal

Cotton Softener

Stearic Acid	15
Paraffin Wax	5
Water	100–400
Ammonia	1

Heat the stearic acid, wax, and water and, while agitating vigorously, run in the ammonia in a thin stream.

Cotton-Cord Glaze Finish

Water	50 gal
Potato Starch	15 lb
Castile Soap	½ lb
Paraffin Wax	4 lb
Beeswax	1 lb
Lard Oil	2½ pt

Mix the starch in cold water and pour into the boiler when the water is at the boiling point. Add the other ingredients and boil 15 minutes. Use when cold.

Cotton-Warp Sizing

Tapioca Flour	14– 20 lb
Ground Animal Glue	1⅓– 3 lb
50% Sulfonated Tallow	3– 5 lb
Paraffin Wax	3– 5 lb
Water	90–150 gal

Dressing for Prepared Yarns of Upholstery Cottons

Double-Bleached Montan Wax	10
Stearic Acid	15
Soap	2.5
Gelatin	2.5
Soda Ash	0.3
Starch	10
Water	400

Emulsion for Finishing Dyed Rayon

Paraffin Wax	10
Stearin	10
Rosin	6

To a fused mixture of the ingredients is added a solution of:

Sodium Hydroxide (40°Bé.)	2
Ammonia (25%) Diluted 1:2.5 with Water	6

The mixture is then heated to 80°C. and water is added to make 100 parts.

Felt-Hat Stiffener

Bright-Drying Carnauba-Wax Emulsion	90
Shellac (Ammonia Water Solution)	10

Increasing the Flexibility and Luster of Cotton and Rayon Braid

A mixture of glyceryl tristearate and Flexo Wax C* gives high luster and flexibility to cotton and rayon braids, shoe laces, etc. The blended waxes are melted and the braid is run through in continuous lengths and then over brushes on polishing rollers. The addition of oil-soluble colors to the wax mixture will make the coating suitable for different colored braids.

Luster Emulsion for Starching

Stearic Acid	1 kg
White Beeswax	200 g
Borax	60–75 g
Water	4 kg
Citronella Oil for Perfume	5 g

Of this emulsion, 10 spoonfuls are used for 1 kg of starch.

Primer for Rubber Cloth

Hard Soap	1.7
Light Carnauba Wax	8.7
Shellac	1.3
Borax	1.4
Water	86.9

Rotproofing Canvas

Japan Wax	28
Paraffin Wax	22
Rosin Soap	12
Starch	33
Alum Solution (5%)	5

Sash-Cord Impregnant

Paraffin Wax (130–132°F.)	8
Rosin	4
Rosin Oil	1
Carnauba Wax	1

Sizing for Polishing Cotton Cordage

Trisodium Phosphate	2	oz
Irish-Potato Starch	4	lb
Japan Wax	1¾	lb
Paraffin Wax (M.P. 127°)	1¾	lb
Narobin*	6	oz
Mineral-Oil Softener	14	oz

Mix thoroughly and make up to 10 gal with water. Use sufficient water to dissolve the starch and heat, with constant stirring, until all the ingredients are thoroughly mixed; then allow to cool and use cold.

The trisodium phosphate is used primarily as a water softener.

* Glyco Products Co.

Stiffening and Waterproofing for Cotton Goods

Batavia Dammar A/D	5
Paraffin Wax	5
Morpholine	1
Oleic Acid	2
Water	100

Melt the paraffin. Sift in the dammar and mix the two thoroughly, keeping the temperature at about 125°C. Allow to cool to 90–95°C. Add the oleic acid and stir in. Then stir in the morpholine, keeping the temperature at 90–95°C. Heat the water also to 90–95°C. and add it to the mixture slowly with thorough agitation.

Strips of cotton cloth may be stiffened by impregnating with this mixture and drying at 210°F.

String Waxing

Formula No. 1

Pull the string through melted paraffin and rub off with leather when cooled.

No. 2

Venice Turpentine	500
Vegetable Wax	400
Ozokerite	300
Spermaceti	250

Textile Gloss Oil

Formula No. 1

Paraffin Wax	20
Lanette Wax Sx*	10
Water	70

Heat together and stir until dispersed.

No. 2

Paraffin Wax	20
White Petrolatum	2
Glycosterin**	10
Water	180

Heat together and stir until dispersed.

Water Repellent for Textiles

Formula No. 1

Paraffin Wax (130°F.)	140
Stearic Acid	60
Naphthenic Acids	10
Gelatin	20
Aluminum Formate (30% Solution)	200
Water	570

Melt the wax and stearic acid and add the naphthenic acids, following with 150 g of boiling water. Stir until an emulsion is formed. Dissolve the gelatin in 180 g of water. Add the wax mixture to the gelatin solution at 60°C. Run through a colloid mill, if desired. Finally add the aluminum formate solution very slowly and with much stirring. Now add the remaining water and filter if necessary.

No. 2

Rubber Cement	46
Ester Gum	22
Cumar	2
Paraffin	31
Naphtha	100

* Du Pont.
** Glyco Products Co.

No. 3

Paraffin Wax	40
Oleic Acid	5
Trigamine*	3
Water	112

The Trigamine is dissolved in the water and the oleic acid is added, stirring thoroughly until completely homogeneous. This is then heated to 65–70°C. and the wax, previously melted, is added with rapid agitation. Stirring is continued until the emulsion is cold. This emulsion can be diluted readily with warm water.

No. 4

Rubber	2½
Xylene	90

Allow to stand until swollen. Then mix slowly until dissolved and add:

Hydrogenated Fish Oil	7½

This may be brushed or sprayed.

No. 5
(White)

Asbestine	20
Lithopone	10
Paraffin Wax (135°F.)	20
Petrolatum (140°F.)	72
Aluminum Stearate	8

With this formula which is designed to give fairly open constructions a 35–45% fill should be used based on the original weight of the fabric. The closeness of the fabric will be the determining factor for the degree of fill in the final analysis.

No. 6
(Sunproof)

Hard Amorphous Wax (148°F.)	22
Petrolatum (130°F.)	70
Magnesium or Aluminum Stearate	8

No. 7
(For Acetate)

Paraffin Wax	5 g
Soap	5 g
Ammonium Hydroxide (0.880)	30 cc
Water	970 cc

Heat to 55°C. and stir vigorously until a uniform emulsion results. Soak the rayon in this emulsion at 45°C. for 15 minutes; remove; centrifuge and treat for 10 minutes at 20°C. with a 5° Tw. solution of aluminum acetate.

No. 8
(Canvas)

Gilsonite	10
Asphaltum	2
Neutral Degras	4
Crude Beeswax	1
Lead Oleate	3
Kerosene	31
Gasoline	41

No. 9
(Canvas)

Aluminum Oleate	2–3
Paraffin Wax	7–8
Linseed-Oil Varnish	100

* Glyco Products Co.

Apply by brush on both sides of the canvas.

No. 10
(Canvas)

Ammonia (0.880)	1
Linseed Oil	5
Hard Paraffin Wax	8
Stearic Acid	8
Bleached Montan Wax	8
Glue	10
Soft Soap	15
Water	200

No. 11
(Brown)

Amorphous Wax (145–165°F.)	40
Paraffin Wax (133°F.)	50
Gilsonite Selects	20
Petrolatum (140°F.)	15
Red Iron Oxide	5
Extender, Repeller Clay or Asbestos	55

No. 12
(Tan)

Amorphous Wax	45
Paraffin Wax	45
Yellow Ochre	30
Extender	30
Petrolatum	18

No. 13
(Olive)

Amorphous Wax	25
Paraffin Wax	25
Petrolatum	16
Asphalt (190°F.)	2
Yellow Ochre	10
Chrome Green	2
Extender	20

No. 14
(Canvas)

Raw Linseed Oil	1 gal
Crude Beeswax	13 oz
White Lead	1 lb
Rosin	12 oz

Boil and apply warm to the upper side of canvas, wetting the canvas with a sponge on the underside before applying.

No. 15
(Canvas)

Gilsonite	80
Stearine Pitch	62
Scale Wax	34
Mineral Oil	10
Creosote Oil	10
Copper Linoleate	9

Melt together.

Apply at 300°F. Scrape off excess while hot.

No. 16
(Canvas)

Beeswax	25
Glyceryl Stearate	5
Stearine Pitch	102
Cooper Oleate	15
Castor Oil	48
Naphtha	50

No. 17
(Canvas)

Dark or Amber Petrolatum	8½ lb
Yellow Refined Beeswax	1½ lb
Dry Earth Pigment (Ochre, Sienna, or Umber)	5 lb
Volatile Mineral Spirits (Painters' Naphtha)	5 gal

No. 18
(Tarpaulin)

Crude Fuel Oil	60
Hard Paraffin Wax	20
Soft Paraffin Wax	5
Montan Wax	10
Wood Pitch	10

The hardness of the coating may be varied by changing the

relative proportions of the first three ingredients.

No. 19

(Tarpaulin)

Neutral Petroleum Asphalt	25
Paraffin Wax (124–126°F.)	20
Petrolatum (135–140°F.) (Amorphous Wax)	25
Venetian Red	35

Water- and Alkali-Resistant Textile Finish

Ceresin Wax	90
Bitumen (M.P. above 100°)	45
Tricresyl Phosphate	45
Turpentine	820

Wax Size

Formula No. 1

Japan Wax	100
Soap	10
Water	40

Boil and stir until homogeneous. Dilute with boiling water and stir before use.

This is used for treating paper-cloth.

No. 2

Water	59.60
Beeswax	9.40
Japan Wax	11.00
Stearic Acid	11.00
Oleic Acid	3.00
Caustic Soda	3.00
Soluble Olive Oil	3.00

Yarn Finish

Potash Carbonate	2
Rosin	7
Soft Paraffin Wax	10
Bleached Montan Wax	10
Water	65

WATERPROOFING

Antirust Waterproof Oil

Raw M. Rosin	100 lb
Red Lead	16 lb
Manganese Dioxide	4 lb
China-Wood Oil	32 gal
Menhaden Fish Oil	8 gal
Beeswax	1 lb
Kerosene (46°)	16 gal
Wood Turpentine	32 gal
Heavy Naphtha (48°)	16 gal

Melt the rosin at about 320°F., pull the kettle from the fire and sprinkle in the driers, previously mixed together, stirring them well into the melted rosin. When the red lead and black oxide are all in, put the kettle back on the fire and cook at 350—400°F. for 15 minutes, or until a sample on glass shows a transparent amber color.

As soon as the drying salts are taken up, pull the kettle from the fire and heat quickly to 575°F., stirring frequently and beating down the foam if necessary. As soon as the temperature reaches 575°F. pull from the fire and immediately chill back with the 8 gal of fish oil. Stir very frequently while the mixture is cooling to 480°F., at which point add the beeswax in small pieces and stir until dissolved.

When the wax has dissolved, thin first with the kerosene then follow with the turpentine and heavy naphtha.

This waterproof oil is of a medium amber color, has a good luster, and dries in 8–10 hours.

It contains 35.71% oil. The oil is used as an antirust, waterproof varnish on metals and as a paint-grinding oil.

Greaseproof, Waterproof Coating

An excellent greaseproof coating with good waterproof properties is obtained by the application of a 1/1000 in. film of a special wax mixture, using a modified waxing machine. The formula is as follows:

Glyco Wax A*	400
Flexo Wax C*	100

The waxes are melted together with intermittent slow stirring until clear. This may be done in the reservoir of the waxing machine, directly before use, since it will keep indefinitely. The approximate coverage is 30,000 sq. ft. In actual practice, a 15-point board thus requires about 200 lb of the wax mixture. Being stable to heat, the waxes will tolerate the prolonged heating required in their application. Good flexibility is maintained even at low temperatures, and the melting point, approximately 135°F., is sufficient for the maximum temperatures to which containers made from the coated board are subjected.

This wax mixture is nontoxic and will not support or induce bacterial or mold growth. The coated board will withstand 7 days' continuous contact with lard before being penetrated; its resistance to mineral or vegetable oils and greases is, in some cases, even higher.

* Glyco Products Co.

Moistureproofing Varnish

Sorbitol Alkyd Resins	1.50
Colorless Paraffin Oil	0.21
Paraffin Wax	0.42
Dissolve in 1 gal Toluol	

This mixture gives a coating which, when dried, is odorless. It may be used for coating glassine paper, etc.

Moisture-Vaporproof Lacquers for Paper

Chlorinated Rubber	50
Diamyl Phthalate	20
Pale East India Chips	30
Paraffin Wax	10
Toluol	165

Mix the ingredients by cold cutting.

The mixture is cloudy. Applied to paper, it imparts excellent moistureproofness. It is odorless, tasteless, and nontoxic.

Paperboard Proofing

The proofing of paperboard is of great importance, particularly when it is used in the manufacture of food containers. In most cases, both grease- and moisture-proofing is necessary and in some cases the moisture in the packaged product has also to be retained.

Proofing of this type has not as yet found wide application due to the fact that the coating must be absolutely odorless. The odor of rubber, while not pronounced ordinarily, is readily absorbed by many substances particularly if allowed to stand above ordinary room temperature.

Emulsified waxes may be used to good advantage with sodium silicate to give the desired surface effect.

A typical mixture of this kind may be prepared as follows:

Paraffin Wax Emulsion	800 g
60% Latex	300 g
Sodium Silicate (Water Glass)	100 g

The paraffin-wax emulsion is prepared by emulsifying 150 g of paraffin in 650 g of water with the aid of heat, stirring, and emulsifying agents. When cool, the sodium silicate is stirred into the emulsion, made slightly alkaline with ammonia, and the latex then added with mild stirring.

A coating of this type is not much more viscous than water, but can be readily applied to paper on an inclined bed-coating or any other similar machine.

If the coating is to be applied by means of a doctor blade, the solution may be thickened by the addition of bentonite or similar substances.

Sizing and Waterproofing Emulsion

Paraffin Wax	40.00
Oleic Acid	3.75
Stearic Acid	3.75
Trigamine*	3.00
Water	273.50

The Trigamine is dissolved in the water and the oleic acid is added, stirring thoroughly until completely homogeneous. This is then heated to 60–70°C. and the wax and stearic acid, previously melted, added with rapid agitation. Stirring is con-

* Glyco Products Co.

tinued until the emulsion is cold.

This emulsion can be diluted readily with warm water.

Straw-Hat Coating

Low-Viscosity Ethyl Cellulose	20
Japan Wax	5
Toluol	60
Ethyl Alcohol	15

This solution may be applied by means of a pressure spray gun or with a brush. It makes the straw hat completely water resistant and rainproof and will not yellow on standing.

Textile Waterproofing

Formula No. 1

(Impregnol)

Paraffin Wax (48–52°C.)	17.0
Bone Glue	2.5
Alumina	3.0
Formic Acid	7.0
Rosin	0.7
Caustic Potash (50°Bé.)	0.2
Water	To suit

No. 2

(Hydrophobol)

Bleached Montan Wax	6.3
Paraffin Wax (48–50°C.)	6.3
Stearic Acid	3.2
Formic Acid	12.0
Alumina	8.0
Zirconium Oxychloride	10.0
Water	To suit

No. 3

(Textal)

Paraffin Wax	26
Bone Glue	11½
Sodium Tetralinsulfonate	2

Add water to suit before use.

No. 4

(Persistol)

Methylol Stearamide	38
Paraffin Wax	2
Emulsifier	4
Water	56
Preservative	1/10

Warm together and mix until uniform.

Water- and Ink-Resistant Lacquers

A 5% solution of Flexo Wax C* in toluol can be incorporated into nitrocellulose-lacquer mixtures, giving excellent water and ink resistance. It also exhibits marked antiblock effects.

Waterproofing Composition

Formula No. 1

Paraffin Wax	20
Acrawax C*	4
Ester Gum	8

No. 2

Modified Hydrocarbon Wax	17
Carbon Tetrachloride	10
Ethylene Dichloride	10
Benzol or Naphtha	60

Digest until dissolved.

* Glyco Products Co.

No. 3

Petroleum Jelly	33
Rosin	11
Beeswax	56

Waterproofing Shoes

Natural Wool Grease	8
Dark Petrolatum	4
Paraffin Wax	4

Melt the ingredients together by warming them carefully and stirring thoroughly. Apply the grease when it is warm but never hotter than the hand can bear.

Waterproofing Substantive Emulsion

Paraffin Wax	32.0
Armac HT* (Cationic Emulsifier)	5.6
Span 40** (Nonionic Emulsifier)	1.2
Tween 40 (Nonionic Emulsifier)	1.2
Water	60.0

* Armour and Co.
** Atlas Powder Co.

index

A

B

C

D

G

P

Y

Z

www.ingramcontent.com/pod-product-compliance
Lightning Source LLC
LaVergne TN
LVHW011534200926
848484LV00001B/260

* 9 7 8 0 8 2 0 6 0 1 5 6 4 *